AF251367

A Handbook Series on
Electromagnetic Interference and
Compatibility

Volume 5

Electromagnetic Control in Components and Devices

Michel Mardiguian

Interference Control Technologies, Inc.
Gainesville, Virginia

Interference Control Technologies, Inc.
Route 625, Gainesville, VA 22065
TEL: (703) 347-0030   FAX: (703) 347-5813

Library of Congress Catalog Card Number: 88-80524
ISBN: 0-944916-05-8

Preface

Electromagnetic interference (EMI) can be attacked from any of several angles, such as addressing the equipment that creates the interference, the propagation paths of the unwanted energy or, finally, the equipment that suffers from it. However, almost invariably, the prime source of the EMI and its ultimate receptor are electrical components (exceptions being natural EMI sources and electromagnetic effects on living creatures or flammable substances). Before EMI propagation can take place through the conductor or space separation, enhanced or not by some wiring antenna effects, the energy which produces the unwanted effects has been **created** by an electrical part. And after the propagation has occurred, an electrical part at the end of the coupling paths exhibits an undesirable (not meaning "unpredictable") response.

Recognizing that a sound strategy to cure EMI must encompass the electrical component which is either the cause or the effect, it would be tempting to classify components as EMI sources (the villains) and EMI victims. The engineer would then be equipped with a twofold list of the components he must fear and those he must protect. Unfortunately, things are not so simple. Many components alternately can be sources or victims. In addition, some components have been developed which are neither sources nor victims but which instead suppress EMI. In turn, an EMI suppression part can generate some other forms of noise, so an attempt to categorize components as mentioned above would result in a perplexing, impenetrable catalog of sources and victims.

Instead, this book addresses each component individually, giving it a fair trial such that the designer can know what to expect in terms of its creation of, or reaction to, EMI. This way the prime building blocks of his system will be EMI characterized.

Of the three books I have written so far for Interference Control Technologies, this one is the most ambitious and, by and large, the largest in size (if not in depth). Its making would not have been possible without the thorough review of editors Jeff Eckert and Christine Miller.

Michel Mardiguian
February 14, 1988

Other Books in the 12-Volume Series

VOLUME 1
Fundamentals of Electromagnetic Compatibility

VOLUME 2
Grounding and Bonding

VOLUME 3
Electromagnetic Shielding

VOLUME 4
Filters and Power Conditioning

VOLUME 5
EMC in Components and Devices

VOLUME 6
EMI Test Methodology and Procedures

VOLUME 7
EMC in Telecommunications

VOLUME 8
EMI Control Methodology and Procedures

VOLUME 9
United States Commercial Standards

VOLUME 10
European and International Commercial Standards

VOLUME 11
Military EMC Standards

VOLUME 12
Supporting Military EMC Standards

Table of Contents

Chapter 5   Transformers and Magnetic Coupling Components

Chapter 6   Electromechanical Devices

Chapter 1

The Component in the General EMI/EMC Subject

EMI problems occur at many layers of a system hierarchy, from the environment down to the most elementary device in the smallest subassembly. The EMI manifestations can range from the most complex level (coupling among several elaborate systems, or between a system and its natural environment) to the lowest element.

Figure 1.1 illustrates nine levels of EMI situations. The number is arbitrary; wider or narrower categories could be made. However, the figure shows a scale based on functional identification. On the right, a few examples of what each layer can be are listed.

The first and second layers from the bottom are the ones with which this handbook from the EMC collection deals. Cards, components and devices/chips are small (or not so small) building blocks performing a function which is part of a machine, unit or equipment operations. For instance, PCBs and motherboards are the highest size of any major section in an individual box or equipment. At the bottom, the first layer contains the individual components which can be the smallest removable, sizable or repairable items. These include LSI modules, dual-in-line or chip carrier packages, relays, transformers, etc.

An enormous spread exists in the magnitude of the size, complexity, cost and repairability along the nine levels of EMI/EMC. Consequently, the higher the level, the more lead time, cost and engineering activity (design or retrofit) are necessary to implement an EMC solution. Addressing EMI at the earliest levels where susceptibility or emission take place (the component level) is a sound approach.

Some components can be generators of conducted or radiated EMI. Others can be victims of conducted or radiated EMI. Some are both sources and victims. Additionally, some parts are EMI-suppressing parts. In turn, EMI suppressors can generate other

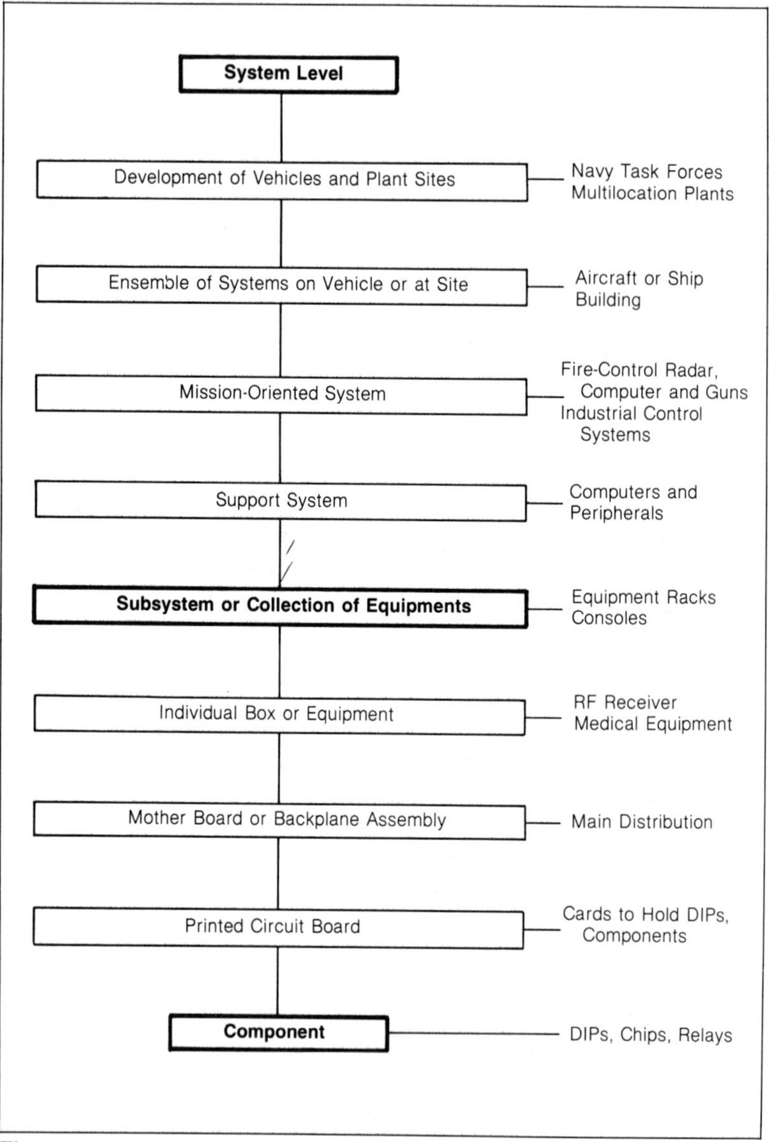

Figure 1.1—The Many Levels of EMI Manifestations

forms of EMI as a side effect. The law of energy conservation dictates that unless it is transformed into heat, light or movement, electromagnetic energy can be transformed only into electromagnetic energy.

For example, a gas tube surge suppressor will create an abrupt voltage drop which, in turn, creates EMI. Or a filtering inductor will radiate a magnetic field which induces noise in nearby circuits. So it seems that trying to classify components' interference is just like opening Pandora's box. Attempting to rigorously classify parts which defy any such classification into EMI categories would be more academic than real. Instead, the chapters of this book address all EMI aspects of parts, arranging them by functional families.

As a prerequisite to this book, it is helpful for the reader to have a background in EMI principles and basic EMI coupling mechanisms. If this background is not the case, a brief review of EMI fundamentals follows.

As seen in Fig. 1.2, interference occurs when three things are present: a source, or generator of electromagnetic energy, a victim, or receptor of electromagnetic energy, and a coupling path between the two.

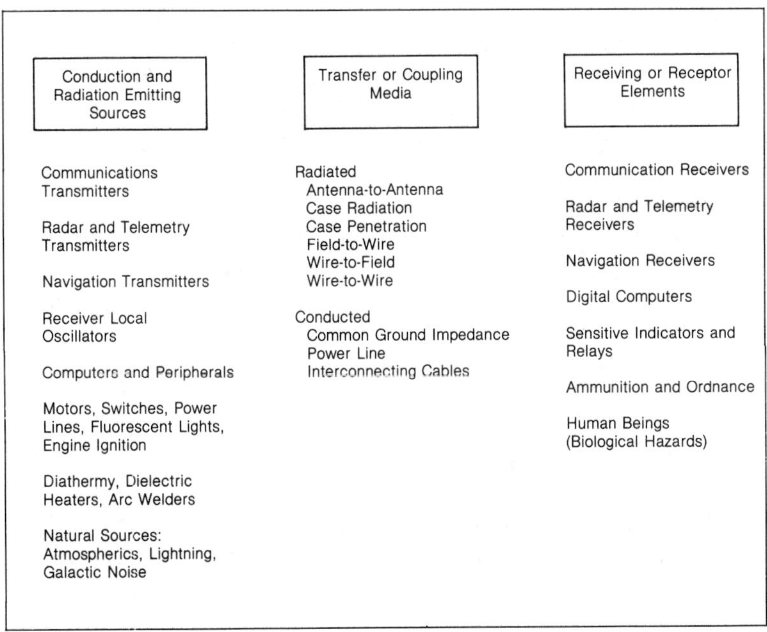

Figure 1.2—Three Basic Elements of an Emitting-Susceptibility Situation

Components can be **sources** of EMI by the wanted and unwanted signals they generate. The sources can be characterized by the amplitudes and waveforms (or spectrum) of their signature. Components can be **victims** of EMI by the unwanted energy they receive. The victims can be characterized by their sensitivity and bandwidth. Eventually, components such as EMI-fix parts can be put at either end of the coupling path to make the sources less emitting or the victims less vulnerable.

The magnitude of EMI signals can be expressed in these forms:

Voltage	= V or dB above 1 V (dBV)
	μV or dB above 1 μV (dBμV)
Current	= A, μA or dB above 1 μA (dBμA)
Power	= W, mW or dB above 1 mW (dBm)
E-field	= V/m, μV/m or dBμV/m
H-field	= T, A/m, μA/m or dBμA/m
Radiated power density	= W/m^2, mW/cm^2 or dBm/cm^2

There are five kinds of coupling paths by which a source creates EMI or a victim receives EMI. These paths are illustrated in Figs. 1.3 through 1.7 and enumerated below.

1. **Common impedance coupling** is a phenomenon by which a common impedance (generally return or power bus) is shared between the emission source and the victim. The $I \times R$ or $L\omega I$ drop in this path will affect the victim by creating a common-mode (CM) voltage in series in the circuit loop.

2. **Common-mode induction or radiation** is a phenomenon by which a radiating source induces a common-mode voltage in the loop formed by circuit(s) and a common ground. If the field is predominantly electric, this CM voltage appears transversely, line to ground. If the field is predominantly magnetic, the CM voltage appears in series in the ground loop and its value is:

$$V_{CM} = -(\Delta B/\Delta t) \times \text{area} \qquad (1.1)$$

3. **Differential-mode induction or radiation** is direct radiation pickup in a wire pair or between a trace and its return. The induced voltage appears directly across the victim's terminals.

4. **Wire-to-wire coupling (crosstalk)** occurs when two different circuits having a parallel run exhibit a mutual

capacitance C_{1-2} and a mutual inductance M_{1-2}. Therefore, if the source (culprit) circuit carries a voltage V_c and a current I_c, the victim circuit will see:

    a. A voltage capacitively coupled across its impedance equal to $V_{cap} = R_v C_{1-2}\, dV_c/dt$, where R_v is the parallel combination of victim near-end and far-end resistances and

    b. A voltage, magnetically coupled in series, equal to $V_{mag} = M_{1-2}\, dI_c/dt$

5. **Power-source coupling** is a phenomenon by which the power bus is disturbed by an EMI generator and conveys this disturbance into other users.

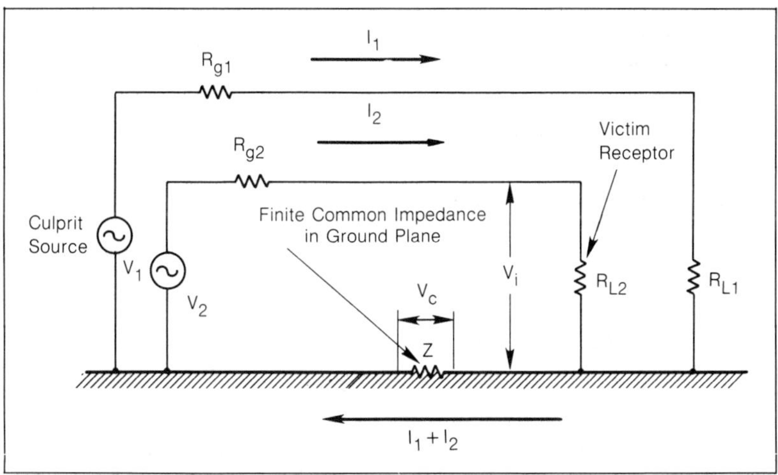

Figure 1.3—Coupling by Common Ground Impedance

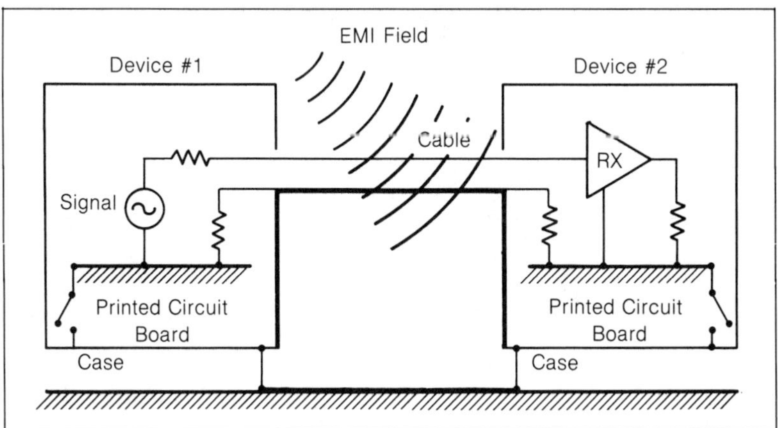

Figure 1.4—Field Induction via Ground Loops

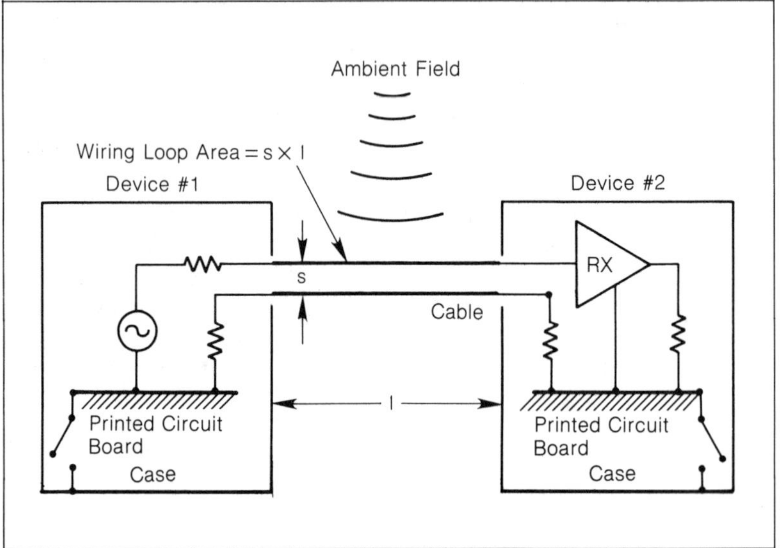

Figure 1.5—Field Induction via a Conductor Pair

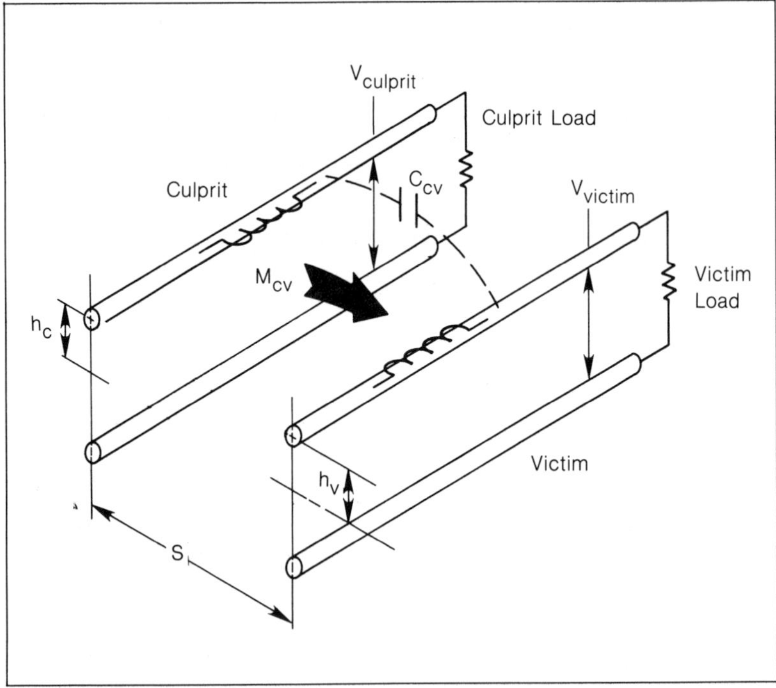

Figure 1.6—Cable-to-Cable Capacitive and Inductive Coupling

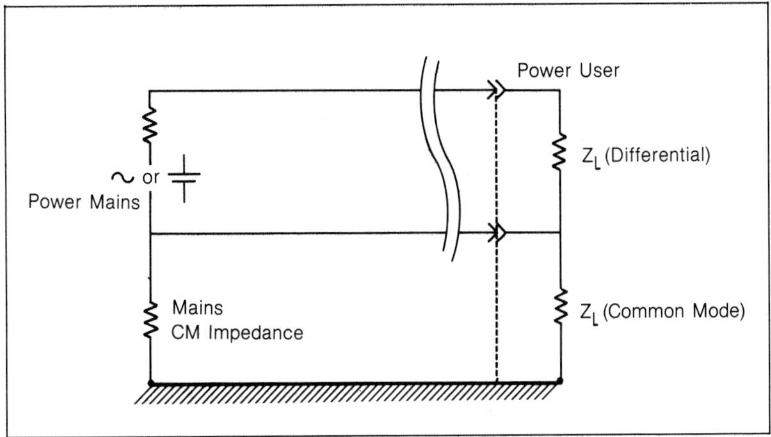

Figure 1.7—Power Coupling

Chapter 2

Resistors, Capacitors, Inductors

2.1 Resistors

2.1.1 Characteristics

The different kinds of resistors considered here are carbon composition, deposited carbon composition film, pyrolitic carbon films, metal film, wirewound, microelectronic and special-purpose. Determination of the type of resistor to be used is made by considerations of resistance, wattage, cost, compactness, precision, distributed capacitance, distributed inductance, life and internal noise.

Composition resistors may be of the pellet or filament types. The pellet type is made of finely divided carbon with a binder pressed into a slug with leads imbedded in each end. The slug is then enclosed in a phenolic or other case, and the resistor's body is molded. In some cases, the resistor is enclosed in a ceramic tube with cement covering both ends. The filament type has the carbon and binder mixture coated on the outer surface of a glass tube, and the leads are inserted therein. A phenolic tube is then molded around the resistor body.

Carbon or metal fixed-film resistors are usually made by depositing a controllable thickness of resistive material in a continuous film onto a base. The resistor body is then covered with a plastic or epoxy. The geometry of film resistors enhances their high-frequency characteristics, and they may be used up to about 400 MHz.

Any covering on the resistor body acts as a thermal barrier as well as protection against moisture. Thus, dissipated energy is conducted primarily by the leads. Special metal jackets are made to help heat-energy leave the resistor body. Bifilar winding of a wirewound resistor reduces internal inductance because adjacent turns carry currents in opposite directions. However, adjacent turns may exhibit appreciable shunt capacitance. Capacitive currents may have adverse effects on RF applications. The Ayrton-Perry winding is preferred, since each resistor is constructed of two parallel windings in opposite directions; the turns cross at points of no potential difference. A typical Ayrton-Perry resistor wound on a cylindrical spool exhibits one percent of the inductance of a conventional spool-wound power resistor (Fig. 2.1).

A composition resistor can exhibit an ac resistance lower than its dc value. It is primarily due to the shunting effect of distributed capacitance that results from the large number of conducting particles mixed with the dielectric material. To reduce this effect, resistors with a minimum of dielectric are used so the dielectric constant and associated loss factors are minimized. Decreasing the resistor cross section and increasing the resistor length, as in the filament type of resistor, decreases this problem. Because of the greater amount of dielectric used, higher values of resistance exhibit a greater percentage of change in value.

Finally, skin effect occurring at high frequencies causes current flow to be concentrated at the surface with little current flowing in the rest of the cross section. **Because current is not evenly distributed through the entire cross section of the con-**

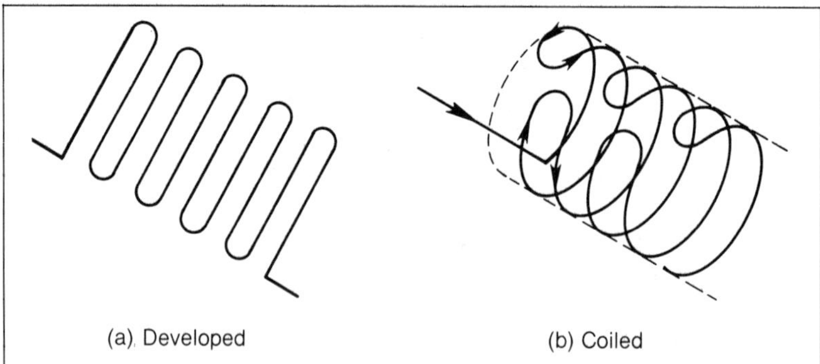

Figure 2.1—Wirewound Resistor with Inductanceless Design

ductor, skin effect results in increased effective resistance at RF over that of the dc value (see Table 2.1).

Table 2.1—Ratio of RF Resistance to Dc Resistance of a 1 MΩ Axial Lead, Carbon-Composition Resistor

Radio	Manufacturer No.1			Manufacturer No. 2		
Frequency	1/2 W	1 W	2 W	1/4 W	1/2 W	1 W
10 kHz	1.00	1.00	1.00	1.00	1.00	1.00
100 kHz	0.89	0.85	0.75	1.00	1.00	1.00
1 MHz	0.54	0.46	0.37	0.92	0.89	0.90
10 MHz	0.21	0.15	0.12	0.65	0.60	0.67
100 MHz	0.07	0.04	0.04	0.32	0.28	0.36

For resistors other than wirewound, better high-frequency performances are achieved when the ratio of body length to cross section is maximum (this ratio decreases capacitive reactance which is more a problem than the increase in self-inductance) and when a minimum amount of dielectric binder (for composition-type) is used.

Film-type resistors have the best high-frequency performance. Most resistance values remain constant up to 100 MHz before they start decreasing.

2.1.2 Equivalent Circuits

The equivalent circuit of a resistor depends upon manufacturing processes, techniques and raw materials used. One equivalent circuit is shown in Fig. 2.2a. The shunt capacitance, C, is of the order of 0.1 to 0.3 pF for a typical 1 W composition resistor. Figure 2.2b is the equivalent circuit of a resistor located near the return circuit and operating at a frequency where the capacitance of the return circuit is significant and where the capacitance, Cd, is low. For composition resistors, the inductance may be negligible.

Wirewound resistors have a relatively large series inductance and distributed capacitance. They are also influenced by skin effect and exhibit an increasing resistance as the frequency increases. The equivalent circuit of a wirewound resistor is shown in Fig. 2.3.

Table 2.1 shows the effective resistance at different radio frequencies of some general-purpose, axial-lead, carbon-composition resistors and 1 MΩ resistance value. The resistor produced by manufacturer No. 1 is described as a hot-molded fixed resistor. The resistance element is a carbon composition film on a glass body.

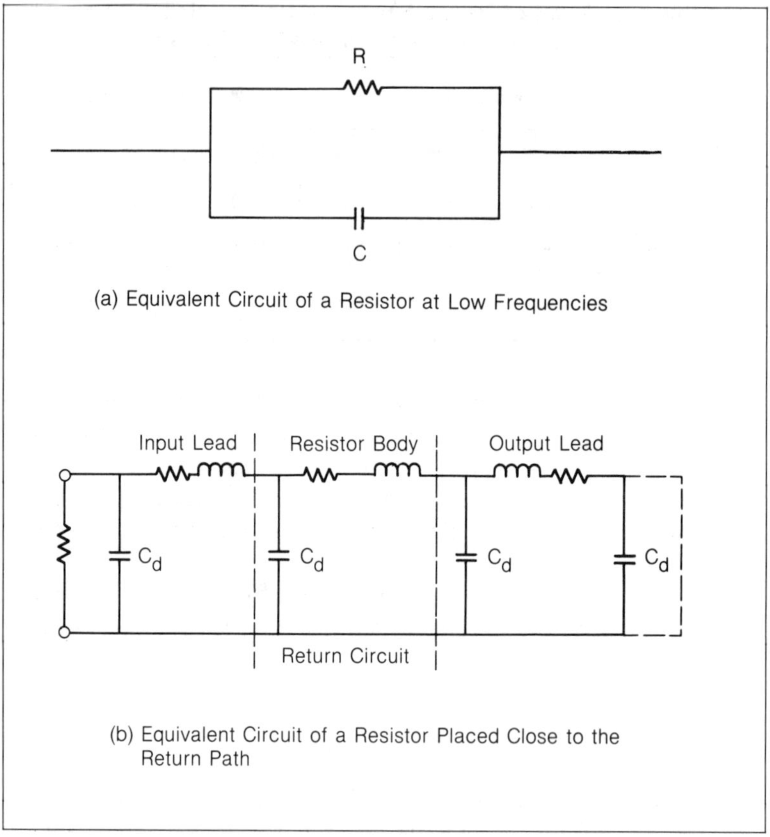

R
C
(a) Equivalent Circuit of a Resistor at Low Frequencies
Input Lead
Resistor Body
Output Lead
Cd
Cd
Cd
Cd
Return Circuit
(b) Equivalent Circuit of a Resistor Placed Close to the Return Path

Figure 2.2—Composition Resistor Equivalent Circuits

Lead
Lead
Resistor Body

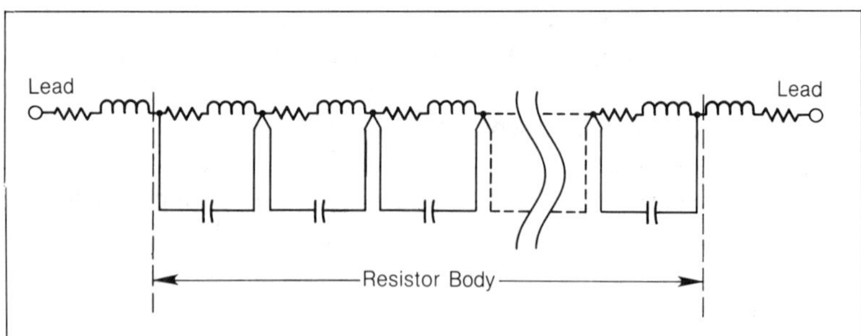

Figure 2.3—Equivalent Circuit of a Wirewound Resistor

The manufacturer's published frequency characteristics include both inductance and capacitance effects.

When both the skin effect and the reactive parts are considered, their combined effect gives apparent resistor values as in Fig. 2.4. This figure shows that for a 1 W hot-molded carbon composition resistor, the skin effect would double the resistor's value at about 100 MHz. On the other hand, the shunt capacitance takes over at a frequency which depends on the resistor's value. A 1 MΩ resistor starts to show a decrease in its impedance at around 500 kHz, even before the inductive and skin effects show up. By contrast, a 1 kΩ resistor shows an apparent increase before the $1/C\omega$ effect takes over.

Up to approximately 10 MHz, proper spacing and short leads minimize the effects of self and mutual inductance, while various capacitances and dielectric losses are negligible.

A conductor of uniform cross section, diameter d and distance D from a return circuit consisting of a conductor of the same dimen-

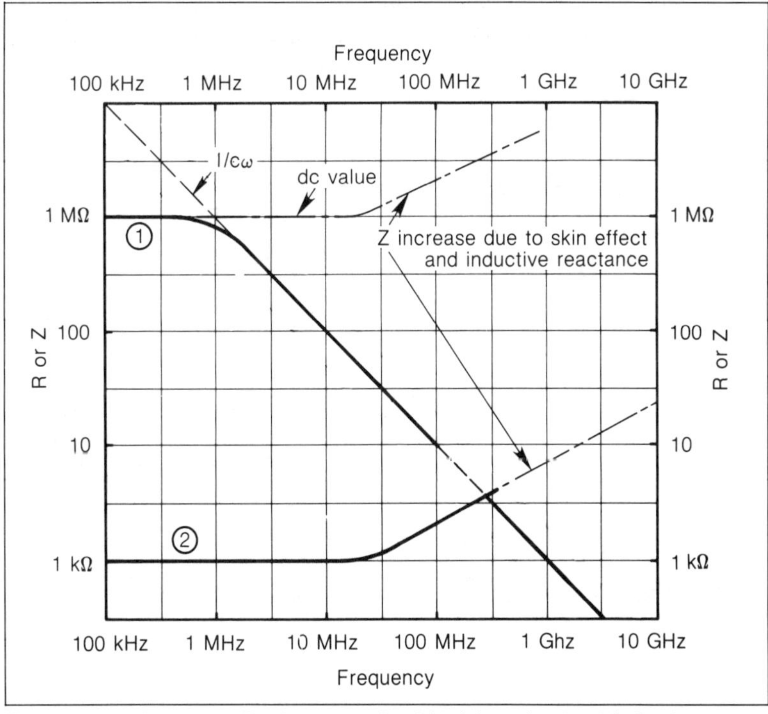

Figure 2.4—Apparent Resistance of 1 MΩ (1) and 1 kΩ (2) Resistors, 1 Watt, Carbon Composition

sions has an inductance L per unit length of:

$$L = (4 \ln (D/d)+1) \text{ nH/cm} \qquad (2.1)$$

The reactance X_L, at frequency f_{Hz} is:

$$X_L = 6.28 \, f \, (4 \ln (D/d)+1) \, 10^{-9} \, \Omega/\text{cm} \qquad (2.2)$$

As a rule of thumb, for a typical configuration where D/d ranges from 5 to 10, the above equations yield about 10 nH/cm, i.e., 0.06 Ω/cm/MHz.

When Eqs. (2.1) and (2.2) are applied to a 1 W, RC 32 resistor representing one manufacturer's form of construction, the resulting values occur as shown in Fig. 2.5.

The inductance and capacitance of helix form resistors exhibit broadband effects and parallel resonance at certain frequencies. Therefore, these resistors are often limited to power frequencies and, generally, frequencies below 100 kHz. But proper design can use these characteristics to filter pulses or reduce undesired frequencies.

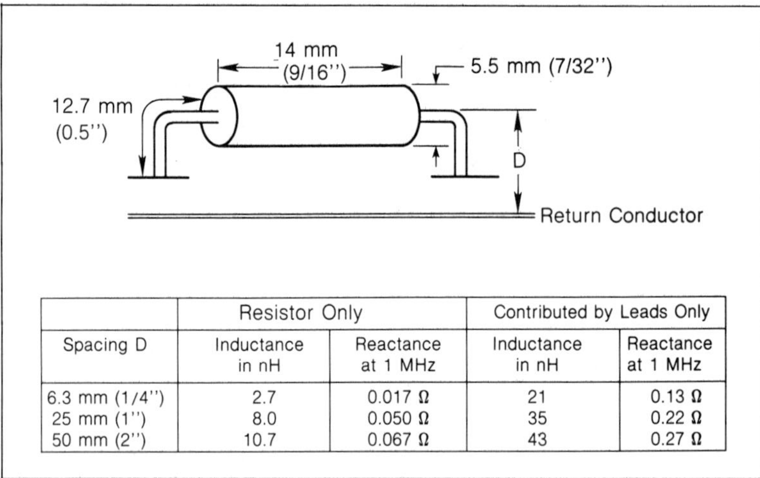

Spacing D	Resistor Only		Contributed by Leads Only	
	Inductance in nH	Reactance at 1 MHz	Inductance in nH	Reactance at 1 MHz
6.3 mm (1/4'')	2.7	0.017 Ω	21	0.13 Ω
25 mm (1'')	8.0	0.050 Ω	35	0.22 Ω
50 mm (2'')	10.7	0.067 Ω	43	0.27 Ω

Figure 2.5—Reactance of a 1 W Resistor at 1 MHz

2.1.3 Leadless (Surface-Mount) Resistors and Microelectronic Resistors

As far as skin effect and parasitic inductance are concerned, surface-mounted (see Figure 2.6) resistors have a definite advantage; since the lead inductance represents about three-fourths of the total inductance, the surface-mounted package exhibits only one-fourth of the inductive reactance of a conventional mounting. This is confirmed by actual measurement, along with the fact that the rectangular cross section of some surface-mount packages also reduces the skin effect.

The microelectronic resistors consist of integrated (monolithic MSI, LSI and VLSI) and film deposit (hybrid) devices, none of which is really a miniaturization of its discrete counterpart. For the integrated type, the primary material is silicon, precisely the base region of a diffused transistor. Since extremely high- or low-impurity doping cannot be used, the sheet resistance used is in the 200 Ω/square range, resulting in practical resistor values ranging from a few tens of ohms to about one MΩ, and satisfying a large percentage of applications.

From the EMI standpoint, the integrated type exhibits a lower self-inductance due to its generally flat shape. On the other hand, the parasitic capacitance is aggravated by the ϵ_r of the substrate. Also, nonlinear effects due to parasitic semiconductor junctions may

Figure 2.6—Typical Surface-Mount Resistor (courtesy of General Electric)

appear. For these reasons, accuracy and high-frequency performances of such resistors cannot be excellent.

Film deposit (hybrid) resistors require more manufacturing processes but offer excellent accuracy, especially at high frequencies. They have lower parasitic capacitance and their thinness precludes skin effect. For instance, the typical features of modern nickel chrome film resistors (chip networks) are:

Film thickness: 1 μm or less
Sheet resistivity: 1 to 1,000 Ω/sq.
Tolerances: 0.005 to 2 percent
Noise figure: -30 to -50 dB [3 percent to 0.3 percent]
    (excess noise above thermal noise
     of the same ideal resistor)
High-frequency capability: 100 MHz to > 1 GHz

2.1.4 Intrinsic Resistor Noise

All resistors generate **thermal** and **current** noise. Thermal noise (also called **Johnson** noise) is due to the thermal agitation in the atoms as soon as the component's temperature departs from the absolute zero ($-273.16°C$). This voltage, which appears even if no current is forced through the resistance, is expressed by its rms value, valid at any frequency:

$$V_t = \sqrt{4RKTB} \qquad (2.3)$$

where, R = resistance component affected by thermal agitation
     K = Boltzmann's constant = 1.374×10^{-23} W/°K/Hz
     T = absolute temperature in degrees Kelvin
     B = noise bandwidth in Hz

Eq. (2.3) is shown in graphic form in Fig. 2.7 for different values of resistance and bandwidth. For a 50 Ω resistance at 20°C, Eq. (2.3) results in 0.9 nV in one hertz or 0.9 nV/$\sqrt{\text{Hz}}$.

In addition to the thermal noise, noise is also produced by the current flowing through a resistor. The corresponding noise voltage is approximately equal to:

$$V_c = I \sqrt{k/f} \qquad (2.4)$$

where, V_c = rms V/Hz of bandwidth at frequency f
    k   = noise quality constant of proportionality
            dependent upon type of resistor
    I   = current through the resistor (dc or rms A)
    f   = frequency in Hz

This noise exists only when current is flowing and is due to the nonhomogeneous nature of some resistors (and semiconductor devices as well) made of many small particles molded together: the more homogeneous the material, the less current noise. For instance, on a scale from the less noisy to the more noisy, we find wirewound resistors, film (deposited carbon) resistors and composition carbon resistors.

Also, the ratio of dissipated versus rated power is important: for a given dissipated power, the resistor with the highest power rating will exhibit the least noise.

Different authors and many studies have documented this current-generated noise. All data show noise decreasing when frequency increases, although some studies report a $1/\sqrt{f}$ dependency;[1] others, a $1/f$ dependency.

For instance, Ott[2] indicates that a 10 kΩ carbon composition resistor shows a noise voltage of 10^{-7} V/mA/$\sqrt{Hz}$ at 10 kHz, which is above the sole thermal noise of this resistance at 20°C. But the same noise drops down to 10^{-8} V/mA/$\sqrt{Hz}$ around 300 kHz where it becomes inferior to the thermal noise, which predominates from then on.

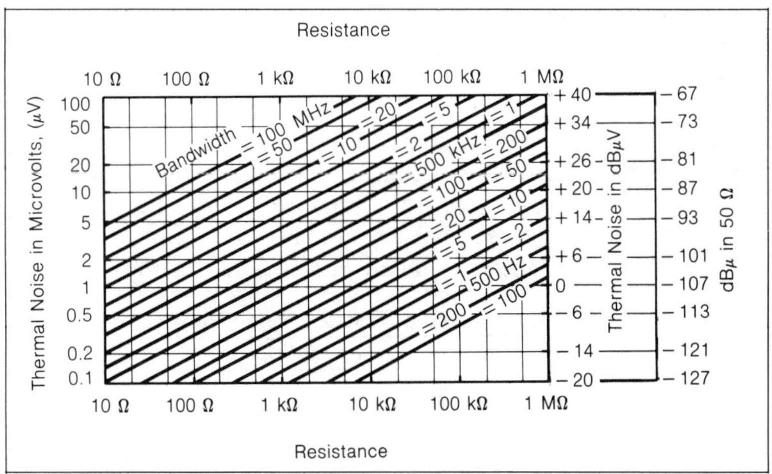

Figure 2.7—Thermal Noise Voltage of Resistors as a Function of Bandwidth at Room Temperature

Since the two noise voltages given by Eqs. (2.3) and (2.4) are independent, the total noise voltage is found by their rms addition:

$$V_{total} = \sqrt{V_t^2 + V_c^2} \qquad (2.5)$$

Another cause of EMI in resistors may be the imperfect contact between the leads and the resistor body or the presence of microscopic fractures in the resistor body. These two defects can go unnoticed during the manufacturing process or can show up later with aging, thermal overrun, etc. They do not reach a level where they cause circuit malfunction, but they may cause steady or intermittent EMI which is difficult to pinpoint. Moreover, being located generally at a junction of different conductors, these defects have a nonlinear VI curve and tend to create audio rectification in the presence of a modulated sine wave (see Section 4.1.3).

Strong electromagnetic fields can affect resistors, usually causing a change in resistance due to heating. While composition resistors exhibit only the heating effect, spiral film and ordinary wirewound resistors also behave as inductors which can couple energy into associated circuits.

In the case of variable resistors or potentiometers, noise may result from several causes: foreign materials formed on the resistance due to wipers, dust particles or chemical contamination, formation of oxide films on contact surfaces, mechanical imbalance and triboelectric effect from the wiper sliding on the element which causes a self-generated voltage and thermoelectric effect from external or frictional heat.

Noise can be generated when high current densities exist at the wiper-to-element interface, such as occurs when a wiper makes contact with the high spot on the surface of a composition or a ceramic carbon metal (cermet) element. As the surface heats, the contact may change quickly, often producing arcing and white noise. The wirewound variable resistor wiper can make or break contact with adjacent turns and cause small arcs. In precision variable resistors, the noise level is reduced to 100 mV or less. Table 2.2 shows typical

Table 2.2—Variable Resistor Noise Voltage

Resistance in Ohms	Resistor Current		
	1 mA	0.1 mA	0.01 mA
1,000	0.01 to 0.03 V	0.002 V	0.0002 V
10,000	0.15 to 0.25	0.02	0.002
100,000	1.0 to 4.0	0.3	0.035

noise voltage for resistor currents varying from 10 μA to 1 mA.

As a summary, Table 2.3 illustrates some common EMI situations with resistors.

Table 2.3—Effects of Resistor Characteristics

Resistor Characteristics	Example	Problem
Direct end-to-end capacitance	Attenuator	High-frequency signals fed through to a point not grounded for the signal frequency
Total capacitance (end-to-end and lead-to-ground)	Feedback amplifier plate load resistor	Phase shift to signal components as a function of frequency
Resistance varies with frequency	Some amplifiers, some measurement methods	Boella and skin effects
Inductance	Shunt resistors in attenuators	Change of effective impedance with frequency. Important in low value resistors below 100 MHz.
Inductance	Any resistor	Phase shift, change of effective impedance. Important about 100 MHz.
Susceptibility to R-F fields	Composition and metal film resistors	Can change resistance and overheat
Susceptibility to R-F fields	Ordinary spiral-wound resistors	Induced voltage, proportional to the numbers of turns and field strength, is transferred to circuitry

2.2 Capacitors

Capacitance is formed between any two physically separate conductive objects. An intentional capacitance is a capacitor constructed of two metal plates or foils separated by a controllable dielectric. Its capacitance C is:

$$C \cong \frac{A\epsilon}{t} \text{ F for } \sqrt{A/t} > 10 \qquad (2.6)$$

where, A = area of each plate or foil in m^2

ϵ = permittivity of dielectric medium = $\epsilon_0 \epsilon_r$

ϵ_r = dielectric constant relative to air

ϵ_0 = absolute permittivity of air = 8.85×10^{-12} F/m

t = thickness of medium or separation of plates or foil in meters

Equation (2.6) applies when the least dimension associated with the area is much greater than the thickness, corresponding to a negligible capacitance of the fringing field.

Since the ratio A/t has the dimension of a length, when A/t is in centimeters and inches, Eq. (2.6) becomes:

$$C = \frac{0.0885\epsilon_r A}{t} \text{ pF (for dimensions in cm)} \qquad (2.7)$$

$$C = \frac{0.225\epsilon_r A}{t} \text{ pF (for dimensions in inches)} \qquad (2.8)$$

Figure 2.8 is a plot of Eq. (2.8) for various dielectric constants ranging from 1 to 1,000.

For a cylindrical (coaxial) capacitor, the formula is:

$$C = \frac{0.55\epsilon_r l}{\ln d_2/d_1} \text{ pF} \qquad (2.9)$$

where, l = length of the cylinder in cm
           d_2= outer diameter
           d_1= inner diameter

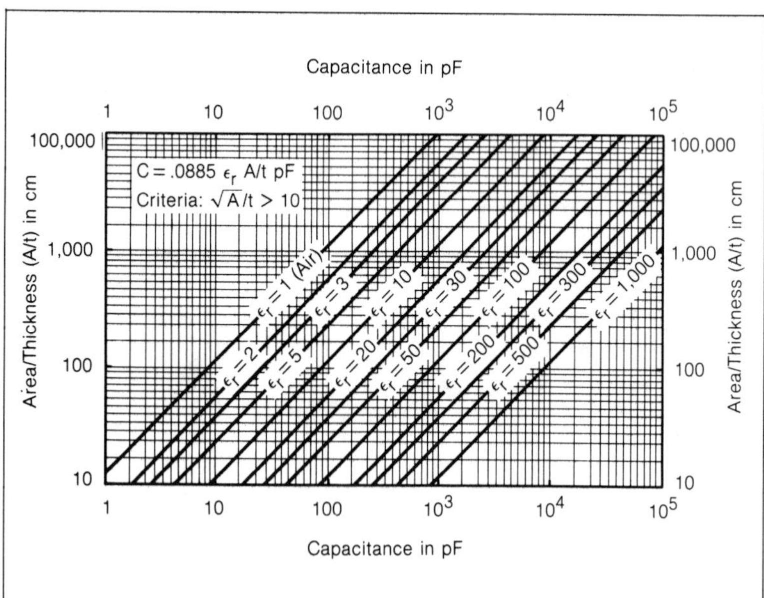

Figure 2.8—Capacitance of Parallel-Plate Capacitor

2.2.1 Principal Electrical Characteristics

The following are a few parameters which are useful in the application of capacitors:

Dissipation Factor (DF) is defined as:

$$\frac{\text{Equivalent Series Resistance (ESR)}}{\text{Capacitive Reactance } (X_c)}$$

It relates directly with the heat losses in the capacitor parasitic series resistance (including connection, dielectric and plate material). Sometimes DF is expressed as capacitor **Efficiency,** i.e.:

$$\frac{\text{Power Lost}}{\text{Power Stored}}$$

Figure of Merit (QM) and Power Factor (PF) are also used to characterize the loss terms:

$$QM = \frac{1}{DF} = \frac{X_c}{ESR}$$

$$\text{and,} \quad PF = \frac{ESR}{Z} = \frac{ESR}{\sqrt{X_c^2 + (ESR)^2}} \cong \frac{ESR}{X_c} \text{ or DF}$$

Insulation Resistance (IR) relates to the insulating properties (therefore to the leakage current) of the particular dielectric used. At normal operating range, IR ($> 1{,}000$ MΩ) is usually high enough to be disregarded. As the amount of dielectric increases (larger C) the IR decreases. IR also decreases with frequency.

Dielectric Absorbtion (DA) is the measure of the voltage gradient which develops after a capacitor has been charged then quickly discharged. This recovery voltage is caused by the dielectric "giving up" electrons that were held during the charging period.

$$DA = \frac{V_{recovery}}{V_{charge}}$$

Polarized capacitors are prone to this phenomenon. Therefore, nonpolar dielectric (most plastics) should be used in designs requiring fast discharge or where an accurate threshold depends on a known percentage of charge from a capacitor.

Surge Voltage is the maximum nonpermanent voltage (including surges and transients) that the capacitor can withstand. For electrolytics and impregnated paper capacitors, this surge voltage is generally 110 to 130 percent of the maximum working voltage.

The equivalent circuit of most capacitors over a wide frequency range is shown on Fig. 2.9a. Figure 2.9b is the equivalent circuit

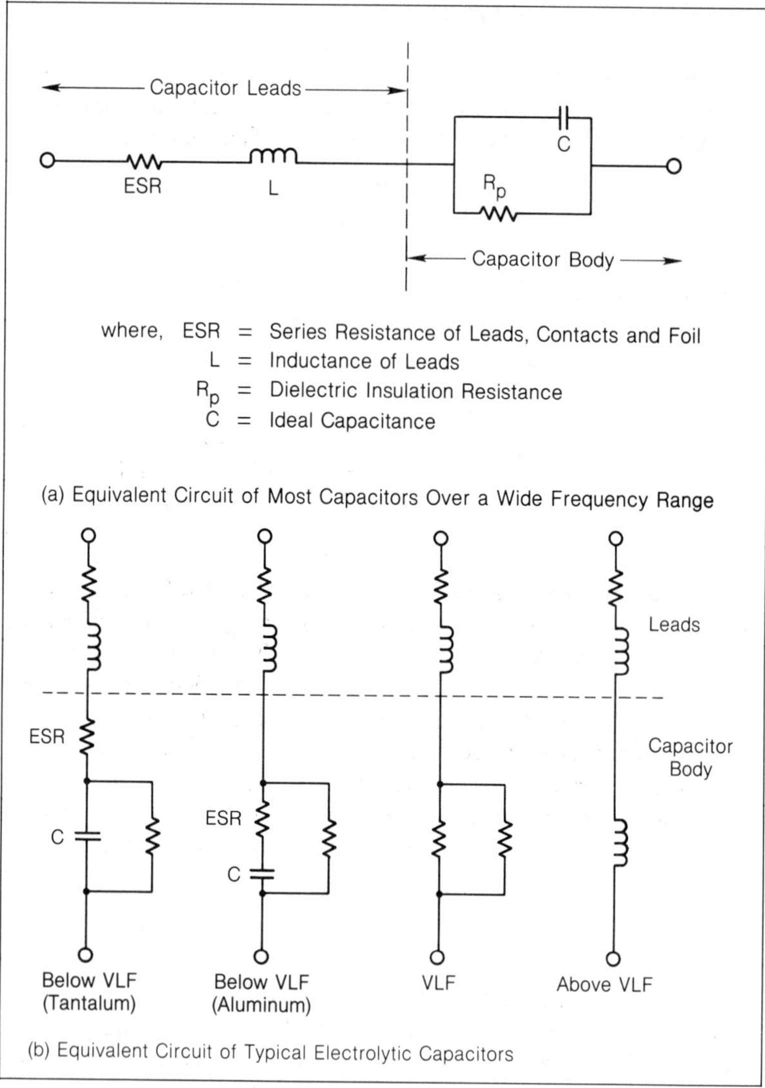

Figure 2.9—Capacitor Equivalent Circuits

of electrolytic capacitors.

In most capacitors, the lead inductance combines with the capacitance to give a terminal impedance, which generally is small at the resonant frequency and increases on either side of this frequency. Above resonance, the capacitance appears as an inductor. Inductance and series resistance limit the rate of change of current during sudden charge or discharge such as for transients. The Equivalent Series Resistance (ESR) affects the dissipation factor and may cause problems in ac, high-precision and timing circuits. For most applications, series resistance is considered constant and independent of frequency.

The total inductance of a capacitor is the sum of the internal electrode inductance and the external lead inductance. Table 2.4 lists typical values of internal (electrode) inductance for various types of capacitors.

Table 2.4—Typical Internal Inductance of Various Capacitors

Capacitor Type	Inductance
Porcelain and ceramic fixed capacitors	1.4 nH
Wet-anode tantalum capacitors	25 nH
Solid tantalum capacitors	20 nH
Foil tantalum, tubular case, with leads	50 nH
Foil tantalum, rectangular case, lug terminals	23 nH

As already shown in Section 2.1, Eq. (2.1), the lead inductance (external) can be determined from the self-inductance of a straight wire, i.e., about 10 nH/cm. Internal and external inductances may be added to determine the total series inductance.

Operating frequency is the most decisive criterion when choosing a capacitor. Figure 2.10 shows the typical frequency domain of the various capacitor technologies. The upper frequency boundary is generally limited by self-resonant frequency and HF dielectric losses.

The self-resonant frequency is determined by many factors, including physical size, dielectric properties, capacitance, lead inductance and inductance of the plates. Figure 2.11 shows lead-length effects for a 0.05 μF capacitor with 6.35 mm leads. Note the complete degradation of performance above about 5 MHz. Above this frequency, the capacitor behaves as an inductor. If this capacitor

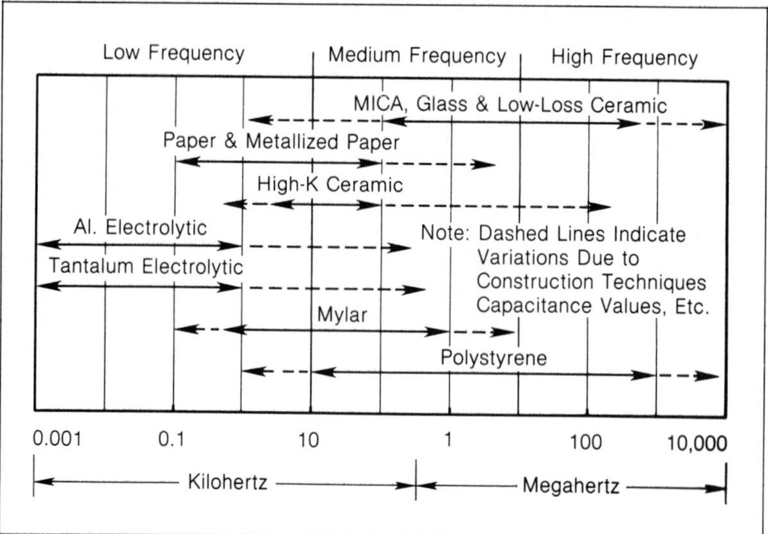

Figure 2.10—Approximate Usable Frequency Ranges for Various Types of Capacitors (From Ott, Ref. 2)

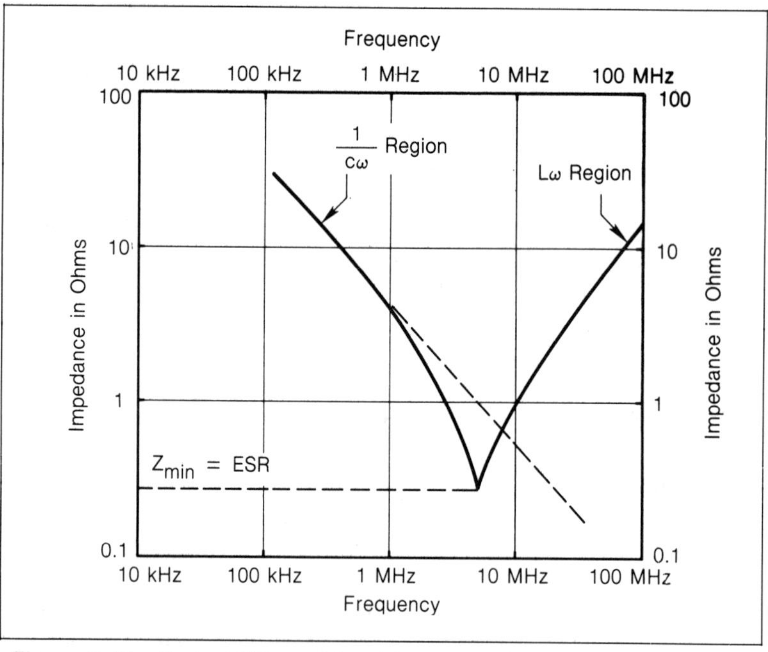

Figure 2.11—Typical Effect of Lead Length on Capacitor [0.05 μF with 6.35 mm (1/4 inch) Leads]

were used as a single-stage filter in a 50 Ω line, the insertion loss would degenerate above 4 MHz as shown in Fig. 2.12. A contrast can be made with the performance of a 0.05 μF feed-through capacitor in which the associated series inductance is very small. Above 50 MHz, the dielectric and series resistive loss protect the capacitor's performance as a filter.

The superior performance of a feed-through capacitor is due to its "coaxial" nature (Fig. 2.13). With one electrode as the center

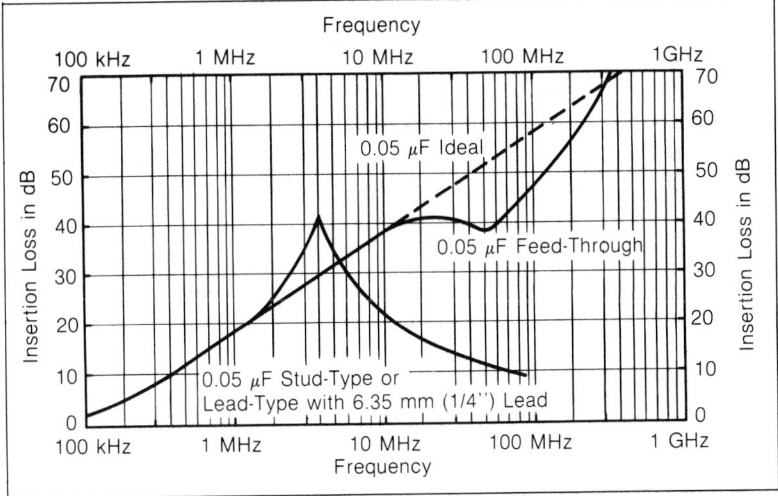

Figure 2.12—Comparison of Filtering Performance of Feed-Through and Lead-Type Capacitors with Ideal Capacitors, as Measured in a 50 Ω/50 Ω System

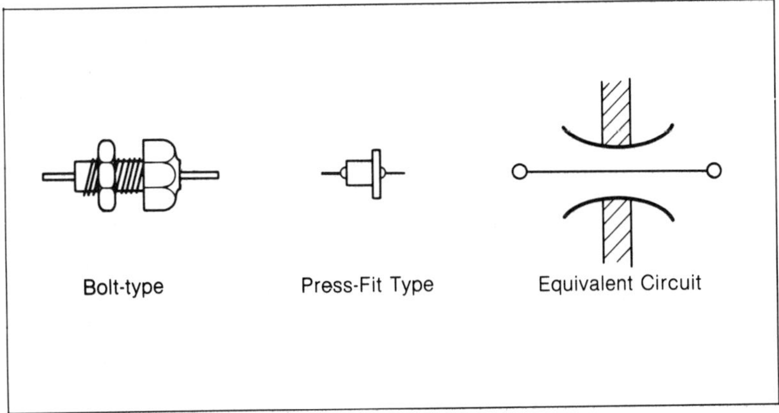

Figure 2.13—Feed-Through Capacitors

conductor, its self-inductance is minimal and, in the case of a filter, does not deteriorate the insertion loss (to the contrary, it might improve it). The other electrode, being tubular and directly pressed, soldered or screwed against the mounting surface, shows no parasitic inductance. The impedance of such a capacitor decreases as frequency increases, with practically no lower bound. The insertion loss, in the case of a filter, is bound by the contact resistance of the thread or fitting. If this is, for instance, equal to 1 mΩ, the insertion loss measured in a 50 Ω/50 Ω system, will be asymptotic to 20 log = 25 Ω/1.10^{-3} Ω = 88 dB.

Shunt conductance is caused by current leakage and voltage stress across the dielectric. Current leakage is usually small in solid dielectrics but may be a problem in both high-precision capacitors and some electrolytic capacitors. Shunt conductance is affected by both the instantaneous and longer-duration application of voltage stress. The effects are energy loss, heating and change in power factor.

Absorption of energy results in the reappearance of voltage on the capacitor after it has been discharged. Dielectric absorption causes a voltage stress that is delayed because of the time required to displace charges from the dielectric. In high-voltage circuits, a means of discharging capacitors should be provided to prevent danger to personnel.

Starting from the low-frequency end, the electrolytic capacitor used in power supplies and low-frequency filtering is first considered. This capacitor is made of two aluminum foils: a positive foil called "anode" and a negative foil called the "cathode." These foils are etched to increase their effective contact area to the electrolyte in between. When used with ac or pulsed voltages, neither the ripple voltage nor current should exceed the maximum rated values, otherwise premature aging or deterioration will occur.

As seen in Fig. 2.14a, typical ESR for modern electrolytic capacitors can be as low as 3 to 10 mΩ. Figure 2.14b shows the same data for a metalized polypropylene capacitor.

Tantalum capacitors offer attractive space and weight reduction under some conditions of operation. Tantalum capacitors come in three types: solid, foil and wet anode. Solid slug types are made by sintering. This method forms a spongy slug of metal that has a large effective surface area and is extremely small for its capacitance and voltage rating. Feed-through capacitors of the solid-tantalum type are effective up to 5 GHz.

Foil-type tantalum capacitors can be made in voltage ratings up

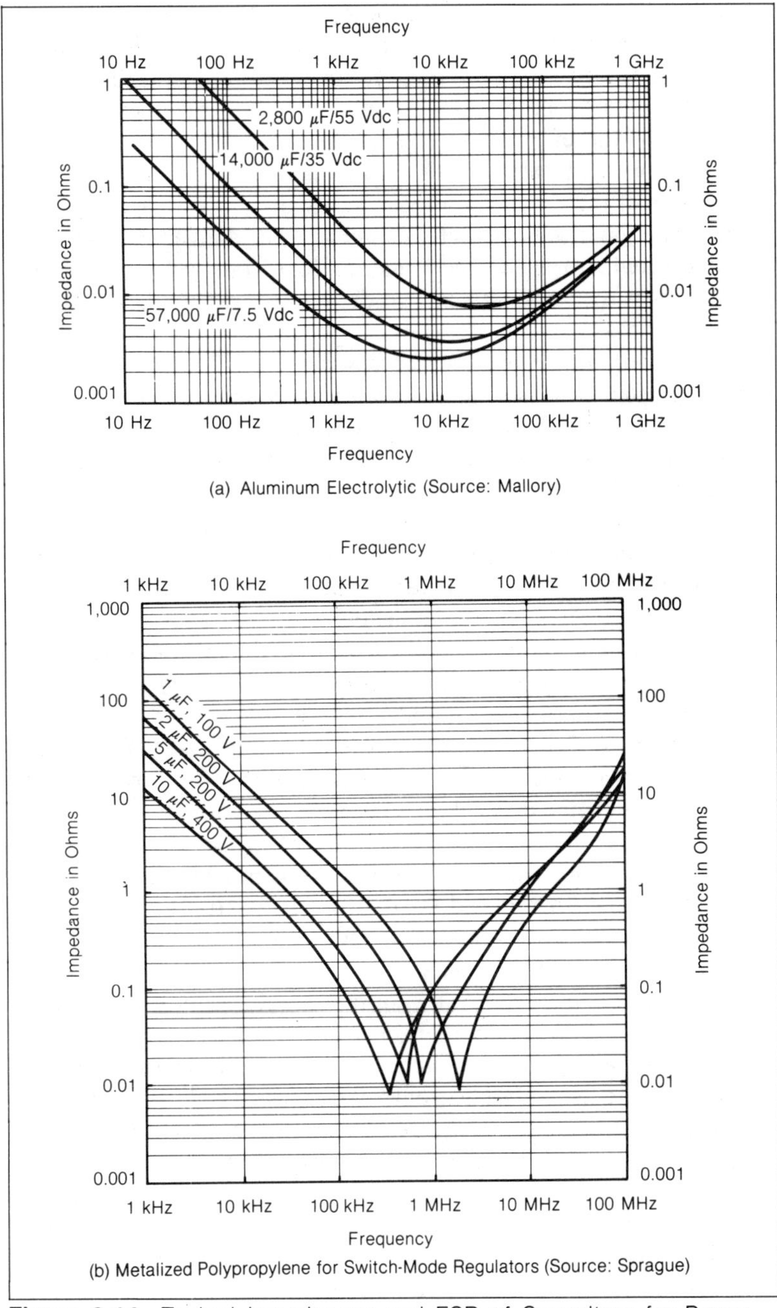

Figure 2.14—Typical Impedances and ESR of Capacitors for Power Applications

to 300 V compared with about 50 V for the solid and 125 V for the wet-type. The foil-type is limited to audio and low RF applications because of high internal inductance. Sometimes a tantalum-foil capacitor is shunted with a smaller paper or ceramic capacitor to extend the effective bypass range.

The original wet-type tantalum capacitor is no longer used because of the danger of electrolyte leakage. In current wet-type construction, the electrolyte is a gel, and the danger of leakage and consequent corrosion is no longer a factor. The wet-type construction offers the smallest size and the highest capacitance of all tantalum types.

The tantalum capacitor has a further advantage in that low-temperature performance is greatly superior to the aluminum-electrolytic-type. Tantalum capacitors also have longer shelf life and less current leakage, especially at high temperatures and after long periods of idle time.

Aluminum electrolytics, tantalum and other polarized capacitors are often packaged in a tubular metal can which is connected to the cathode, i.e., (the negative) terminal. Also, metallized paper or plastic types have one of the leads connected to the outside foil. In both cases, the lead being connected to the electrical zero reference will cause the can or outside foil to act as a shield, thus shunting any stray capacitance.

With two-stud style electrolytic capacitors, the case is theoretically floating with respect to one or the other electrode. However, there is an undetermined resistance between the cathode and the can. Therefore, if the cathode terminal is not at the chassis potential in the application, the capacitor should be mounted in such a way that the can is **insulated from the mounting surface** (these capacitors are sometimes equipped with a transparent PVC insulating sleeve for that reason).

For applications requiring high-frequency decoupling and small, rugged capacitors, ceramic capacitors are the most suitable. Ceramic dielectric capacitors and filters using ceramic (barium titanate) dielectric are useful because of their small size and low weight compared with capacitors using more conventional materials such as paper or mica. However, some ceramic capacitors with a very high ϵ_r (for instance $\epsilon_r = 1,000$ or more) are extremely sensitive to temperature. A significant degradation of equipment performance results when they are used as a bypass operating at a local temperature of $-40°C$ or at an elevated temperature of $+100°C$.

For example, according to one manufacturer's specifications, a nominal 1,000 pF ceramic disc feed-through type capacitor will have a capacitance of 330 pF at −40°C. While the operating temperature range is −55°C to +85°C, the capacitor is rated for a temperature of 20°C ± 1°C. Continuous working voltage rating is 500 Vdc, but the capacitance is stated for a 0.5 to 5.0 V range. With 500 V applied and a feed-through current of 25 A, the effective attenuation as a bypass capacitor is reduced by 20 dB at +25°C. Thus, operating conditions must be considered when a capacitor or filter incorporating a ceramic dielectric is used.

Ceramic capacitors are available as axial leads, radial (disc), multilayers, chips and arrays. Disc types are probably the most widely used in discrete assemblies. Three types of resonances can occur in disc-type capacitors: low-frequency resonance due to long leads, medium-high frequency resonance when discs are connected in parallel (due to internal leads) and high-frequency resonance due to resonant cavity effects in high-dielectric capacitors.

Figure 2.15 shows the resonant frequency of some ceramic

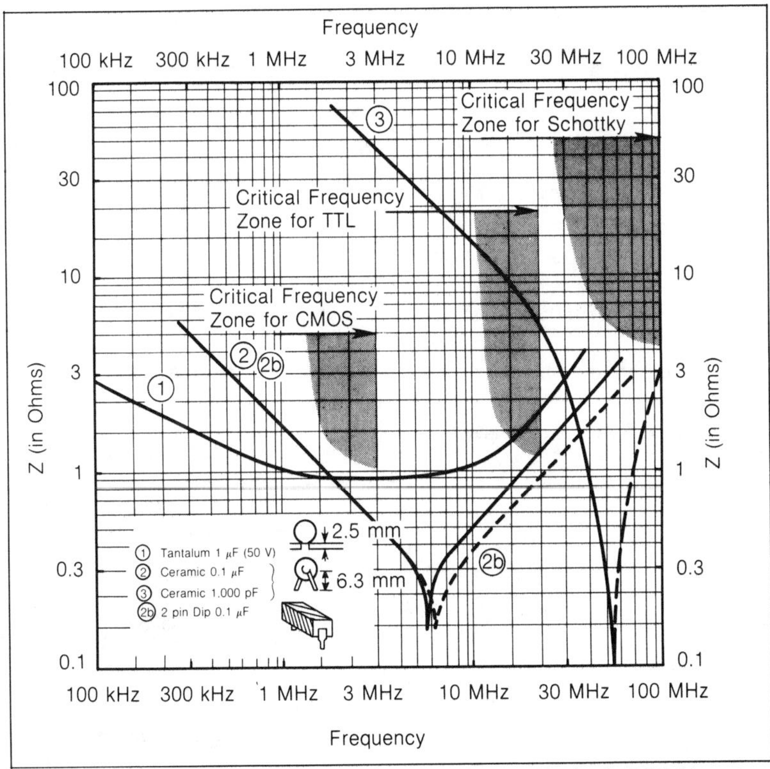

Figure 2.15—Impedance Frequency of Some Typical PCB-Mount Capacitors. By contrast, the occupied bandwidths of a few popular logic technologies are shown. When this bandwidth exceeds significantly the capacitors resonant frequency, its decoupling will be inefficient.

capacitors, depending on their size, value and lead type. More application aspects of these capacitors will be covered in Chapter 8, "PCBs."

For ceramic and metallized paper or plastic capacitors used in power-line filters, an additional requirement exists: international safety requirements, like those specified in IEC Recommendation 161, stipulate that capacitors used for interference suppression on power mains should meet the safety ratings of class X for line-to-line and class Y for line-to-ground. Since a failure of a class Y capacitor causes an electrical hazard, these capacitors have the most stringent requirement.

A class Y capacitor must withstand for one minute a stress voltage of:

$$2,250 \text{ Vdc and } 1,500 \text{ Vac (50 Hz) plate-to-plate}$$

$$(1,500 \text{ V} + 2 \times \text{ line voltage}) \text{ plate-to-case}$$

For class X capacitors, these values are:

$$4.3 \times \text{ line voltage, plate-to-plate}$$

$$(1,500 \text{ V} + 2 \times \text{ line voltage}) \text{ Vac plate-to-case}$$

To summarize the EMI aspects associated with capacitors, Table 2.5 shows the type of problems, their causes and remedies.

Table 2.5—EMI Problems, Causes and Corrections in Capacitors

Type of Noise	Causes	Result and Correct
Effective inductance of leads is too large even with short leads	Leads of capacitor behave as inductors	Use parallel capacitors connected by short conductors with the return as close as possible to the other conductor; use coax conductors or a flat-strap conductor close to the return conductor. Use coaxial or surface mount capacitors.

(continued next page)

Table 2.5—(continued)

Type of Noise	Causes	Result and Correct
Internal Spike	Combination of temperature and voltage stresses causing breakdown of dielectric	Permanent damage to most capacitors; stress should be decreased by using several capacitors in series. Self-healing capacitors like foil tantalum might help
Random noise on RF capacitors, e.g., silver mica, silver ceramic	Flutter or scintillation caused by random and sudden change in capacitance due to improper adhesion of silver to the dielectric, allowing intermittency	Some areas of the capacitor are out of the circuit; proper manufacturer will avoid this
Electrolytic capacitor scintillation	Voltage surges greater than working voltage or exceeding temperature rating	Keep voltage sources and the temperature below the ratings
Noise pulses during charging and discharging at VLF	Release of dielectric stress, particularly with polystyrene and quartz dielectrics	Avoid low frequency use of polystyrene and quartz dielectrics; liquid dielectrics may minimize effect
Noise pulses	Plastic dielectrics such as polyethylene continue to polymerize even after being put in use. Stresses eventually cause noise pulses	Use other dielectrics
Possible noise in polyethylene, quartz and mica dielectric capacitors (particularly small capacitance units	Light, radioactive particles, and X-rays produce ionic action on the dielectric	Shielding or other means
Passing or absorption of energy at certain frequencies	Resonant frequency caused by capacitance, inductance, and physical qualities	Proper design and shielding

2.2.2 Leadless (Surface-Mount) Capacitors and Other Special Mountings

As for resistors, leadless capacitors have the advantage of a virtually null lead inductance, so the only remaining inductance is the internal foil inductance (see Table 2.4). A lower inductance is especially attractive for high-frequency capacitors because as the parasitic resonance is moved higher in frequency the useful frequency domain of the capacitor is extended.

Leadless ceramic capacitors are available in values ranging from 0.5 pF up to 10 μF, in sizes not exceeding 6 × 5 × 2 mm. Leadless electrolytics are available in values from 0.1 μF up to several hundred μF, in sizes not exceeding 12 × 4 × 4 mm.

Other special realizations with low or no lead inductances are: the capacitive busbars, the BitGuard® and the capacitor arrays.

The capacitive busbar like the "Q-Pac" from Rogers Corp. was designed to provide a low-impedance decoupling in PCB applications. It is made of two parallel copper stripes separated by a thin dielectric layer with a high ϵ_r. These stripes can be placed underneath the device to filter or used to distribute the supply voltage with a highly capacitive line. Applications of this technique are shown in Chapter 8 (PCBs).

The BitGuard®, introduced in 1982 by AVX, is a ceramic capacitor chip designed to be incorporated directly into the substrate of an integrated circuit so that its decoupling action is available right at the supply pins (Fig. 2.16). Its parasitic inductance is in the order of 1 to 2 nH.

The capacitor array is a multiple-port capacitor made of a ceramic block with a series of plated through holes and a metallized edge. Each wire or lead to be filtered is inserted into a metallized hole, constituting one electrode. The other electrode is made by the metallized rim which can be wave soldered to the common ground in the application (Fig. 2.17). Arrays with up to 128 holes and values from 30 pF up to about 1 μF are available by this process.

2.2.3 Microelement Capacitors

Like, and even better than with resistors, capacitors can be fabricated by integrated circuit technologies. They, too, can be found as integrated (monolithics) and film deposit (hybrid) types. In-

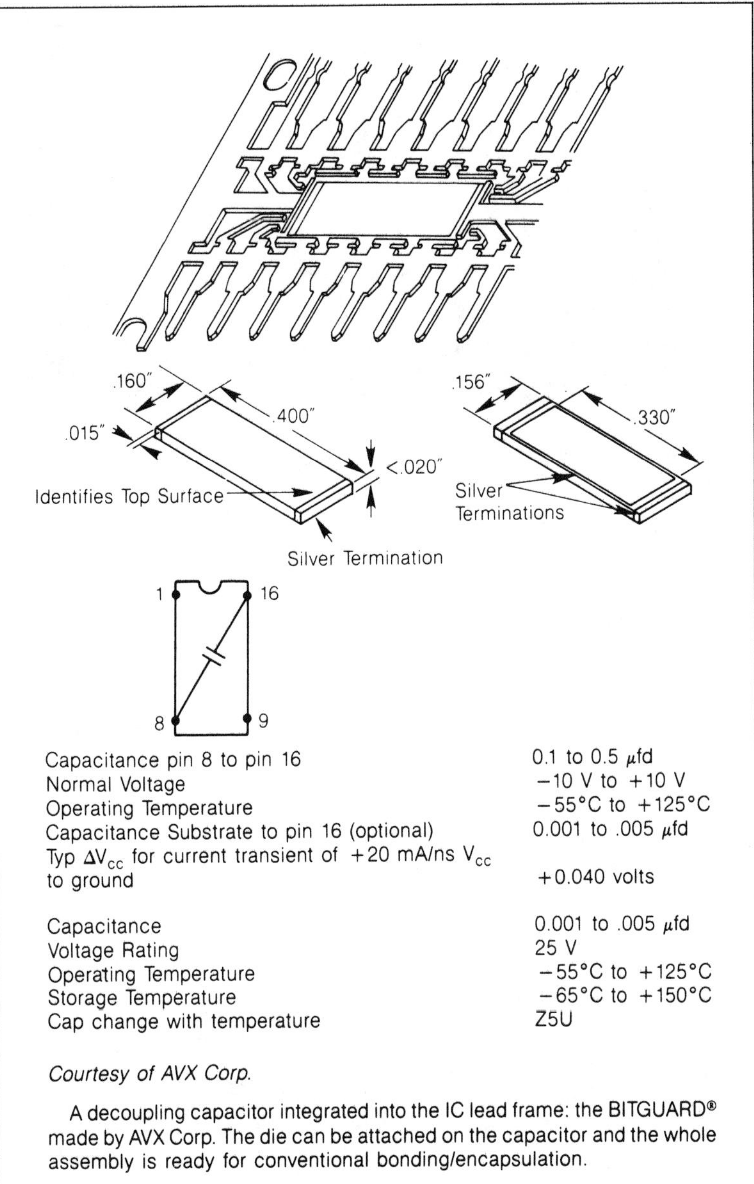

Capacitance pin 8 to pin 16	0.1 to 0.5 μfd
Normal Voltage	-10 V to $+10$ V
Operating Temperature	$-55°C$ to $+125°C$
Capacitance Substrate to pin 16 (optional)	0.001 to .005 μfd
Typ ΔV_{cc} for current transient of $+20$ mA/ns V_{cc} to ground	$+0.040$ volts
Capacitance	0.001 to .005 μfd
Voltage Rating	25 V
Operating Temperature	$-55°C$ to $+125°C$
Storage Temperature	$-65°C$ to $+150°C$
Cap change with temperature	Z5U

Courtesy of AVX Corp.

A decoupling capacitor integrated into the IC lead frame: the BITGUARD® made by AVX Corp. The die can be attached on the capacitor and the whole assembly is ready for conventional bonding/encapsulation.

Figure 2.16—A decoupling capacitor integrated into the IC lead frame: the BITGUARD® made by AVX Corp. The die can be attached on the capacitor and the whole assembly is ready for conventional bonding/encapsulation.

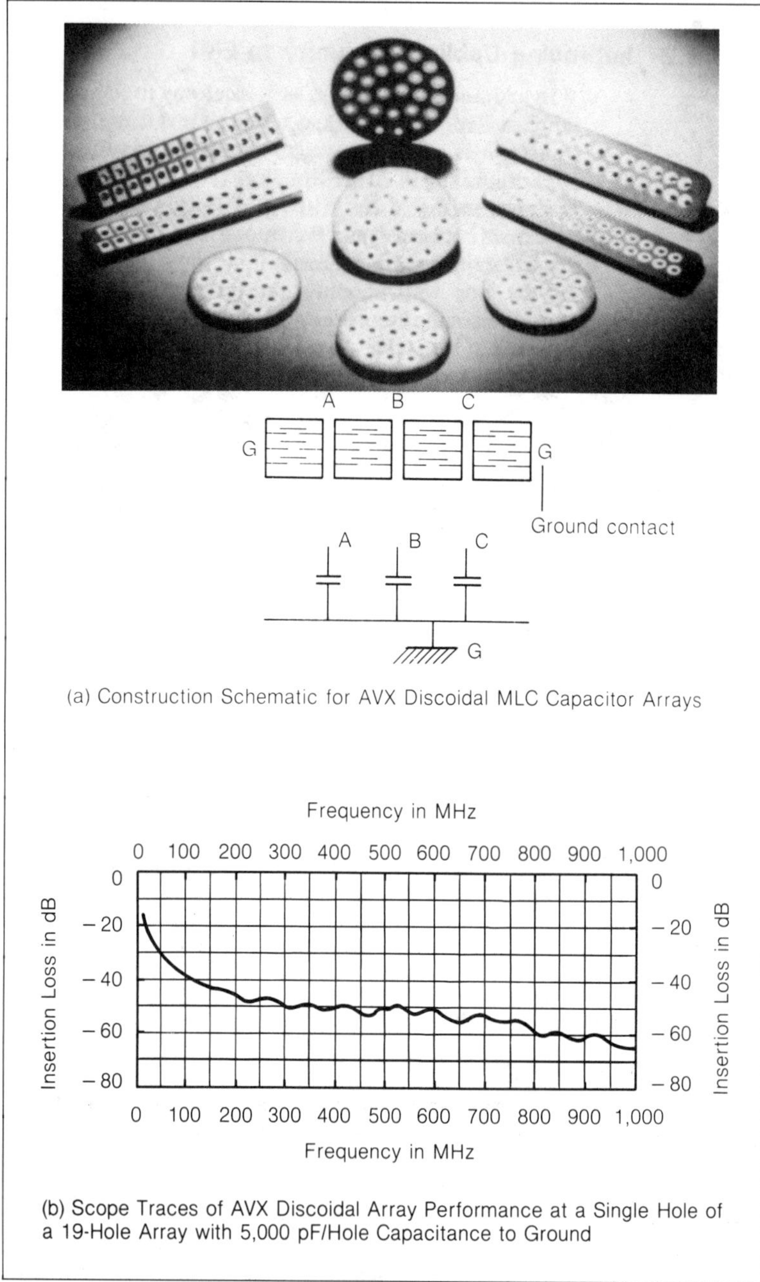

Figure 2.17—Ceramic Capacitor Arrays by AVX Corp.

tegrated capacitors are available in both bipolar and unipolar (metal-oxide) technologies. For bipolar, because it uses a collector-base or emitter-base junction, the capacitance is dependent on the thickness of the depletion region, which is dependent on the applied voltage. Therefore the accuracy is poor and parasitic effects may occur with the C value being modulated by an EMI signal. Also, this capacitor is polarized.

A MOS capacitor more closely resembles a conventional capacitor since it consists of a thin layer of SiO_2 between two conductive or semiconductive surfaces. Such a capacitor has lower parasitic effects than its bipolar counterpart.

No large values of C can be obtained in this way, but with a dielectric constant $\epsilon_r = 12$ for the SiO_2 and a junction thickness of 10 μm, the corresponding capacitance is about 1 nF/cm^2. Practical values from a few pF to a few thousand pF are obtained.

Unlike their discrete counterparts, such capacitors have virtually no parasitic inductance and can be used up to hundreds of GHz. Thin film (hybrid) capacitors can use many different dielectric materials; hence, they have the capability of larger values of C and can show high Q factors. They are nonpolarized and immune to voltage fluctuations, an advantage against EMI. On the other hand, they require more manufacturing steps.

2.3 Inductors

An inductor can be formed by a portion of a conductor and its return, a single conductor coil or, usually, a multiturn coil. The inductance of different straight wires as a function of length is shown in Fig. 2.18. An intentional inductor exhibits inductance, resistance, capacitance between turns, and capacitance between turns and ground, shields, and other circuits.

Figure 2.19 shows the equivalent circuit of a typical inductor. An air-core inductor may be wound on a nonmagnetic metal or insulator core. Magnetic cores are made with steel or iron alloy in sheet, strip, wire or powder form.

Current flowing through an inductor coil causes magnetic flux to produce an EMF proportional to the rate of change of flux linkage.

Self-inductance is:

$$L = N\Phi/I \tag{2.10}$$

where,

L = self-inductance in henrys
N = number of turns, loops, or linkages
Φ = magnetic flux or webers (1 T × 1 m^2)
I = current in amperes

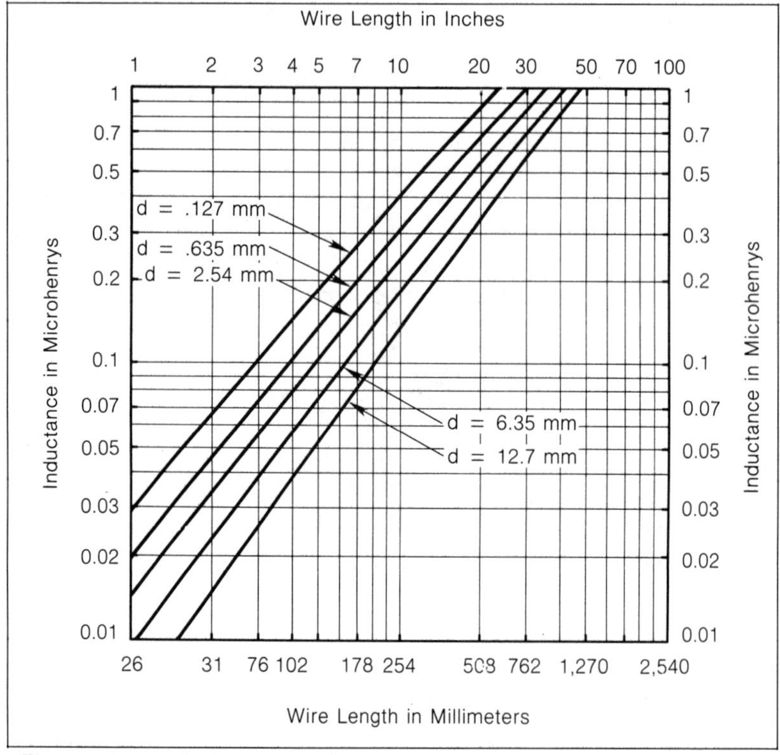

Figure 2.18—Self-Inductance of a Straight Round Wire at High Frequencies

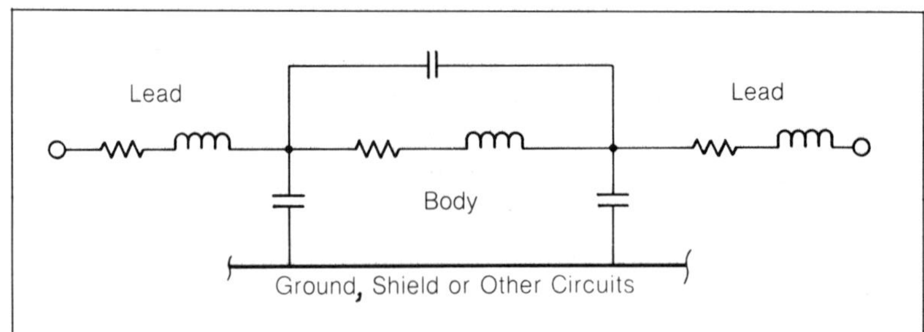

Figure 2.19—Equivalent Circuit of an Inductor

Since on the other hand the flux is $\Phi = \mu HS = \mu S \times 4\pi NI/l$ for a cylindrical coil, the value of L for a bobbin-type inductance is:

$$L = \mu_r 4\pi N^2 S/l \times 10^{-9} \text{ H} \tag{2.11}$$

where,

S = area of one loop in cm^2
l = length of the bobbin core in cm
μ_r = relative permeability of core material
    ($= 1$ for air)

or since,

$S = \pi D^2/4$
$$L = \mu_r \, 10^{-8} N^2 D^2/l \text{ H} \tag{2.12}$$

So the inductance is proportional to the square of the number of turns and to the relative permeability of the core material.

2.3.1 Inductor Parameters

The important parameters of an inductor, as far as EMI is concerned, are distributed capacitance, effective inductance, Q factor, self-resonant frequency and saturation current.

Distributed Capacitance—A small capacitance between turns of the windings is unavoidably produced by their proximity. The combination of all these small capacitances across the whole inductor, comprising the turn-to-turn capacitance and eventually the layer-to-layer capacitance, is called the distributed capacitance.

For one layer, the distributed capacitance is approximated by:

$$Cp = 0.0088 \; \epsilon_r A/t \text{ pF} \tag{2.13}$$

where, A = layer area in mm^2
             t = dielectric thickness in mm
             ϵ_r = insulation dielectric constant

If several layers of winding are used (see Ref. 3),

$$Cp \, (\text{tot}) = \frac{4Cp}{3N} \left(1 - \frac{1}{N} \right) \tag{2.14}$$

where Cp is the distributed capacitance of the first layer, per Eq. (2.13).

The value of Cp is generally validated by measurements, according to the formula:

$$Cp = \frac{C_1 - 4C_2}{3}$$

(2.15)

where, C1 = capacitance in pF required to resonate the inductor at frequency f

C2 = capacitance in pF required to resonate the inductor at frequency 2 × f

The distributed capacitance is one of the key features in parasitic effects of the inductor, since not only does it govern its parallel resonance (Fig. 2.20) but also its dielectric losses and instability of characteristics. High-precision circuitry can lose appreciable precision and stability when temperature and humidity cause dielectric losses from the insulation and support of the inductors.

Several techniques to reduce it are available, some of them traceable back to the pioneers in the early days of radio (Lee De Forest). They include minimum amount of dielectric in the coil support and winding insulation, maximum separation between the in-

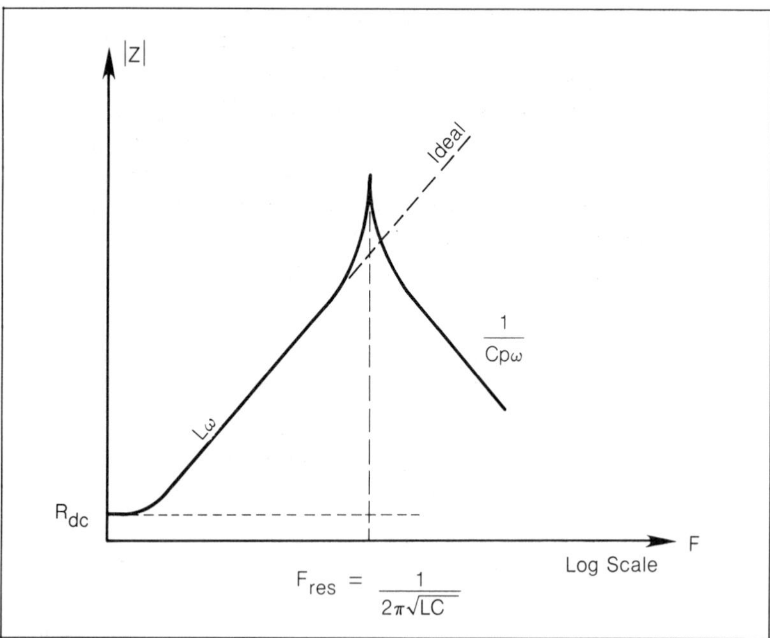

Figure 2.20—Impedance vs. Frequency of Ideal and Actual Inductors

2.30

put and output leads and special winding techniques as shown in Fig. 2.21.

Effective Inductance—The value which departs from the true nominal value, which is defined as follows:

$$L_e = \frac{L}{1 - (LCp\ \omega^2)\ 10^{-18}} \tag{2.16}$$

where, L_e = effective inductance in μH
   L = nominal inductor value, as measured at 1 kHz, for instance
   Cp = distributed capacitance

It appears that from 1 kHz up to the resonance, L_e is always greater than L, generally by a few percent.

The **Q factor**, or figure of merit of an inductor, is the ratio of its reactance to its resistance **at a given frequency** (since X_L, R and even L are frequency dependent).

The formula for Q is:

$$Q = \frac{\omega L_e}{R} \times 10^{-6} \tag{2.17}$$

for L_e in μH, and R in Ω.

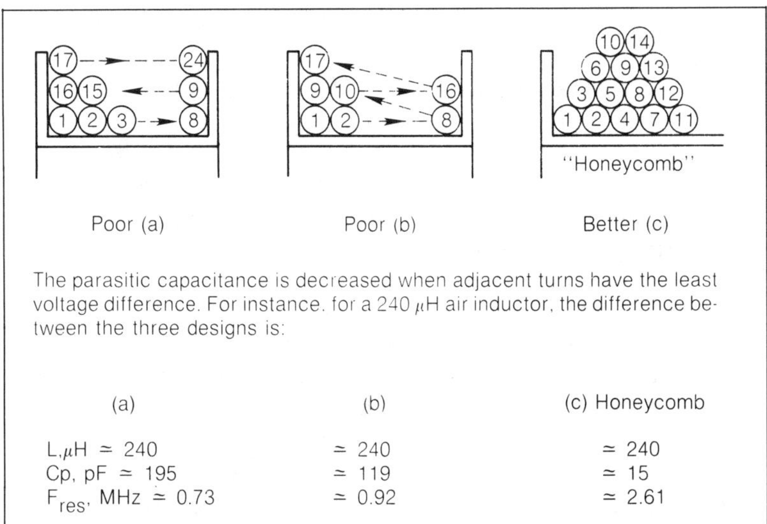

The parasitic capacitance is decreased when adjacent turns have the least voltage difference. For instance. for a 240 μH air inductor, the difference between the three designs is:

	(a)	(b)	(c) Honeycomb
L,μH	$\simeq$ 240	$\simeq$ 240	$\simeq$ 240
Cp, pF	$\simeq$ 195	$\simeq$ 119	$\simeq$ 15
F_{res}, MHz	$\simeq$ 0.73	$\simeq$ 0.92	$\simeq$ 2.61

Figure 2.21—Reduction of Parasitic Capacitance in Bobbin-Type Inductor

For values of $Q > 10$, the reciprocal of Q is a close approximation of the circuit power factor, i.e.:

$$\frac{1}{Q} \cong \frac{P_{active}}{P_{apparent}} \text{ or } \frac{watts}{volts \times amperes}$$

Self-resonant frequency is the minimum frequency at which X_L and X_C are equal.

Saturation current, for magnetic core inductors, is the current above which the core permeability decreases due to saturation, generally expressed as the value of dc current which causes L to decrease by five percent. Following are some typical values for actual inductors:

Induct.	Q at Freq.	R_{DC} (Ω)	I_{MAX} (A)	F_0, Self-Reson. (MHz)
Air Induct. 10 μH	60 @ 7.9 MHz	0.65	0.55	40
100 μH	45 @ 2.5 MHz	4.9	0.2	8.5
1,000 μH	50 @ 0.79 MHz	17	0.11	2

2.3.2 EMI Considerations in Inductor Applications

Besides self-resonance, the usual EMC problems involving inductors are:

EMI Generation
- H-field coupling
- E-field coupling
- Counter-EMF when switching
- Ringing on transients
- Core saturation (increases leakage flux)

EMI Susceptibility
- Skin effect
- Core saturation (modifies L)
- Resonant frequencies
- Mutual inductance

H-field coupling is due to the portion of the field which is not "confined" within the inductor volume. Air cores or open type magnetic cores are more likely to create EMI in their vicinity due to high leakage flux.

Mutual inductance (inductive coupling) occurs when magnetic flux lines of one element link with an area of another element. Energy is thereby coupled from one element to the other and can cause interference to other circuits as suggested in Fig. 2.22. Mutual coupling decreases with distance, and the least coupling occurs when the inductor axes are at right angles to the area of the victim circuit. Conductors made from toroids have considerably reduced external magnetic fields since nearly all the field is inside the magnetic toroid material.

Good practices to minimize magnetic leakage are similar to those used for transformers (see Chapter 5). Unfortunately, they work against what should be done to minimize parasitic capacitance; the designer has to decide on some tradeoff.

When both capacitance and leakage flux have to be kept low, the solution is an electromagnetic shield. The magnetic shield can be made of ferrous metal or a high permeability material around the inductor bobbin. A reduction of low-frequency magnetic leakage can also be made by a good-conductivity closed ring around the

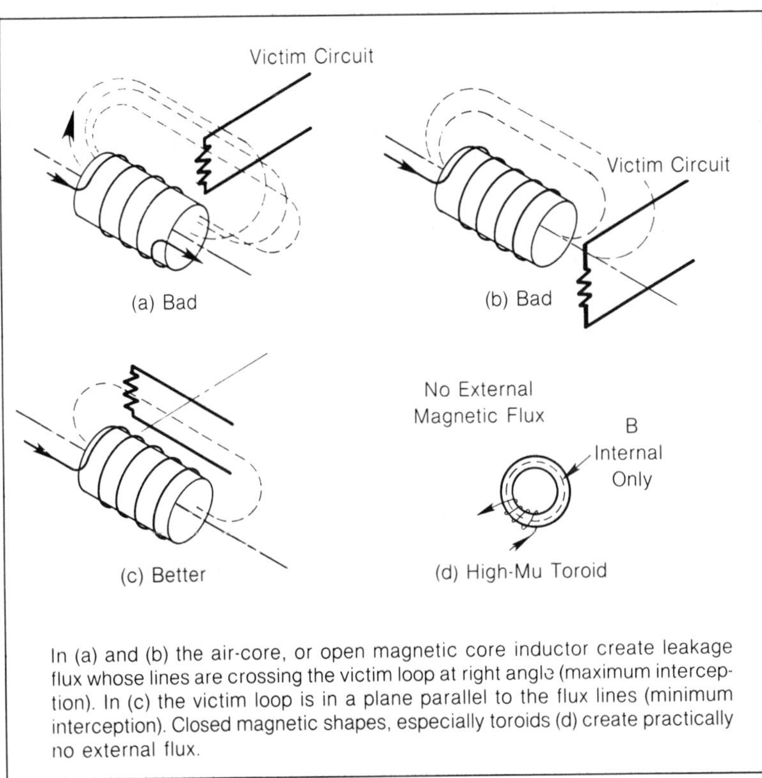

In (a) and (b) the air-core, or open magnetic core inductor create leakage flux whose lines are crossing the victim loop at right angle (maximum interception). In (c) the victim loop is in a plane parallel to the flux lines (minimum interception). Closed magnetic shapes, especially toroids (d) create practically no external flux.

Figure 2.22—External Magnetic Fields for an Air-Core Inductor and Torroid Inductor

inductor; the existing leakage field will induce a voltage in this short-circuited loop, which in turn will force a current to flow. Hence, a flux will develop which will more or less neutralize the leakage flux. The electrostatic shield prevents electric coupling to the nearby circuit in the cases where the inductor generates high values of dV/dt. This shield needs to be a casing made of good-conductivity metal such as copper, aluminum, zinc or tin-plated steel. Contrary to the magnetic shield, the electrostatic shield cannot be left floating (it would be useless) and must be referenced to the low-potential terminal of the inductor.

Since both electric and magnetic shields interfere with the reactive behavior of the inductor, they unavoidably affect the value of L. So for a shielded inductor, **the value of L should be measured with the shield in place** or calculated from the curves of Fig. 2.23.

Too often neglected, there is a simple, inexpensive recipe to

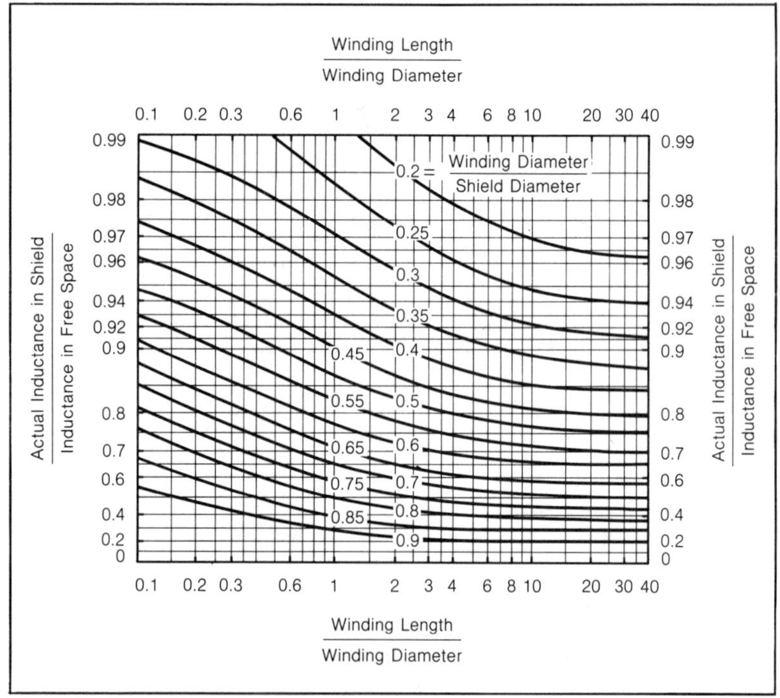

Figure 2.23—Determination of Inductance Decrease When a Solenoid Is Shielded.[1] Note: these curves are accurate provided the radius of the shield (if cylindrical) or half its side (if square) is at least twice the coil radius.

minimize inductive coupling from (and to) an inductor: this is to take care of the respective orientation of the source and victim loops and to keep their axes at right angles.

Core saturation, which creates both emission and susceptibility problems, can be avoided by proper core shape, cross-section and eventually using gapped or slotted cores. If the inductor has a constant (or at least unipolar) current flowing through, the corresponding saturation can be neutralized to some extent by putting a permanent magnet in series in the magnetic circuit to provide an H-field "offset" in the opposite direction. This reverse pre-magnetization will move the actual saturation current to higher values.

From the previous discussion it can be concluded that although capacitors can be quasiperfect elements (the coaxial, feed-through capacitor), the quasiperfect inductor has yet to be invented. The closest configuration to a perfect inductor is the one-turn ferrite bead or sleeve (see Section 5.3, "Ferrites and Baluns"), but these components are limited to low values of L.

While they have EMI problems of their own, inductors are the keystone of many EMI filters. In such devices, inductors are sometimes used as common-mode chokes. The inductor is made as a bifilar winding, and the two leads of the line to be filtered are wound in the same direction. With such an arrangement (Fig. 2.24) the differential-mode current (the normal line current, for instance) does not "see" the inductance since the two wires create fluxes which cancel each other. As a result, the choke does not saturate for the normal line current. By contrast, the common mode EMI currents are flowing in both wires in the same direction, so they create in the core two fluxes having the same orientation. But flux Φ_1 created by current I_1 induces in wire No. 2 a current $I_{1\text{-}2}$ which

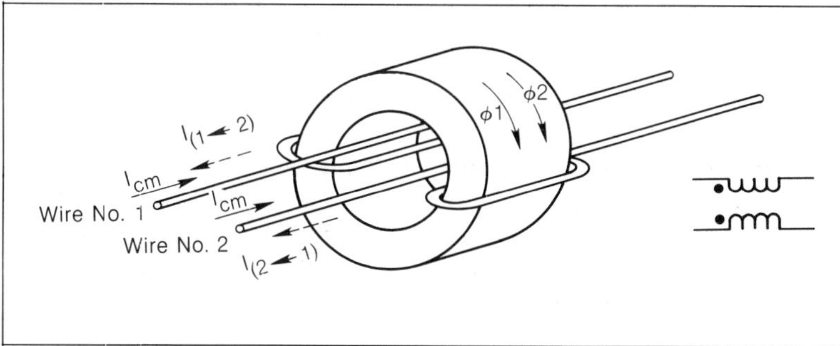

Figure 2.24—Principle of the Common-Mode Choke

tends to oppose current I_2. All the same, the flux Φ_2 created by current I_2 induces in wire No. 1 a current $I_{2\text{-}1}$ which is opposed to current I_1.

Another type of inductor specific to EMI is the RF blocking choke to be inserted in grounding or earthing conductors to avoid circulation of ground-loop currents. These chokes have a typical value of 20 to 1,000 μH so that they do not represent a safety hazard at 50/60 Hz, but oppose a significant impedance in the 10 kHz to a few MHz range. Above this range, their parasitic capacitance and the *in situ* stray capacitance generally spoil their efficiency.

Such chokes, because of their mounting in safety ground wires, are subject to some specific requirements such as IEC 85 and VDE 0550. Their wire size must be at least 1 mm^2 for rated currents of 16 A or less, then be at least equal to the size of power-line wires for currents above 16 A. These chokes could temporarily become a hazard if the installation is subject to large surges of earth return currents such as those occurring during lightning stroke, sudden common mode gradients, etc.

In that situation, the LdI/dt surge created by the inductor is clamped by a varistance or other suppression component mounted across the coil. Another approach, used by the British manufacturer Belling & Lee, is to use a highly saturable core so that when the current exceeds a few amperes, and especially with large dI/dt, the magnetic circuit saturates. Consequently, the inductor is left with a $\mu_r = 1$ and behaves temporarily like an air choke with a few hundred nH of inductance. Table 2.6 summarizes EMI problems and solutions associated with inductors.

2.3.3 Inductors in Microelectronics

Of the three passive elements (R, L and C), inductances are the least compliant with miniaturization or integration. Of the elements which can be formed in a silicon integrated semiconductor circuit, the inductor is simply **not** available.

Therefore, all that can be done is to mimic a real inductor by forming a circular or square conductor spiral on the substrate. Concerning EMI, this limitation means a high leakage field. Sometimes a tiny chip of ferrite with miniature winding is used as an add-on rather than to form it as a film.

Table 2.6—EMI Problems, Causes and Corrections in Inductors

Problem	Possible Cause	Corrections
Passing or stopping desired signals	Resonance due to inductive and capacitive characteristics	Design and shielding
Power dissipation over the ratings	Skin effect, conductor resistance, eddy current and hysteresis losses	Use of stranded conductors, non-magnetic cores and toroidal geometry winding
Varying effective inductance	Distributed capacitance in coil causes apparent decrease in inductance	Minimize distributed capacitance and/or permit high eddy current effect to reduce effective inductances
Interference by mutual inductance	Proximity of two coils, one coil and a conductor, or any two circuit elements	Separation, shielding and keeping object axes at right angles
Transients	Current applied or cut off to the inductor	Same as for electromagnetic relays

2.4 References

1. Jordan, E.C., ed., *Reference Data for Radio Engineers: Radio, Electronics, Computer, and Communications*, Seventh Edition, (Indianapolis: Howard W. Sams & Co., Inc., 1985).
2. Ott, H.W., *Noise Reduction Techniques*, (New York: John Wiley & Sons, 1976).
3. Mazda, F., *Electronic Engineer's Reference Book*, (Butterworths, 1983).

Chapter 3

Insulators and Conductors

This chapter reviews the topics of insulators and conductors as potential sources of EMI and some of the control techniques that are employed to suppress interference. Details regarding both the conductor physics and EMI control methods are covered in Volume No. 2, *Grounding,* and Volume 3, *Shielding,* of this handbook series.

Although insulators and conductors are not actually "components," their contribution to EMI/EMC, both as causes and possible remedies, make it necessary to review their dc, ac and transient behavior. Even though there is not a clear line of definition between what is an insulator and what is a conductor, the regions of Table 3.1 are generally accepted.

Table 3.1—Classification of Conductive and Insulating Materials Based on Their Resistivity

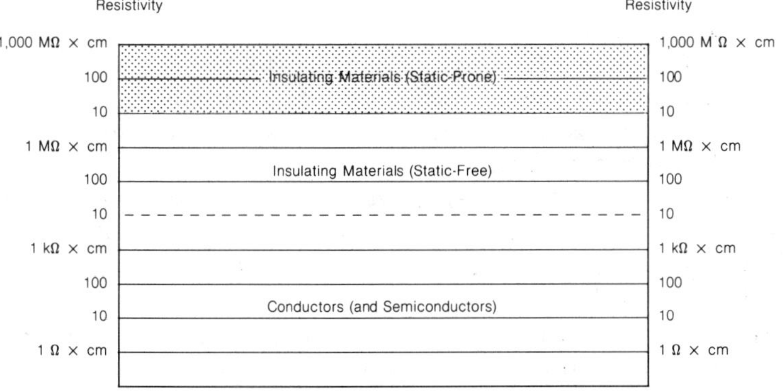

3.1 Insulators

Insulators permit small leakage currents to flow despite their low conductivity, leading to voltage surges and changes in current. The insulation around a cable illustrates this phenomenon. When a corona occurs, insulation breaks down and small currents creep along the surface.

Dielectric loss in insulators, called loss tangent or power factor, causes dissipation of power, loss of linearity and coupling with other circuits. It is especially important in coaxial lines and capacitor construction. This problem is eliminated by selecting proper materials such as polyethylene or Teflon® for the frequency range of interest.

Surface tracking, a condition in which small currents creep across an insulator, is an important source of EMI. It is caused by surface contamination of the insulation by moisture or solid conductive particles, by chemical degradation of the insulation material or by momentary overvoltage. When a discharge occurs across a low impedance circuit, significant energy may be released and quick, catastrophic degradation occurs. If the circuit impedance is high, a slow discharge occurs over a long time interval. The arc changes paths continuously during this interval, extinguishing in one section but igniting in another. These changes are fast and variable, producing a broad spectrum of interference.

Corona differs from a discharge across insulation because it is highly concentrated at one point, usually shows a visible glow and has a humming sound caused by highly ionized air around the conductor. Special cases of corona may develop in gaseous pockets or voids within solid insulation, usually showing no glow. These cases also cause EMI since they develop slowly and lead to insulation failure.

Internal breakdown of the insulation produces small changes in current as occurs in surface tracking. Small internal paths are overheated, causing chemical change that results in lower local resistivity and accelerated damage. High-voltage gradients sometimes occur across small mechanical voids in an insulator. Gaseous discharges also occur in the voids, with current changes generally leading to degradation.

These effects are simulated during high-potential tests and life-tests which are performed on capacitors and electrical wires using insulating materials. With modern insulating materials, temporary breakdown of the insulation can occur during a voltage transient of short duration and have no effect on the long-term reliability of the insulation barrier, because of the "self-healing" effect.

This effect occurs, for instance, in metallized paper or plastic capacitors. It also occurs on power lines: utility companies consider that on multiconductor buried lines and building wiring, although short transients in theory could reach greater values, they are "clamped" to about 6 to 8 kV by the self-limiting effect of insulation breakdown.

Another effect of a recoverable insulation rupture is the distortion of the ac sine wave on high-voltage power lines. A degraded insulator may still perform well enough for 90 percent of the voltage amplitude but will arc every time near the peak of the sine wave. As shown in Fig. 3.1, a frequency spectrum which now extends to higher frequencies than the mere 50 or 60 Hz causes the power line to generate conducted and radiated EMI.

Finally, there has been a recent outgrowth of problems created by the propensity of insulating materials to build up static charges, resulting in a subsequent electrostatic discharge which can create interference problems and even destroy microelectronic devices. Covering the ESD problem is beyond the scope of this chapter, but a brief summary of the static build up by insulators is given. (See Refs. 1 through 3 for a deeper explanation).

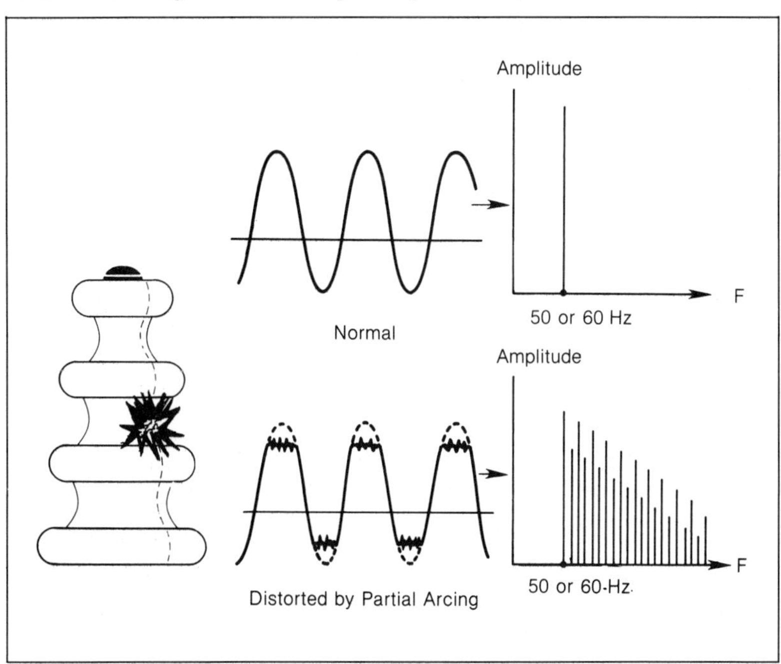

Figure 3.1—Degraded Insulator Causing Partial Arcing and EMI Noise

Any material is made of atoms. Unless submitted to certain external influences (heating, rubbing, electrical stress, etc.), the atom is in equilibrium; that is, the amount of negative charges represented by the electrons orbiting around the nucleus is exactly balanced by an equal number of positive charges, or protons, aggregated in the nucleus. Therefore, the net electric charge seen from the outside is zero.

In metals, the mobility of electrons is such that the conditions of equilibrium will always exist, i.e., no significant static field will exist between different zones of the same piece of metal. With nonconductive materials, however, the lesser mobility of electrons does not provide such rapid recombination of charge imbalance. If heated or rubbed strongly (which also creates heat), a nonconductor will free up electrons. A nonconductive material which gives up an electron will become positively charged. Such unbalanced atoms are called positive ions. A nonconductive material which takes extra electrons will become negatively charged, and its atoms with an excess of electrons are called negative ions.

In conductors, the recombination is instantaneous. In insulators, the high resistance does not allow quick recombination, and charges can build up and remain until a high gradient of field causes arcing. The ability of insulators to acquire electrostatic charges is often shown as a triboelectric scale (Table 3.2).

Table 3.2—Triboelectric Series

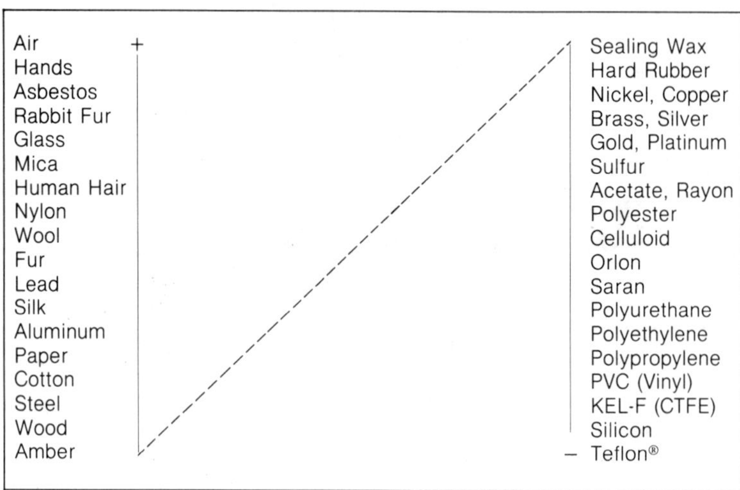

Aguet relates the propensity toward electrostatic charge to the dielectric constant of the materials.[4] He indicates the surface charge density ρs:

$$\rho s = 15.1^{-6} (\epsilon_{r1} - \epsilon_{r2}) \text{ C/m}^2 \qquad (3.1)$$

where, ϵ_{r1} and ϵ_{r2} are the relative permittivity of the two materials.

Besides the well-known phenomenon of carpets and synthetic textiles causing human ESD of 10 kV or more, static charges can develop in rubber or textile belts and conveyors and their pulleys/rollers, cooling fans with plastic rotor blades, moving paper (printers, copiers) and with the rapid flow of gas, liquid or granules against an insulating material or ungrounded conductor. Examples of insulating materials contacting ungrounded conductors include cleaning with an air gun, PVC "skin-packing" with hot air blast, cleaning with a solvent, fuel lines, loading or dumping grain, in silos, rocket exhaust nozzles, radomes, fiberglass hoods and tips, thermal blankets (spacecraft) and continuous friction of a cable harness against an insulated structure.

The static charge can be especially critical when a cable harness is attached to a moving piece of equipment. Some modern insulators used for wiring insulation, such as KAPTON®, have an extremely high resistivity and can keep their charge for a very long period. If they connect to a high impedance circuit, several kilovolts can appear at the victim's input.

Table 3.3 gives some parameters for common insulating materials. The propagation data allow for calculation of propagation delay when conductors are embedded in an homogeneous insulating medium. A simple rule of thumb is that compared to the velocity V_o in the air, the velocity V_x in an insulating material is reduced as:

$$V_x = \frac{V_o}{\sqrt{\epsilon_r}} \qquad (3.2)$$

A side effect of modern dielectric materials, which also contributes to EMI, is the fact that their outstanding insulation resistance allows the use of less insulation thickness around the wires and inside the connectors. As a result, mutual capacitance between wires in a bundle (or between connector pins) increases, and crosstalk is aggravated accordingly.

For instance, in many ways Teflon is an ideal insulating material for wires in cables. It offers a high insulation with good mechanical protection in extremely thin conductor coatings, allowing a high density of wires within a cable or harness. This in turn may increase circuit-to-circuit coupling within a cable because of closer spacing. Capacitive coupling is a function of the distance between conductors, the length of the conductors and the dielectric constant of the material between conductors. Since capacitance is proportional to dielectric constant, the capacitance will increase by a factor of two when the air between two wires is replaced with Teflon.

Ionic atmosphere, created by field gradients in the region of conductance, can promote the deterioration of the insulation and eventually result in a change from the original characteristics.

Finally, Table 3.4 shows typical noise problems with insulators and their solutions.

Table 3.3—Important Parameters of Common Insulating Materials

Material	Average Dielectric Constant ϵ_r	Volume Resistivity $\Omega \times cm$	Propogation	
			Delay ns/m	Speed m/ns
Plexiglass® (Polycarbonate)	3	10^{15} to 10^{16}	5.88	0.17
Glass	4 to 8	10^{11} to 10^{13}	6.5 to 9.4	0.14 to 0.1
Epoxy glass	4.4	10^{7} to 10^{9}	7	0.143
PVC	3	10^{14} to 10^{15}	5.88	0.17
Paper	3 (typical)		5.88	0.17
Polyethylene	2.3	10^{15} to 10^{17}	5	0.2
Polystyrene	2.5	10^{18} to 10^{19}	5	0.2
Teflon®	2	10^{17} to 10^{18}	4.7	0.21
Kynar®	8	10^{14}	9.4	0.1
Kapton®	3.5	10^{17}	6.2	0.16
Pure water	78		29.5	0.034

Table 3.4—EMI Problems, Causes and Corrections in Insulators

Problem	Cause	Preventive Measure
Surface tracking causing dI/dt charges which produce wide band of frequencies	Surface contamination, chemical degradation of insulation, momentary over-voltage	Protection from contamination, use of proper material, proper voltage design
Surface tracking resulting in catastrophic degradation	Low impedance discharge circuit releasing high energy	Protection from contamination, use of proper material, proper voltage design
Surface tracking with small but relatively persistent EMI	High impedance discharge circuit which limits energy to circuit component tolerances	Protection from contamination, use of proper material, proper voltage design
Glow discharge which may be visible and audible but causing electrical noise because of dI/dt effects	Corona with high voltage across one conductor to ground or to another conductor	Prevention of voids within insulators, protection from contamination, prevention of sharp points at high potentials, low voltage gradient design
Static voltage appearing on insulated wires and harnesses	Static buildup by friction, or dry air flow, etc.	Electrostatic shield, conductive lubricant over cable jacket

3.2 Conductors

A conductor is any material which readily permits an electric current to flow when subjected to a difference in potential. It is usually desirable to use a low-resistance metal such as copper as the conductor in an electronic circuit. Copper is inexpensive and relatively stable in the normal ambient temperature range. It is easily soldered but will corrode when exposed to the atmosphere. For this reason, copper is sometimes protected with a plating or coating of tin, silver or gold.

The selection of a conductor size is generally related to the maximum allowable voltage drop in the conductor or heating effect by power (I^2R) loss. Concerning EMI, there are other relevant parameters to consider in a conductor: ac resistance, external resistance, internal inductance, distortion of steep wave front by skin effect and possible nonlinearities due to boundary phenomenon and corrosion.

The total impedance of a conductor is given by:

$$Z = R_{(dc\ or\ ac)} + j\omega x\ (L_{ext} + L_{int}) \qquad (3.3)$$

3.2.1 External Inductance

External self-inductance is by far the greatest cause of parasitic (undesired) effects in conductors. It is the phenomenon by which any change of current creates a changing flux which, in turn, induces a reverse electromotive force into the original conductor. Since a conductor cannot carry a current without some kind of return path, the self-inductance can only be defined for a given geometry, i.e., the distance between the conductor in question and the return plane (or conductor).

As seen in Fig. 3.2, the larger the wire diameter, the lower the external inductance L_{ext}. However, L_{ext} does not change significantly with diameter since a natural logarithm ln is involved. A ten-times increase in diameter barely reduces the external inductance by a factor of two. Disappointing results generally occur, for example, when one tries to eliminate ground noise problems just by increasing the jumper wire size.

The equations governing external inductance are:

For one wire, and $h < l_w$ (or $D < 2l_w$ in Fig. 3.2),

$$L_{ext} = 0.2 \ln\left(\frac{4h}{d}\right) \mu H/m \qquad (3.4)$$

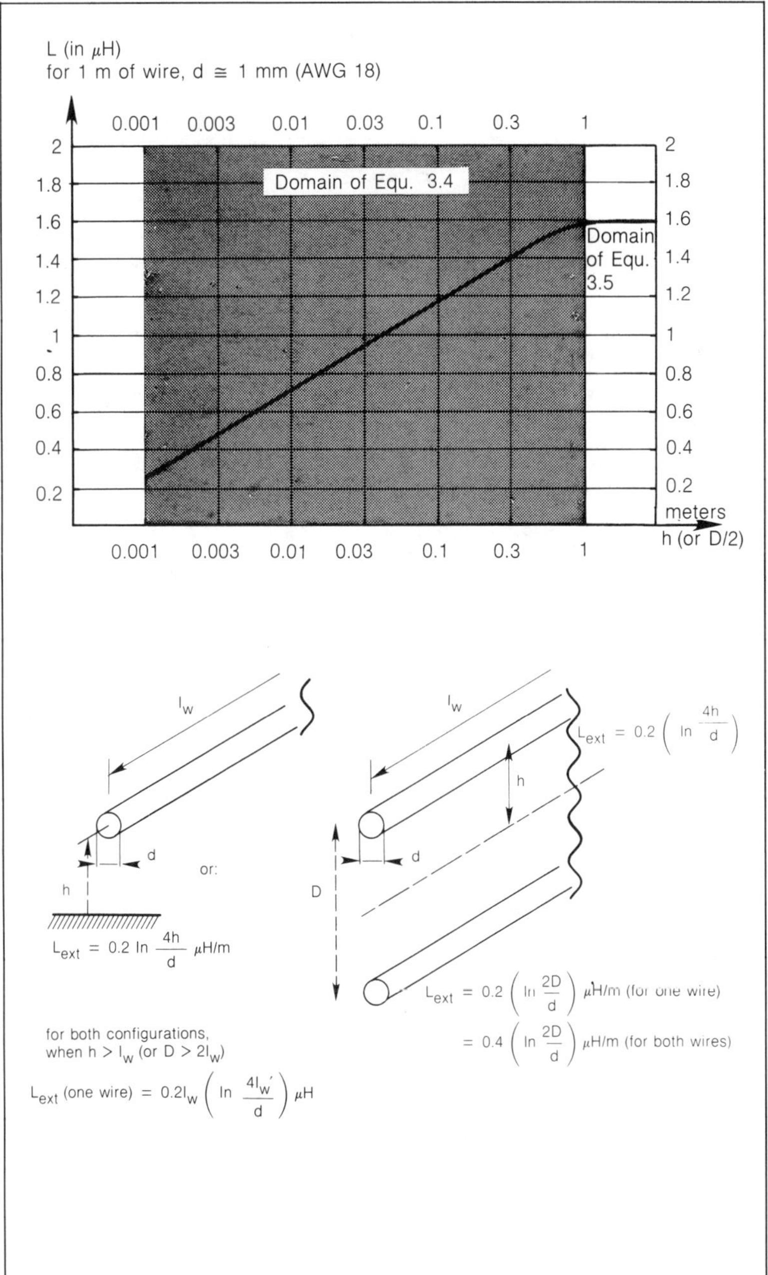

Figure 3.2—External Self-Inductance of Round Solid Conductors

For $h > l_w$ (or $D > 2l$),

$$L_{ext} = 0.2\, l_w \ln\left(\frac{4l_w}{d}\right) \mu H \tag{3.5}$$

where, h = height above ground plane, in meters
      d = wire diameter, in meters
      l_w = wire length in meters

The first equation shows that increasing h for a given diameter d increases L.

It seems that this increase could go on forever, but beyond a certain height (exactly when h becomes equal to l_w, the length of the wire sample), the flux produced by the current in the wire is no longer uniform in the loop and no longer increases with h. At this point Eq. 3.5, which is the free-space inductance of the wire, applies.

For a rough approximation and for ordinary types of configurations, L_{ext} can be approximated by 1 $\mu H/m$ or 10 nH/cm.

3.2.2 Internal Inductance

Internal inductance is due to the effects of magnetic field created by the current within the conductor itself. It, in turn, is closely associated with skin effect since both phenomena are interactive. At low frequency, since the current density is uniform, the H-field distribution inside a perfect conductor appears as in Fig. 3.3a. When frequency increases due to current concentration on a peripheral "skin," the H-field is also leaving the center core (Fig. 3.3b) and the internal inductance L_i decreases accordingly. At dc and low frequencies (quasi-uniform current density) the value of L_i is given by:

$$L_{i(o)} = \frac{\mu}{8\pi}\ H/_m \times l_w \tag{3.6}$$

For a copper wire, in the air, since $\mu = 4\pi \times 10^{-7}$ H/m,

$$L_{i(o)} = 0.05\ \mu H/m \times l_w \tag{3.7}$$

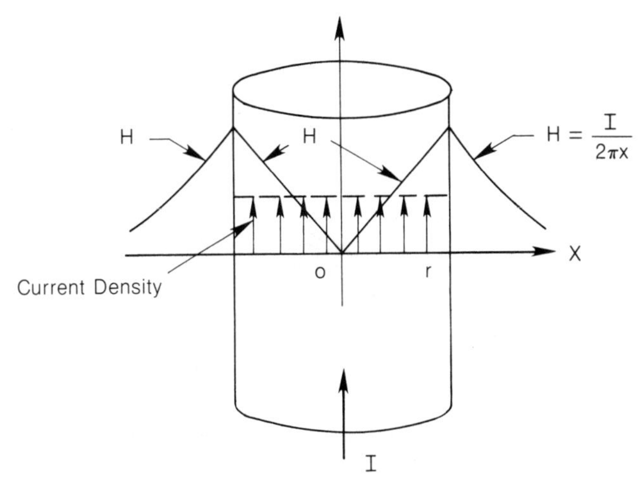

(a) Current and Internal H-Field Distributions at Low Frequency

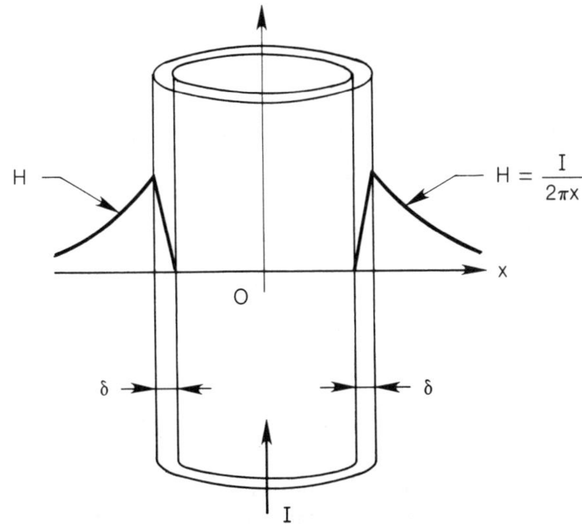

(b) Current and Internal H-Field Distributions at High **Frequency**
(Skin Depth Region)

Figure 3.3—Internal Inductance (L_i) **(continued next page)**

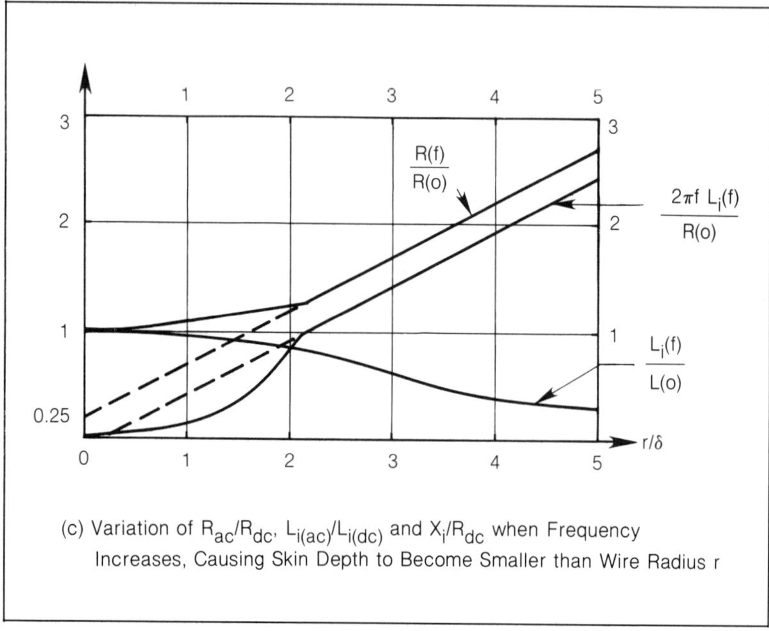

(c) Variation of R_{ac}/R_{dc}, $L_{i(ac)}/L_{i(dc)}$ and X_i/R_{dc} when Frequency Increases, Causing Skin Depth to Become Smaller than Wire Radius r

Figure 3.3—(continued)

As frequency increases, L_i decreases and becomes asymptotic to a value $L_{i(HF)}$ which is approximately equal to $0.25 \times L_{i(o)}$.[5] See Fig. 3.3c.

The frequency above which L_i levels off has been calculated for a few typical examples:

1. Strip line, 1 mm spacing, any width: 435 kHz
2. Typical coaxial cable, 3.5 mm O.D.: 65 kHz
3. Parallel wire pair, wire diameter 1 mm (AWG 18), separation 2 mm: 27 kHz

The value of 50 nH/m of Eq. (3.7) indicates that L_i represents only a few percent of the external inductance. Therefore, L_i is generally neglected in the calculation of the total inductance of a wire. However, its contribution, jointly with that of skin effect, is taken into account in the design of optimal conductor cross section and configuration at high frequencies.

It is also demonstrated that in the high-frequency region, the inductive part L_i and the resistive part R_{ac} are equal. This equality is due to the fact that as L_i varies as $1/\sqrt{f}$, both R_{ac} and ωL_i are increasing as $\sqrt{f}$ (see skin effect below).

3.2.3 AC Resistance

"Skin effect" is the phenomenon by which, due to the magnetic field produced by the current in the conductor itself, the current leaves the center and concentrates near the conductor surface as frequency increases. Consequently, the effective cross section available for the current to flow through decreases.

For any conductor, a fictitious thickness is defined where everything behaves "as if" the current were concentrated in a skin depth given by:

$$\delta \cong \frac{0.066}{\sqrt{\sigma_r \, \mu_r \, F_{MHz}}} \text{ in mm} \tag{3.8}$$

where, $\quad \sigma_r$ = relative conductivity = 1 for copper
$\quad\quad\quad \mu_r$ = relative permeability

From Ref. 5 the ac resistance for a cylindrical conductor is:

$$R_{ac} = R_{dc}\left(\frac{d}{4\delta} + \frac{1}{4}\right) \tag{3.9}$$

where d is the wire diameter and δ the skin depth, in similar units. Therefore, for a copper wire at room temperature and for d/2 < δ,

$$R_{ac} = R_{dc}\,(3.78 \times d \sqrt{F} + 1/4) \tag{3.10}$$

with, $\quad$ d $\quad$ = diameter in mm
$\quad\quad\quad$ F $\quad$ = frequency in MHz

or, $\quad\quad$ R_{ac} $\quad$ = $R_{dc}\,(0.096 \times d\sqrt{f} + 1/4)$ $\tag{3.11}$

with, $\quad$ d $\quad$ = diameter in inches
$\quad\quad\quad$ f $\quad$ = frequency in Hz

If the material is not copper, Eqs. (3.10) and (3.11) can be used provided that the first term is multiplied by a factor:

$$\sqrt{\mu_r \times \sigma_r} \quad \text{or} \quad \sqrt{\mu_r/\rho_r}$$

where μ_r is the relative permeability and σ_r and ρ_r are the relative conductivity and resistivity, respectively.

Table 3.5 shows the values of resistance and skin depth for few conductors. When looking at this table, realize that at high frequency, and speaking **only in resistive terms**, a hollow tube has the same ac resistance as a solid cylinder. Additionally, the ac resistance can be decreased by changing the aspect ratio, i.e., offering more peripheral area for the same cross-section (rectangular conductors have less ac resistance than round ones). The ac resistance for any shape of conductor can be calculated from Eq. (3.10 or 3.11) by letting d = perimeter of cross section/π.

Table 3.5—Skin Depth for a Few Typical Conductors

	Absolute conductivity σ (siemens/m)	δ (mm)
Copper	5.8×10^7	$66/\sqrt{f_{Hz}}$
Aluminum	3.7×10^7	$83/\sqrt{f_{Hz}}$
Brass	1.6×10^7	$127/\sqrt{f_{Hz}}$

f	50 Hz	10^4 Hz	1 MHz	100 MHz	1,000 MHz
δ at 20°C Copper, $\mu_r = 1$	9.4 mm	0.66 mm	66 μm	6 μm	1.9 μm
Iron, $\mu_r = 100$	2 mm	150 μm	15 μm	1.5 μm	0.475 μm

3.2.4 Summary of Impedance Terms

Considering the impedance of various conductors, when all parameters are combined, the impedance of a round conductor at all frequencies is therefore given by:

$$Z = R_{(dc\ or\ ac)} + j\omega L$$

$$\text{with,} \quad L = L_{ext} + L_{int} = \frac{\mu\, l_w}{2\pi} \left[\ln\left(\frac{4\, l_w}{d}\right) + 1/4 \right] \qquad (3.12)$$

for a wire height (or separation) greater than its length l_w.

The constant 1/4 accounts for L_{int}, which will always be equal ($Li_{(o)}$) or less ($Li_{(HF)}$) than this value and is often neglected.

For a round copper wire :

$$L_{(1\ \text{wire})} = 0.2\ l_w \left[\ln \left(\frac{4\ l_w}{d} \right) + 1/4 \right] \mu H \qquad (3.12a)$$

Table 3.6 shows the value of impedances, and Table 3.7 the value or L and R for standard round wires.

When the frequency increases:

1. Current migrates towards periphery.
2. Eventually, current migrates towards sharp angles.
3. When penetrating inside a conductor, the current density varies, as 1/e (one neper) per skin depth, translatable into an 8.8 dB decrease per skin depth.
4. At certain frequencies, inner portions of the conductor become useless or even detrimental (contribute to L_i, do not help for R_{ac}.)

Skin effect also causes distortion of high-frequency signals, especially digital pulses or square wave signals in general. Even for a perfectly matched line and load configuration, with subnanosecond signals, a few meters of cable can already show uneven propagation delays simply by skin effect.

Since R_{ac} is highest for the highest frequencies, any complex spectrum will show more attenuation of its high frequency components than for the fundamental. Fig. 3.4b shows an example of this distortion.[5]

For any line, assuming no reflection mismatch, the distortion can be calculated from Fig. 3.4a where $T_{0.5}$, called the "cable time constant," is defined as follows:

$T_{0.5}$ = time after which a step input reaches 50 percent of its value at the output

$T_{0.5}$ can be calculated from the cable loss (picked at any frequency):

$$T_{0.5} = \frac{\alpha^2 l_w^2}{\pi f}\ \text{seconds} \qquad (3.13)$$

with
α = attenuation in Np/m
= 2.3/20 $\times$ attenuation in dB/m
l_w = cable (wire) length in m
f = frequency in Hz

If f is entered in MHz, T will result in μs.

Table 3.6—Impedance of Straight Circular Copper Wires**

FREQ.	AWG# = 2, D = 6.54 mm				AWG# = 10, D = 2.59 mm				AWG# = 22, D = 0.64 mm			
	l = 1cm	l = 10cm	l = 1 m	l = 10 m	l = 1cm	l = 10cm	l = 1 m	l = 10 m	l = 1cm	l = 10cm	l = 1 m	l = 10 m
10 Hz	5.13 μ	51.4 μ	517 μ	5.22 m	32.7 μ	327 μ	3.28 m	32.8 m	529 μ	5.29 m	52.9 m	529 m
20 Hz	5.14 μ	52.0 μ	532 μ	5.50 m	32.7 μ	328 μ	3.28 m	32.8 m	529 μ	5.29 m	53.0 m	530 m
30 Hz	5.15 μ	52.8 μ	555 μ	5.94 m	32.8 μ	328 μ	3.28 m	32.9 m	529 μ	5.30 m	53.0 m	530 m
50 Hz	5.20 μ	55.5 μ	624 μ	7.16 m	32.8 μ	329 μ	3.30 m	33.2 m	530 μ	5.30 m	53.0 m	530 m
70 Hz	5.27 μ	59.3 μ	715 μ	8.68 m	32.8 μ	330 μ	3.33 m	33.7 m	530 μ	5.30 m	53.0 m	530 m
100 Hz	5.41 μ	66.7 μ	877 μ	11.2 m	32.9 μ	332 μ	3.38 m	34.6 m	530 μ	5.30 m	53.0 m	530 m
200 Hz	6.20 μ	99.5 μ	1.51 m	20.6 m	33.2 μ	345 μ	3.67 m	39.6 m	530 μ	5.30 m	53.0 m	530 m
300 Hz	7.32 μ	137 μ	2.19 m	30.4 m	33.7 μ	365 μ	4.11 m	46.9 m	530 μ	5.30 m	53.0 m	531 m
500 Hz	10.1 μ	219 μ	3.59 m	50.3 m	35.3 μ	425 μ	5.28 m	64.8 m	530 μ	5.31 m	53.2 m	533 m
700 Hz	13.2 μ	303 μ	5.01 m	70.2 m	37.7 μ	500 μ	6.66 m	84.8 m	530 μ	5.32 m	53.4 m	537 m
1 kHz	18.1 μ	429 μ	7.14 m	100 m	42.2 μ	632 μ	8.91 m	116 m	531 μ	5.34 m	53.9 m	545 m
2 kHz	35.2 μ	855 μ	14.2 m	200 m	62.5 μ	1.13 m	16.8 m	225 m	536 μ	5.48 m	56.6 m	589 m
3 kHz	52.5 μ	1.28 m	21.3 m	300 m	86.3 μ	1.65 m	25.0 m	336 m	545 μ	5.71 m	60.9 m	656 m
5 kHz	87.3 μ	2.13 m	35.6 m	500 m	137 μ	2.72 m	41.5 m	559 m	571 μ	6.39 m	72.9 m	835 m
7 kHz	122 μ	2.98 m	49.8 m	700 m	189 μ	3.79 m	58.1 m	783 m	609 μ	7.28 m	87.9 m	1.04 Ω
10 kHz	174 μ	4.26 m	71.2 m	1.00 Ω	268 μ	5.41 m	82.9 m	1.11 Ω	681 μ	8.89 m	113 m	1.39 Ω
20 kHz	348 μ	8.53 m	142 m	2.00 Ω	533 μ	10.8 m	165 m	2.23 Ω	1.00 m	15.2 m	207 m	2.63 Ω
30 kHz	523 μ	12.8 m	213 m	3.00 Ω	799 μ	16.2 m	248 m	3.35 Ω	1.39 m	22.0 m	305 m	3.91 Ω
50 kHz	871 μ	21.3 m	356 m	5.00 Ω	1.33 m	27.0 m	414 m	5.58 Ω	2.20 m	36.1 m	504 m	6.48 Ω
70 kHz	1.22 m	29.8 m	498 m	7.00 Ω	1.86 m	37.8 m	580 m	7.82 Ω	3.04 m	50.2 m	704 m	9.06 Ω
100 kHz	1.74 m	42.6 m	712 m	10.0 Ω	2.66 m	54.0 m	828 m	11.1 Ω	4.31 m	71.6 m	1.00 Ω	12.9 Ω
200 kHz	3.48 m	85.3 m	1.42 Ω	20.0 Ω	5.32 m	108 m	1.65 Ω	22.3 Ω	8.59 m	142 m	2.00 Ω	25.8 Ω
300 kHz	5.23 m	128 m	2.13 Ω	30.0 Ω	7.98 m	162 m	2.48 Ω	33.5 Ω	12.8 m	214 m	3.01 Ω	38.7 Ω
500 kHz	8.71 m	213 m	3.56 Ω	50.0 Ω	13.3 m	270 m	4.14 Ω	55.8 Ω	21.4 m	357 m	5.01 Ω	64.6 Ω
700 kHz	12.2 m	298 m	4.98 Ω	70.0 Ω	18.6 m	378 m	5.80 Ω	78.2 Ω	30.0 m	500 m	7.02 Ω	90.4 Ω

Table 3.6—(continued)

FREQ.	AWG# = 2, D = 6.54 mm				AWG# = 10, D = 2.59 mm				AWG# = 22, D = 0.64 mm			
	l = 1cm	l = 10cm	l = 1 m	l = 10 m	l = 1cm	l = 10cm	l = 1 m	l = 10 m	l = 1cm	l = 10cm	l = 1 m	l = 10 m
1 MHz	17.4 m	426 m	7.12 Ω	100 Ω	26.6 m	540 m	8.28 Ω	111 Ω	42.8 m	714 m	10.0 Ω	129 Ω
2 MHz	34.8 m	853 m	14.2 Ω	200 Ω	53.2 m	1.08 Ω	16.5 Ω	223 Ω	85.7 m	1.42 Ω	20.0 Ω	258 Ω
3 MHz	52.3 m	1.28 Ω	21.3 Ω	300 Ω	79.8 m	1.62 Ω	24.8 Ω	335 Ω	128 m	2.14 Ω	30.1 Ω	387 Ω
5 MHz	87.1 m	2.13 Ω	35.6 Ω	500 Ω	133 m	2.70 Ω	41.4 Ω	558 Ω	214 m	3.57 Ω	50.1 Ω	646 Ω
7 MHz	122 m	2.98 Ω	49.8 Ω	700 Ω	186 m	3.78 Ω	58.0 Ω	782 Ω	300 m	5.00 Ω	70.2 Ω	904 Ω
10 MHz	174 m	4.26 Ω	71.2 Ω	1.00 kΩ	266 m	5.40 Ω	82.8 Ω	1.11 kΩ	428 m	7.14 Ω	100 Ω	1.29 kΩ
20 MHz	348 m	8.53 Ω	142 Ω	2.00 kΩ	532 m	10.8 Ω	165 Ω	2.23 kΩ	857 m	14.2 Ω	200 Ω	2.58 kΩ
30 MHz	523 m	12.8 Ω	213 Ω	3.00 kΩ	798 m	16.2 Ω	248 Ω	3.35 kΩ	1.28 Ω	21.4 Ω	301 Ω	3.87 kΩ
50 MHz	871 m	21.3 Ω	356 Ω	5.00 kΩ	1.33 Ω	27.0 Ω	414 Ω	5.58 kΩ	2.14 Ω	35.7 Ω	501 Ω	6.46 kΩ
70 MHz	1.22 Ω	29.8 Ω	498 Ω	7.00 kΩ	1.86 Ω	37.8 Ω	580 Ω	7.82 kΩ	3.00 Ω	50.0 Ω	702 Ω	9.04 kΩ
100 MHz	1.74 Ω	42.6 Ω	712 Ω	10.0 kΩ	2.66 Ω	54.0 Ω	828 Ω	11.1 kΩ	4.28 Ω	71.4 Ω	1.00 kΩ	12.9 kΩ
200 MHz	3.48 Ω	85.3 Ω	1.42 kΩ	20.0 kΩ	5.32 Ω	108 Ω	1.65 kΩ	22.3 kΩ	8.57 Ω	142 Ω	2.00 kΩ	25.8 kΩ
300 MHz	5.23 Ω	128 Ω	2.13 kΩ	30.0 kΩ	7.98 Ω	162 Ω	2.48 kΩ	33.5 kΩ	12.8 Ω	214 Ω	3.01 kΩ	38.7 kΩ
500 MHz	8.71 Ω	213 Ω	3.56 kΩ	50.0 kΩ	13.3 Ω	270 Ω	4.14 kΩ	55.8 kΩ	21.4 Ω	357 Ω	5.01 kΩ	64.6 kΩ
700 MHz	12.2 Ω	298 Ω	4.98 kΩ	70.0 kΩ	18.6 Ω	378 Ω	5.80 kΩ	78.2 kΩ	30.0 Ω	500 Ω	7.02 kΩ	90.4 kΩ
1 GHz	17.4 Ω	426 Ω	7.12 kΩ		26.6 Ω	540 Ω	8.28 kΩ		42.8 Ω	714 Ω	10.0 kΩ	

* AWG = American Wire Gage
  D = wire diameter in mm
  l = wire length in cm or m
  μ = microhms
  m = milliohms
  Ω = ohms

Non-Valid Region for which l ≥ λ/4

* * Values derived from free-space inductance. when the conductor is far from its return circuit or plane

Table 3.7—Resistance and Inductance
of a Straight Wire from AWG No. 0 to 34

Diam	AWG	R/L	10 cm	20 cm	30 cm	50 cm	70 cm	1 m	2 m	3 m	5 m	7 m	10 m	20 m	30 m
8.2	0	Res/μ	32	64	96	160	225	322	645	967	1,613	2,259	3,227	6,455	9,683
		Ind/n	62	153	254	474	710	1,086	2,450	3,918	7,042	10,329	15,470	33,712	53,001
6.5	2	Res/μ	51	102	153	56	358	512	1,026	1,539	2,565	3,592	5,132	10,265	15,398
		Ind/n	67	162	267	497	743	1,132	2,543	4,057	7,273	10,654	15,933	34,639	54,392
5.2	4	Res/μ	81	162	244	408	570	816	1,631	2,447	4,080	5,712	8,161	16,322	24,484
		Ind/n	71	171	281	520	775	1,179	2,635	4,196	7,505	10,978	16,397	35,567	55,783
4.1	6	Res/μ	129	259	389	648	907	1,297	2,594	3,892	6,488	9,083	12,976	25,954	38,931
		Ind/n	76	180	295	543	807	1,225	2,728	4,335	7,737	11,303	16,860	36,494	57,174
3.2	8	Res/μ	206	412	618	1,031	1,443	2,062	4,126	6,189	10,316	14,443	20,634	41,269	61,904
		Ind/n	80	189	309	566	840	1,271	2,821	4,474	7,969	11,627	17,324	37,422	58,566
2.6	10	Res/μ	328	655	983	1,640	2,296	3,280	6,561	9,842	16,404	22,967	32,810	65,621	98,432
		Ind/n	85	199	323	589	872	1,318	2,913	4,613	8,200	11,952	17,788	38,349	59,957
2	12	Res/m	0.5	0.1	1.5	2.5	3.5	5	10	15	25	36	51	103	156
		Ind/n	90	208	336	612	905	1,364	3,006	4,753	8,432	12,277	18,252	39,277	61,348
1.6	14	Res/m	.08	1.6	2.4	4	5.6	8	16	24	40	57	82	165	248
		Ind/n	94	217	350	635	937	1,410	3,099	4,892	8,664	12,601	18,715	40,204	62,739
1.3	16	Res/m	1	2	3	6	8	12	25	39	65	91	131	263	395
		Ind/n	99	226	364	659	969	1,457	3,191	5,031	8,896	12,926	19,179	41,132	64,131
1	18	Res/m	2	4	6	10	14	20	41	62	104	146	209	418	628
		Ind/n	103	235	378	682	1,002	1,503	3,284	5,170	9,128	13,250	19,643	42,059	65,522
0.8	20	Res/m	3	6	9	16	22	32	66	99	166	232	332	666	999
		Ind/n	108	245	392	705	1,034	1,549	3,377	5,309	9,360	13,575	20,107	42,987	66,913
0.6	22	Res/m	5	10	15	26	36	52	105	158	264	370	529	1,060	1,590
		Ind/n	113	254	406	728	1,067	1,596	3,470	5,448	9,592	13,900	20,570	43,914	68,305
0.5	24	Res/m	8	16	24	42	58	83	168	252	421	589	842	1,685	2,529
		Ind/n	117	263	420	751	1,099	1,642	3,562	5,587	9,823	14,224	21,034	44,842	69,696
0.4	26	Res/m	13	26	39	66	93	133	267	401	669	937	1,340	2,680	4,021
		Ind/n	122	273	434	774	1,132	1,688	3,655	5,726	10,055	14,549	21,498	45,769	71,087
0.3	28	Res/m	21	42	63	106	148	213	425	639	1,065	1,491	2,131	4,263	6,394
		Ind/n	127	282	448	798	1,164	1,735	3,748	5,865	10,287	14,873	21,962	46,697	72,478
0.25	30	Res/m	34	67	101	168	236	338	677	1,016	1,694	2,372	3,389	6,776	10,168
		Ind/n	131	291	461	821	1,197	1,781	3,841	6,005	10,519	15,198	22,425	47,624	73,870
0.2	32	Res/m	54	107	161	268	376	538	1,077	1,616	2,694	3,772	5,389	10,779	16,168
		Ind/n	136	300	475	844	1,229	1,828	3,933	6,144	10,751	15,523	22,889	48,552	75,261
0.16	34	Res/m	85	170	256	428	599	856	1,713	2,570	4,284	5,998	8,569	17,139	25,709
		Ind/n	140	310	489	867	1,261	1,874	4,026	6,283	10,983	15,847	23,353	49,479	76,652

$\mu = \mu\Omega$ m = milliohm   n = nanohenry

If we take the example of Fig. 3.4b, the coaxial cable was RG8 with 25 m length. Manufacturers' data for this cable indicate a loss at 1,000 MHz of 30 dB/100 m.

$$\text{Therefore, } \alpha l_w \text{ for 25 m} = \frac{30 \times 25}{100} \times \frac{2.3}{20} = 0.86 \text{ Np}$$

$$T_{0.5} = \frac{(0.86)^2}{\pi \times 1,000} = 0.24 \times 10^{-3} \text{ } \mu s = 0.24 \text{ ns} \qquad (3.13a)$$

The curve in Fig. 3.4a shows that a step front V_o at the input of this coax will reach 0.5 V_o after 0.24 ns, 0.82 V_o after 2.4 ns and 0.9 V_o after 7.2 ns. Therefore, the delay distortion introduced by the sole skin effect is not negligible.

Litzendraht (litz) wire was designed to reduce skin effect. It is composed of several strands of enamel-coated wire, individually insulated, interwoven and connected in parallel at each end. Skin effect combined with the effect of flux linkages between a conductor and its return above 1 GHz makes conduction within a solid con-

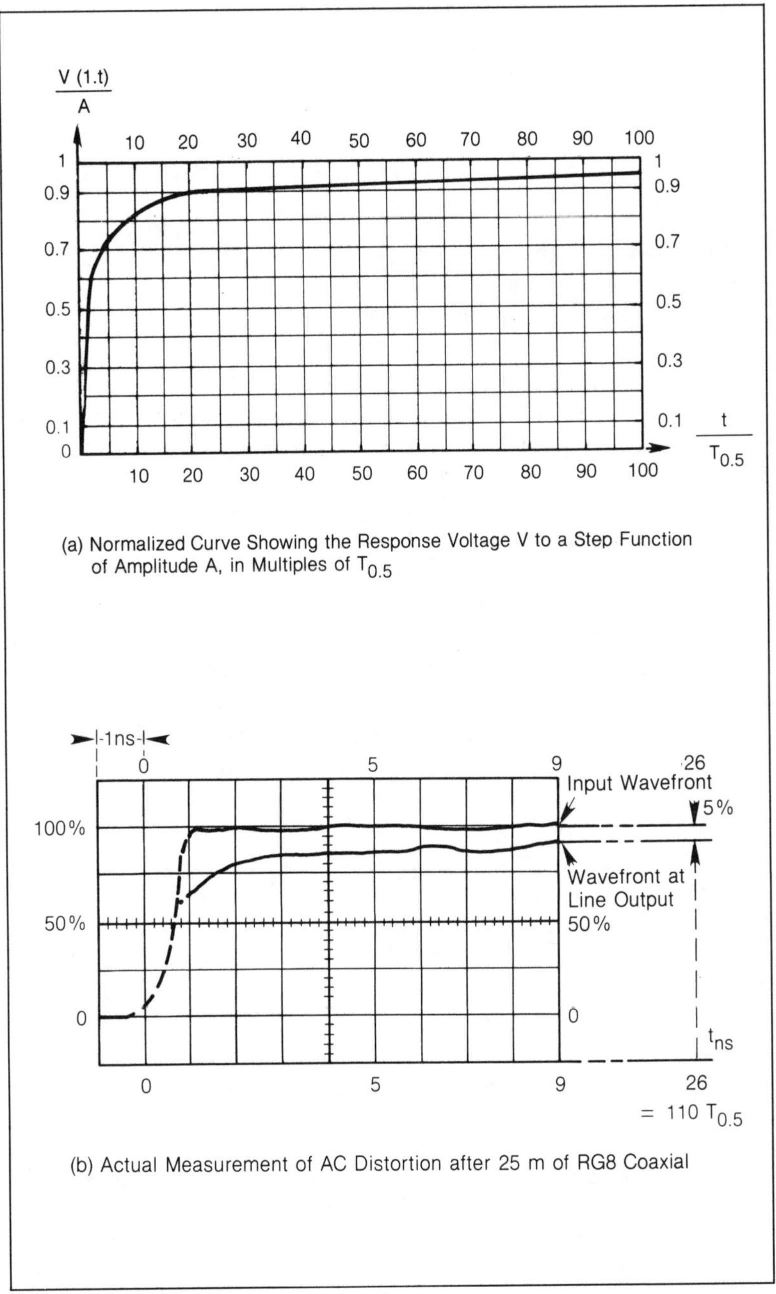

(a) Normalized Curve Showing the Response Voltage V to a Step Function of Amplitude A, in Multiples of $T_{0.5}$

(b) Actual Measurement of AC Distortion after 25 m of RG8 Coaxial

Figure 3.4—Distortion of Step Fronts by Skin Effects

ductor difficult. Accordingly, a transition is often made to waveguide for signal transmission at about 1 GHz (especially 10 GHz). Waveguide, as a conductor, is effective up to about 300 GHz.

3.2.5 Other Considerations in Conductors

Parallel wiring systems may be loaded up to a point where the wire fuses or where corona is probable, whereas arcing limits the power in coaxial and waveguide systems.

In general, parallel lines exhibit less loss than equivalent coaxial lines. However, coaxial lines have less tendency to radiate. Shielded wire is also relatively more expensive. The attenuation offered by shields is due to energy reflection at the boundary and absorption within as external signals pass through.

Braided sleeving is sometimes used as a bond across a vibration mount because of its flexibility. However, because of its construction it has a higher inherent inductance than a strap made of solid copper sheet. Braided sleeving also makes a convenient cable shield, but it exhibits a limited effectiveness.

For low-frequency, and especially low-impedance applications, an overall insulating cover should be used so that the braid can be grounded at one point and insulated from ground over the rest of its length. As a high-frequency circuit shield it can be bare, but when it is unprotected the eventual corrosion between intersecting strands will degrade shielding effectiveness.

Another phenomenon shows up with braids and stranded wire if they are exposed to corrosion, weather effects, etc. Oxide films form diodes at junctions or connections and create nonlinear effects. These show at least as distortion (a perfect sine wave input will show up distorted at the end and, therefore, will have harmonic contents and can cause EMI) or even audio rectification if the incident power is large enough.

3.2.6 Impedances of Other-Than-Round Conductors

Impedance of Conductor Straps—For the same cross-sectional area, the previous section showed that one way to reduce the RF impedance of a conductor is to increase its periphery. The dc resistance is:

$$R_1 = \frac{1{,}000\ l_{mm}}{\sigma A_{mm}{}^2} = \frac{1{,}000\ l_{mm}}{\sigma w_{mm} t_{mm}} = 17.2\ l_{mm}/w_{mm}\ t_{mm} \text{ for copper}$$

where, l_{mm} = strap length in mm   (3.14)

w_{mm} = strap width in mm

t_{mm} = strap thickness in mm

The inductance of the same strap is:

(3.15)

$$L_{\mu H} = 0.0002\, l_{mm} \left[\ln\left(\frac{2\, l_{mm}}{w + t} \right) + 0.5 + 0.2235\left(\frac{w + t}{2\, l_{mm}} \right) \right]$$

The impedance of a copper strap is obtained by converting Eq. (3.15) to inductive reactances and adding quadratically the resistance. This impedance is tabulated in Table 3.8 for 3 cm to 10 m lengths of three sizes of copper straps: 0.3 mm × 10 mm, 1 mm × 10 mm and 2 mm × 50 mm.

Impedance of Metal Ground Planes—The dc resistance R_{dc} of a metal ground plane is:

$$R_{dc} = \frac{\rho l_{mm}}{A} \qquad (3.16)$$

$$= \frac{1{,}000\, l_{mm}}{\sigma l t_{mm}}$$

$$R_{dc} = \frac{1{,}000}{\sigma t_{mm}} \ \Omega/sq \qquad (3.17)$$

where, ρ = resistivity of the metal
σ = $\sigma_c \sigma_r$ conductivity of the metal = $1/\rho$
σ_c = 5.80×10^7 mhos/m for copper
σ_r = conductivity relative to copper
= 0.17 for cold-rolled steel
t = metal thickness in mm

Thus Eq. 3.17 becomes:

$$R_{dc} = \frac{17.2}{\sigma_r t_{mm}} \ \mu\Omega/sq \qquad (3.18)$$

The RF impedance of a metal ground plane is:[6]

$$Z_{RF} = \frac{369\sqrt{\mu_r F/\sigma_r}}{1 - e^{-t/\delta}} \qquad \mu\Omega/sq \qquad (3.19)$$

Table 3.8—Impedance of Copper Straps**

FREQ.	0.3 mm × 10 mm Strap				1 mm × 10 mm Strap				2 mm × 50 mm Strap			
	$l=3$ cm	$l=10$ cm	$l=1$ m	$l=10$ m	$l=3$ cm	$l=10$ cm	$l=1$ m	$l=10$ m	$l=15$ cm	$l=50$ cm	$l=1$ m	$l=10$ m
10 Hz	173 μ	574 μ	5.74 m	57.4 m	52.4 μ	172 μ	1.72 m	17.2 m	30.1 μ	108 μ	180 μ	1.90 m
20 Hz	174 μ	574 μ	5.74 m	57.5 m	53.3 μ	172 μ	1.73 m	17.3 m	34.4 μ	130 μ	201 μ	2.36 m
30 Hz	175 μ	574 μ	5.75 m	57.5 m	54.1 μ	172 μ	1.73 m	17.5 m	38.8 μ	151 μ	232 μ	2.98 m
50 Hz	176 μ	575 μ	5.75 m	57.6 m	55.8 μ	173 μ	1.76 m	17.9 m	47.4 μ	195 μ	312 μ	4.40 m
70 Hz	178 μ	575 μ	5.76 m	57.9 m	57.5 μ	175 μ	1.79 m	18.6 m	56.0 μ	238 μ	404 μ	5.93 m
100 Hz	181 μ	576 μ	5.79 m	58.3 m	60.0 μ	177 μ	1.86 m	19.9 m	69.0 μ	304 μ	549 μ	8.29 m
200 Hz	189 μ	581 μ	5.92 m	60.9 m	68.5 μ	192 μ	2.24 m	26.4 m	112 μ	522 μ	1.05 m	16.3 m
300 Hz	198 μ	589 μ	6.14 m	65.0 m	76.9 μ	215 μ	2.75 m	34.7 m	155 μ	739 μ	1.57 m	24.3 m
500 Hz	215 μ	614 μ	6.79 m	76.6 m	93.8 μ	275 μ	3.97 m	53.1 m	242 μ	1.17 μ	2.61 m	40.5 m
700 Hz	233 μ	615 μ	7.66 m	91.3 m	111 μ	346 μ	5.30 m	72.5 m	328 μ	1.61 μ	3.65 m	56.7 m
1 kHz	259 μ	721 μ	9.25 m	116 m	136 μ	462 μ	7.37 m	102 m	457 μ	2.26 μ	5.22 m	81.1 m
2 kHz	345 μ	1.04 m	15.5 m	210 m	220 μ	874 μ	14.4 m	201 m	889 μ	4.44 μ	10.4 m	162 m
3 kHz	432 μ	1.43 m	22.4 m	309 m	304 μ	1.29 m	21.5 m	302 m	1.32 m	6.62 μ	15.6 m	243 m
5 kHz	605 μ	2.25 m	36.7 m	510 m	473 μ	2.15 m	35.8 m	503 m	2.84 m	11.0 μ	26.1 m	405 m
7 kHz	779 μ	3.11 m	51.0 m	712 m	642 μ	3.00 m	50.2 m	704 m	3.05 m	15.3 μ	36.5 m	567 m
10 kHz	1.04 m	4.40 m	72.7 m	1.01 Ω	895 μ	4.29 m	71.7 m	1.00 Ω	4.34 m	21.9 μ	52.2 m	810 m
20 kHz	1.91 m	8.75 m	145 m	2.02 Ω	1.74 m	8.57 m	143 m	2.01 Ω	8.66 m	43.6 μ	104 m	1.62 Ω
30 kHz	2.77 m	13.1 m	217 m	3.04 Ω	2.58 m	12.8 m	215 m	3.01 Ω	13.0 m	65.4 μ	156 m	2.43 Ω
50 kHz	4.51 m	21.8 m	362 m	5.07 Ω	4.27 m	21.4 m	358 m	5.03 Ω	21.6 m	109 μ	261 m	4.05 Ω
70 kHz	6.24 m	30.5 m	507 m	7.10 Ω	5.95 m	30.0 m	501 m	7.04 Ω	30.2 m	152 μ	365 m	5.67 Ω
100 kHz	8.84 m	43.7 m	725 m	10.1 Ω	8.42 m	42.8 m	716 m	10.0 Ω	43.2 m	218 μ	522 m	8.10 Ω
200 kHz	17.5 m	87.4 m	1.45 Ω	20.2 Ω	16.9 m	85.7 m	1.43 Ω	20.1 Ω	86.4 m	436 μ	1.04 Ω	16.2 Ω
300 kHz	26.2 m	131 m	2.17 Ω	30.4 Ω	25.3 m	128 m	2.15 Ω	30.1 Ω	129 m	653 μ	1.56 Ω	24.3 Ω
500 kHz	43.5 m	218 m	3.62 Ω	50.7 Ω	42.2 m	214 m	3.58 Ω	50.3 Ω	216 m	1.09 Ω	2.61 Ω	40.5 Ω
700 kHz	60.8 m	305 m	5.07 Ω	71.0 Ω	59.1 m	300 m	5.01 Ω	70.4 Ω	302 m	1.52 Ω	3.65 Ω	56.7 Ω

Table 3.8—(continued)

	0.3 mm × 10 mm Strap				1 mm × 10 mm Strap				2 mm × 50 mm Strap			
FREQ.	l = 3 cm	l = 10 cm	l = 1 m	l = 10 m	l = 3 cm	l = 10 cm	l = 1 m	l = 10 m	l = 15 cm	l = 50 cm	l = 1 m	l = 10 m
1 MHz	86.9 m	437 m	7.25 Ω	101 Ω	84.4 m	428 m	7.16 Ω	100 Ω	432 m	2.18 Ω	5.22 Ω	81.0 Ω
2 MHz	173 m	874 m	14.5 Ω	202 Ω	169 m	857 m	14.3 Ω	201 Ω	863 m	4.36 Ω	10.4 Ω	162 Ω
3 MHz	260 m	1.31 Ω	21.7 Ω	304 Ω	253 m	1.28 Ω	21.5 Ω	301 Ω	1.29 Ω	6.53 Ω	15.6 Ω	243 Ω
5 MHz	434 m	2.18 Ω	36.2 Ω	507 Ω	422 m	2.14 Ω	35.8 Ω	503 Ω	2.16 Ω	10.9 Ω	26.1 Ω	405 Ω
7 MHz	607 m	3.05 Ω	50.7 Ω	710 Ω	590 m	3.00 Ω	50.1 Ω	704 Ω	3.02 Ω	15.2 Ω	36.5 Ω	567 Ω
10 MHz	867 m	4.37 Ω	72.5 Ω	1.01 kΩ	843 m	4.28 Ω	71.6 Ω	1.00 kΩ	4.32 Ω	21.8 Ω	52.2 Ω	810 Ω
20 MHz	1.73 Ω	8.74 Ω	145 Ω	2.02 kΩ	1.69 Ω	8.57 Ω	143 Ω	2.01 kΩ	8.63 Ω	43.6 Ω	104 Ω	1.62 kΩ
30 MHz	2.60 Ω	13.1 Ω	217 Ω	3.04 kΩ	2.53 Ω	12.8 Ω	215 Ω	3.01 kΩ	13.0 Ω	65.3 Ω	156 Ω	2.43 kΩ
50 MHz	4.33 Ω	21.8 Ω	362 Ω	5.07 kΩ	4.21 Ω	21.4 Ω	358 Ω	5.03 kΩ	21.6 Ω	109 Ω	261 Ω	4.05 kΩ
70 MHz	6.07 Ω	30.5 Ω	507 Ω	7.10 kΩ	5.90 Ω	30.0 Ω	501 Ω	7.04 kΩ	30.2 Ω	152 Ω	365 Ω	5.67 kΩ
100 MHz	8.67 Ω	43.7 Ω	725 Ω	10.1 kΩ	8.43 Ω	42.8 Ω	716 Ω	10.0 kΩ	43.2 Ω	218 Ω	522 Ω	8.10 kΩ
200 MHz	17.3 Ω	87.4 Ω	1.45 kΩ	20.2 kΩ	16.9 Ω	85.7 Ω	1.43 kΩ	20.1 kΩ	86.3 Ω	436 Ω	1.04 kΩ	16.2 kΩ
300 MHz	26.0 Ω	131 Ω	2.17 kΩ	30.4 kΩ	25.3 Ω	128 Ω	2.15 kΩ	30.1 kΩ	130 Ω	653 Ω	1.56 kΩ	24.3 kΩ
500 MHz	43.3 Ω	218 Ω	3.62 kΩ	50.7 kΩ	42.1 Ω	214 Ω	3.58 kΩ	50.3 kΩ	216 Ω	1.09 kΩ	2.61 kΩ	40.5 kΩ
700 MHz	60.7 Ω	305 Ω	5.07 kΩ	71.0 kΩ	59.0 Ω	300 Ω	5.01 kΩ	70.4 kΩ	302 Ω	1.52 kΩ	3.65 kΩ	56.7 kΩ
1 GHz	86.7 Ω	437 Ω	7.25 kΩ		84.3 Ω	428 Ω	7.16 kΩ		432 Ω	2.18 kΩ	5.22 kΩ	81.0 kΩ

* Strap dimensions are thickness × width in mm

l = strap length in cm or m
μ = microhms
m = milliohms
Ω = ohms

Value Subject To Error
Where l ≥ λ/20

**Worst-case values derived from free-space inductance, when the conductor is far from its return circuit or plane.

where, μ_r = permeability relative to copper
                   = 200 for cold-rolled steel
     F   = frequency in MHz
     δ = skin depth in mm
                   = $0.066\sqrt{\mu_r\, \sigma_r\, f_{MHz}}$ for any metal

The impedance between two points in a ground plane will approximate the above Ω/sq provided the ground plane is at least as wide as the distance between the two ground points and this distance, d, is short compared with a wavelength, λ.

Therefore, the impedance of a metal ground plane at any frequency including dc is:

$$Z = (R_{dc} + jZ_{RF})\,[1 + \tan|(2\pi d/\lambda)|] \qquad (3.20)$$

$$\cong R_{dc} + jZ_{RF} \text{ for } d \leqslant 0.05\lambda \qquad (3.21)$$

Eq. 3.21 is tabulated in Table 3.9 for both copper and cold-rolled steel from 10 Hz to 10 GHz, for six different thicknesses: 0.03 (about 1 mil), 0.1, 0.3, 1, 3 and 10 mm. Examine Table 3.9 and note that the dc resistance governs at 10 Hz for the thin metals (t = 0.03 mm). However, the RF impedance already applies at 10 Hz for the thick metals because they are several skin depths thick. For the same frequency above about 100 MHz for copper, the RF impedances are the same for different thicknesses. This fact also applies for steel above about 10 MHz because the skin depth is much less than metal thickness.

The impedance of any other metal may be determined with the use of Table 3.9, provided that its relative conductivity and permeability are known. The dc and RF impedances are:

$$R_{dc} = \text{from Table 3.9 (for copper)} \div \sigma_r$$

$$Z_{RF} = \text{from Table 3.9 (for copper)} \times \sqrt{\mu_r/\sigma_r} \qquad (3.22)$$

For example, for aluminum, $\sigma_r = 0.6$ and $\mu_r = 1$.

The values found in the table can be used to approximate the impedance of metal plane samples which are not square.

For instance, the impedance of a rectangular plate with length L and width W can be found by:

$$Z_\Omega = Z_{table} \times L/W \qquad (3.23)$$

provided that L/W does not exceed 10, above which gross error would occur (except for dc) since the element would start behaving like a wire due to external self-inductance.

Similarly, the surface impedance of a short beam or cylinder can be calculated by using the developed area and comparing it to a rectangular plane.

3.2.7 EMI Reduction in Conductors

Various methods exist to minimize EMI in conductors. Some of them are:

1. *Multiple-conductor cables that bundle each conductor with its return conductor.* They carry current in opposite directions, resulting in magnetic field cancellation. The amount of cancellation depends upon the relative equality of the currents and the spacing between conductors.

2. *Two conductors with similar current levels twisted together.* The field generated by one tends to cancel the field of the other if the currents are opposite in direction. The greater the number of uniform twists, the more effective is the cancellation of magnetic fields.

3. *A conductor surrounded by a return path shield, as in a coaxial cable.* Theoretically, it will not have an external magnetic field. Various factors degrading this effect are the lack of homogeneity of the surrounding conductor and its conductivity.

4. *Hollow conductors.* Skin effect at higher frequencies has an increasing effect on electric and magnetic fields. At higher frequencies, hollow conductors should be considered to minimize external fields.

5. *Magnetic shields.* Those shields can be used to minimize magnetic field penetration. Eddy currents that occur in grounded shields create absorption losses that minimize the external magnetic field.

6. *Coated conductors.* Conducted interference can be limited by coating conductors with high-permeability material which magnifies skin-effect losses. There is a large effect in the frequency range from about 25 kHz to 50 MHz.

7. *Ferrite beads.* These beads are tiny cylindrical beads that may be strung on a wire to increase inductance of the wire and thereby cause attenuation of signals due to filter action. Attenuation by inductive reactance and I^2R losses is a function

Table 3.9—Metal Ground Plane Impedances in Ohms/Square

Freq.	COPPER, COND-1, PERM-1						STEEL, COND-0.17, PERM-200					
	$t=0.03$	$t=0.1$	$t=0.3$	$t=1$	$t=3$	$t=10$	$t=0.03$	$t=0.1$	$t=0.3$	$t=1$	$t=3$	$t=10$
10 Hz	574 μ	172 μ	57.4 μ	17.2 μ	5.74 μ	1.75 μ	3.38 m	1.01 m	338 μ	101 μ	38.5 μ	40.3 μ
20 Hz	574 μ	172 μ	57.4 μ	17.2 μ	5.75 μ	1.83 μ	3.38 m	1.01 m	338 μ	102 μ	49.5 μ	56.6 μ
30 Hz	574 μ	172 μ	57.4 μ	17.2 μ	5.75 μ	1.95 μ	3.38 m	1.01 m	338 μ	103 μ	62.3 μ	69.3 μ
50 Hz	574 μ	172 μ	57.4 μ	17.2 μ	5.76 μ	2.30 μ	3.38 m	1.01 m	338 μ	106 μ	86.2 μ	89.6 μ
70 Hz	574 μ	172 μ	57.4 μ	17.2 μ	5.78 μ	2.71 μ	3.38 m	1.01 m	338 μ	110 μ	105 μ	106 μ
100 Hz	574 μ	172 μ	57.4 μ	17.2 μ	5.82 μ	3.35 μ	3.38 m	1.01 m	338 μ	118 μ	127 μ	126 μ
200 Hz	574 μ	172 μ	57.4 μ	17.2 μ	6.04 μ	5.16 μ	3.38 m	1.01 m	340 μ	157 μ	179 μ	179 μ
300 Hz	574 μ	172 μ	57.4 μ	17.2 μ	6.38 μ	6.43 μ	3.38 m	1.01 m	342 μ	199 μ	219 μ	219 μ
500 Hz	574 μ	172 μ	57.4 μ	17.3 μ	7.36 μ	8.27 μ	3.38 m	1.01 m	350 μ	275 μ	283 μ	283 μ
700 Hz	574 μ	172 μ	57.4 μ	17.3 μ	8.55 μ	9.77 μ	3.38 m	1.01 m	362 μ	335 μ	335 μ	335 μ
1 kHz	574 μ	172 μ	57.4 μ	17.5 μ	10.4 μ	11.6 μ	3.38 m	1.01 m	385 μ	403 μ	400 μ	400 μ
2 kHz	574 μ	172 μ	57.5 μ	18.3 μ	16.1 μ	16.5 μ	3.38 m	1.02 m	495 μ	566 μ	566 μ	566 μ
3 kHz	574 μ	172 μ	57.5 μ	19.5 μ	20.3 μ	20.2 μ	3.38 m	1.03 m	623 μ	693 μ	694 μ	694 μ
5 kHz	574 μ	172 μ	57.6 μ	23.0 μ	26.2 μ	26.1 μ	3.38 m	1.06 m	862 μ	896 μ	896 μ	896 μ
7 kHz	574 μ	172 μ	57.8 μ	27.1 μ	30.9 μ	30.9 μ	3.38 m	1.10 m	1.05 m	1.06 m	1.06 m	1.06 m
10 kHz	574 μ	172 μ	58.2 μ	33.5 μ	36.9 μ	36.9 μ	3.38 m	1.18 m	1.27 m	1.26 m	1.26 m	1.26 m
20 kHz	574 μ	172 μ	60.4 μ	51.6 μ	52.2 μ	52.2 μ	3.40 m	1.57 m	1.79 m	1.79 m	1.79 m	1.79 m
30 kHz	574 μ	172 μ	63.8 μ	64.3 μ	63.9 μ	63.9 μ	3.42 m	1.99 m	2.19 m	2.19 m	2.19 m	2.19 m
50 kHz	574 μ	173 μ	73.6 μ	82.7 μ	82.6 μ	82.6 μ	3.50 m	2.75 m	2.83 m	2.83 m	2.83 m	2.83 m
70 kHz	574 μ	173 μ	85.5 μ	97.7 μ	97.7 μ	97.7 μ	3.62 m	3.35 m	3.35 m	3.35 m	3.35 m	3.35 m
100 kHz	574 μ	175 μ	104 μ	116 μ	116 μ	116 μ	3.85 m	4.03 m	4.00 m	4.00 m	4.00 m	4.00 m
200 kHz	575 μ	183 μ	161 μ	165 μ	165 μ	165 μ	4.95 m	5.66 m	5.66 m	5.66 m	5.66 m	5.66 m
300 kHz	575 μ	195 μ	203 μ	202 μ	202 μ	202 μ	6.23 m	6.93 m	6.94 m	6.94 m	6.94 m	6.94 m
500 kHz	576 μ	230 μ	262 μ	261 μ	261 μ	261 μ	8.62 m	8.96 m	8.96 m	8.96 m	8.96 m	8.96 m
700 kHz	578 μ	271 μ	309 μ	309 μ	309 μ	309 μ	10.5 m	10.6 m	10.6 m	10.6 m	10.6 m	10.6 m

Table 3.9—(continued)

	COPPER, COND-1, PERM-1						STEEL, COND-0.17, PERM-200					
Freq.	t = 0.03	t = 0.1	t = 0.3	t = 1	t = 3	t = 10	t = 0.03	t = 0.1	t = 0.3	t = 1	t = 3	t = 10
1 MHz	582 μ	335 μ	369 μ	369 μ	369 μ	369 μ	12.7 m	12.6 m	12.6 m	12.6 m	12.6 m	12.6 m
2 MHz	604 μ	516 μ	522 μ	522 μ	522 μ	522 μ	17.9 m	17.9 m	17.9 m	17.9 m	17.9 m	17.9 m
3 MHz	638 μ	643 μ	639 μ	639 μ	639 μ	639 μ	21.9 m	21.9 m	21.9 m	21.9 m	21.9 m	21.9 m
5 MHz	736 μ	827 μ	826 μ	826 μ	826 μ	826 μ	28.3 m	28.3 m	28.3 m	28.3 m	28.3 m	28.3 m
7 MHz	855 μ	977 μ	977 μ	977 μ	977 μ	977 μ	33.5 m	33.5 m	33.5 m	33.5 m	33.5 m	33.5 m
10 MHz	1.04 m	1.16 m	1.16 m	1.16 m	1.16 m	1.16 m	40.0 m	40.0 m	40.0 m	40.0 m	40.0 m	40.0 m
20 MHz	1.61 m	1.65 m	1.65 m	1.65 m	1.65 m	1.65 m	56.6 m	56.6 m	56.6 m	56.6 m	56.6 m	56.6 m
30 MHz	2.03 m	2.02 m	2.02 m	2.02 m	2.02 m	2.02 m	69.4 m	69.4 m	69.4 m	69.4 m	69.4 m	69.4 m
50 MHz	2.62 m	2.61 m	2.61 m	2.61 m	2.61 m	2.61 m	89.6 m	89.6 m	89.6 m	89.6 m	89.6 m	89.6 m
70 MHz	3.09 m	3.09 m	3.09 m	3.09 m	3.09 m	3.09 m	106 m	106 m	106 m	106 m	106 m	106 m
100 MHz	3.69 m	3.69 m	3.69 m	3.69 m	3.69 m	3.69 m	126 m	126 m	126 m	126 m	126 m	126 m
200 MHz	5.22 m	5.22 m	5.22 m	5.22 m	5.22 m	5.22 m	179 m	179 m	179 m	179 m	179 m	179 m
300 MHz	6.39 m	6.39 m	6.39 m	6.39 m	6.39 m	6.39 m	219 m	219 m	219 m	219 m	219 m	219 m
500 MHz	8.26 m	8.26 m	8.26 m	8.26 m	8.26 m	8.26 m	283 m	283 m	283 m	283 m	283 m	283 m
700 MHz	9.77 m	9.77 m	9.77 m	9.77 m	9.77 m	9.77 m	335 m	335 m	335 m	335 m	335 m	335 m
1 GHz	11.6 m	11.6 m	11.6 m	11.6 m	11.6 m	11.6 m	400 m	400 m	400 m	400 m	400 m	400 m
2 GHz	16.5 m	16.5 m	16.5 m	16.5 m	16.5 m	16.5 m	566 m	566 m	566 m	566 m	566 m	566 m
3 GHz	20.2 m	20.2 m	20.2 m	20.2 m	20.2 m	20.2 m	694 m	694 m	694 m	694 m	694 m	694 m
5 GHz	26.1 m	26.1 m	26.1 m	26.1 m	26.1 m	26.1 m	896 m	896 m	896 m	896 m	896 m	896 m
7 GHz	30.9 m	30.9 m	30.9 m	30.9 m	30.9 m	30.9 m	1.06 Ω	1.06 Ω	1.06 Ω	1.06 Ω	1.06 Ω	1.06 Ω
10 GHz	36.9 m	36.9 m	36.9 m	36.9 m	36.9 m	36.9 m	1.26 Ω	1.26 Ω	1.26 Ω	1.26 Ω	1.26 Ω	1.26 Ω

*   t is in units of mm

μ = microhms

m = milliohms

Ω = ohms

NOTE: Do not use table at frequencies in MHz above $5/l_m$ since the separation distance in meters. l_m of two grounded equipments will exceed 0.05λ where error becomes significant.

of frequency because there is no dc current in the bead. As an example, four ferrite beads can increase the inductance of a two-inch piece of wire by 15 times and can cause a 6 dB attenuation over a wide frequency range (see Section 5.3). The cost is less than that of resistors or RF chokes.

8. *Lossy flexible sleeving.* The sleeving, made up from ferrite particles in a suitable binder, is a continuation of the ferrite bead and rod concept. Above about 10 MHz, conducted EMI is converted to heat, and attenuation becomes very significant above about 50 MHz (see Section 5.3).

9. *Bare conductors operating at high-voltage potentials.* These can produce large electrostatic fields at sharp corners or points that ionize the surrounding atmosphere, resulting in a broadband, white-noise corona. Minimum bend radius for a high-voltage conductor should be 10 times its outside diameter and connection points and other irregularities should be rounded. Contamination across a component can also lead to corona by reducing the required breakdown potential.

3.3 Connectors

Connectors are not active devices; their function is to provide electrical continuity between two circuits while ensuring enough insulation around them. For this reason, they are included in the conductors/insulator part of this handbook. The many types of connectors which are used can be categorized as low-frequency power, low-frequency signaling, high-frequency power or high-frequency signaling. They can also be categorized by their hardware style, such as wire ends, ribbon cable, PCBs, coaxial BNC, "N," TNC, etc.

As far as noise is concerned, the connector features which are of interest are crosstalk, characteristic impedance, contact impedance and insertion loss, transfer impedance, shielding and the presence of filter pins or varistor pins. All connectors become limiting factors in circuits operating at high frequencies.

3.3.1 Crosstalk in Connectors

Crosstalk is the mechanism by which two nearby circuits are coupled by capacitance or inductance (see Chapter 1, and also Volume 8 of this handbook series), and connectors are not exempt from it. In fact, when 60 dB or more isolation is required among different channels of the same equipment, the connectors can very well be the limiting factor.

For instance, it is a wise practice in critical equipment to separate the different cabling by families based on at least 30 dB of isolation between classes. To be sure not to defeat that concept at connector level by more than 1 dB, crosstalk in connectors should stay below 50 dB.

Crosstalk is defined as :

$$X_{dB} = 20 \log \frac{V_{noise\ on\ victim\ pin}}{V_{noise\ on\ culprit\ pin}} \qquad (3.24)$$

Figure 3.5 shows the capacitive crosstalk versus frequency (or rise time) in a typical "sub D" connector. Figure 3.6 illustrates pin-to-pin differential capacitance in circular connectors. Crosstalk increases when centerline spacing decreases. The capacitive crosstalk is generally the predominant mode in connectors because it is enhanced by the ϵ_r of the insulating material, while magnetic coupling is not. However, if a connector carries significant current in some pins (more than 30 mA/V of drive, corresponding to 30 Ω or less of culprit circuit load), the inductive crosstalk may be a problem as well (see Fig. 3.7). In the latter case, it is sometimes expressed as a coupling impedance or transfer impedance, relating to the culprit pin current:

$$Z_{crosstalk} = \frac{V_{noise\ on\ victim\ pin}}{I_{noise\ in\ culprit\ pin}} \qquad (3.25)$$

If the victim pin is surrounded by more than one "culprit" pin, the crosstalk is aggravated by:

$20 \log N$ (for coherent culprit signals)

$10 \log N$ (for noncoherent culprit signals),

where N is the number of adjacent pins within the same distance of the victim pin.

Crosstalk in connectors can be reduced by careful segregation of culprits and victim pins, for instance, avoiding some pin assignments where large power contacts could be next to small signal ones. For example, in a same connector, contacts carrying 10 W (or 40 dBm) will be used along with contacts going to analog circuits with a sensitivity of 1 mV/kΩ (or -60 dBm). It is clear that at least 100 dB of isolation is needed in that connector which, depending on the frequency of concern, may not be easy to achieve. Crosstalk can also be reduced by the insertion of ground pins between culprit and victim contacts. Figure 3.8 shows the influence of ground pin distribution over crosstalk in a connector.

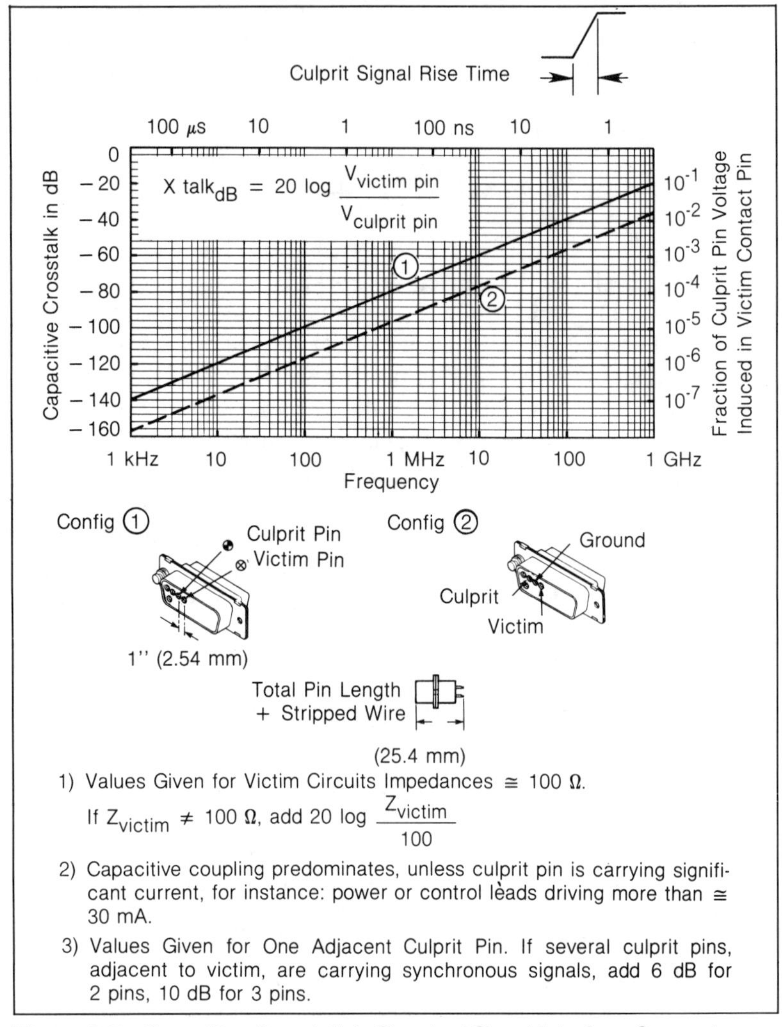

Figure 3.5—Capacitive Crosstalk in Standard Signal Interface Connectors

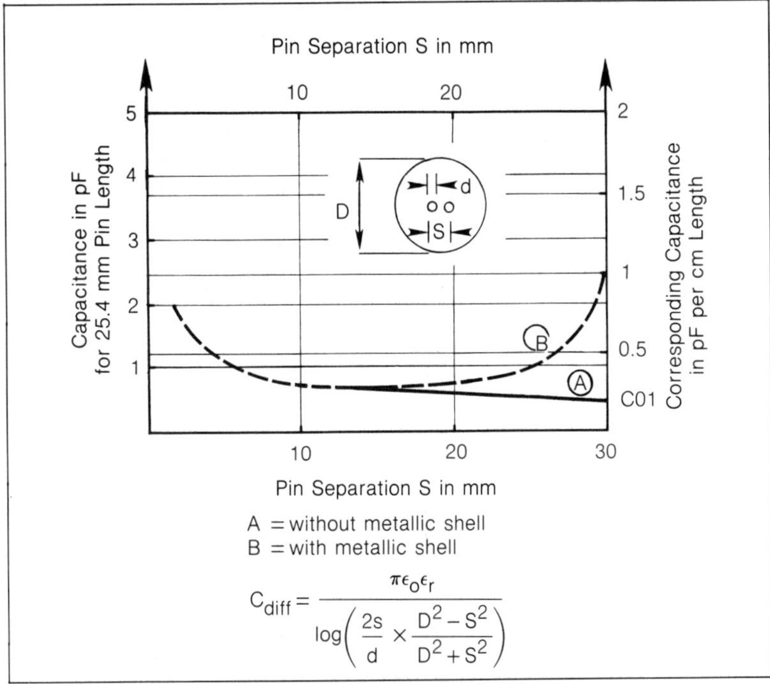

$$C_{diff} = \frac{\pi\epsilon_0\epsilon_r}{\log\left(\dfrac{2s}{d} \times \dfrac{D^2 - S^2}{D^2 + S^2}\right)}$$

Figure 3.6—Pin-to-Pin Differential Capacitance in Circular Connectors (Each Pin Floated from Shell)

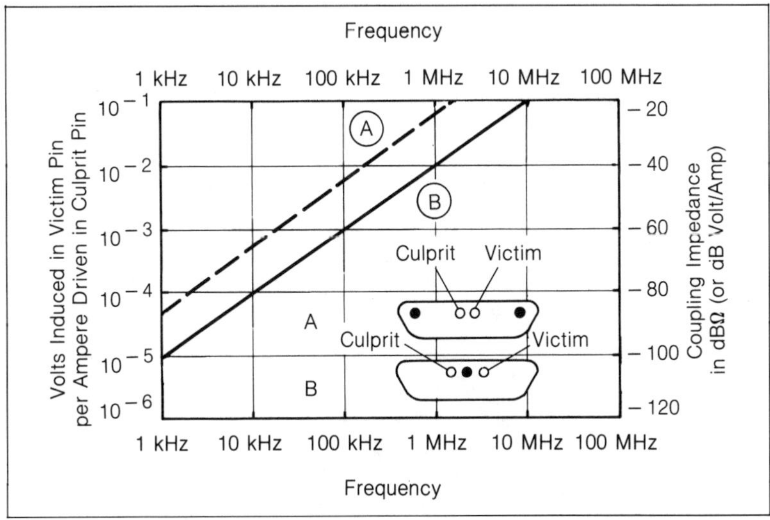

Figure 3.7—Pin-to-Pin Inductive Coupling in Type D Connectors in Function of Culprit Pin Current. The voltage shown is the voltage at either end of victim circuit if they have similar resistance.

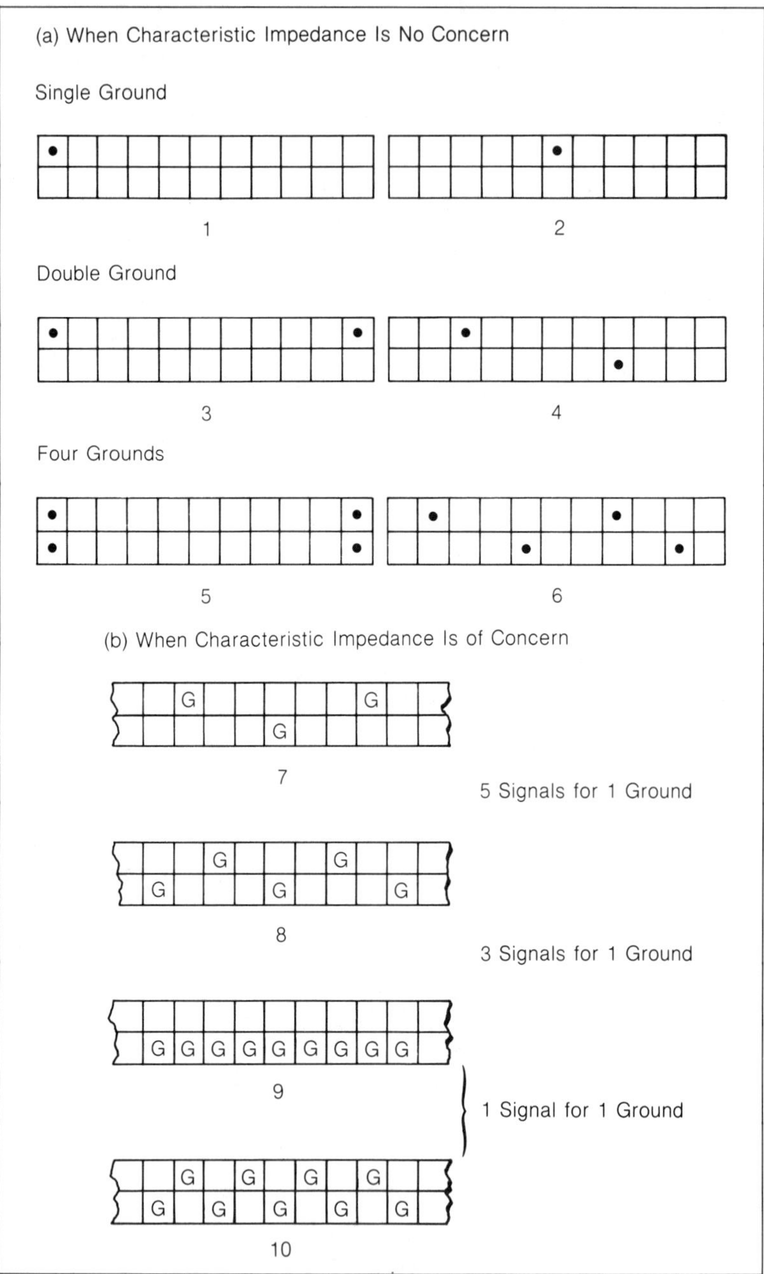

Figure 3.8—Comparison of Crosstalk versus Ground Pin Arrangement for PCB and Flat Cable Connectors. Performance improves constantly from pattern #1 to #10 (Unmarked Locations Are Signals).

Crosstalk in connectors can be reduced by careful segregation of culprits and victim pins, for instance, avoiding some pin assignments where large power contacts could be next to small signal ones. For example, in a same connector, contacts carrying 10 W (or 40 dBm) will be used along with contacts going to analog circuits with a sensitivity of 1 mV/kΩ (or -60 dBm). It is clear that at least 100 dB of isolation is needed in that connector which, depending on the frequency of concern, may not be easy to achieve. Crosstalk can also be reduced by the insertion of ground pins between culprit and victim contacts. Figure 3.8 shows the influence of ground pin distribution over crosstalk in a connector.

3.3.2 Characteristic Impedance and Mismatch

When two lines of characteristic impedance Z_0 are mating via a connector, the combination ideally should have the same characteristic impedance to avoid line impairment. At low frequencies this factor has little importance, but when the connector length starts to represent a non-negligible fraction of the signal wavelength, a connector characteristic impedance different from the Z_0 of the line will create reflections and ringing. Taking a connector whose pins and body length represent about 5 cm, and given a velocity in connector insulation (generally resin or dialyl-Phtalate) of about 15 cm/ns, the connector length represents a trip time of 0.3 ns. Considering (as a rule of thumb) that 1/10 of trip delay (or 1/10 wavelength) is the criterion beyond which impedance matching must be satisfied, this connector must be matched for signals with rise times equal to or less than 3 ns or for frequencies above $1/\pi \times 3$ ns, i.e. 100 MHz.

Regardless of connector length, the other factor dictating the need for or against matching is the length of the mating lines. Since a line typically will be longer than the connector itself, it ends up that for any line longer than $\lambda/10$, the connector needs to be matched. Although this fact has been known for many years with RF coaxial connectors, it has to be considered as well for high-speed logic PCB and flat cable connectors.

Figure 3.9 shows the effects of impedance mismatch on a typical PCB connector, having a pin inductance of about 20 nH.[7]

For RF or HF connectors, impedance mismatch is rated in terms of maximum voltage standing wave ratio (VSWR) versus frequency. The maximum signal amplitude variation as a function of VSWR is the amplitude of the VSWR, per se. For example, depending upon the length of cable, a connector rated with a VSWR of 2:1 at a particular frequency could exhibit a 6 dB peak-to-peak variation in signal or EMI amplitude. Thus, connector VSWR becomes especially important at frequencies for which an associated cable length approaches or exceeds $\lambda/10$.

Another aspect associated with mismatch or line impairment is the line-to-ground imbalance which can be introduced by the connector for a balanced line (wire pair or twinax). If a wire pair has a line-to-ground or line-to-shield symmetry of three percent, due to the unevenness of each wire-to-shield capacitance and inductance, the connector must have a balance equal to or better than this.

For instance, if a balanced line, terminated into a well-balanced differential receiver, is used to avoid common-mode interference (see Volume 8, *EMI Control Methodology and Procedures,* of this handbook series), the balancing at the connector must be considered as well. This balancing is usually done by ensuring that the two balanced pins are at an equal distance from the connector shell or from surrounding ground pins.

3.3.3 Contact Impedance and Insertion Loss

EMI problems associated with connectors are similar to those manifested by mechanical switches. One principal problem is poor contact which may result directly in arcing or in overheating which eventually leads to arcing. Poor connector contact also invites driven-circuit voltage variations due to contact impedance modulation of the driving current source. Common impedance coupling from outside sources can exist in connector grounding paths. For unbalanced (coaxial) connectors with the shell both a shielding item and a part of the signal return path, the contact impedance and insertion loss should be addressed both at pin level and at shell-to-shell level.

The achievable contact dc resistance is in the milliohm region, ranging from about 0.1 mΩ for the best, up to about 10 mΩ for the poorest. Although contact resistance is directly proportional to the inverse of the lateral pressure, excessive applied pressure is not desirable because of contact wear. Depending on the rate of use of the connector, contact quality or durability is usually specified for a maximum number of insertions ranging from 50 insertions for less frequently used connectors up to 500 or more for frequently used connectors.

Low-impedance contact can occur in either of two ways: by simple contact under pressure in which the pins are pressed together to break or wipe away film and tarnish, or breakdown contact in which the pins have a film which is not ruptured by connector mating.

Arcing forms channels of molten metal which can degrade a low-resistance contact due to reduced cross section. A capacitive effect

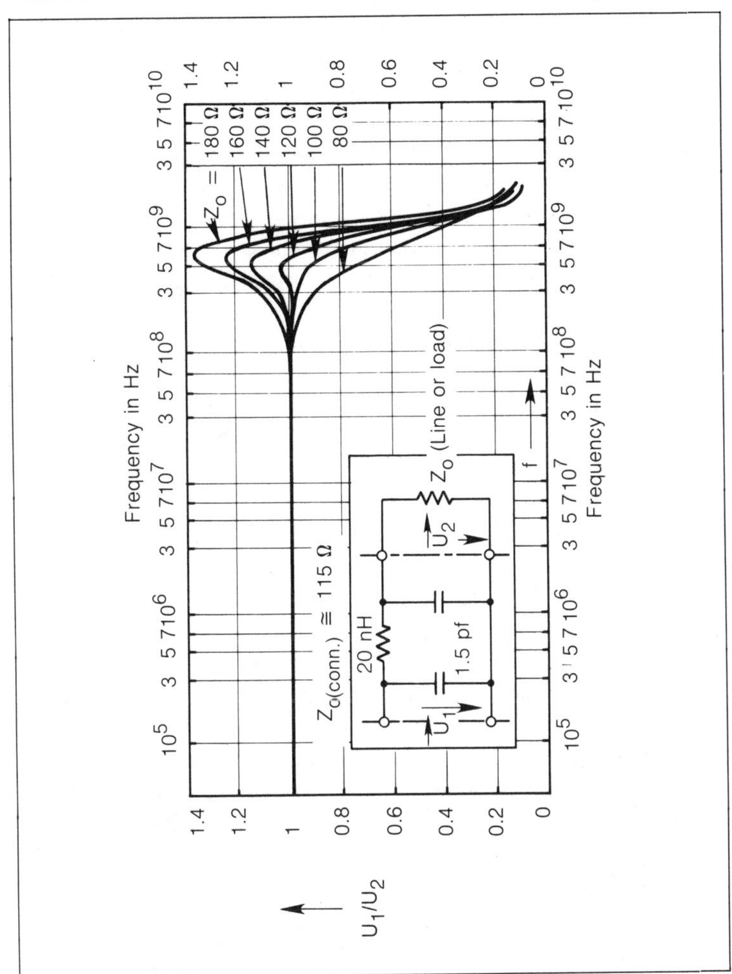

Figure 3.9—Effect of Connector-to-Line Mismatch for a Typical PCB Connector versus Frequency

can exist if the voltage level between the pins is not enough for the dielectric oxide film breakdown. High-resistance contact can be produced on some metal oxides on the surface of the connectors.

To mitigate those problems, contact plating is used. Plating yields the advantages of increased tarnish and corrosion resistance. For low-cost, average-voltage/current application, tin-plated contacts are used, but for the signal connectors, gold is the most widely used plating. Certain hard-gold alloy platings are preferred for electrical conductivity corrosion resistance as well as for wearability.

For an underplate for the gold alloy, 2.5 microns (0.1 mil) of ductile nickel (elongation of not less than 5 percent) provides the best overall combination from a performance aspect for a sustained period of time.

Nonhermetically sealed connectors contain a copper-based alloy, while hermetically sealed connectors usually consist of an iron-nickel alloy. The gold underplating for hermetic seals should be copper over the iron-nickel substrate followed by a plating of nickel. If gold is plated directly to the copper plating, the gold will be diffused into the copper, thereby establishing a seal leakage as well as a corrosion mechanism.

The finer the microfinish of the contacts' mating surfaces, the better the corrosion resistance characteristics and the less the insertion and withdrawal forces. Microfinishes as low as 0.25 microns are achievable. Gold platings of 0.75 microns are generally used for 500-insertion ratings, while 0.25 microns is adequate for 50-insertion types.

The best design combines low spring pressure and good peripheral electrical contact. Contacts should maintain a low-impedance bond in the link position. Thus, requirements exist for testing connectors to detect and isolate high-impedance contacts:

1. Ensure that the applied voltage is about one-tenth the normal working voltage, so that the normal film and tarnish accumulation is not broken down by the test voltage.
2. Use the highest test frequency to which the driven loads may be susceptible. It is sufficient to conduct dc resistance measurements when the loads are audio and low-frequency circuits.
3. Submit the connector to low-level, low-impedance tests, then perform an environmental test of mechanical forces, wear and corrosion.

A common cause of faulty connector contacts is damage during mating. Adequate contact floating will permit insertion without binding and prevent wedging by correct pin layout. Properly placed guide pins will reduce bending, gouging and abrasion due to misalignment. With guides, alignment occurs without trial-and-error scraping of pins across the female contact to find the alignment position. Protective coverings should extend over the male pins to reduce pin damage. Pin overdesign for an extra low length-to-diameter ratio provides ruggedness. Inexpensive protective plastic caps should be used during handling and storage. The back ends of connectors should be potted to decrease entry of moisture, fumes, contaminants and foreign objects. Clamps prevent wires being pulled and twisted from their contacts. Use of low-fatigue, high-resilience spring materials ensures good contact pressure over a long time.

Connector insertion loss which incorporates both the contact resistance and contact high-frequency impedance is specified by the manufacturer at a few key frequencies. A connector loss of 0.1 to 0.3 dB in the 100 to 1,000 MHz range is typical of an RF connector. Although it seems like a minute factor, a certain cable link having 10 such connectors in series will have some additional 1 to 3 dB loss which must be accounted for in either the signal budget or the measured data, depending on the application.

Another EMI effect of a poor contact occurs with corrosion, generally brought about or aggravated by aging, the worst-case being a connector not frequently used and located in a moist atmosphere. Although this fact is often forgotten or overlooked, a corroded layer over a metal support is a nonlinear junction, i.e., the voltage drop in one direction is not the same as the voltage drop in the other. Two undesirable effects result: rectification of a strong CW interference, since the average value of the CW noise pickup is no longer null once it has gone through this nonlinear junction (see "Audio Rectification" in Chapter 4 of this book), and distortion of intentional signal due to the same nonlinear effects.

3.4 Transfer Impedance and Shielding Effectiveness

Improperly shielded connectors may invite penetration of ambient EMI (susceptibility) or leakage of inside HF signals outside (emission). For example, shell discontinuities can spoil the effect of a good cable shield, that of a good box shielding or, most typically, both.

The connector shell becomes an integral part of the overall system shielding. A system with a box shield better than 60 dB and cable shields better than 50 dB will never benefit from these if the connector shell provides only 30 dB, thus becoming the weakest link in the chain.

The connector aspect of shielding is covered in Volume 3, *Shielding*, of this handbook series. Connector shell shielding can be either characterized by its shielding effectiveness (SE) in dB or its transfer impedance in ohms (see Fig. 3.10).

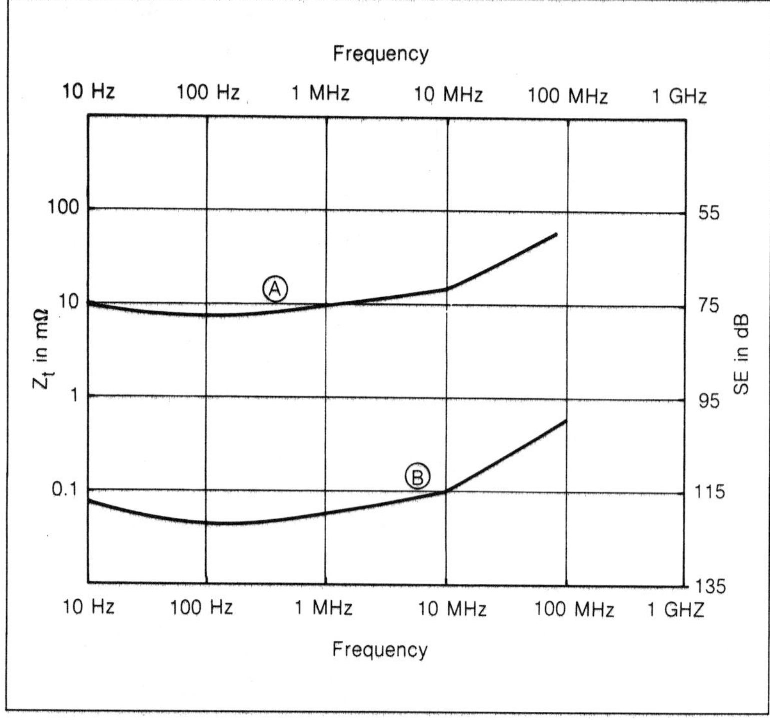

Figure 3.10—Example of a Typical A and Hardened B Circular Connector Backshell Shielding—Effectiveness and Transfer Impedance (source: SOCAPEX Connectors, France)

3.5 Floated Connectors

In the attempt to break ground loops, special BNC, TNC or N connectors are used where the coaxial shell of the contact is electrically isolated from the mounting flange. Signal return continuity is thus provided while still isolating the electrical zero voltage from the chassis ground. This solution is used to break low-frequency ground loops. It has, of course, the penalty of creating a ring-shaped, high-frequency leakage through the mounting panel.

3.6 Filter Pin Connectors and Varistor Pins

Filters, the topic of Volume 4, *Filtering*, of this handbook series, offer significant possibilities for controlling conducted interference. Generally, EMI filters are employed as lumped elements in various portions of circuits and input-output wiring of equipments. In recent years, however, filters have been miniaturized to such an extent that some can now be built into the cable-pin assembly. Figure 3.11 illustrates one type of miniature multipin connector employing π-type filters in each pin. Because of the limitation of the obtainable shunt capacitance and series inductance that can be constructed in the pin, filters of this small type, typically about 3.175 mm × 9.525 mm in size, exhibit little or no attenuation below 1 MHz. Typical attenuation offered by these filter pins in a 50 Ω system is about 20 dB at 10 MHz and up to 80 dB at 100 MHz.

Another filter-pin connector of a somewhat larger body dimension is designed to carry 5 A. Thus, for low dc working voltages, capacitances up to about 1 μF are achievable in the larger pins. Figure 3.12 shows some typical dimensional data for these connectors and associated insertion loss versus frequency. Many of these filters exhibit cutoff frequencies of the order of 100 kHz when measured in a 50 Ω system per MIL-STD-220A. Also connectors have been developed which incorporate transient protection devices such as Transzorbs® or varistors in the contact pin itself (see Chapter 7).

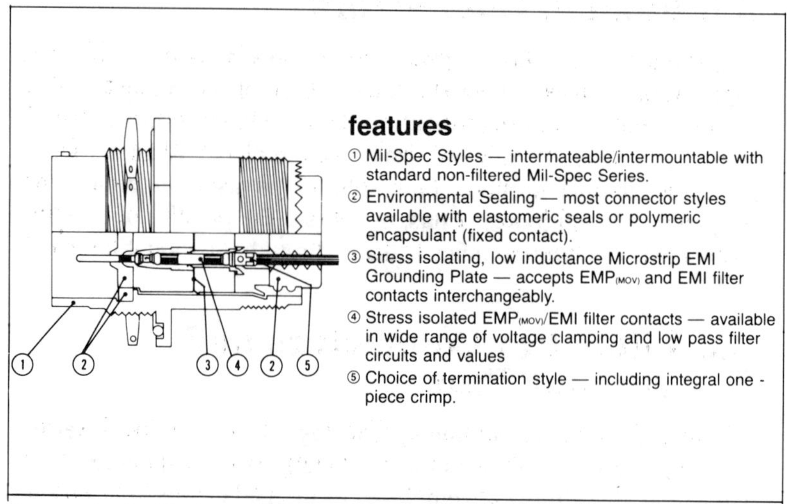

Figure 3.11—Typical Miniature Multipin Filter Connector (courtesy of Amphenol Canada Corp.)

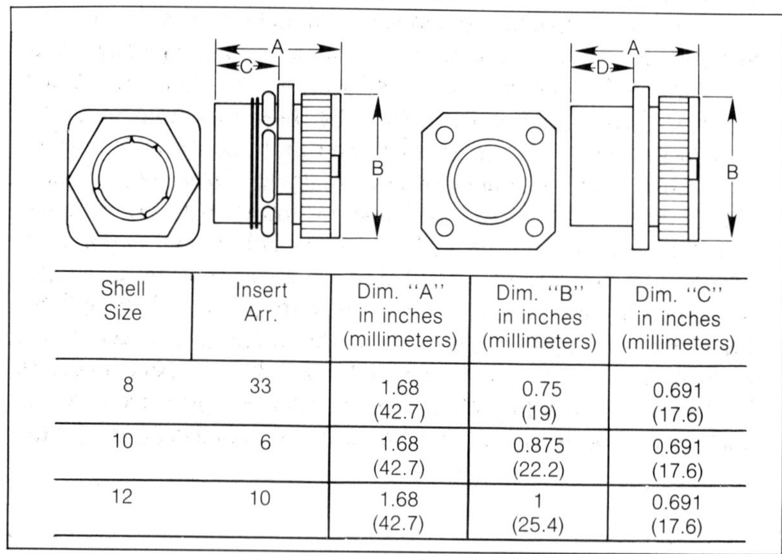

Shell Size	Insert Arr.	Dim. "A" in inches (millimeters)	Dim. "B" in inches (millimeters)	Dim. "C" in inches (millimeters)
8	33	1.68 (42.7)	0.75 (19)	0.691 (17.6)
10	6	1.68 (42.7)	0.875 (22.2)	0.691 (17.6)
12	10	1.68 (42.7)	1 (25.4)	0.691 (17.6)

Figure 3.12—Insertion Loss versus Frequency for a 5A Connector (continued next page)

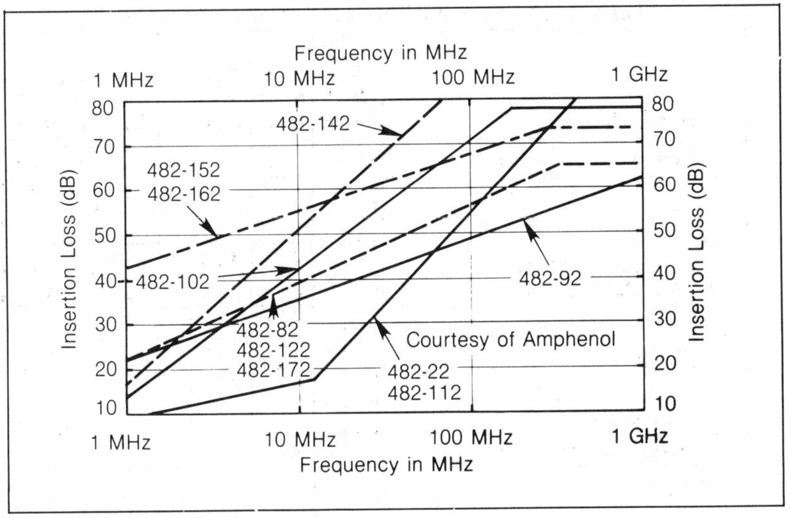

Figure 3.12—(continued)

3.7 Coaxial Connectors

For applications above 10 kHz, and more typically above 10 MHz, connectors are of the unbalanced-line, coaxial type in order to mate with coaxial cables.* Connectors of this type maintain a 360° low-impedance integrity of the outer shield through the connector interface to the mating connector assembly. A low-impedance shield is extremely important since this impedance exists in the return-wire path of the associated coaxial cable. Thus, the outer-cable shield impedance of the termination is of paramount consideration in the performance of coaxial connectors.

Specifying the correct coaxial connector for a specific requirement will simultaneously address most EMI considerations. To do this, use Table 3.10. Work down the characteristics in the left-hand column, making note of the connector series that have the desired characteristics, and progressively eliminate those series that do not. When the checklist is complete, the selection will have been narrowed down to types that will fulfill the requirements.

---

*An exception to this application is the "TWINAX" connector which looks like an ordinary coaxial one but, in fact, contains two balanced pins to match with balanced shielded pairs.

Table 3.10—Coaxial Connector Selection Chart*

Characteristics	SMA	SMAA	SMB	SMC	TNC	BNC	N	SC	C	UHF	HN	LT
COUPLING Threaded	X	X		X	X		X	X		X	X	X
Bayonet						X			X			
Quick Disconnect			X									
CABLE Crimp	X	X	X	X	X	X	X	X	X	X		
ATTACHMENT Clamp	X		X	X	X	X	X	X	X	X	X	X
Solder	X	X	X	X	X					X		
IMPEDANCE 50 Ω	X	X	X	X	X	X	X	X	X		X	X
70 Ω			X	X	X	X	X					
FINISH (BODY)												
Passivated Stainless Steel	X	X	X	X	X		X					
Gold Plated	X	X	X	X								
Silver Plated					X	X	X	X	X	X	X	X
Nickel Plated					X	X	X	X	X	X	X	X
SIZE 0.320" Maximum		X	X	X								
0.400" Maximum	X											
0.625" Maximum					X	X						
0.827" Maximum							X	X	X	X	X	
1.500" Maximum												X
VOLTAGE 250 VRMS		X	X	X								
RATING 500 VRMS	X		X	X	X	X		X				
1500 BRMS							X		X		X	
5000 VRMS												X
SEALING Moisture	X	X			X	X	X	X	X		X	X
MAXIMUM OPERATING FREQ.												
0.3 GHz										X		
2.5 GHz											X	
4.0 GHz			X			X						X
10.0 GHz				X								
11.0 GHz							X	X	X			
12.4 GHz	X	X										
15.0 GHz					X							
18.0 GHz	X	X										
VSWR AT MAX. OPERATING FREQ.												
1.25:1		X					X	X				
1.30:1		X			X	X	X	X				
1.35:1									X	X	X	X

Table 3.10—(continued)

Characteristics	S M A	S M A A	S M B	S M C	T N C	B N C	N	S C	C	U H F	H N	L T
1.40:1	X											
1.50:1			X									
1.60:1				X								
1.70:1				X								
QPL MIL-C-39012	X		X	X	X	X	X	X	X			
BODY MATERIAL												
Beryllium copper	X	X										
Stainless Steel	X	X			X		X					
Brass			X	X	X	X	X	X	X	X	X	X
Aluminum					X		X				X	

*Courtesy of Electronic Products Magazine, May 15, 1972; pp. 148.

3.8 References

1. Testone, A., *Static Electricity in the Electronics Industry*, (Massachussetts: Testone Enterprises).

2. Mardiguian, M., *Electrostatic Discharge*, (Gainesville, Virginia: Interference Control Technologies, Inc., 1986)

3. Ramo, S.; Whinnery, J.S.; and T. VanDuzer, T., *Fields and Waves in Communication Electronics*, (New York: John Wiley & Sons, 1967).

4. Ianovici, M., *Compatibilite Electromagnetique*, (Switzerland: EPFL, 1985).

5. Metzger, G. Vabre, J.P., *Electronique des Impulsions*, (Paris: Masson & Cie, second edition, 1975). U.S. Edition: *Transmission Lines with Pulse Excitation* (New York: Academic Press).

6. White, Donald R.J., *Electromagnetic Shielding Materials and Performance*, (Gainesville, Virginia: Interference Control Technologies, Inc., 1975), p. 1.12.

7. Ott, H.W., *Noise Reduction Techniques*, (New York: John Wiley & Sons, 1976).

3.9 Bibliography

1. Nowacki, H., "Connector System for Data Processing," (*Electronic Engineering*, September, 1981).

2. Southard, R.K., "High Speed Signal Pathways," (*Electronic Engineering*, September, 1981.

Chapter 4

Analog and Logic Active Devices

Although most EMI cases seem to occur because of installation parameters such as site, cabling, grounding, etc., the real crux of the problem, whether it is the source or the victim, is actually an active device.

This chapter deals with the EMI characteristics of active electronic devices, first from the susceptibility point of view, then (if applicable) from the emission point of view.

While the whole world is moving toward more digital-type processing and less analog, this move does not affect the fundamental understanding of the elementary EMI mechanisms at the active component level. At this level, EMI is related to the **unintended** behavior of diodes, transistors, etc., and many EMI features will come into play indistinctively in analog or logic circuits.

On the other hand, the trend to ultra-integration allows an entire functional block, such as an audio amplifier, RF receiver, multiplexer, microprocessor, video amplifier, modem etc., to be included in a single chip. Susceptibility cannot be tested at this lower level, but the designers of these ICs need to take care of the parasitic behavior of the inner components when their elementary cells are created and implemented. In a sense, they have to accomplish electromagnetic compatibility "inside the chip."

4.1 Definitions, Susceptibility Thresholds

EMI can occur **inside** or **outside** the normal range of frequency where the device is intended to perform. In this respect, and at the risk of over simplifying, it can be said that all circuits or components designed to receive and process an electrical signal can be classified into two categories: tuned devices and baseband devices.

The first category of devices, which behave as passband elements, involves the intentional reception of a modulated signal at (or around) a certain frequency, like radio communication receivers, audio amplifiers, modems, etc. Tuned devices have a given set of properties (some of them being, unfortunately, seldom characterized and documented) such as: center frequency, bandwidth, selectivity (or shape factor, i.e., the ratio of the 60 dB to the 3 dB bandwidth) and out-of-band spurious responses.

The second category of devices, which behave as low-pass elements, involves the reception of any frequency from dc (or VLF) up to a certain cutoff frequency. This category includes instrumentation amplifiers, digital logic, video amplifiers and the like.

Their characteristics relevant to EMI are cutoff frequency (which can be derived from the rise time or the time constant) and rejection slope beyond the cutoff frequency. Examples include 6, 12 or 18 dB/octave (corresponding respectively to 20, 40 or 60 dB/decade since dB/decade = $3.32 \times$ dB/octave) and out-of-band spurious response.

In addition to these frequency aspects, both categories need to be characterized in terms of input impedance (real and complex) and symmetric (balanced) or asymmetric (unbalanced) nature of their input.

4.1.1 In-Band Susceptibility

Regardless of the tuned or baseband nature of the device, the first parameter to start with is the in-band susceptibility. Presuming that the transfer function or gain within the passband is relatively flat, it is helpful to establish some measure of latent in-band susceptibility of receptors to electromagnetic exposure in order to rate them for classification purposes. This rating then requires that a scoring system calculate and assess relative vulnerability to EMI environments. Such a scoring technique has been developed and will now be discussed.

Sensitivity (N) and bandwidth (B) appear to be the two most pertinent parameters required for a rating system. The greater the sensitivity (a lower number) and bandwidth, the greater the tendency for EMI in a receptor. If sensitivity is defined as equal to internal noise, the bandwidth term will be in the numerator and the sensitivity term will be in the denominator.

4.1.1.1 Susceptibility Index for Analog (Linear) Devices

The sensitivity can be based on the intrinsic noise of the device; that is, a statement is made that the "minimum discernible signal" (MDS) occurs when signal equals internal noise. Therefore, a figure of merit or relative score can be defined by:

$$RS = \frac{\text{device bandwidth}}{\text{device input noise}}$$

Since in practically any receiving equipment, sensitivity and selectivity are built into the first-stage input block, i.e., a monolithic or hybrid electronic circuit, the RS rating of the first-stage amplifier, eventually with its selectivity filters, will dictate the RS rating of the entire equipment.

If an interfering source is coherent* (such as a broadband transient or pulse emission), then a receptor susceptibility **voltage-to-noise** ratio RS_c (subscript c stands for coherent) is proportional to bandwidth:

$$RS_c = \frac{B}{N_v} = \frac{B}{\sqrt{4RFKTB}} = \sqrt{\frac{B}{4RFKT}} \qquad (4.1)$$

where,
$$B = \text{bandwidth in Hz}$$
$$N_v = \text{internal noise voltage}$$
$$R = \text{resistive component of the equivalent input impedance in ohms}$$
$$F = \text{equivalent noise figure of receptor}$$
$$KT = 4 \times 10^{-21} \text{ W/Hz bandwidth, from Boltzmann's constant, at } 20°C \text{ ambient.}$$

If the EMI source is noncoherent, e.g., bandwidth-limited white

---

*A coherent signal or EMI emission has a specified frequency and phase relation between its environmental frequency components, whereas a noncoherent emission has a random phase and often a random amplitude between increments.

noise such as an unmodulated arc discharge, the RS voltage-to-noise ratio is proportional to the square root of bandwidth:

$$RS_n = \frac{\sqrt{B}}{M_v} = \frac{\sqrt{B}}{\sqrt{4RFKTB}} = \frac{1}{\sqrt{4RFKT}} \qquad (4.2)$$

Higher values of RS correspond to greater tendencies for EMI susceptibility. Since potential EMI sources may be either coherent or noncoherent, Eq. (4.1) is chosen as the more damaging of the two equations; it becomes the basis for latent receptor susceptibility. Replacing KT by its value at 20°C yields:

$$RS_v = \sqrt{\frac{B}{4R \times 4 \times 10^{-21}F}} = 0.8 \times 10^{10} \sqrt{B/RF}$$

$$= 198 + 10 \log (B/RF) \text{ dB} \qquad (4.3)$$

Sometimes it is more useful to calculate RS directly from the rated (power) sensitivity, N_p, whereupon Eq. (4.1) becomes:

$$RS_p = RS_v{}^2 = \frac{B^2}{N_v{}^2} = \frac{B^2}{4R \, N_p} = 20 \log B - N_{dBW} - 10 \log 4R$$

$$= 20 \log B - (N_{dBm} - 30 \text{ dB}) - 10 \log 4R$$
$$= 24 \text{ dB} + 20 \log B - 10 \log R - N_{dBm} \qquad (4.4)$$

where,

N_{dBW} = sensitivity in units of dBW

N_{dBm} = sensitivity in units of dBm = N_{dBW} + 30 dB

Equation (4.4) is plotted in Fig. 4.1 with resistance and sensitivity as parameters. To select the pertinent curve in the figure, first use the companion table.

Illustrative Example 4.1

Determine the RS receptor susceptibility rating of a receiver having a sensitivity of –104 dBm, an input impedance of 50 Ω and a bandwidth of 1 MHz. For this situation, Eq. (4.4) or Fig. 4.1 is used:

$$RS = 24 \text{ dB} + 20 \log 10^6 \text{ Hz} - 10 \log 50 - (-104 \text{ dBm})$$
$$= 24 \text{ dB} + 120 \text{ dB} - 17 \text{ dB} + 104 \text{ dBm} = 231 \text{ dB}$$
$$(4.5)$$

If, instead of sensitivity, the data had been given to us as a noise figure of 10 dB or a noise factor of 10, then Eq. (4.3) would yield:

$$RS = 198 + 10 \log (10^6/50 \times 10) \text{ dB}$$
$$= 198 + 33 = 231 \text{ dB} \qquad (4.6)$$

For further practice with the ranges of RS, a few different ones will now be computed.

Sensitivity and Impedance Levels for Determining Curve to Use

Sensitivity N_{dBm}	Equivalent Input Impedance in Ohms												
	10	30	50	100	300	600	1k	3k	10k	30k	100k	300k	1M
−160		--	A	A	AB	B	B	BC	C	CD	D	DE	E
−150	A	AB	B	B	BC	C	C	CD	D	DE	E	EF	F
−140	B	BC	C	C	CD	D	D	DE	E	EF	F	FG	G
−130	C	CD	D	D	DE	E	E	EF	F	FG	G	GH	H
−120	D	DE	E	E	EF	F	F	FG	G	GH	H	HI	I
−110	E	EF	F	F	FG	G	G	GH	H	HI	I	IJ	J
−100	F	FG	G	G	GH	H	H	HI	I	IJ	J	JK	K
−90	G	GH	H	H	HI	I	I	IJ	J	JK	K	KL	L
−80	H	HI	I	I	IJ	J	J	JK	K	KL	L	LM	M
−70	I	IJ	J	J	JK	K	K	KL	L	LM	M	MN	N
−60	J	JK	K	K	KL	L	L	LM	M	MN	N	NO	O
−50	K	KL	L	L	LM	M	M	MN	N	NO	O	OP	P
−40	L	LM	M	M	MN	N	N	NO	O	OP	P	PQ	Q
−30	M	MN	N	N	NO	O	O	OP	P	PQ	Q	QR	R
−20	N	NO	O	O	OP	P	P	PQ	Q	QR	R	RS	S
−10	O	OP	P	P	PQ	Q	Q	QR	R	RS	S	ST	T
0	P	PQ	Q	Q	QR	R	R	RS	S	ST	T	TU	U
10	Q	QR	R	R	RS	S	S	ST	T	TU	U	UV	V
20	R	RS	S	S	ST	T	T	TU	U	UV	V	VW	W

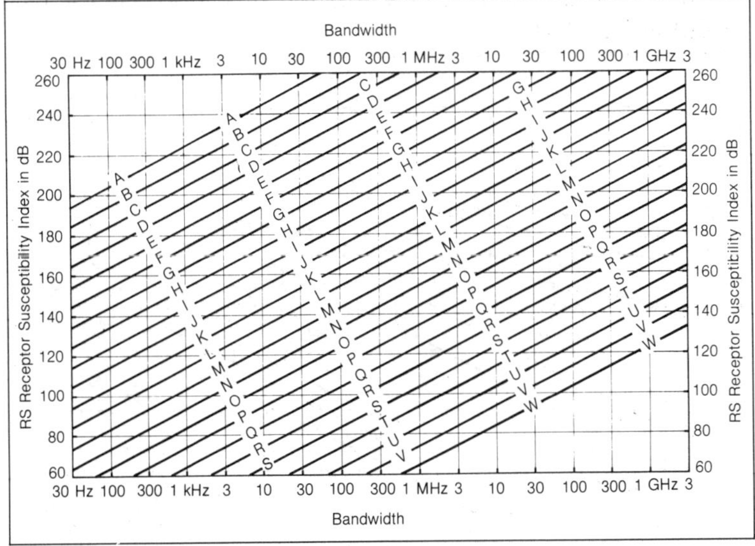

Figure 4.1—Susceptibility Index Based on Device's Input Resistance and Bandwidth

Illustrative Example 4.2

Determine the RS receptor susceptibility rating of the following amplifiers:

Amplifier Type	$N_{dB\mu V}$	Z Level	N_{dBm}	Bandwidth	RS
TWT Amplifier	+37	50 Ω	−70	2 GHz	263 dB
IF Amplifier	+37	50 Ω	−70	1 MHz	197 dB
Crystal-Video Rcvr	+75	1 kΩ	−45	3 MHz	169 dB
Video Amplifier	+106	50 Ω	−1	1 MHz	128 dB
Audio Amplifier	+100	600 Ω	−18	10 kHz	94 dB
Sensor Amplifier	+60	1 kΩ	−60	100 Hz	94 dB
Digital Data RX	+126	150 Ω	+14	1 MHz	108 dB

From such approach, an RS level ranking can be created in terms of susceptibility value judgments corresponding to all types of receptors. One such rating appears in Fig. 4.2 for typical receivers and amplifiers. The RS rating scores shown range from > 230 dB (extremely susceptible) to < 80 dB (rarely susceptible). A number of interesting observations now can be made.

Amplifiers which have a voltage sensitivity independent of bandwidth (such as baseband video amplifiers driven from a second detector with about 0.2 V threshold) have an RS slope of 20 dB/decade of receptor bandwidth. N_{dBm} is independent of bandwidth in Eq. (4.4).

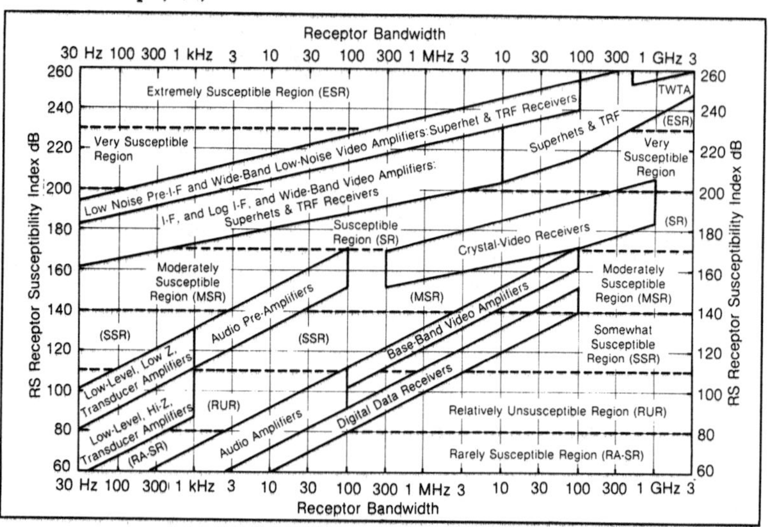

Figure 4.2—Susceptibility Index of Various Types of Receiving Devices

Amplifiers and receivers which have noise-limited front ends have an RS slope of about 10 dB/decade of receptor bandwidth. N_{dBm} decreases at 10 dB/decade and 20 log B increases at 20 dB/decade in Eq. (4.4).

The spread in RS susceptibility of receivers and amplifiers is in excess of 200 dB — an enormous range! Broadband amplifiers and receivers are more susceptible than their narrowband counterparts and low-noise (2 to 20 dB) receivers. For example, IF and wideband video amplifiers are very susceptible receptors.

Low-level (0.1 to 1 mV), narrowband transducer amplifiers are relatively unsusceptible. However, without considering the way it is installed, a low-impedance, baseband transducer amplifier is more susceptible than high-impedance amplifiers at low frequencies. Assuming all amplifiers and receivers exhibit a 100 dB rejection to out-of-band EMI emission, wideband, low-noise devices still show moderately susceptible characteristics.

The RS mentioned before was based on the intrinsic noise level of the device. A result was that, for a given bandwidth, the analog devices with the lowest input resistance appear the most susceptible since their noise level, Nv, is small. This fact seems to contradict experience where high-impedance analog amplifiers do exhibit a higher vulnerability. The vulnerability occurs because once the amplifier is wired to an actual equipment, any component will be exposed to ground-loop couplings and stray capacitance coupling where more induced EMI voltage will appear across larger resistances. Therefore, it is useful to show, in addition to the RS index due to intrinsic noise only, another figure of merit which would take into account the device's input resistance.

$$\text{RS (in situ)} = \text{RS (intrinsic)} \times \text{R input}$$

$$= \sqrt{\frac{B}{4RFKT}} \times R = \sqrt{\frac{B \times R}{4FKT}} \qquad (4.7)$$

Notice that while the intrinsic RS decreases (less susceptible) at 10 dB/decade when R increases, the "when-installed" RS increases (more susceptible) at 10 dB/decade.

4.1.1.2 Review of Susceptibility Scores of Some Typical Receivers

Communications-Electronics (CE) Receivers

CE receivers are the first of four categories of man-made receptors. These are generally the most susceptible to EMI since their RS rating varies from 160 dB to over 260 dB (see Fig. 4.2). They are only summarized here with regard to their RS susceptibility scores.

Broadcast CE Receivers

The broadcast bands cover:

	RS Rating	Value Judgment
HF Amplitude Modulation (150-1,605 kHz)*	195 dB	Susceptible
VHF Frequency Modulation (88-108 MHz)	215 dB	Very Susceptible
VHF Television:		
Lower Bands (54-88 MHz)	230 dB	Extremely Susc.
Upper Bands (174-216 MHz)	230 dB	Extremely Susc.
UHF Television (470-890 MHz)	225 dB	Very Susceptible

*The AM band starts at 535 kHz in the United States.

In addition to man-made sources of electromagnetic emissions, AM broadcast receivers are susceptible to broadband atmospheric noise. FM and TV receivers, while immune to atmospheric noise, are quite susceptible to automobile ignition noise if their associated antennas are located near roads and are not located well above the ground, e.g., less than 10 m.

Communications CE Receivers

Communications equipments are the greatest in number and most varied of all CE types. They occupy portions of the spectrum interlaced between other activities from about 10 kHz to about 2 GHz. Many of the CE receivers here are of the land-mobile type. Above 2 GHz, point-to-point communication is generally of the relay type. RS ratings for communications receivers vary from 150 dB at VLF (moderately susceptible) to about 235 dB UHF (extremely susceptible).

Since many CE receivers are of the land-mobile type, they are susceptible to automobile ignition and nearby industrial-area noise. This susceptibility is a lesser problem for the military since all vehicles employ ignition suppression devices. Co-channel and adjacent-channel interference is more of a problem for the military because of the perishability of combat frequency assignments.

Relay Communications CE Receivers

Point-to-point communications generally consist of one or more of the following four types:

	RS Rating	Value Judgment
Common Carrier, Microwave Relay (2.1—11.7 GHz Interspersed)	$\cong$ 245 dB	Extremely Suscept.
Satellite Relay (2.4—16 GHz Interspersed)	$\cong$ 225 dB	Very Susceptible
Ionospheric Scatter (400—500 MHz)	$\cong$ 220 dB	Very Susceptible
Tropospheric Scatter (1.8—5.6 GHz Interspersed)	$\cong$ 230 dB	Extremely Suscept.

It is interesting to note that microwave relay receivers are the most susceptible (RS rating of 245 dB). Yet, pragmatically, microwave relay links are relatively immune to EMI because of both the enormous off-axis interference rejection of their antennas (typically 60 to 70 dB) and high S/N ratios (typically about 50 dB) of the links. This illustrates what can be done for good EMI control in system equipment design.

Navigation CE Receivers

Navigation receivers in this classification exclude radars since radars are covered in the next section. The RS rating of navigation receivers varies from about 170 dB (moderately susceptible for certain VLF receivers) to about 230 dB (extremely susceptible) for certain UHF types. Typical navigation receiver types included are:

> VOR (VHF Omni-Range): 108—118 MHz
> TACAN (Tactical Air Navigation)
> Marker Beacons: 76.6—75.4 MHz
> ILS (Instrument Landing System)
>   Localizer: 108—118 MHz
>   Glide Path: 328.6—355.4 MHz
> Altimeter: 4.2—4.4 GHz
> Direction Finding: 405—415 kHz
> Loran C: 90—110 MHz
>   A: 1.8—2.0 MHz
> Maritime: 285—325 kHz; 2.9—3.1 GHz; 5.47—5.65
>   GHz
> Land: 1638—1708 kHz

Radar CE Receivers

Radar is used in intermittent portions of the spectrum from about 225 MHz to 35 GHz. It is used in many ways (the Department of Defense is the largest user of the powerful radars) including air-traffic control, air and surface search, harbor surveillance, mapping, tracking and fire control, police speed monitoring and weather.

Except for narrowband doppler police and CW radars, radar receivers are usually broadband and operate at higher frequencies. Hence, their RS rating is usually about 230 to 240 dB (extremely susceptible). This is a bit ironic since radars are significant offending sources of EMI because of their high effective radiated powers (+95 dBm to +135 dBm).

Spread-Spectrum Receivers

Spread-spectrum (SS) communication techniques use the non-discrete spectrum of a randomized, binary coded message. The most used techniques are direct-sequence SS (DSSS) and frequency-hop SS (FHSS). In the first type, the keyed-in data is simply mixed with a pseudorandom generator, and the resulting signal is sent out. In the second type, a similar signal is mixed with an RF carrier whose frequency itself is varied pseudorandomly, then sent out. In both cases, the spectrum envelope is determined by the pseudo-random coding rather than the original modulating signal. The immunity to EMI of an SS system is determined by the despreader (SS receiver). The effect of despreading is shown on Fig. 4.3 for a DSSS and a narrowband EMI. The result of the despreading is that the original signal which was stretched over a frequency band $2 \times Bs$ given by the randomizer is now recomposed into its initial bandwidth dependent of the information rate or baud rate (Bd). The similar process spreads the discrete interference over the bandwidth $2 \times Bs$. Consequently, the demodulator which accepts only the bandwidth $2 \times Bd$ receives only a portion of the CW interference. Then the signal to interference ratio is:

$$S/I = \frac{Ps}{Pi} \times \frac{Bs}{Bd} \qquad (4.8)$$

where, Ps = Signal Power
Pi = EMI Power

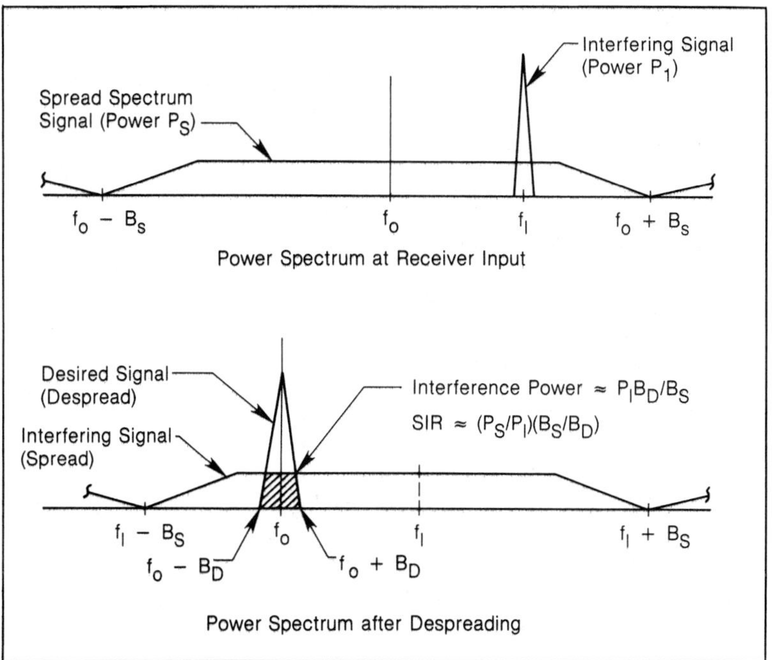

Figure 4.3—Narrowband Interference to a Direct Sequence Spread Spectrum Receiver (from Hopkins & Carvey, Ref. 6)

Bs/Bd is usually defined as the SS processing gain (PG). Therefore,

$$S/I = \frac{Ps}{Pi} \times PG \qquad (4.9)$$

For FHSS, the effect is similar to DSSS, except that, as shown on Fig. 4.4, for a given observation time, the desired signal shows a discrete spectrum of all the carrier hops. Then the receiver compresses this spectrum around a center frequency Fo while the EMI power is stretched over a bandwidth $2 \times Bs$, and the same equations (4.8) apply.

Figure 4.5 shows the probability of error of a binary DSSS channel for an S/I ratio of −20 dB; i.e., the EMI power received at victim input being 20 dB above signal power.[1]

The case of broadband EMI is more complex to model. To the worst extreme, the EMI spectrum could occupy the same bandwidth 2 × Bs with an equal or greater power density than the desired signal, but this is generally not the case for continuous interference. A more realistic case is the one of fortuitous or intentional jamming of an SS receiver by a multifrequency source, which can be itself another or several SS systems. A given number (n − 1) of users utilizing identical power will affect the nth user in a similar manner than n − 1 simultaneous narrowband EMI sources, with the possible added effect of false sychronization at the despreader, loss of confidentiality etc. To summarize, we may say that the RS rating of an SS receiver generally is equal to that of a conventional receiver having the same bandwidth 2 × Bd than the demodulator, improved (i.e. divided) by the processing gain PG of the despreader.

Amplifiers

A previous section summarized CE receivers and stated that they are generally the most susceptible of the man-made receptor

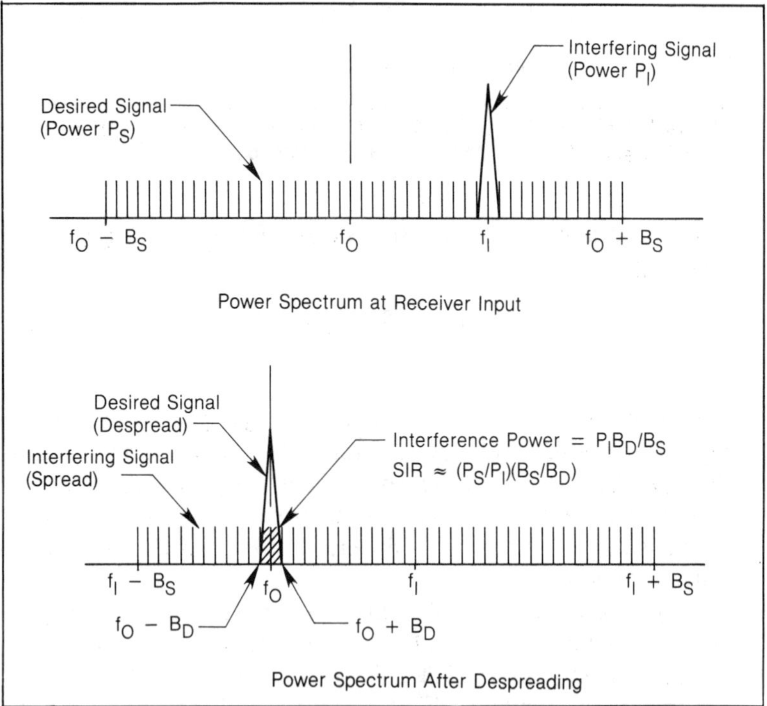

Figure 4.4—Narrowband Interference to a Frequency-Hopping Spread Spectrum Receiver (from Hopkins & Carvey, Ref. 6)

categories shown in Fig. 4.2. There exist a few exceptions such as low-noise, wideband video amplifiers that exhibit higher RS susceptibility ratings than receivers. However, many applications of such amplifiers are intended to improve the sensitivity of receiver front-ends and therefore become a part of the CE receiver. On the other hand, low-noise, wideband video amplifiers are used in other applications.

Intermediate frequency (IF) amplifiers are always part of CE receivers by definition since they are the natural devices to the superheterodyne process. Also, either video (not low-noise, wideband types) and/or audio amplifiers are always found in superheterodyne receivers. In contrast to IF amplifiers, however, video and audio amplifiers constitute a separate class of potential

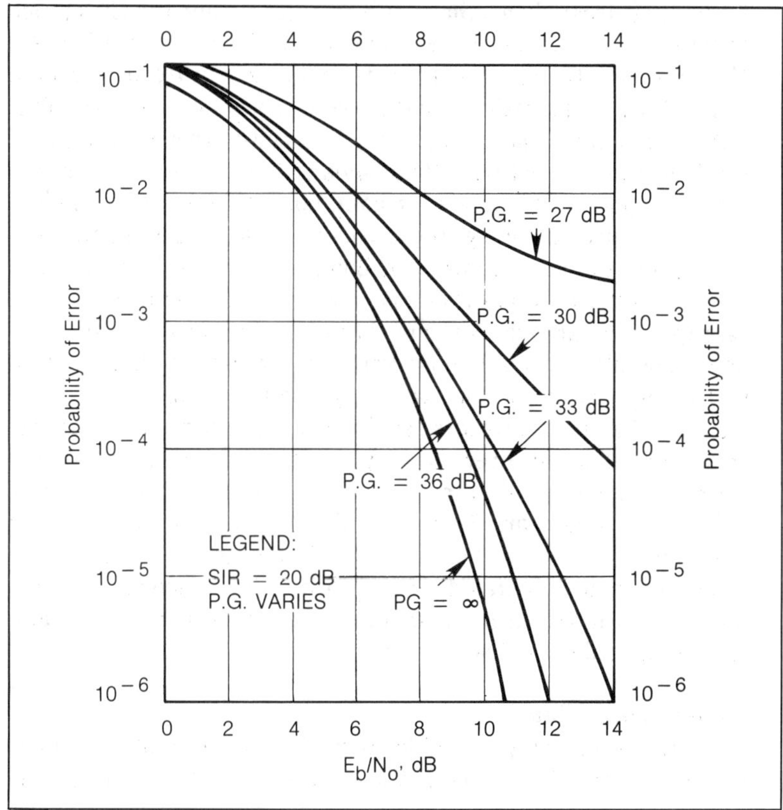

Figure 4.5—Performance of Binary DSSS Channel with Interference. The abcissa is scaled in energy per bit/thermal noise energy density, in consistent units, which is equivalent to the receiver S/N ratio.

EMI victims. Whether or not they appear in CE receivers, they are shown as a distinct category of man-made receptors in Fig. 4.2.

IF Amplifiers

IF amplifiers constitute another input terminal pair to which a receiver may be susceptible. Culprit emissions ordinarily enter a receiver input terminal by the antenna and sometimes by the lead-in RF transmission line pickup (leakage). If the input signals exist at intermediate frequencies, the receiver front-end rejection may or may not be adequate to prevent a susceptible situation. Assuming such rejection to IF emissions is adequate, it may still be possible for these emissions to bypass the receiver front end and be picked up by the IF input cable. An example is when the receiver front end is located on a mast with an antenna, and the IF is piped down to the deck where the remainder of the receiver is located. Here the long IF coaxial cable acts as a pickup **antenna** and the IF amplifier (and remainder of the receiver) now may be susceptible.

An IF amplifier may be more or less sensitive than its associated receiver overall sensitivity. The amount depends upon the loss of the RF input cable; the loss or gain, noise figure and bandwidth of the premixer circuitry; the conversion loss or gain of the converter; and the noise figure and bandwidth of the IF amplifier.

Figure 4.2 shows that low-noise, pre-IF amplifiers may have RS ratings comparable to receivers. These ratings typically vary from about 190 dB (susceptible) for small bandwidths to about 240 dB (extremely susceptible) for very large bandwidths.

Figure 4.2 also shows less-sensitive IF amplifiers which typically accommodate receivers having either premixer RF gain, first-converter gain or pre-IF amplifiers. Such IF amplifiers may have RS ratings which are 10 to 30 dB less than low-noise, pre-IF amplifiers. However, their susceptibilities are still significant, especially when operated under conditions in which they are physically removed from their receiver front ends by a long inter-connecting cable as discussed above.

Video Amplifiers

There are many types of video amplifiers. One simple classification of such amplifiers is (1) low-level video amplifiers and (2) baseband, high-level video amplifiers. A few examples of each classification are:

Low-Level Video Amplifiers	High-Level Video Amplifiers
Receiver Front-End Noise Figure Improvement Crystal-Video Receiver Preamplifiers Nontunable, Low-Noise Receivers Low-Noise Oscilloscope Amplifiers	Receiver Post-Detector Amplifiers Digital Data Receivers Telemetry Data Amplifiers A/D and D/A Converters

The distinguishing features of the video amplifier classification are that low-level types are generally noise limited by either typical (50 Ω, 72 Ω or other) input impedances or high input impedances (100 kΩ, 1 MΩ or other). On the other hand, high-level video amplifiers are generally characterized by requiring a minimum voltage to perform. For example, receiver post-detector video amplifiers typically need a level of about 0.2 V because of the second detector characteristics. A second example is digital data receivers which may require a minimum level of 2 V to sense a mark or "1-bit."

Figure 4.2 shows that low-noise video amplifiers have RS ratings comparable to receivers and pre-IF amplifiers, e.g., from about 190 to 240 dB. Unless protected by a low- and high-pass filter, the wideband, low-noise video amplifier is highly susceptible to EMI emissions and intermodulation. The susceptibility occurs because the amplifiers generally operate over a broad-frequency base band having a low-frequency cutoff of perhaps 100 Hz to 1 kHz where many broadband EMI emissions are the highest.

Low-noise video amplifiers can even have passbands of 1 GHz or more due to recent advances in solid state circuitry, where gain-bandwidth products of active devices may be of the order of 3 GHz or more. Here, Fig. 4.2 can be extrapolated to yield RS ratings of 250 to 260 dB -- an extremely susceptible situation.

Finally, when the low-noise, wideband video amplifier has a high low-frequency cutoff, it acquires other names, such as traveling wave tube amplifiers (TWTAs) shown in the figure. Low-noise TWTAs have the highest RS ratings of all (250 to 270 dB). Fortunately, the lower broadband EMI noise and CE transmitter spectrum activity in the 10 GHz and higher frequency regions create fewer problems with the TWTAs than their latent RS ratings would indicate. From 500 MHz to 10 GHz, however, they are extremely susceptible to EMI.

Another EMI problem associated with low-level video amplifiers is their out-of-band rejection performance (see Section 4.1.3). Unless protected by additions, low-pass filters and high-pass filters, such amplifiers may exhibit only 60 dB rejection to out-of-band emissions at a few decades beyond the amplifier cutoff frequencies. Thus, for strong out-of-band EMI, video amplifiers have out-of-band RS ratings of perhaps 110 dB (190 dB − 80 dB = somewhat susceptible) to 180 dB (240 dB − 60 dB = susceptible).

Consequently, a 30 MHz low-level video amplifier can be very susceptible to a high-level UHF TV transmission or to a nearby VHF land-mobile transmitter.

The high-level video amplifiers pose an altogether different susceptibility situation. Figure 4.2 indicates that RS ratings may vary from about 90 dB (relatively unsusceptible) for 100 kHz baseband bandwidths to about 160 dB (moderately susceptible) for 30 MHz bandwidths (high data rates or short pulse widths). For similar bandwidths this level corresponds to 80 to 130 dB lesser susceptibilities than their low-level video amplifier counterparts. This notwithstanding, the high-level, wideband video amplifiers such as used in high-clock-rate digital receivers and computers can be susceptible to EMI.

Audio Amplifiers

There are also many types of audio amplifiers. One simple classification of such amplifiers somewhat parallels that for video amplifiers, except that 100 kHz is arbitrarily defined here as the upper frequency limit for audio amplifiers. The classification suggested is: (1) low-level audio amplifiers and (2) high-level audio amplifiers. A few examples of each classification are:

Low-Level Audio Amplifiers	High-Level Audio Amplifiers
Audio Preamplifiers	Telephone-Line Repeaters
Low-Noise Oscilloscope Amplifiers	Public Address Amplifiers
Low-Level Sensor Amplifiers	Hi-Fi Power Amplifiers
Biomedical Instruments	Chart Recorder Drivers

The distinguishing feature of the audio amplifier classification is that low-level types are often noise limited with either low input impedances (approximately 10 Ω) or high input impedances (10 kΩ, 100 kΩ or other). If not noise limited, they have input sensitivities less than about 10 mV. On the other hand, high-level audio amplifiers are generally characterized by requiring a minimum voltage, such as about 100 mV as in the case of many hi-fi audio power amplifiers, to perform.

Figure 4.2 shows that audio preamplifiers have RS ratings ranging from about 100 dB (somewhat susceptible) to 170 dB (susceptible). If the preamplifier is used for either low-level, high-impedance transducer or oscilloscope applications, Figure 4.2 shows that it may be 30 to 40 dB more susceptible than its lower impedance counterpart.

4.1.1.3 Susceptibility Index for Digital (Two-State) Devices

Since a digital device is a "threshold" or two-state device, the minimum discernible signal has nothing to do with the input resistance of the device. Rather, it is related to the minimum voltage (noise margin) required to trigger it. Therefore, an intrinsic rating of susceptibility RS could be established such as for analog devices by using:

$$\text{RS intrinsic} = B/N_V$$
$$= 0.32/(\tau_r \times N_V)$$

where $\quad B = $ Bandwidth in Hz $= \dfrac{1}{\pi \tau_r} = \dfrac{0.32}{\tau_r}$

$\tau_r = $ Rise time

$N_v = $ Worst-case dc noise margin in volts

All the same, an "in situ" rating taking into account the device's input resistance can be defined which also takes into account the device's immunity **and** its vulnerability to EMI coupling:

$$\text{RS (in situ)} = \frac{B}{N_v} \times R$$

$$= \frac{0.32}{N_v \tau_r} \times R \tag{4.10}$$

In the latter, we see that the RS is inversely proportional to the $N_v \times \tau_r$ (volt $\times$ second) product necessary to trigger the device. When RS intrinsic and in situ are calculated for the most popular logic families, Table 4.1 results and the family which has the higher number is the most susceptible.

Table 4.1—Intrinsic and In-Situ Susceptibility Ratings for the Principal Logic Families. The Values in parenthesis are when the effective input impedance is taken into account, instead of resistance only.

	Nv volts	τ_r sec	Rin ohms	Susceptibility Rating (intrinsic) (volts $\times$ sec)-1	Susceptibility Rating (in-situ)	Rel. index in positive terms RSp
CMOS	1 v	10^{-7}	10^6 (3.10^5)	0.3×10^7 (130 dB)	0.3×10^{13} (10^{12})	125 (120 dB)
HCMOS	1 v	10^{-8}	10^6 (3.10^4)	0.3×10^8 (150 dB)	0.3×10^{14} ($.10^{12}$)	135 (120 dB)
TTL	0.4 v	10^{-8}	3.10^3	0.75×10^8 (158 dB)	2.2×10^{11}	113
STTL	0.3 v	3×10^{-9}	3.10^3	0.3×10^9 (170 dB)	10^{12}	120
ECL 10k	0.1 v	2×10^{-9}	2×10^3	0.15×10^9 (164 dB)	0.3×10^{12}	115

In all fairness, the 10^6 input resistance of HCMOS and CMOS starts to drop above 50 kHz due to the shunting effect of the 3 to 5 pF input capacitance, reaching an input impedance of 10 to 15 kΩ at 3 MHz (CMOS) and 1 to 1.5 kΩ at 30 MHz (HCMOS). Therefore, the value in parenthesis is based on a more pragmatic value, i.e., the equivalent input impedance averaged on the whole bandwidth.

4.1.2 Susceptibility Criteria

The receptor susceptibility rating concept, introduced in Section 4.1, was based on the notion of signal equals noise as part of the basis for scoring. As it develops, S = N may not be an acceptable criterion for susceptibility threshold for these reasons:

1. Different detection systems are affected differently by the same interference or noise spectral intensity levels.

2. Different readout and display systems may respond differently to the same detection systems.
3. Different user requirements impose different interpretations
to any given level of receptor performance.
4. Some receptors either perform correctly or malfunction entirely. A few examples of these "black and white" situations
are: an electroexplosive device (EED) either detonates or it
does not (e.g., for explosive belts, an associated fuel wing tank
is dropped, a canopy and pilot are jettisoned, a missile is
launched, etc.), a carrier-operated relay is closed or it is not,
or any two-state output device from the receptor is either in
state 0 or 1.

The problem, however, develops because most receptor output devices are not two-state. In fact, they are not three-state or n-state, but rather they represent a continuum of degradation (a very large or infinite state) from "white to black," i.e., from perfect performance to complete malfunction. Consequently, it is necessary to relate the status of performance versus S/(N + I) by referring to either the input of the receptor or to its output.

Figure 4.6 shows that for both analog and digital devices, there is a certain nonlinearity of the EMI effect; *that is, the effect is not*

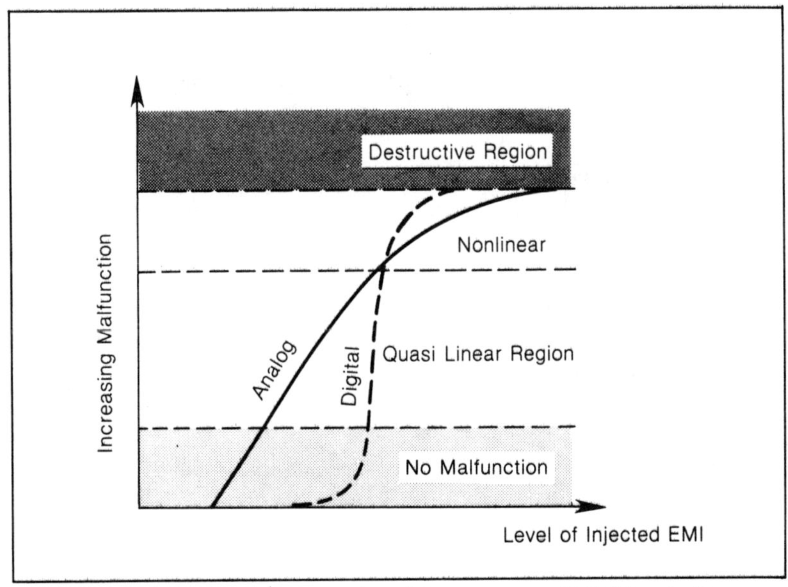

Figure 4.6—Conceptual View of EMI Effects versus EMI Stimuli, Showing Their Nonproportionality (from R. Rode, Ref. 3)

always proportional to the cause. Figure 4.6 reveals three regions: the quasilinear region (above the threshold of sensitivity), the nonlinear region (>> the threshold of sensitivity) and the destructive region.

Thus, it is necessary to introduce the concept of receptor susceptibility criteria in order to relate any given interference level referenced to the input of a receptor to the impact upon the user's satisfaction with the degree of performance. The next section summarizes this concept for voice intelligibility and articulation, digital-error acceptance, TV picture displays, radar displays and other outputs.

Voice Intelligibility

The problem of specifying an operational performance measure for voice communications systems is complicated by the random nature of received voice signal variations in message content and differences in hearing and understanding abilities from one receiver operator to another.

Articulation Score

One performance measure used for voice systems is the **articulation score** obtained by using trained speakers and listeners to determine the percentage of words scored correctly by the listener out of the total number of words contained in the test.

The procedure for an articulation test consists of a speaker (or standardized voice generator such as a tape-recorded voice) reading a set of selected words or syllables over a communications system (which may be subjected to interference). The listener panel interprets what it hears. Various levels of interference may be introduced into the receiver system along with the selected words. The percentage of words interpreted correctly by the listener indicates the intelligibility level or articulation score for the particular set of conditions tested.

The resulting empirical data can be translated into suitable electrical characteristics (such as signal-to-interference ratio) which, in turn, can be used in an EMI prediction process to determine voice system performance.

Figure 4.7 shows the relationship between signal-to-interference ratio and articulation score for different combinations of desired and interfering signal conditions. All of the cases illustrated are for co-channel interference conditions. One significant factor that is

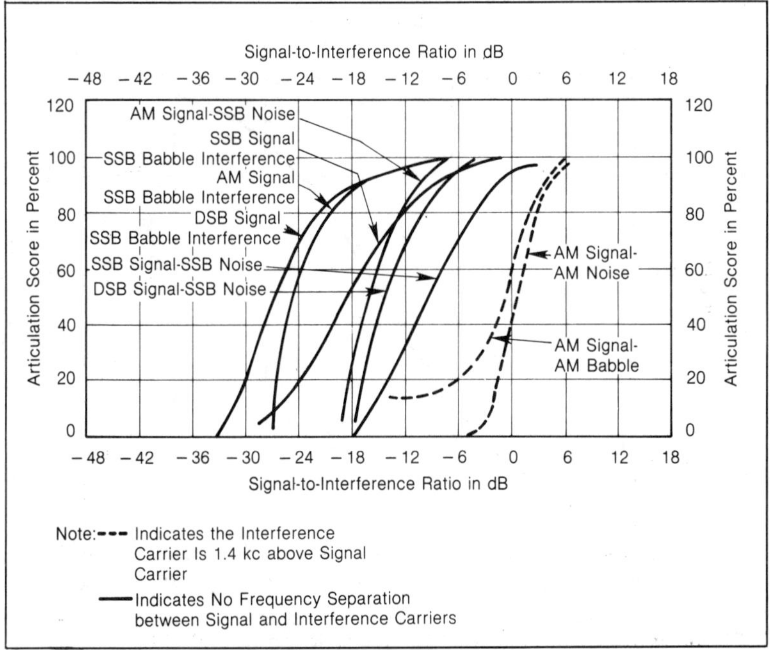

Figure 4.7—Voice System Performance

clear in the figure is that there is a fairly rapid transition from good to poor performance versus S/I ratio. One also sees that AM reception is relatively immune to co-carrier EMI (a message buried — 24 dB below a broadband interference is still 70 percent understood) while a slight 1.4 kHz offset creates the worst situation, mainly due to intermodulation (it takes now an S/I of + 3 dB to get a 70 percent intelligibility).

Articulation Index

Another method for specifying performance of voice communications systems is the articulation index. Figures 4.8 through 4.10 show the relationship between articulation index and signal-to-interference ratio for a number of different types of interfering signals.

The curves may be grouped into one of two classes: CW interference or modulated interference. Considering the many uncertainties that exist in an EMI prediction, the variations among the different curves within a group are relatively insignificant and all

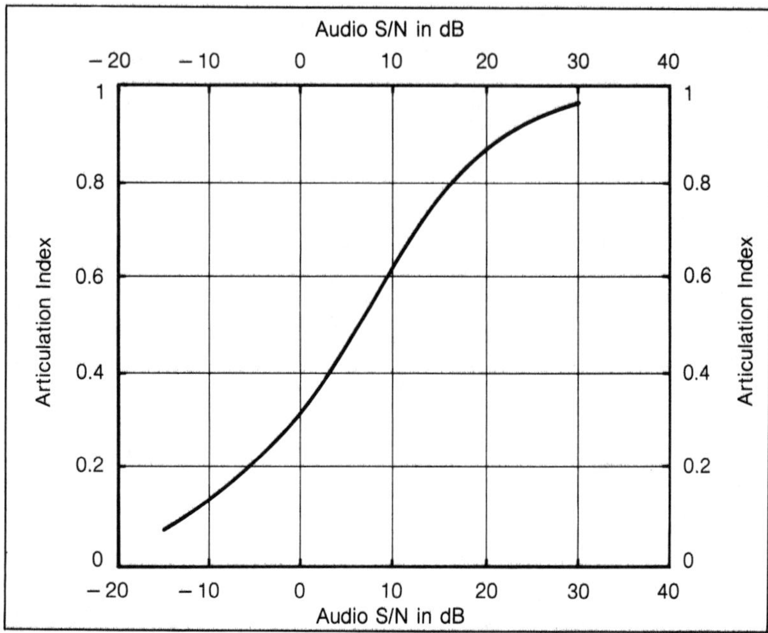

Figure 4.8—Audio Performance, White Noise Interference

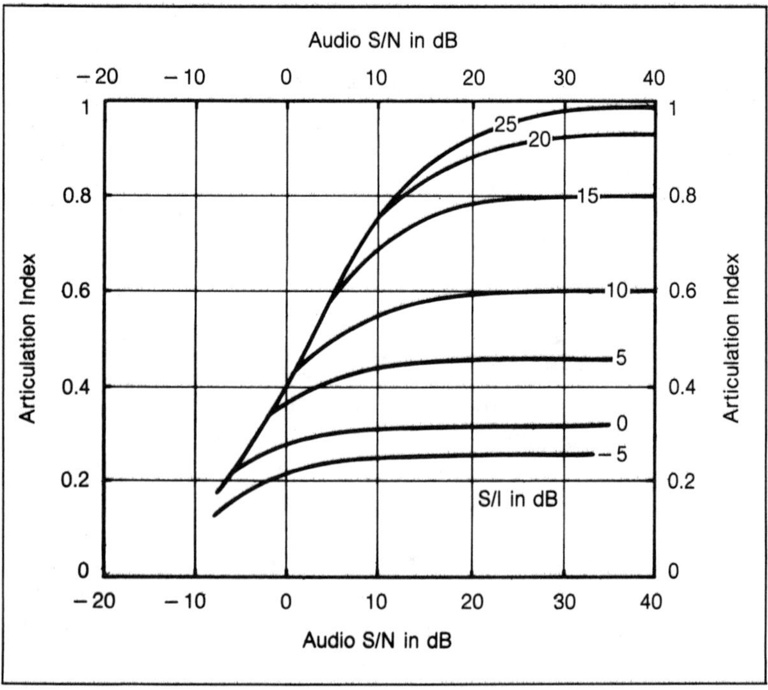

Figure 4.9—Audio Performance, White Noise-Pulse Interference 200 pps

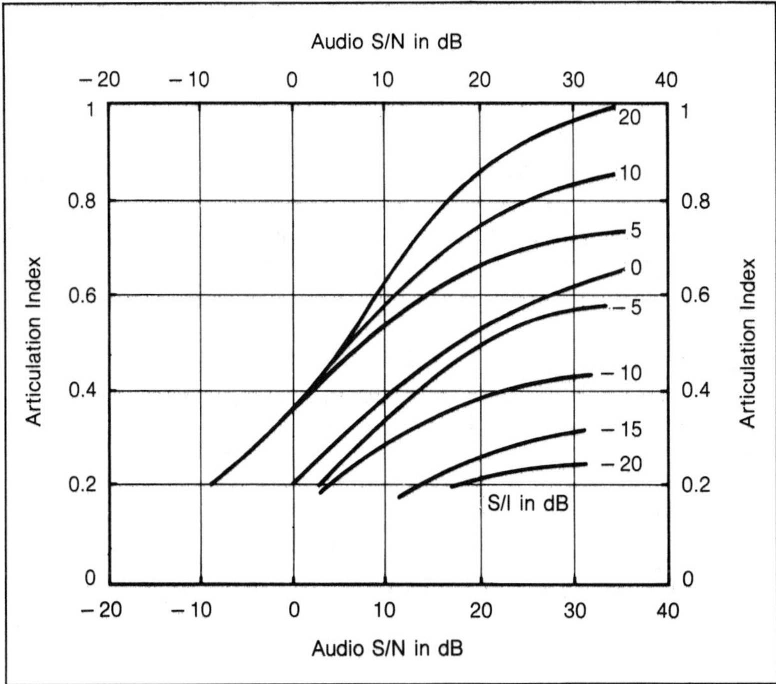

Figure 4.10—Audio Performance, White Noise-500 Hz Tone Interference

of the curves within a group may be approximated by the average curve shown in the figures.

This shows the combined effect of a pre-existing background noise (S/N) and a discrete interference (S/I). Except for single-frequency tones, which do not severely degrade performance, the articulation index relationships shown in the figures exhibit similar trends.

Error Acceptance Criteria for Digital Circuits

The evaluation of performance for a digital system consists of calculating the probability of error (bit-error rate or BER). Two basic types of errors are false acceptance (mistaking interference or noise for the signal) and false dismissal (not recognizing the presence of the signal). For on-off binary transmission, false acceptance is equal to the probability of false alarm, and false dismissal is equal to one minus the probability of detection.*

The relative occurrence of false acceptance or dismissal can be determined from the probability densities for signal, interference and noise at the receiver output.

The relationship between the basic decision process and the two types of errors is illustrated in Fig. 4.11. The density function designated IN(x) refers to the output probability distribution density when interference and noise are present while SIN(x) is the output distribution density when a signal, interference and noise (S + I + N) are present. Decision regions are defined so that when the output exceeds a certain threshold T, the decision is **"signal present,"** whereas if the output is less than T the decision is **"no signal present."**

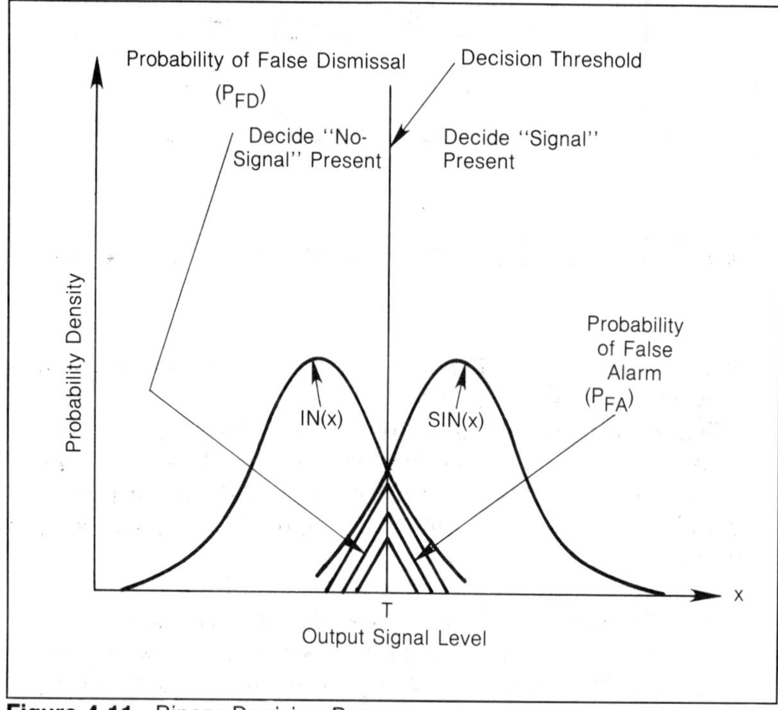

Figure 4.11—Binary Decision Process

Probability of false alarm* is the conditional probability of deciding that a signal is present when no signal was transmitted. **Probability of detection is the conditional probability of deciding that a signal is present, given that a signal was transmitted. (see Ref. 2).

Figure 4.12 illustrates the relationship between signal-to-noise ratio and error rate. Many types of interference result in noise-like signals at the receiver output, and for these situations Fig. 4.12 provides a good approximation of the error rate. One especially significant factor that is evident in the figure is that there is a rapid transition from good to poor performance. The binary pulse amplitude modulation (PAM) signal shown is a random NRZ. The S/N ratio is signal-to-gaussian noise power ratio in the minimum theoretical bandwidth (one-half the symbol rate) and the interference is a discrete sinewave. For instance, assuming S/N =

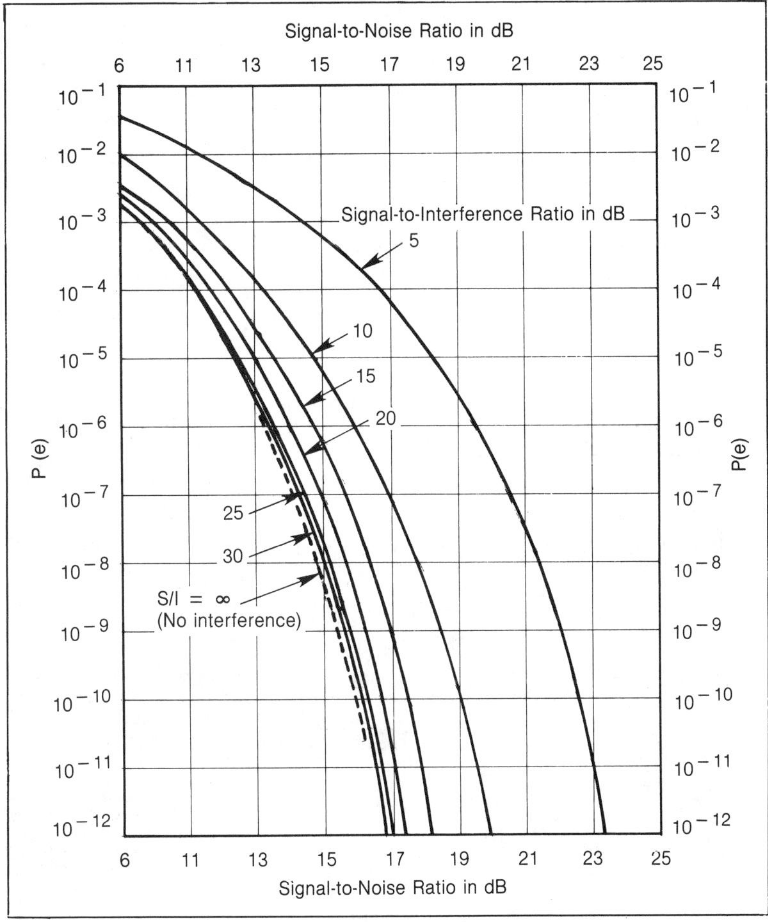

Figure 4.12—Two-Level PAM Performance in an Additive Gaussian Noise and Sinusoidal Interference Environment

15 dB, the probability of error $P_{(e)}$ is 10^{-8} without I, but if an EMI 10 dB below the signal (S/I = 10) is present, $P_{(e)}$ is degraded to 10^{-5}, which is generally unacceptable.

Picture Communication Systems

Television and facsimile systems transmit information that is eventually displayed in the form of a picture, an important form of communication. Aside from the many TV sets currently in use, this form of communication is experiencing increased use by law enforcement and criminal justice agencies for transmitting mug shots and lineups to neighboring agencies. The picture phone is another example of picture communication which is expected to become widely used.

Interference can degrade picture transmission by introducing dots, lines or bars; causing the picture to be blurred; or causing the receiver to lose sync and roll. Figure 4.13 shows typical effects of different types of interference on TV. Television receivers (particularly color receivers) are relatively sensitive to interference as seen in Fig. 4.14. For example, with pulse interference such

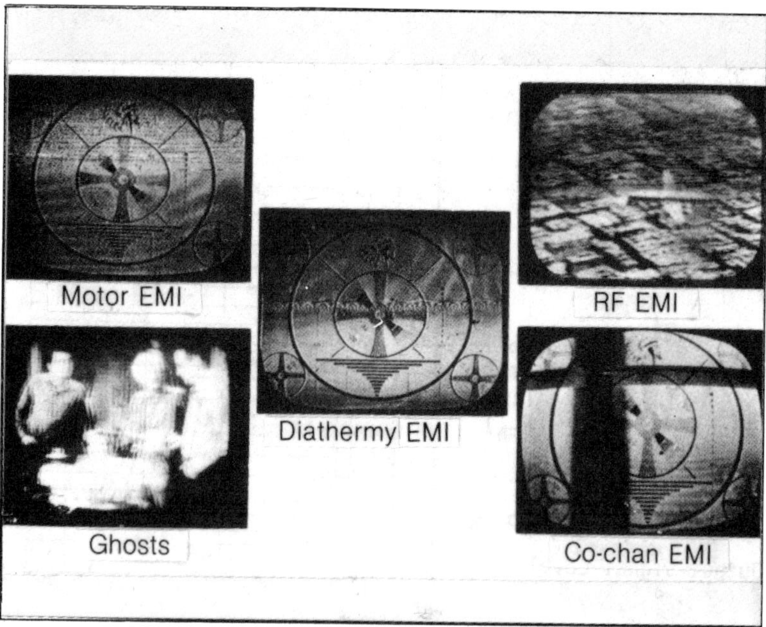

Figure 4.13—Typical TV Interference

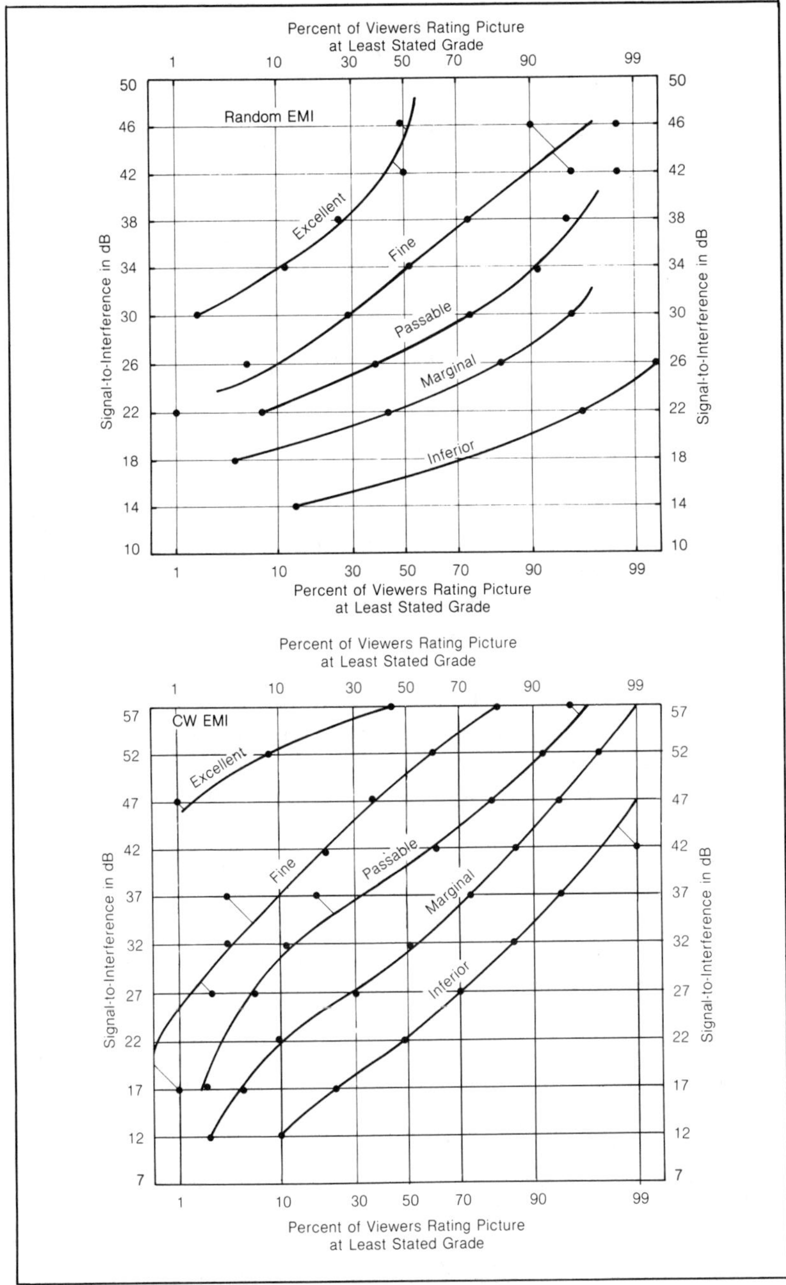

Figure 4.14—Scoring of Two Kinds of TV Interference, by a Sample of Viewers. Top: Random Noise EMI. Bottom: Co-channel CW-EMI, 604 Hz offset.

as would be produced by a radar, an S/I < 15 dB starts to create "snow" in the picture.

4.1.3 Out-of-Band Susceptibility

Out-of-band performance of analog and digital devices is defined in terms of the slope or reduction in sensitivity versus frequency.

Above cutoff frequency, an amplifier gain degenerates and exhibits a slope in the stopband of X dB per decade. This slope is related to the slope in units of dB/octave by:

$$X \text{ dB/decade} = [\log_2 (10)] \; X \text{ dB/octave} \qquad (4.11)$$

$$= 3.32 \times X \text{ dB/octave} \qquad (4.12)$$

Parasitic shunt capacitance often accounts for performance degradation to produce the stopband. It corresponds to a one-stage (N = 1) lowpass filter (see Ref. 3) which exhibits a rolloff of 6 dB/octave or 20 dB/decade. Hereafter, the term dB/decade will be used to describe the stopband performance of all transfer functions of circuits.

Many EMI problems exist out-of-band to the victim and not in-band. Therefore, the out-of-band performance of receivers, analog and digital circuits is very important. Unless the designer makes a special effort, the out-of-band rolloff due to parasitic capacitance corresponds to 20 dB/decade or behaves as an N = 1, one-stage filter. A question is raised regarding amplifier out-of-band performance if the designer intentionally forces a faster rolloff.

In many respects, the performance of an amplifier in the stopband is similar to that of a passive filter, especially when faster rolloff is designed into the amplifier. Figure 4.15 shows the most common response of that of a maximally flat or Butterworth filter for N = 1 to five stages. Beyond the cutoff region, i.e., in the stopband, the slope equals 20 × N dB/decade. The X-axis corresponds to the EMI frequency f_{EMI} in multiples of amplifier cutoff frequency f_{CO}. Thus, the ratio of f_{EMI}/f_{CO} is formed to determine the X-axis value. For any number of stages, N, the Y-axis gives the performance relative to passband sensitivity or NIL in units of dB.

Figure 4.15 shows that all rejection responses are truncated at some "default" value like 50 dB for n = 1 stage etc., a reminder to the user that active amplifiers, logic and passive filters are not

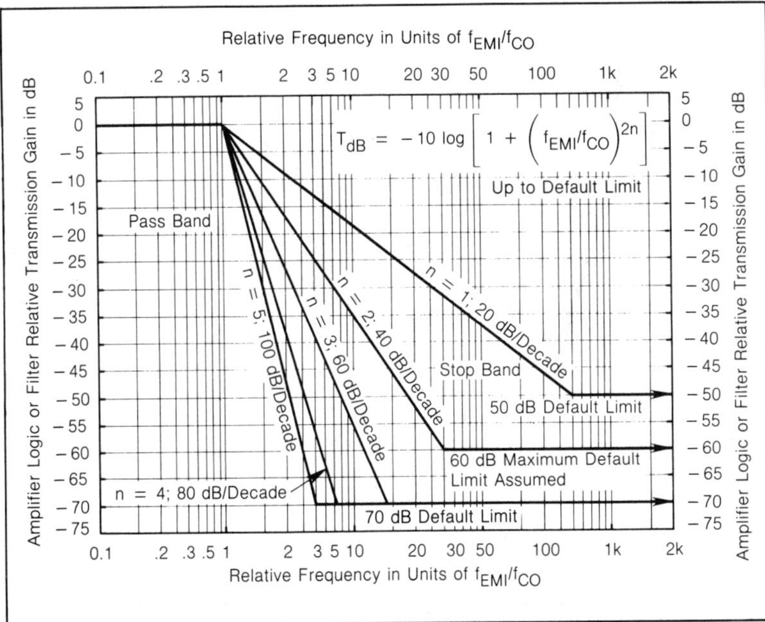

Figure 4.15—Victim Response to Out-of-Band EMI in Function of Its Rejection Slope

perfect devices and that there are technical limitations in the realization of filters. If the user knows that the out-of-band susceptibility response is greater than (e.g., -70 dB) or less than (e.g., -50 dB) the truncated value, then the known measured data should be used. Otherwise, -60 dB is used for $N = 2$, -70 dB for $N = 3$.

Figure 4.16 shows the rejection of a device near f_{CO}, as function of its shape factor SF.

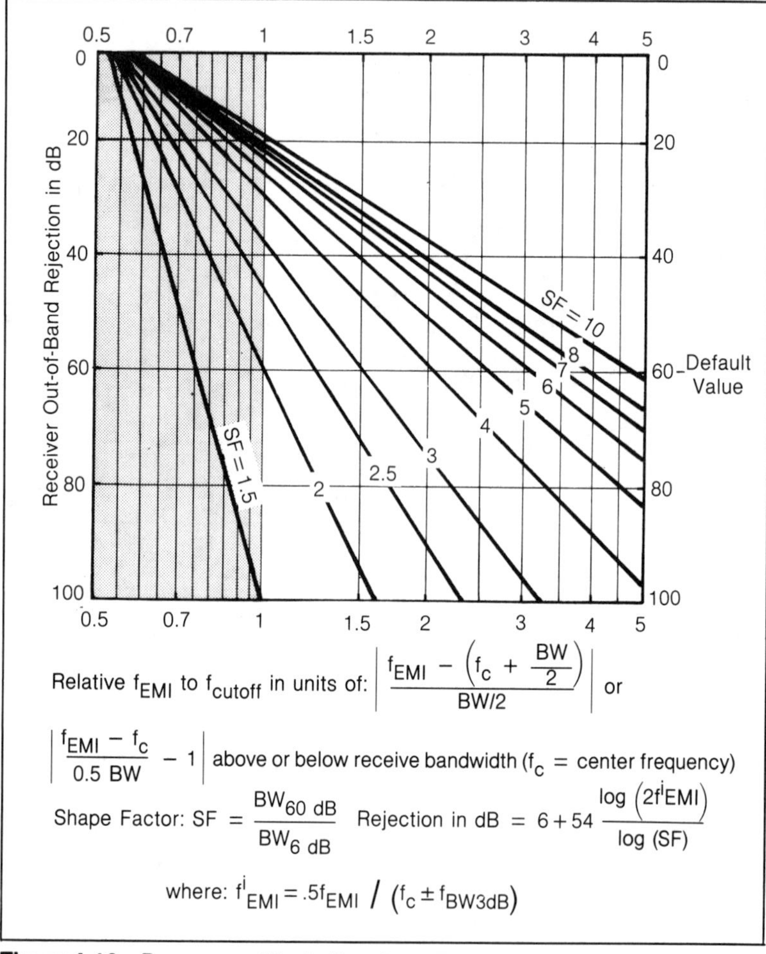

Relative f_{EMI} to f_{cutoff} in units of: $\left| \dfrac{f_{EMI} - \left(f_c + \dfrac{BW}{2}\right)}{BW/2} \right|$ or

$\left| \dfrac{f_{EMI} - f_c}{0.5\ BW} - 1 \right|$ above or below receive bandwidth (f_c = center frequency)

Shape Factor: $SF = \dfrac{BW_{60\ dB}}{BW_{6\ dB}}$ Rejection in dB $= 6 + 54\ \dfrac{\log\left(2f^{i}_{EMI}\right)}{\log(SF)}$

where: $f^{i}_{EMI} = .5 f_{EMI} \ / \ \left(f_c \pm f_{BW3dB}\right)$

Figure 4.16—Response of Radio Receivers Expressed in Function of Their Selectivity (Shape Factor)

4.1.3.1   Audio Rectification Region in Amplifiers and Logic

In addition to out-of-band slope, performance of analog and digital devices is also defined in terms of the sensitivity to audio rectification.

For the out-of-band region ($f_{EMI}/f_{CO} \gg 1$) in amplifiers and digital logic, Figure 4.17 indicates that the **ideal** response continues indefinitely. This situation does not happen in real life due to a phenomenon known as **audio rectification**. A frequency region is reached in the stopband in which antiresonance responses take place as suggested in case No. 3 of Fig. 4.17. The responses constitute a number of RF **windows** in which the far out-of-band responses are much less than both the ideal and the 60 dB down default values previously suggested in Fig. 4.15.

The audio rectification phenomenon is explained by the fact that any active device exhibits a degree of nonlinearity. When the victim, for example, is overdriven by an out-of-band emission, the RF carrier is rectified at an emitter base junction and the carrier sees a low capacitive reactance to ground as in a superheterodyne second detector. The modulation envelope is stripped off or recovered and processed through the amplifier. Audio rectification explains why radar can jam a computer, or a citizens band (CB) transmitter can cause EMI in a stereo amplifier. It is a recurring EMI problem in all active electronic devices.

The only way to determine the audio rectification response in an amplifier is to measure its out-of-band susceptibility performance. However, if this measure is not made or is otherwise

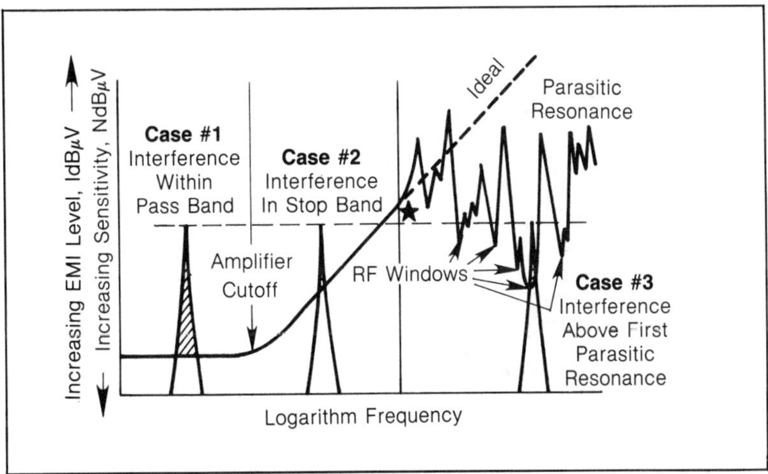

Figure 4.17—Audio Rectification Phenomena in Amplifers and Logic Circuits

unknown, a default model can be used. Figure 4.18 shows the default model* for audio rectification in an amplifier. For the near-in response corresponding to $1 \leqslant f_{EMI}/f_{CO} \leqslant 10$, the amplifier or logic behaves as a lowpass filter. For $f_{EMI}/f_{CO} \geqslant 10$, however, the attenuation may depart dramatically.

The lowest frequency at which audio rectification exists is:

in relative terms:
$$f_{ar}/f_{CO} = 100/\sqrt{f_{CO}}, \text{ for } f_{CO} \text{ in MHz}$$

or in absolute terms:
$$F_{Audio - Rect} = 100\sqrt{f_{CO}} \text{ MHz} \tag{4.13}$$

For values of $f_{EMI}/f_{CO} > f_{ar}$, the response becomes equal to the horizontal line corresponding to N = 1, N = 2 or N $\geqslant$ 3, as applicable. The discontinuities in the figure range from 0 dB (e.g., at f_{EMI}/f_{CO} = 10 and N = 1) to 40 dB (e.g., from the −60 dB

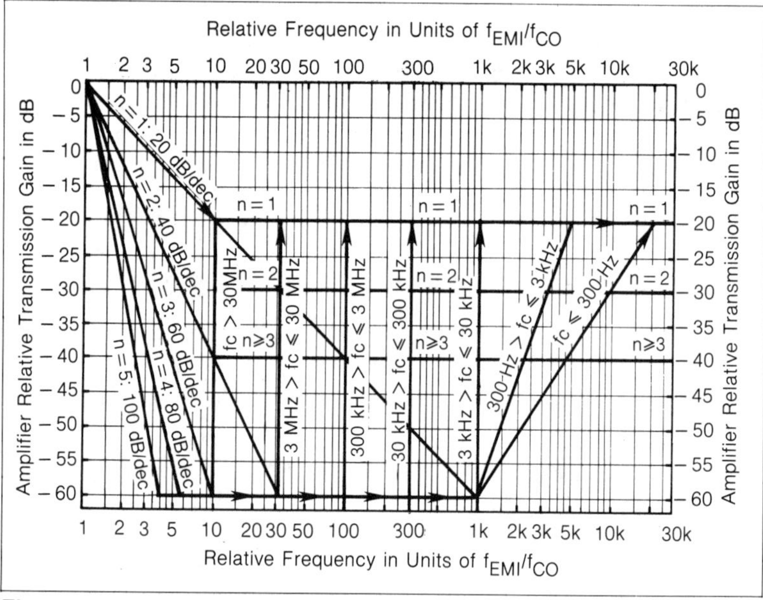

Figure 4.18—Default Model for Audio Rectification

---

*Note: This default math model has been derived from engineering estimates since very little published data are available. It is subject to substantial error since it is also a function of emission level and modulation type.

minimum default line to -20 dB corresponding to N = 1). Since jump discontinuities do not exist in physics, a quantization of the discontinuity is suggested to reflect the effect of emission level at the input terminals. Three differential (normal) mode voltage levels, V_d, are used:

1. For $V_d \geq 500$ mV, use the figure directly.

2. For 10 mV $\leq V_d \leq$ 500 mV:
   $$AR_{dB} = A_{dB} - \Delta AR_{dB} \log ([1 + 0.0184 (V_d - 10)]),$$
   for V_d in mV where, A_{dB} = attenuation due to filtering action only.

   ΔAR_{dB} = difference in dB between the filter curve portion and applicable AR levels in figure.

3. For $V_d \leq 10$ mV, use attenuation due to filtering action only.
   Example: compute the device response for the following:

Operational Amplifier
   Bandwidth = 200 kHz
   Diff. input sensitivity = 1 mV = 60 dBμV
   Rejection slope above F_{CO} = 40 dB/decade

EMI Input at Victim Port
   VHF TV, Channel 9
   F_{EMI} = 200 MHz
   V_{EMI} = 0.2 V = 106 dBμV

The calculation routine is:
   1. Given f_{EMI}/f_{CO} = 200/0.2 = 1,000, Figure 4.18 shows that for n = 2, rejection may be as poor as -30 dB

   2. Since $V_{EMI} <$ 500 mV,
      $$AR_{dB} = A_{dB} - \Delta AR_{dB} \text{ Log } (1 + 0.0184(V_d - 10))$$
      $$AR_{dB} = 60 \text{ dB} - (60 - 30) \log [1 + 0.0184 (200 \text{ mV} - 10)]$$

In the above equation, 60 dB represents the filter action only, and (60 − 30) represents degradation due to worst-case audio rectification.

$$AR_{dB} \cong 40 \text{ dB}$$

3. 106 dBμV $-$ 40 dB $=$ 66 dBμV, i.e., larger than N_{victim}

4. Check for frequency of EMI modulation vs. victim bandwidth. Here TV modulation is usually 150 to 300 kHz, so no reduction for bandwidth.

4.1.3.2 Inside-Circuit Analysis of Audio Rectification

Audio rectification is the mechanism through which out-of-band or microwave signals are converted to in-band signals. The process is essentially envelope detection.

All semiconductor devices are built around pn junctions which have a nonlinear current-voltage characteristic. Figure 4.19 illustrates how this characteristic acts to rectify a sine-wave signal. A continuous sinusoidal voltage is applied to the detection with a dc bias voltage. The current flows in a series of nonsymmetrical pulses. The average value of the current pulses is higher than the

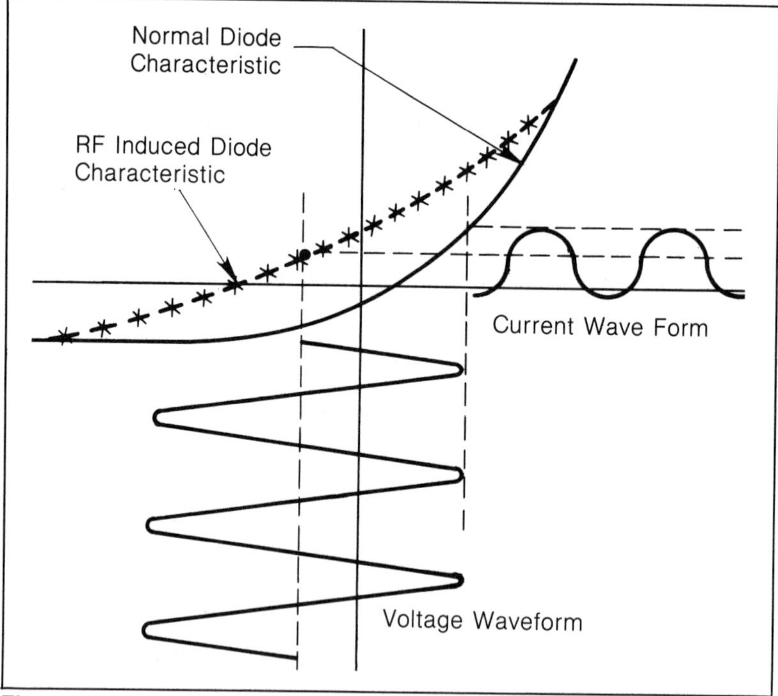

Figure 4.19—Illustration of Rectification in p-n Junction

dc value of the current which would exist if only the bias voltage were present. Therefore, it appears that the signal has caused a dc offset current to flow. As the voltage level changes, the offset current also changes.

A locus of average current versus dc bias voltage can be plotted. It is shown as a series of Xs in Fig. 4.19. This illustration may be interpreted as an **RF-induced** diode I-V characteristic. It is exactly what a low-frequency circuit "sees" of the diode (Fig. 4.20). For a given bias voltage, the dc current flow is given by the RF-induced characteristic if RF is present. If the diode is biased with a load line, standard load-line analysis is used to find the operating point, which is the intersection of the load line and the RF-induced diode characteristic. In general, a diode or transistor junction biased with a load line will experience a decrease in average voltage and an increase in average current when stimulated with RF.

The RF-induced diode characteristics are easily measured in the laboratory (Fig. 4.21). Figure 4.22 (from Larson, Ref. 4) illustrates the I-V characteristics of a 1N914 diode stimulated at 220 MHz. Curve 1 is the familiar diode characteristic: conduction at about 0.7 V and little or no conduction at lower voltages. Curves 2 through 7 illustrate the RF-induced characteristics for power levels from 5 to 380 mW. The greatest change in characteristics occurs near the diode knee, because it is here that the greatest nonlinearity occurs in the diode.

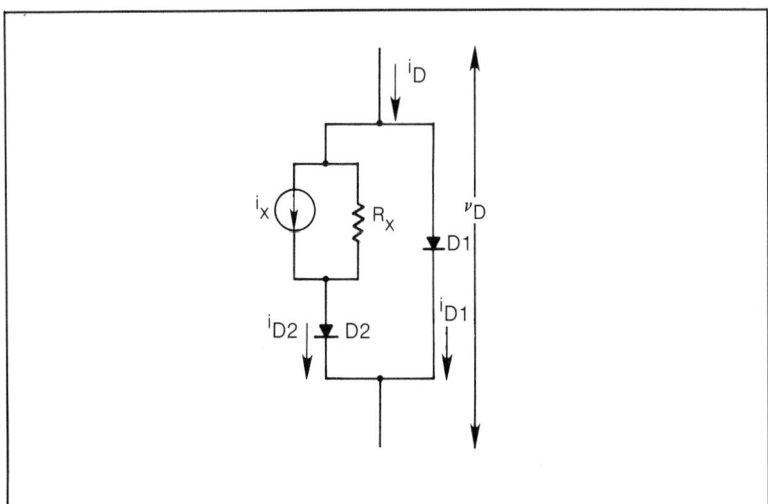

Figure 4.20—Circuit Model of Diode under RF Influence

In transistors, most of the interference effects occur due to rectification in the transistor junctions. A bipolar transistor contains two such junctions, and RF detection can occur in both the base-emitter and base-collector. Concentrating on the RF vulnerability of a transistor's base input which has proven to be the most sen-

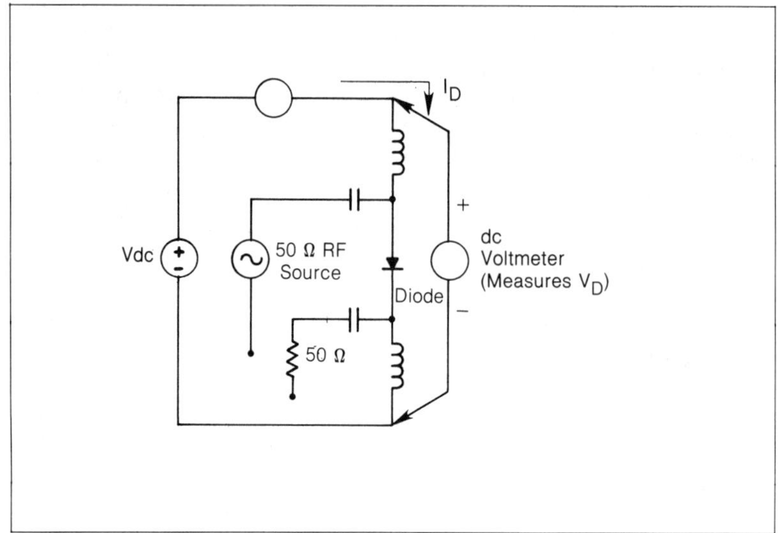

Figure 4.21—Test Setup for Measurement of RF Induced Diode Characteristics

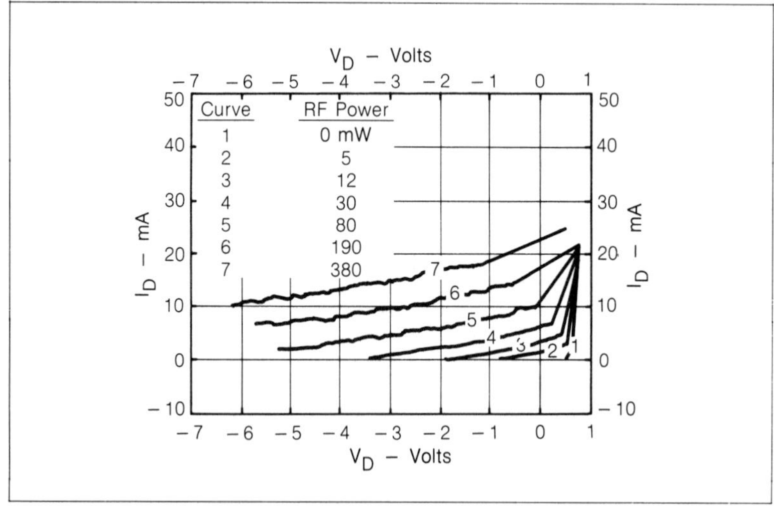

Figure 4.22—RF Induced Diode Characteristics for IN914 Diode at 220 MHz

sitive terminal (Ref. 5), we see that the V_{be}^2 term caused by a sine wave RF input corresponds to a dc shift of:

$$\Delta V = \frac{I_c}{4} \frac{(g_m V_{be}^2)}{(I_c)} = \frac{I_c}{4} \times 1600 \; V_{be}^2$$

where, I_c = collector dc current   (4.14)

          g_m = mutual conductance $\cong 40 \; I_c$ at

                 ambient temperature

       V_{be} = base to emitter voltage

If the signal is modulated, then the peak of the envelope gives maximum shift and the trough gives the minimum shift: the signal is demodulated. We can calculate the CW RF signal to give a certain dc shift from:

$$V_{be} = \frac{1}{20} \sqrt{\frac{\text{shift}}{I_c}} \; \text{peak}$$

$$= \frac{1}{20} \sqrt{\frac{\text{shift}}{2 \, I_c}} \; \text{rms} \qquad (4.15)$$

Knowing V_{be} and the base input impedance (generally a few kΩ paralleled by a capacitance), the necessary EMI current or injected power can be derived for a given permissible ΔV shift.

For instance, from Ref. 5 the results of Fig. 4.23 were obtained for a BC 108 npn transistor. The criteria for upset was a 10 percent change in the dc collector current of 5 mA. Corresponding to the EMI current, V_{EMI} and P_{EMI} are also shown. Since this transistor has a bandwidth of 300 MHz, the test jig was using 50 Ω impedances for the input and output measurements.

4.1.4 Emission Aspects

For an active device, the emission aspect does not pose a threshold problem as does susceptibility. Characterizing emissions from linear or digital devices is merely a measurement or a prediction of their amplitude versus frequency signatures, conducted and radiated.

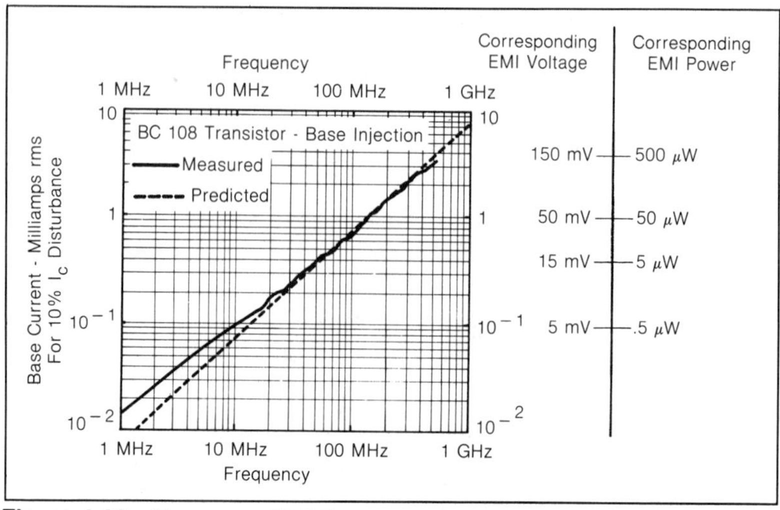

Figure 4.23—Necessary EMI Current to Cause Transistor Malfunction by RF Detection at the Base-Emitter Junction

Measurement of conducted EMI is usually made by supplying the device with a noise-free power supply and connecting its input and output to typical impedances, representative of its actual application(s). Measurement of radiated EMI needs much more caution since the field to be measured is radiated by a component whose dimensions are one or two orders of magnitude less than those of its ancillary connections (wiring, traces, etc.). Usually these measurements are made on a special test jig where everything but the component is shielded inside a thick metal housing.

Prediction of both conducted and radiated EMI includes more or less complex models, from a simple RLC network to sophisticated bipolar or FET equivalent circuits. Prediction model and measurement results for analog and digital components will be addressed in Sections 4.2 and 4.4.4.

4.2 Transistor Noise and EMI Aspects

Since the transistor is the elementary building block of practically every modern electronic component, analog or digital, its intrinsic EMI properties are the first concern.

Low-level transistors can exhibit sensitivity to interference signals and suffer possible destruction. Inherent semiconductor noise limits the minimum detectable and amplifiable signals, and internal resistance and capacitance or transit time limits the maximum amplifiable frequency.

Noise generated by a transistor will appear at the output, but it may be possible to minimize the effects. The three main types of noise generated by a transistor are thermal, shot and flicker noises. Some important definitions concerning noise are:

Thermal noise in a resistance (see Section 4.1):

In volts,
$$V_N = \sqrt{KTBR} = \text{open circuit noise voltage produced by a resistance.}$$

In watts,
$$P_N = KTB, = \text{available noise power (independent of the value of the resistance)} \tag{4.16}$$

Noise Factor is the ratio (greater than one) comparing the actual noise of an active device to that of an ideal noiseless device of the same resistance.

$$\text{NF, or simply F} = \frac{\text{noise power output of actual device}}{\text{noise power putput due to source resistance noise only}} \tag{4.17}$$

If the device has a voltage gain G, an input noise Voltage V_{Ni} (thermal) and an output voltage noise V_{NO} (measured), then:

$$F = \frac{(V_{NO})^2}{(GV_{Ni})^2} = \frac{(V_{NO})^2}{4KTBRG^2} \tag{4.18}$$

where, K = Boltzmann's constant = 1.38×10^{-23} J/°Kelvin
B = Bandwidth in Hz
T = Absolute temperature in °K
R = Input (source) resistance in Ω

Noise Temperature (T_n) is the equivalent temperature of a resistor of the same value as the terminating resistance which generates the same level of noise as the noise source.

Excess Noise Ratio (ENR) is a standard measure of performance. ENR expresses noise power as the noise which is in excess of a resistor at room temperature (290°K):

$$\text{ENR (dB)} = 10 \log \left(\frac{T_N}{290} - 1 \right) \qquad (4.19)$$

where,

$$T = \text{noise temperature (°K)}$$

Spectral Power Density is the measurement of power per unit bandwidth. For example, in a 50 Ω system, 290°K noise temperature equals -174 dBm/Hz.

Peak Factor is the ratio of peak to rms noise power of a particular noise source. It determines the equivalent duty cycle of the noise signal.

White Noise is noise having equal power per unit bandwidth over a specified frequency range.

Pink Noise is noise having equal power per octave bandwidth over a specified frequency range.

Gaussian Noise is noise having an amplitude distribution according to the Gaussian Probability Density function.

Noise Factor F is an attractive parameter and easy to calculate because it is independent of the signal level and the load resistance. However, the signal-to-noise ratio S/N is what really counts in the final process, whether it is audio, video or digital. Relying only on F could lead to a purposeful selection of a very high source resistance; this would result in a very low F, but also in a higher thermal noise with a lower S/N.

Also, the noise factor F becomes less meaningful if the source is not quasiresistive. At the extreme, if the source is purely reactive, the source noise would be zero (no resistance) making the term

F arbitrarily infinite. Therefore, another measure of F, involving S/N, is given by:

$$NF = 10 \log \frac{(S/N)_{in}}{(S/N)_{out}} \qquad (4.20)$$

where,

S and N are power or (voltage)^2 levels

This measurement is made by determining the S/N at the input with no amplifier present and then dividing by the measured S/N at the output with signal source present.

A more recent concept (Ref. 6) has evolved where an active device noise is modeled as equivalent voltage and current sources (Fig. 4.24).

Noise Voltage, or more properly **Equivalent Short-Circuit Input RMS Noise Voltage** is simply that noise voltage V_n which would appear to originate at the input of the noiseless amplifier if the input terminals were shorted. It is expressed in nanovolts per root hertz ($nV\sqrt{Hz}$) at a specified frequency, or in microvolts in a given frequency band. It is determined or measured by shorting the input terminals, measuring the output rms noise, dividing by amplifier gain and referencing to the input; hence the terms **equivalent noise voltage**. An output bandpass filter of known characteristic is used in measurements, and the measured value is divided by the square root of the bandwidth ($\sqrt{B}$) if data is to be expressed per unit bandwidth, or per root

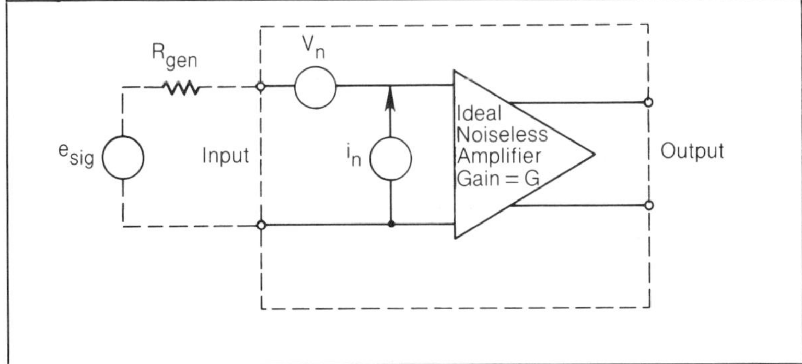

Figure 4.24—Noise Characterization of Amplifier

hertz. The level of V_N is not constant over the frequency band; typically it increases at lower frequencies as shown in Fig. 4.25. This increase is 1/f noise.

Equivalent Open-Circuit RMS Noise Current — Noise current is that noise which occurs apparently at the input of the noiseless device due only to noise currents. It is expressed in picoamps per root hertz (pA/$\sqrt{Hz}$) at a specified frequency or in nanoamps in a given frequency band. It is measured by shunting a capacitor or large resistor across the input terminals so that the noise current will give rise to an additional noise voltage (i_n x R_{in} (or X_{cin}), versus the one due to V_{in} and R_{in} only).

The output is measured, divided by amplifier gain, referenced to input, and the contribution known to be due to V_n and resistor noise is appropriately subtracted from the total measured noise. If a capacitor is used at the input, there is only V_n and i_n X_{cin}. The i_n is measured with a bandpass filter and converted to pA/Hz if appropriate; typically, it increases at lower frequencies for op amps and bipolar transistors but increases at higher frequencies for field-effect transistors.

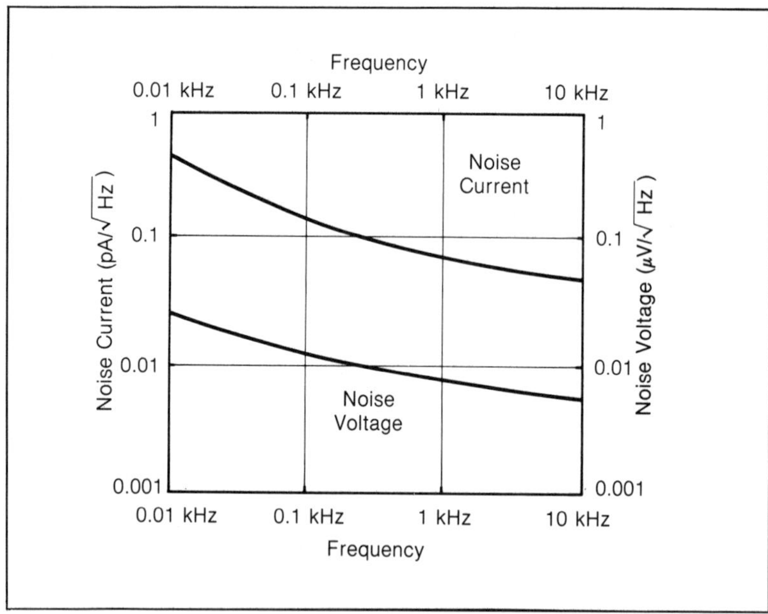

Figure 4.25—Typical Noise Voltage V_n/$\sqrt{B}$ and Noise Current I_n/$\sqrt{B}$ (from Ott, Ref. 1)

Figure 4.25 shows equivalent noise voltage $V_n/\sqrt{B}$ and current $I_n/\sqrt{B}$ versus frequency. The total noise voltage or current over a given frequency band can be found by integrating the square of each segment and computing the square root of the result. For instance, looking at curves 4.25, for a 10 Hz to 10 kHz interval:

$$I_n \ (10 - 10{,}000 \ \text{Hz}) \cong \left[\left(\frac{0.3 \ \text{pA}}{\sqrt{\text{Hz}}} \right)^2 \times 90 \ \text{Hz} + \left(\frac{0.1 \ \text{pA}}{\sqrt{\text{Hz}}} \right)^2 \right.$$

$$\left. \times \ 900 \ \text{Hz} + \left(\frac{0.07 \ \text{pA}}{\sqrt{\text{Hz}}} \right) \times 9{,}000 \ \text{Hz} \right]^{1/2}$$

$$= \left[(0.09 \ \text{pA}^2/\text{Hz}) \times 90 \ \text{Hz} + (0.01 \ \text{pA}^2/\text{Hz}) \times 900 \ \text{Hz} + \right.$$
$$\left. (0.0049 \ \text{pA}^2/\text{Hz}) \times 9{,}000 \ \text{Hz} \right]^{1/2}$$

$$\cong 7.8 \ \text{pA}$$

When the curve slope is fairly constant, multiplying the midband value in $\mu V/\sqrt{\text{Hz}}$ or $\text{pA}/\sqrt{\text{Hz}}$ by the square root of the frequency span gives a sufficient approximation. In this example, integrating $0.009 \ \mu V\sqrt{\text{Hz}}$ at midband over a bandwidth of 10,000 Hz would give:

$$0.009 \ \mu V/\sqrt{\text{Hz}} \times \sqrt{10{,}000} = 0.9 \ \mu V$$

Now we can examine the relationship between V_n and i_n at the amplifier input. When the signal source is connected, the V_n appears in series with the e_{sig} and e_R (the thermal noise in R.) The i_n flows through R_{gen} thus producing another noise voltage of value $i_n \times R_{gen}$. All of these noise voltages add at the input in rms fashion; that is, as the square root of the sum of the squares. Thus, neglecting possible correlation between V_n and I_n, the total input noise is:

$$\overline{V_{NT}}^2 = \overline{V_n}^2 + \overline{e_R}^2 + \overline{I_n}^2 \ R_g^2 \qquad (4.21)$$

Further examination of the NF equation shows the relationship of V_N, i_n and NF:

$$NF = 10 \log \frac{S_{in} \times N_{out}}{S_{out} \times N_{in}}$$

$$= 10 \log \frac{S_{in}\, G_p\, \overline{V_{NT}}^2}{S_{in}\, G_p\, \overline{e_R}^2} \tag{4.22}$$

where,

$$G_p = \text{power gain}$$
$$e_R = \text{thermal noise from signal source resistance } R_g$$

$$= 10 \log \frac{\overline{V_{NT}}^2}{\overline{e_R}^2} \tag{4.23}$$

$$= 10 \log \frac{\overline{V_n}^2 + \overline{e_R}^2 + \overline{I_n}^2\, R_g^2}{\overline{e_R}^2}$$

$$NF = 10 \log \left(1 + \frac{V_n^2 + I_n^2\, R_g^2}{\overline{e_R}^2} \right) \tag{4.24}$$

For small R_{gen}, noise voltage dominates, and for large R_{gen}, noise current becomes important. A clear advantage accrues to FET input amplifiers, especially at high values of R_{gen}, as the FET has essentially zero I_n. Note that for an NF value to have meaning, it must be accompanied by a value R_{gen} as well as frequency. For optimum noise performance, the total noise voltage represented by Eq. (4.21) should be minimized. From Ref. 6, the total equivalent input voltage per square root of bandwidth can be written as:

$$\frac{\overline{V_{nt}}}{\sqrt{B}} = \sqrt{4KTR_g + \left(\frac{\overline{V_n}}{\sqrt{B}} \right)^2 + \left(\frac{I_n R_g}{\sqrt{B}} \right)^2} \tag{4.25}$$

The equivalent input noise voltage due only to device noise can be calculated by subtracting the thermal noise component from Eq. (4.21). The equivalent input device noise then becomes:

$$V_{nd} = \sqrt{V_n^2 + (I_n R_g)^2} \tag{4.26}$$

Figure 4.26 shows a comparison of $V_{nt}/\sqrt{B}$ for a typical operational amplifier, a low-noise bipolar transistor and a junction FET. The curve captioned "thermal noise" is the lower bound of either device's noise, calculated from the source resistance on the X axis. For low sources resistances below 1,000 Ω, a low-noise bipolar is slightly less noisy than any other device. Above few kilohms, the FET is almost an ideal device from noise point of view since it adds virtually no noise to the R_g thermal noise. The noisiest device is the operational amplifier, as will be explained in Section 4.3.

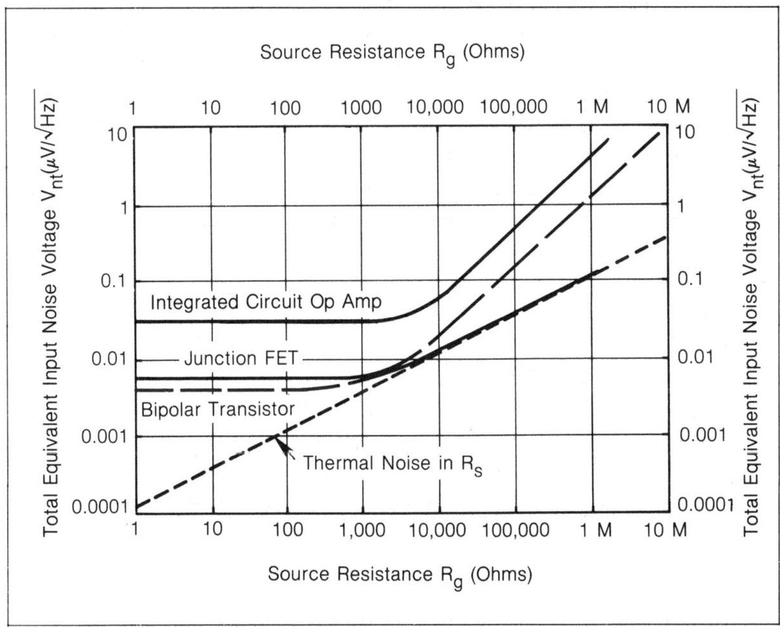

Figure 4.26—Typical Total Equivalent Noise Voltage Curves for Three Types of Devices. (from Ott, Ref. 1)

4.2.1 Source Resistance Optimization

In the latest expression of noise factor NF [Eq. (4.24)], it can be demonstrated that the term $(V_n{}^2 + I_n{}^2 R_g{}^2)/e_R{}^2$ reaches its minimum value when $R_g = V_n/I_n$. This is due to the fact that e_r also includes R_g and making R_g excessively small would increase NF. On the other hand, to maximize S/N would require making R_g infinitely small since, when $R_g = 0$, the 4 KTBR noise of the source resistance would vanish. However, provided that R_g is $\leqslant V_n/I_n$, the total noise voltage is governed by the V_n term over a wide range of frequency. So, as a rule of thumb, for optimum noise performance, use the lowest possible source impedance.

The designer cannot always control the signal source resistance. For instance, sensors and transducers have source resistance imposed by their own technology. Or with telecommunication and radiocommunication, the constraint of matching the source impedance prevails. However the question arises as to whether this source resistance can be adapted to satisfy the above requirement.

We have seen that if $R_g < V_n/I_n$, it is useless to add a series resistor since the total noise is dominated by V_n, the equivalent "shorted-input" noise. If to the contrary, $R_g > V_n/I_n$, the designer can use a step-down transformer to decrease the source resistance seen by the amplifier input. The turns ratio is selected such as the output exhibit an impedance equal to V_n/I_n. The only prejudice is the added thermal noise and possible magnetic domain (Barkhausen) noise of this transformer, plus of course, the impossibility to process dc signals.

Noise Figure of Cascaded Devices

It can be demonstrated that when the first device is a high-gain cell, as generally is the case, the total noise figure is primarily determined by the noise figure NF_1 of the first stage. For instance, good quality analog amplifiers based on low-noise transistors with a 1 to 1.5 dB noise factor can reach a noise factor of 2 or 3 dB over their whole bandwidth.

4.2.2 Noise in Bipolar Transistors

Noise in bipolar transistors is a combination of thermal noise, shot noise and flicker (semiconductor contact) noise. Thermal noise in transistors is believed to be due to thermal agitation causing random motion of electrons and holes in the material.

The equivalent circuit of a thermal noise source is represented by a voltage generator in series with a noiseless resistor as shown in Fig. 4.27. The thermal noise has a flat frequency spectrum (white noise), and the amount of output noise power is limited by the circuit bandwidth to which the noise source is connected.

A constant-current generator, parallel with a noiseless resistor, represents an equivalent shot-noise source circuit as shown in Fig. 4.27. Shot noise is generated by random movement of minority carriers across a junction whenever current is flowing. The current is not uniform due to the random diffusion of minority carriers as well as the random recombination and generation of charges. Shot noise I_{sn} has a flat (white noise) spectrum and is proportional to the number of minority carriers, current flow, temperature and bandwidth.

$$I_{sn}^2 = 2 \ eIB \qquad (4.27)$$

where,

$$I_{sn}^2 = \text{shot-noise current in A}$$
$$e = \text{electron charge} = 1.6 \times 10^{-19} \ C$$
$$I = \text{dc current in A}$$
$$B = \text{bandwidth in Hz}$$

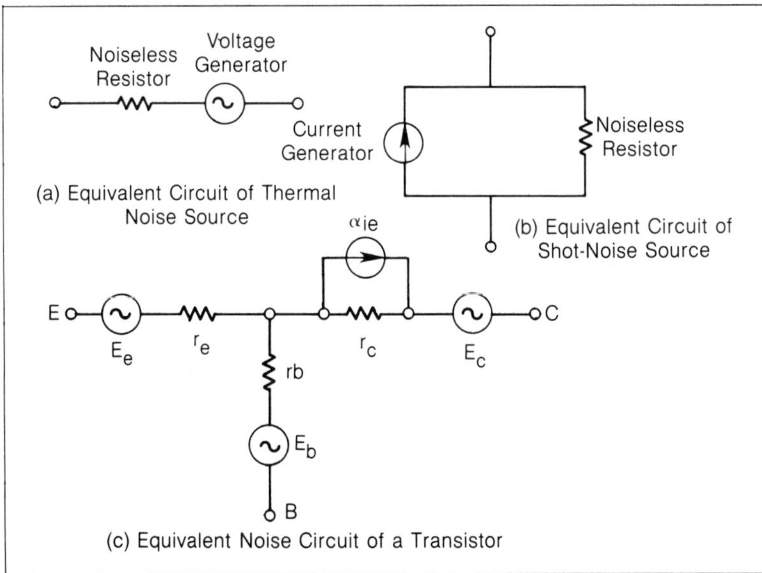

Figure 4.27—Equivalent Transistor Noise Circuits

Flicker noise is called **l/f noise** because it has a power spectrum proportional to l/f. Thus, for a given bandwidth, the noise decreases with an increase in frequency. It is also called **semiconductor noise** and is believed to be due to crystal imperfections and surface effects including trapping of carriers by surface charges with associated surface and internal leakage.

As seen in Fig. 4.28, flicker noise (l/f) usually predominates up to a frequency between 1 and 50 kHz, when thermal and shot (white) noise in emitter and collector junctions become predominant.

Beyond the predominant white-noise region, after frequency $f_2 = f_\alpha \sqrt{1 - \alpha_o}$ (Fig. 4.28), the noise figure increases proportionally to frequency because the output transistor noise (collector junction) increases, and simultaneously the useful signal decreases because of the transistor gain rolloff.

Using the concept of optimal source resistance defined in Section 4.2.1, it is found that the calculated value for optimal source resistance is in the range of values that provides optimum power gain of the transistor. Considering that, in normal application, transistors are used at frequencies below f_α, the alpha cutoff, and assuming $\beta_o >> 1$, the optimum source resistance is:[1]

$$R_{g\ opt} = \sqrt{(2r_b + r_e)\,\beta_o r_e} \qquad (4.28)$$

If, in addition, the base resistance r_b is negligible (not always the case), Eq. (4.28) becomes:

$$R_{g\ opt} \cong r_e \sqrt{\beta_o} \qquad (4.29)$$

This equation is also useful for making quick approximations of the source resistance that produces minimum noise factor. It shows

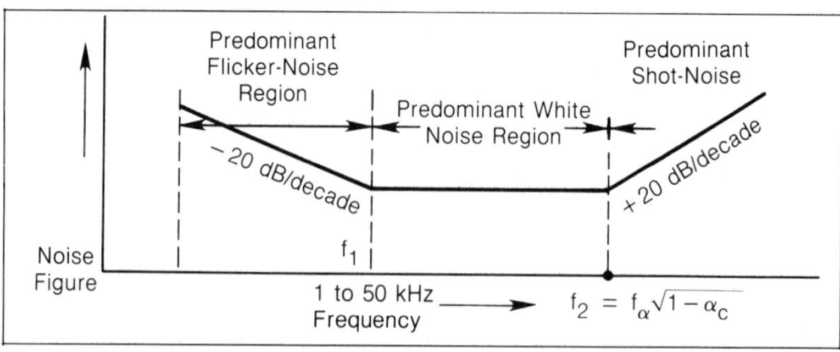

Figure 4.28—Typical Transistor Noise Versus Frequency

that the higher the common emitter current gain β_0 of the transistor, the higher the value of $R_{g\ opt}$.

The equivalent input noise voltage and current, from the model of Fig. 4.27, can be determined (from Ref. 6) by:

Equating R_g to zero:

$$V_n{}^2 = 2\,KTB(r_e + 2r_b) + \frac{2KTB\,(r_e + r_b)^2}{r_e\beta_0}\left[1 + \left(\frac{f}{f_\alpha}\right)^2(1 + \beta_0)\right]$$

$$(4.30)$$

Making R_g very large:

$$I_n{}^2 = \frac{2KTB}{r_e\beta_0}\left[1 + \left(\frac{f}{f_\alpha}\right)^2(1 + \beta_0)\right] \qquad (4.31)$$

4.2.3 Noise with Junction Field Effect Transistors

The noise features of a junction FET are shown on the equivalent circuit of Fig. 4.29. Provided that its contributors are independent variables, the total output noise power is the addition of:

1. Source thermal noise power transferred to the output plus
2. Shot noise power in the FET input (gate) plus

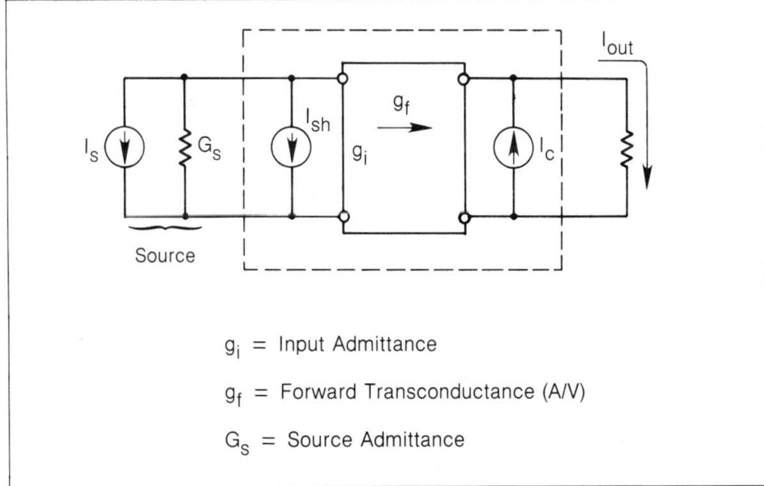

Figure 4.29—Noise Equivalent Circuit of a Junction FET

3. Intrinsic noise power of the output (channel)

Using the admittances in Fig. 4.23, the first contributor can be expressed in terms of source noise current transferred into the output:

$$I_{out}\ (source)\ =\ \frac{g_f}{G_s + g_i}\ \sqrt{4KTB} \qquad (4.32)$$

where,

$$G_s = \text{Source admittance in mhos}$$
$$g_i = \text{Input admittance in mhos}$$
$$g_f = \text{Forward transconductance in mhos}$$

Similarly, the second contributor is translated into its equivalent output noise current:

$$I_{out}\ (shot)\ =\ \frac{I_{sh} \times g_f}{G_s + g_i} \qquad (4.33)$$

where I_{sh}, the FET input gate shot noise, is the same parameter as described already in Section 4.2.2 for bipolar transistors:

$$I_{sh}\ =\ \sqrt{2eI_gB} \qquad (4.34)$$

with,

$$e = \text{Electron charge} = 1.6 \times 10^{-19}\ \text{Coulomb}$$
$$I_g = \text{Gate dc leakage current}$$

The third contributor, expressed in function of its current is simply the thermal noise of the FET channel, equal to:

$$I_c\ =\ \sqrt{4KTBg_f} \qquad (4.35)$$

The noise factor NF of the JFET is actually the effect of the second and third contributors, i.e.:

$$NF\ =\ \frac{I_{out(1+2+3)}}{I_{out(1)}}$$

$$=\ \frac{1 + I_{sh}^2}{4KTBG_s}\ +\ \frac{I_c^2}{4KTBG_s}\left(\frac{G_s + g_i}{G_f}\right)^2 \qquad (4.36)$$

4.50

Given that the input admittance g_i is also equal to $\dfrac{2}{4\,KT}\,I_g$ and replacing I_{sh} and I_c by their expression in Eq. (4.36), leads to:

$$NF = 1 + \frac{g_i}{G_s} + \frac{1}{G_s g_f}(G_s + g_i)^2 \qquad (4.37)$$

If admittances are replaced by resistances:

$$NF = 1 + \frac{R_s}{r_i} + \frac{R_s}{g_f}\left(\frac{1}{R_s} + \frac{1}{r_i}\right)^2 \qquad (4.38)$$

Generally, the input resistance r_i of a FET is much larger than the signal source resistance R_s, at least from dc to a few MHz, and Eq. (4.38) can be approximated by:

$$NF = 1 + \frac{1}{g_f R_s} \qquad (4.39)$$

For those categories of FET where there is no p-n gate junction, like MOSFETs, the shot noise term disappears in Eq. (4.36), and the noise factor becomes:

$$NF = 1 + \frac{I_c^{\,2}}{4KTBG_s}\left(\frac{G_s + g_i}{g_f}\right)^2 \qquad (4.40)$$

or,

replacing I_c by its expression from Eq. (4.35):

$$NF = 1 + \frac{(G_s + g_i)^2}{g_f G_s}$$
$$= 1 + \frac{1}{g_f R_s} + \frac{2}{g_f r_i} + \left(\frac{R_s}{g_f r_i}\right)^2 \qquad (4.41)$$

If, here again, r_i is large, the equation takes the same value as Eq. (4.39).

The total input noise voltage, using V_n, I_n concepts of Fig. 4.24 is found, after Ott (Ref. 6) by using:

$$V_n{}^2 = \frac{4KTB}{g_{fs}}$$

$$I_n{}^2 = \frac{4KTB}{r_{11}}$$

assuming $g_f r_i \gg 1$

4.2.4 EMI Emissions from Transistors

In addition to the previous types of noise, amplifying transistors with HF capabilities (about 100 MHz or greater) may oscillate at spurious frequencies when operated considerably below the device capability. A transistor having usable current gain at 400 MHz, but operated as an amplifier handling frequencies below 80 MHz, tends to react with stray capacitance and inductance within the associated circuitry to oscillate parasitically at unpredictable frequencies. The form is usually that of wideband gaussian noise near or below the operating frequency. The parasitic oscillation is affected by interelement (junction) capacitance which varies, in a varactor fashion, over a wide range with changes of element voltage.

Transistors operated in a linear mode usually will not introduce major interference. When used in a switching mode, they may introduce rapid changes in current in the supply and signal lines and may potentially cause oscillations with fundamental frequencies around 250 kHz to 2 MHz. Inverters may cause serious interference in the range of 15 to 200 MHz with a peak around 60 MHz. Milliampere devices can have switching times in the order of nanoseconds while those of high-amperage devices can be a few microseconds. Thus, the current gradient can be 10^7 A/s. "Soft switching" is a design that controls the rate of change of current as well as the selection of proper dimensions and carrier densities.

To minimize interference, it is best to generate required pulses in the required location rather than send them through long lines where the number of susceptible components involved could significantly increase. Waveform-control techniques can also be used to make devices responsive to certain pulse characteristics only.

For transistors handling upper VHF or UHF power, and for transistors switching amperes and hundreds of volts with only 30 to 100 ns transition times, the generation of EMI is mostly a problem of interconnection between the host circuit and the devices. Powerful radiation and detrimental inductive ringing can come from as little as 1 cm^2 of lead loop area. Such transistors should be mounted in metal cans and output leads formed as flat tabs to minimize parasitic inductance and, eventually, to match with a microstrip configuration on the printed circuit (Fig. 4.30).

4.3 Analog Devices

While there is a definite tendency to use digital circuit elements in new designs, the analog portions of any device (those which act to detect a particular parameter or deliver a stimulus) will continue to be tremendously important to its overall efficacy. Simply because the data to be collected are inherently analog and defiant of direct digitization, analog devices will continue to play an important role in instrumentation, robotics, medical equipment, audio, industrial process, motor vehicle, maritime and airborne electronic controls, etc.

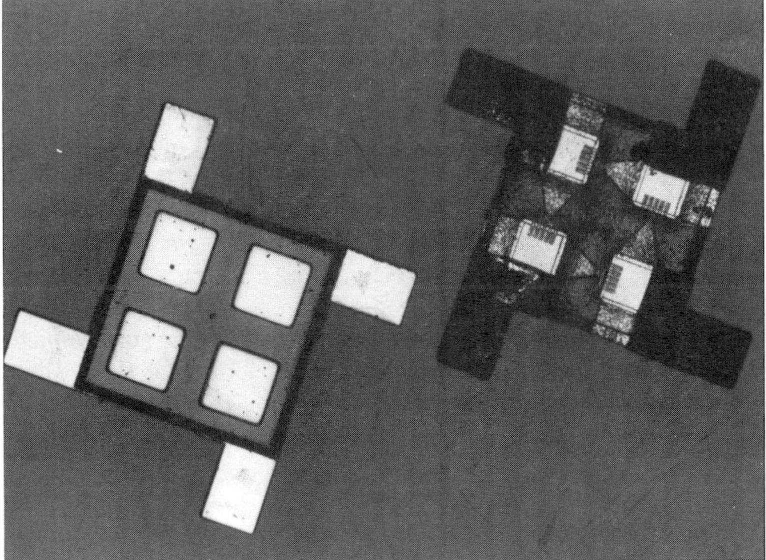

Figure 4.30—Beam-Lead (or Ribbon-Lead) Packages Improve HF Response and Allow Matching with Microstrip PCBs

Analog devices may exhibit either one or both of two EMI problems:

1. EMI energy entering via some "intended" conducted paths: signal input leads, signal output leads and power and ground terminals.
2. EMI induced by direct radiation (from proximate or remote sources) on the device itself. Since practically all analog circuits are mounted on PCBs, macrointegrated hybrids or micro-integrated modules, the relevant coupling mechanisms which end up with these problems will be covered in Chapter 8. This section deals with the intrinsic EMI aspects of analog devices.

4.3.1 In-Band Analog Sensitivity

Since analog amplifiers will often process extremely low level signals, the features of the device's input noise, broadly approached in Section 4.1.1, need to be considered more precisely. As indicated in the previous section, the input stage is the most important one for the noise performance.

The analog device most commonly used to provide gain, summing and other functions is the operational amplifier. While it is still possible to find equipment of fairly recent design made entirely of discrete semiconductors, the majority of new devices clearly exploit advantages offered by the op amp. Although the following discussion is applicable primarily to op amp circuits, general considerations also apply to discrete component design.

Also, operational amplifiers offer the possibility of a true balanced input, a key advantage in EMI reduction techniques. The differential amplifier shown in Fig. 4.31 is the basic scheme from which most of the monolithic operational amplifiers are derived. As already mentioned, the noise figure of a cascade of amplifying stages is determined by the noise figure of the first high-gain device. So, in the case of the differential input, the noise voltages of the two front-end transistors add up randomly, and the resulting noise is $\sqrt{2}$ times larger than for a single transistor.

As was evidenced by Fig. 4.26, op amps exhibit higher noise than discrete transistors. This is true for bipolar op amps versus bipolar transistors, and FET op amp versus FET transistors as well. Op amp noise can be modeled using the general approach described in Section 4.2 and Eq. (4.21), i.e., the rms addition of all the noise contributors, presumably uncorrelated. These noise contributors are modeled in Fig. 4.32 and plotted in Fig. 4.33 as noise voltage

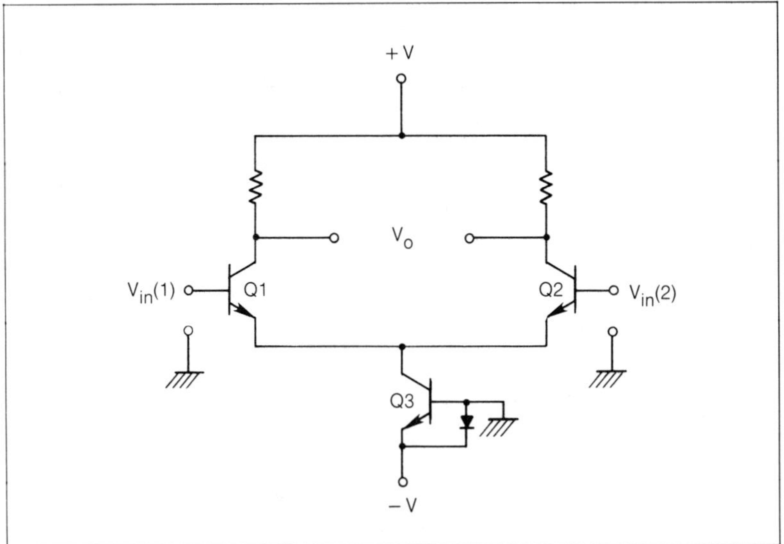

Figure 4.31—Typical Input Circuit Schematic of an IC Operational Amplifier. Transistor Q_3 acts as a constant current source to provide dc bias for the input transistors Q_1 and Q_2 (Reprinted from Ref.1).

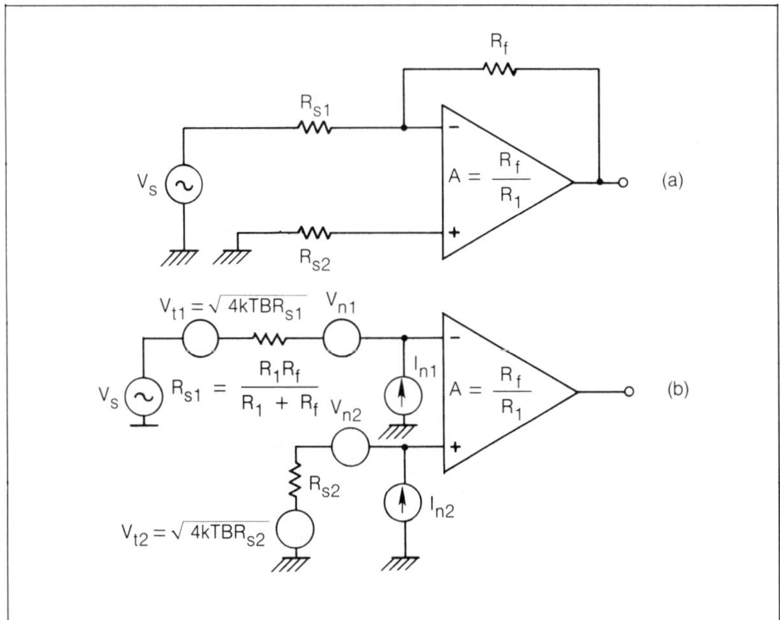

Figure 4.32—(a) Typical Op Amp Circuit; (b) Circuit of (a) with Noise Sources Added (Reprinted from Ref. 1)

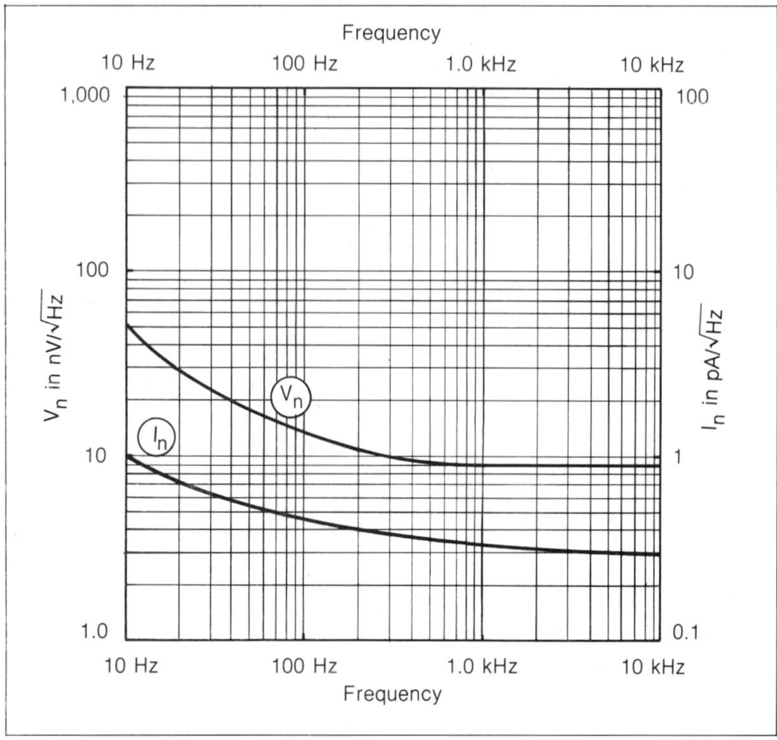

Figure 4.33—Example of Typical Noise Voltage and Current for an Op Amp

and noise current sources, V_n and I_n respectively. The feedback resistor R_f is also transferred as an input parameter. The rms addition expresses as:

$$V_{nt}^2 = 4KTB \, (R_{s_1} + R_{s_2}) + V_{n_1}^2 + V_{n_2}^2 +$$
$$(I_{n_1} \, R_{s_1})^2 + (I_{n_2} \, R_{s_2})^2 \tag{4.42}$$

However, since the inputs of an op amp are made as symmetrical as possible, we can state:

$$V_{n1} = V_{n2} = V_n$$
$$I_{n1} = I_{n2} = I_n$$
$$R_{s1} = R_{s2} = R_s$$

Therefore Eq. (4.42) simplifies as:

$$(V_{nt})^2 = 8\ KTBR_s + 2V_n^2 + 2\ (I_n\ R_s)^2 \qquad (4.43)$$

And the noise factor NF expresses as:

$$NF = 1 + \frac{2V_n^2 + 2\ (I_n R_s)^2}{8KTBR_s}$$

None of this includes the 1/f noise, because in typical applications this contributor only emerges from shot noise and thermal noise below a few kHz. If this frequency domain is the one of concern, the 1/f noise has to be factored into the value of V_n, since it causes V_n, in $V/\sqrt{Hz}$, to increase at low frequencies.

4.3.2   Rejection and Dynamic Properties of Analog Amplifiers in the Presence of EMI

The ideal op amp possesses infinite gain, infinite bandwidth, zero bias current (to generate a functional response at its inputs), zero offset voltage (essentially a perfect match between the input stages) and infinite input impedance. Because of these characteristics, an infinitesimally small input voltage is required at one input with respect to the other to exercise the amplifier output over its full range.

Bandwidth and Slew Rate

The quasi-infinite gain of an op amp starts to fall off rapidly at low frequencies, e.g., 10 Hz, due to many intentional (compensation, stability) and unintentional (input impedance shunted by its own parasitic capacitance) internal features. Therefore, an op amp is characterized by its gain × bandwidth product where:

gain × BW ≡ unity gain BW ≡ $BW_{(3\ dB)}$ × numerical gain

For instance, in the op amp of Fig. 4.34,

$$BW_{(3\ dB)} = \frac{unity\ gain\ BW}{gain} = unity\ gain\ BW \times \frac{R_1}{R_2 + R_1} \cong$$

$$unity\ gain\ BW \times \frac{R_1}{R_2} \qquad (4.44)$$

If, for instance $R_1 = 10$ kΩ and $R_2 = 1$ MΩ, for an op amp having a 10 MHz unity gain BW, the bandwidth (or f_{co}) is equal to:

$$BW_{(3\ dB)} = 10\ \text{MHz} \times \frac{10}{1,000} = 100\ \text{kHz} \qquad (4.45)$$

The 3 dB bandwidth can also be translated into rise time, provided the amplitude of the voltage excursion is set:

$$BW \equiv \frac{1}{\pi \tau_r} \cong \frac{0.32}{\tau_r} \qquad (4.46)$$

In this case:

$$\tau_r = \frac{0.32 \times \text{gain}}{\text{unity gain BW}} = \frac{0.32}{\text{unity gain BW}} \times \frac{R_2}{R_1}$$

In terms of EMI, any unwanted signal appearing *differentially* at the input of an analog amplifier of given gain Gx will be:

1. Amplified by a factor Gx if EMI is in-band
2. Amplified by a factor Gx $\times$ f_{EMI}/f_{co} if out-of-band, f_{co} corresponding to the 3 dB Bandwidth for the given gain Gx.

Note that amplification will still occur until f_{EMI} reaches the unity gain bandwidth. Ideally, the amplification ratio then should become $<$ 1 and decrease forever. However, we have seen in Section 4.1.3 that parasitic resonances and audio rectification do not allow for infinite rejection.

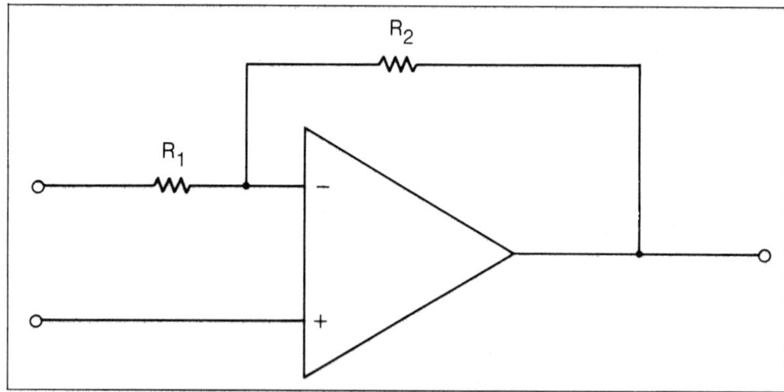

Figure 4.34—Typical Op Amp with Gain-Setting Resistors

Slew-Rate Limiting

Slew-rate limiting affects all amplifiers where capacitance on internal nodes, or as part of the external load, has to be charged and discharged as voltage levels change. Indeed, the effect is clearly seen in digital logic circuitry as limitations in rise and fall time. It is significant in op amps since it determines the difference between the small signal and power bandwidth.

The slew rate limit in most amps is derived from the capability of the internal circuitry to charge and discharge the compensation capacitance network, which may be either internal or external.[7] A general-purpose op amp such as the μA741 has a typical slew rate of 0.5 V/μs. It is also capable of at least a ± 10 V swing into a 2 kΩ load resistance on ± 15 V supplies. The slew rate controls the large signal pulse response, forcing an op amp output to change from -10 V to $+10$ V in a minimum of 40 μs.

Thus, the large signal square wave response cannot be greater than 12.5 kHz (50 percent duty cycle), and at this frequency the output is distorted to a triangular wave. The sine-wave response is affected by slew rate so that the full peak output voltage V_{pk} is available until the frequency F_{max} at which time the maximum dV/dT of the sinewave exceeds the slew rate. Simple differentiation of a sine function, and the maximum rate of change of resultant cosine, lead to the following criterion:

$$\text{slew rate} = 2F_{max}V_{pk} \text{ at the limit.} \qquad (4.47)$$

As an example, the μA741 at nominal 0.5 V/μs slew rate can produce 10 V_{pk} (which is 20 V peak-to-peak or 7.07 V rms) up to a maximum of 8 kHz. The LM318 high-speed op amp with 50 V/μs slew rate could extend the full power bandwidth to 800 kHz. A typical general-purpose op amp output voltage capability-frequency graph is shown in Figure 4.35. Again, note that slew rate is dependent on supply voltage to a large extent.

Common-Mode Rejection

Common-mode rejection is the ability to reject (not react to) voltages appearing on both input leads with the same polarity. If a voltage V_{cm} is applied on both inputs at the same time, due to imperfection of internal symmetry, etc., a small voltage V_o will appear on the output even though the op amp was designed to respond to differential signals only.

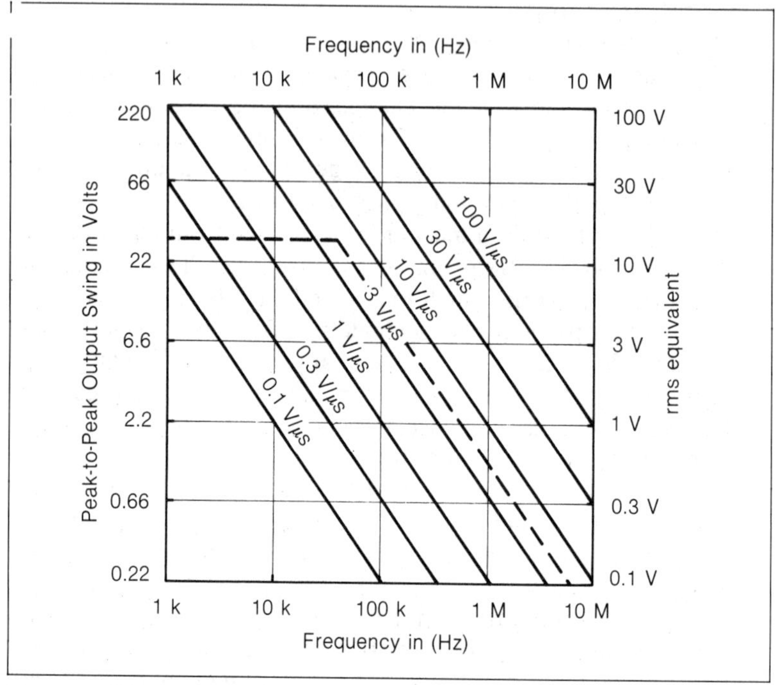

Figure 4.35—Op Amp Output Swing vs. Frequency and Slew Rate. The limitation by slew rate transforms a sinewave input into a triangular output: the peak-to-peak value is clamped to the power supply amplitude. The dotted line shows a practical example for an op amp with a 5 V/µs slew-rate, supplied by 30 V (+15 and −15).

To translate this voltage V_o into the equivalent differential voltage V_D which would cause the same output response, we need to divide V_o by the gain. Then the common-mode rejection ratio (CMRR) is defined as:

$$CMRR = \frac{V_{CM}}{V_o/G} \qquad (4.48)$$

$$or\ in\ dB = 20 \log \left(\frac{V_{CM}G}{V_o} \right)$$

The CMRR is usually very high (80 to 120 dB) for a good-quality op amp. Unfortunately, CMRR is generally specified at low frequency, such as 50 or 60 Hz, beyond which it rolls off rapidly. Due to the extreme sensitivity to any change in the impedance values of the internal matched pairs of amplifiers, the CMRR also varies with temperature and supply voltage. Figure 4.36 shows some typical values of CMRR.

Example: Assume an amplifier having a 100 dB CMRR at 1 kHz and a 40 dB differential gain. What is its reaction to a 1 kHz common-mode EMI?

$$G = 40 \text{ dB } (10^2 \text{ in terms of voltage})$$
$$\text{CMRR} = 100 \text{ dB } (10^5)$$

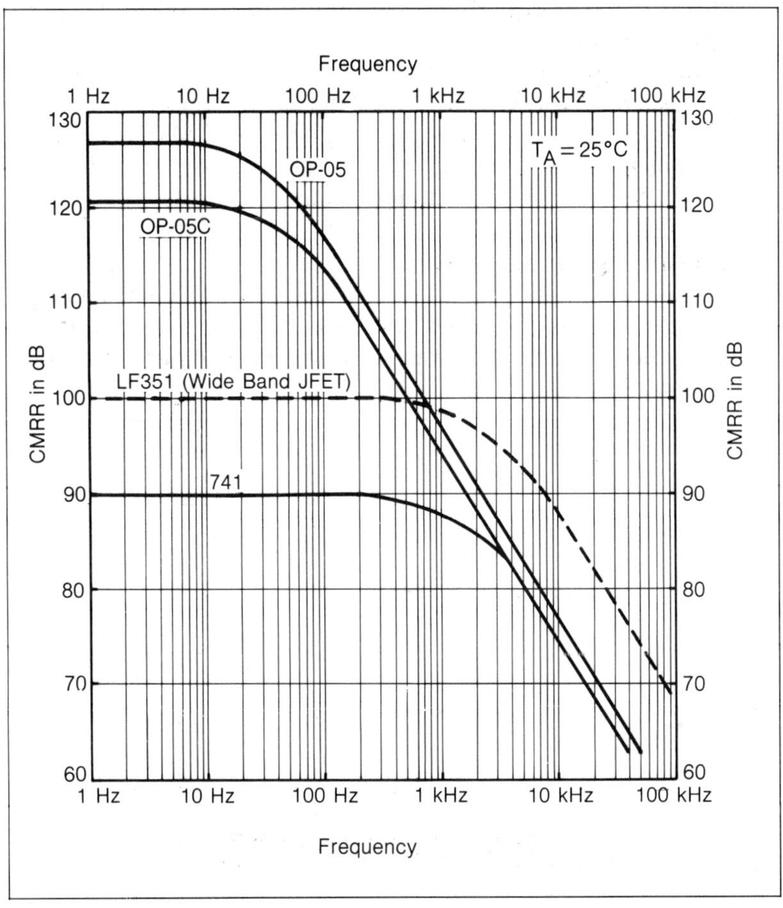

Figure 4.36—Typical Values of CMRR vs. Frequency for Op Amps (Reprinted from Precision Monolithics)

so,
$$\frac{V_{CM}}{V_o} \times G = 10^5$$

$$(4.49)$$

$$V_o = \frac{V_{CM} \times G}{10^5} = V_{CM} \times 10^{-3}$$

Therefore, a 10 V common-mode EMI will create an output error of 10 mV.

Symmetric (Balanced) versus Asymmetric (Unbalanced) Configuration

In many applications, (noninverting amplifier) op amps are not used with a truly floated differential input, but rather with a single supply and one grounded input. In this case, the CM rejection concept is meaningless since any voltage between the input pin and ground (Fig. 4.37) is sensed and amplified.

Common-Mode Slew Rate

Just like the normal-mode slew rate, there is also a common-mode slew rate. If V_{CM} exceeds a certain value of V/s, an additional dissymmetry results and unwanted conversion to differential signal takes place.

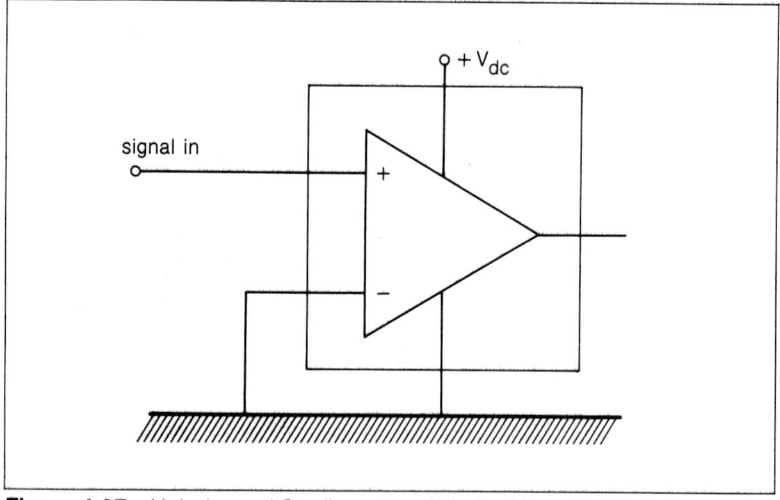

Figure 4.37—Unbalanced Configuration Offering Virtually No CMR (Noninverting Amp)

Power Supply Rejection Ratio (PSRR)

Although sometimes mixed with the CMRR, especially for single-supply devices, there is a definite difference between the effects of injected noise on power supply terminals and signal input common-mode rejection. Ripple, noise and voltage drift appearing on supply inputs can cause fluctuations of the desired output.

The PSRR is defined as:

$$ PSRR = \frac{V_{ps}}{V_o/G} = \frac{V_{ps}}{V_o} \times G \qquad (4.50) $$

where, V_{ps} = Noise superimposed over V_{dc}

V_o = Voltage output fluctuation caused by V_{PS}

Typical values of PSRR are shown in Fig. 4.38.

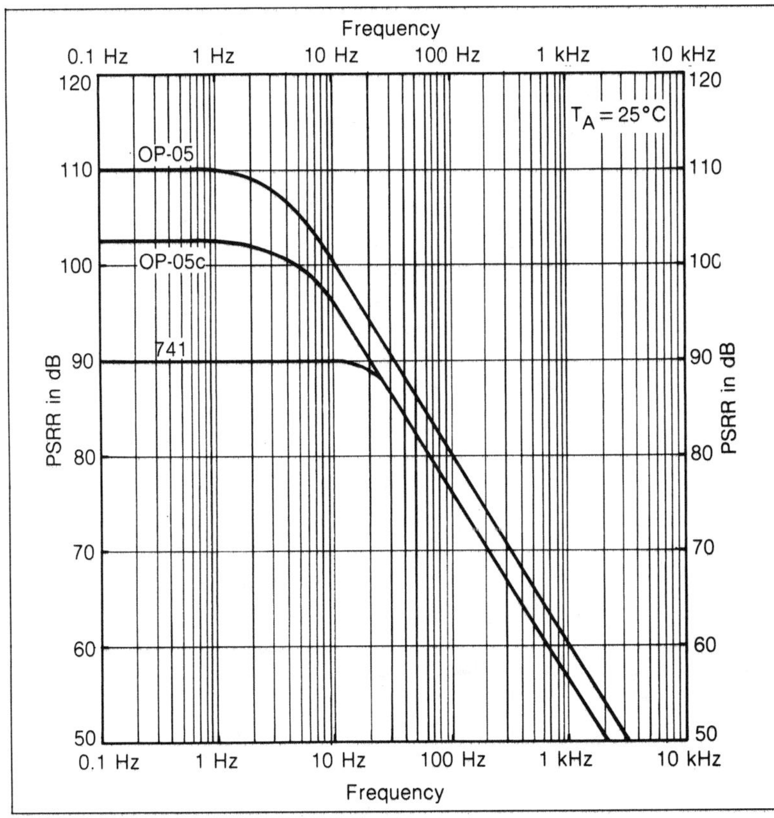

Figure 4.38—Typical Values of PSRR vs. Frequency for Op Amps (Reprinted from Precision Monolithics)

For instance, if the supply regulator exhibits 100 mV of ripple at 120 Hz (full-wave rectification residues), the PSRR from Fig. 4.38 is 74 dB or 5,000:1. Therefore, for a 40 dB gain amplifier this will result in an output noise:

$$V_o = \frac{V_{ps} \times G}{PSSR} = \frac{0.1 \text{ V} \times 100}{5,000} = 2 \text{ mV} \qquad (4.51)$$

which, in turn, will be equivalent to a differential input noise of:

$$\frac{2 \text{ mV}}{100} = 20 \ \mu V \qquad (4.52)$$

If, for instance, the op amp was designed to attain a sensitivity of 1 μV, the equivalent input error caused by the PSRR is 20 times (or 26 dB) the objective. Hopefully, PSRR can be readily improved by built-in or additional RC or LC decoupling, as shown in Fig. 4.39.

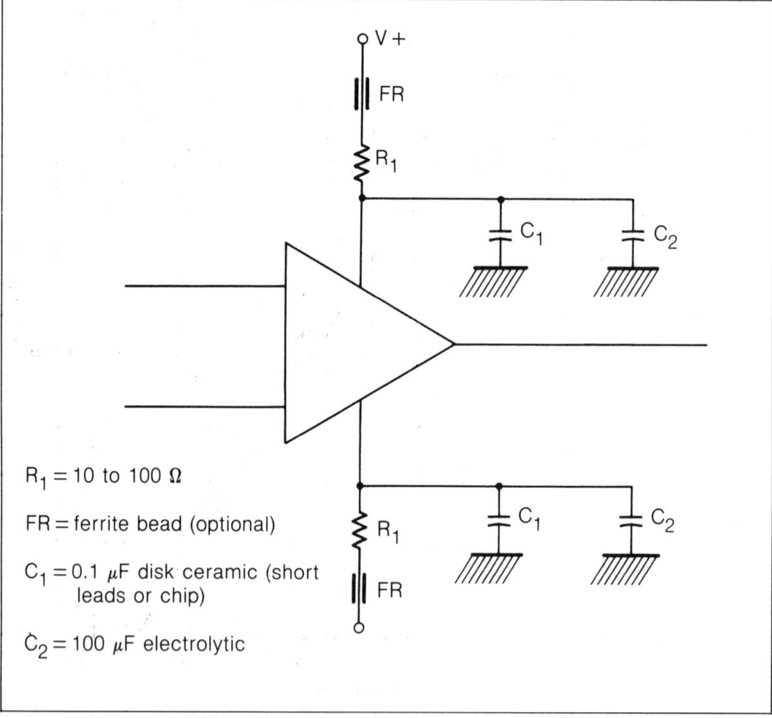

Figure 4.39—Power Supply Decoupling

Common-Mode Range

Whatever the common-mode range, the absolute value of V_{CM} cannot exceed the value of the power supply voltages. Otherwise, malfunctions or permanent deterioration might occur. An exception is isolation amplifiers where several kV of common-mode voltage can be tolerated.

4.3.2.1 Circuit Analysis of CM and DM Rejection

The fundamental functional element of the operational amplifier is the differential pair. This element responds to the differential rather than common-mode signal because the geometries and characteristics of the two transistors, which constitute the differential pair, are so closely matched that their response to a common-mode signal is essentially identical and is cancelled out in succeeding stages.

Regardless of how closely matched they may be, they can never be exactly identical except by accident. Because of manufacturing variables, there will be unavoidable and minute differences in geometries and characteristics in these transistors. These differences, however small, are significant when a strong electromagnetic field is applied to the device.

Consider the circuit of Fig. 4.40. The configuration here is that of a simple analog Schmitt trigger. For simplicity, the power supply is single-ended, with 30 V magnitude. Assume that the reference signal at the inverting input resistor lead is +15 V. If V_{in} has a value of 15 V_{dc}, plus a few mV, then the output is guaranteed to

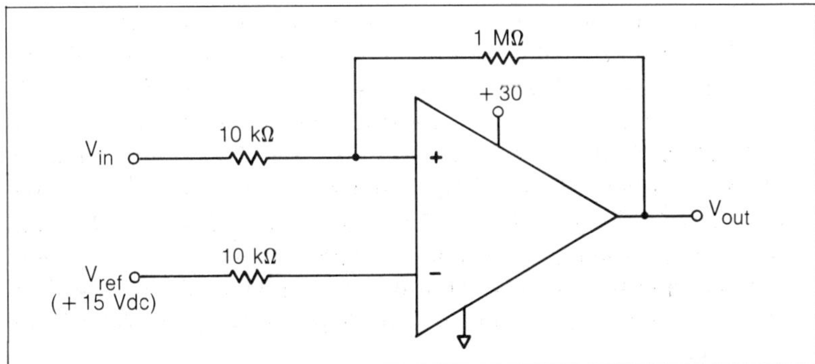

Figure 4.40—Analog Schmitt Trigger

4.65

be saturated near $+30$ V_{dc}. If the voltage is any value less than $+15$ V_{dc} minus a few mV, then the output is guaranteed to be saturated near the power supply common.

If this circuit were placed in a test chamber, and if V_{in} were fixed at 14 V_{dc}, then V_{out} would be nearly 0 V with respect to the power supply common in the quiescent state.

When exposed to a CW field of sufficient intensity and proper frequency, the voltage at V_{out} will begin to pulse toward $+30$ V_{dc}. In a CW field, these pulses will be random in their frequency and duration. As field strength is increased even past this level, the output will saturate at one extreme or the other, and the output will remain at this value no matter what the value of V_{in}.

At this point, there is a complete malfunction of the device due to the radiated electromagnetic field. When the field is removed, the circuit will function normally. Obviously, the behavior is due to an interaction between active elements of the amplifier and the impinging field. Field strengths required to provoke this behavior are strongly frequency-dependent. Usually there will be a band of frequencies which require relatively small field strengths to stimulate anomalous behavior, while very strong field strengths at other frequencies will have no effect.

A more thorough analysis will determine the RF susceptibility and effective input impedance as a function of frequency for typical op amp configurations. To realistically represent the HF impedances, the resistors are all considered to be shunted by a "built-in" stray capacitance of 2 pF. Additionally, the input terminals of the amplifier itself are considered to include a "built-in" stray capacitance to ground of 5 pF. The equivalent circuit is given in Fig. 4.41a.

At low frequencies this circuit can be simplified as shown in Fig. 4.41b. At high frequencies, we can eliminate the amplifier and combine the input stray capacitances with the feedback stray capacitors, C_{f1} and C_{f2}, arriving at the equivalent circuit shown in Fig. 4.41c. At low frequencies (below the unity gain cutoff frequency of the amplifier), the two inputs are maintained at the same potential by the amplifier gain and the feedback to the inverting input. Thus, the input terminals can be considered to be virtually shorted, and the differential input impedance is found by adding the two input resistors (along with their parallel stray capacitors) in series.

For instance, for an LM 118 op amp this "low frequency" region includes frequencies to approximately 10 MHz. Assume that the

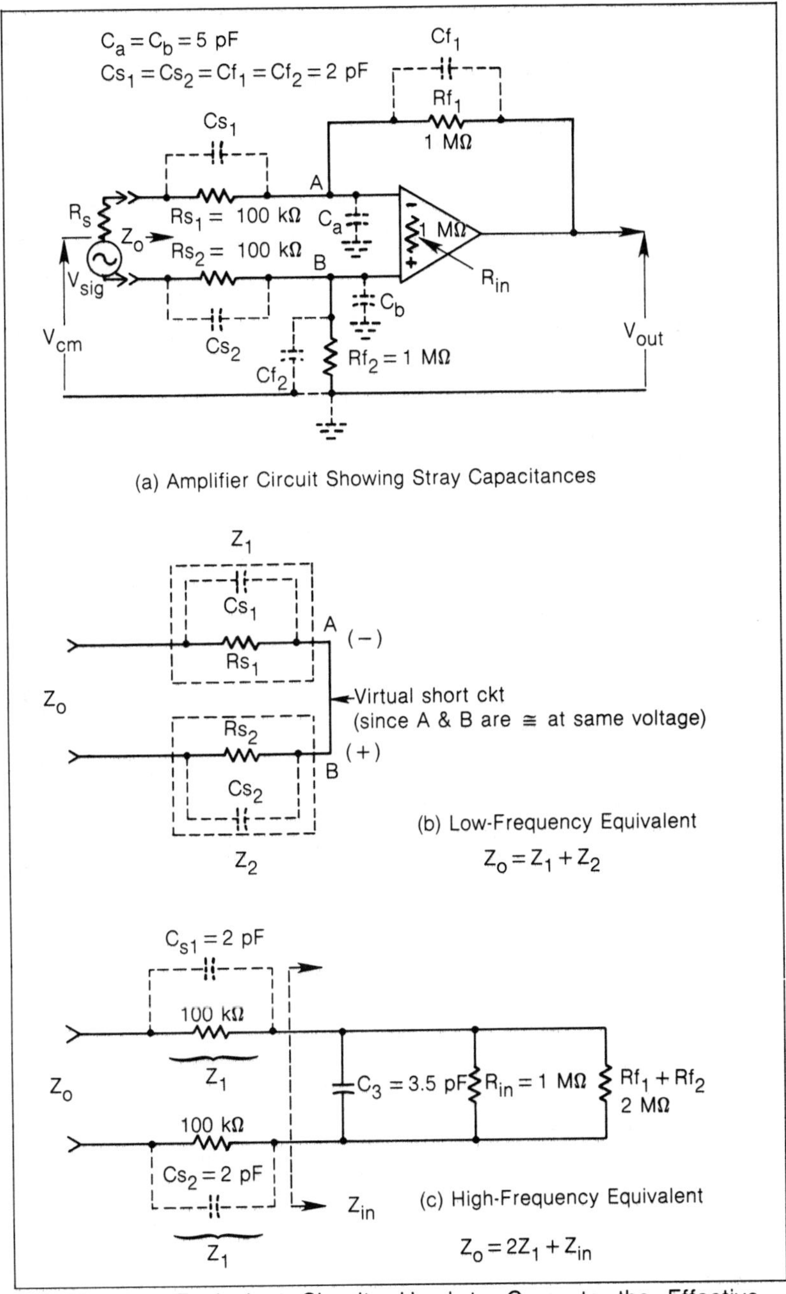

(a) Amplifier Circuit Showing Stray Capacitances

(b) Low-Frequency Equivalent

$$Z_o = Z_1 + Z_2$$

(c) High-Frequency Equivalent

$$Z_o = 2Z_1 + Z_{in}$$

Figure 4.41—Equivalent Circuits Used to Compute the Effective Differential-Mode Input Impedances

op amp is used in the configuration of Fig. 4.41, which provides a design gain of about 20 dB. The differential-mode input impedance is calculated by combining the complex impedances at half-decade frequencies from 1 kHz to 10 GHz. The resulting magnitude of Z_{in} differential is shown in Fig. 4.42.

The common-mode input impedance Z_{CM} is calculated using the equivalent circuit of Fig. 4.43. Common-mode signals see each input resistance in series with the feedback network, and the two resulting branches are parallel (Fig. 4.43b). The resulting module of $|Z_{CM}|$ is shown on Fig. 4.42.

Based on these models, the table and curves of Fig. 4.44 were constructed. Both DM and CM sensitivity are shown for "small" signals and "large" signals. For small signals (50 mV full-scale output), the criterion was the necessary input voltages (DM and CM) to cause an effect equivalent to an in-band 100 μV differential input.

For large signals (5 V full-scale output) the criterion was the necessary input voltage to cause an effect equivalent to an in-band 10 mV differential input. A few remarks are in order.

1. A conservative value of 0.5 percent for the input resistors' tolerance has been taken, resulting in an overall imbalance

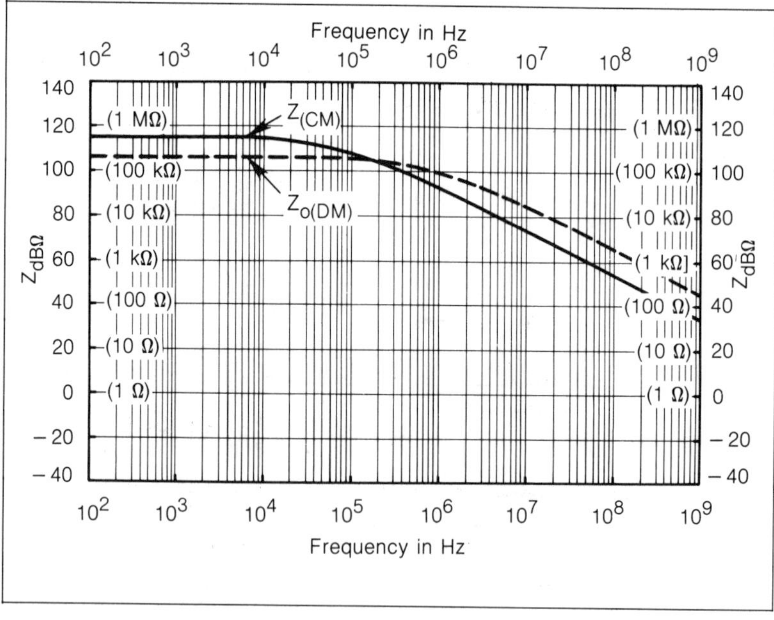

Figure 4.42—Differential-Mode and Common Mode Input Impedances vs. Frequency for the Typical Op Amp Circuit of Fig. 4.41

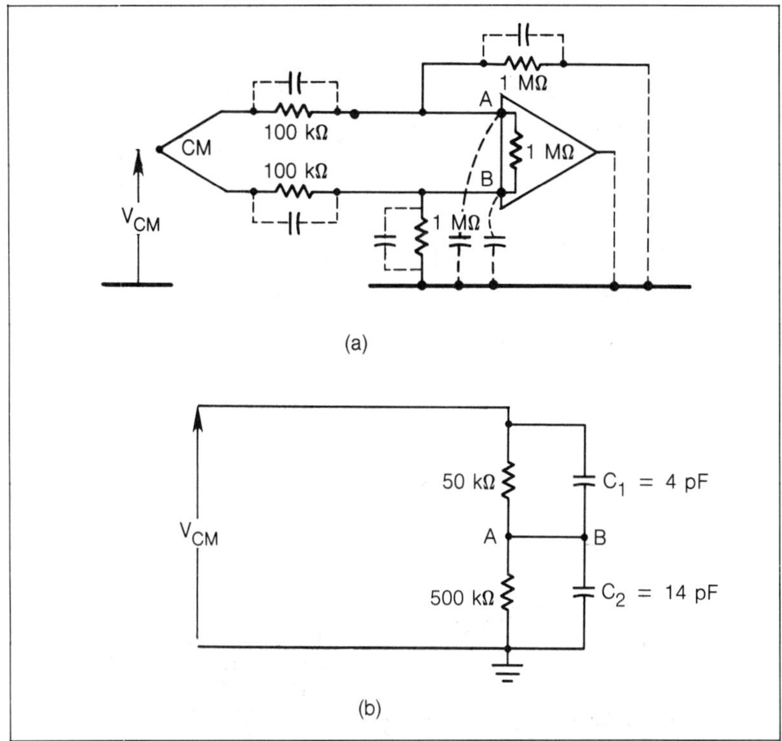

Figure 4.43—Basic (a) and Simplified (b) Equivalent Circuits Used to Calculate CM Input Impedance and CM Rejection. Note that CM voltage appearing with same magnitude on both inputs, these can be "commonned." Also, from an input point of view, the load (being low impedance) can be assimilated to a grounded output.

of 1 percent, therefore, the in situ CMR cannot exceed 40 dB (even though the op amp itself may have a 100 dB CMR).

2. The CMRR for large and small signals is practically the same, i.e., CMRR is independent of the gain.

3. The CMRR approaches 0 dB in the VHF (30 to 300 MHz) region; that is, a common-mode EMI voltage has the same effect as the same voltage applied differentially.

4. The in-band DM rejection is, of course, 0 dB since that is where the amplifier is designed to function. Then, the frequency beyond which DM rejection exists depends on the gain, since the gain × bandwidth product comes into play.

5. From low to high frequencies, CM rejection degrades while

(a)

Frequency in Hz	10^3	10^4	10^5	10^6	10^7	10^8	10^9	10^{10}
CMR (dB)								
Large Signals (Small Gain) $= 20 \log V_{cm}/10\ mV$	40	36	28	20	10			
Small Signals (High Gain) $= 20 \log V_{cm}/100\ \mu V$	40	36	28	(20)	(10)	(6)		
DMR (dB)								
Large Signals (Small Gain) $= 20 \log V_{DM}/10\ mV$	0	0	0	0	6	20	20	20
Small Signals (High Gain) $= 20 \log V_{DM}/100\ \mu V$	0	0	6	26	46	60	60	60

(b)

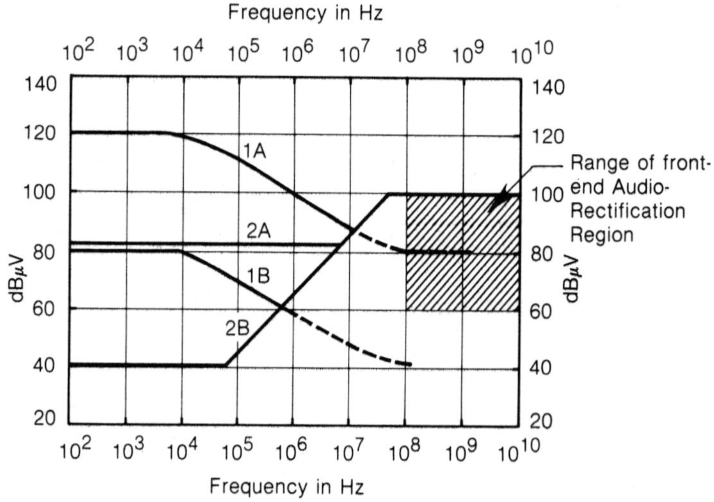

1A = CM sensitivity large signal
2A = DM sensitivity large signal
1B = CM sensitivity small signals
2B = DM sensitivity small signals

Figure 4.44—CM and DM Sensitivity of a Typical Op Amp (LM/18) Mounted in a Typical Gain Setting Network like Fig. 4.40

DM rejection improves. When they become equal, DM rejection takes over since the sensitivity to a CM input cannot be worse than the one to a DM input, all other conditions being the same. Beyond about 100 MHz, EMI rejection levels off to a value derived from the McDonnell Douglas studies (see Section 4.3.3 and Ref. 8). An RF level of about 0.03 to 0.1 V is necessary to cause an undesired response, up to $\cong 10$ GHz. The aforementioned study shows that the necessary power increases about 100 times when EMI frequency changes from 1 to 10 GHz. Since, at the same time the in situ impedance of the op amp and its associated resistor network drops from 100 Ω to 10 Ω, a deduction can be made that necessary EMI voltage has varied as $\sqrt{F}$. All this does explain for the fact that parasitic inductances cause resonances and that the input port becomes severely unmatched to its normal input wiring beyond a few hundred MHz. Since, at the same time the in situ impedance of the op amp and its associated resistor network drops from 100 Ω to 10 Ω, a deduction can be made that necessary EMI voltage has varied as $\sqrt{F}$. All this does explain for the fact that parasitic inductances cause resonances and that the input port becomes severely unmatched to its normal input wiring beyond a few hundred MHz.

For CM sensitivity, a 0.5 percent resistor tolerance is assumed (1 percent mismatch overall). The table in Fig. 4.44a is the ratio between the necessary EMI amplitude to cause a given output change and the inband differential input which would cause the same change. The curves in Fig. 4.44b are the EMI voltage impressed in CM or DM mode to cause an output change equivalent to a 10 mV (large signals) or 100 μV (small signals input. For small signals (lesser bandwidth, more gain), the CMR figures in parenthesis are calculated ones, corresponding to the dotted lines of curves B, since CMR cannot be less than DMR.

4.3.2.2 Audio Rectification and Other RF Side Effects in Operational Amplifiers

When the susceptibility of an operational amplifier is considered, radiated susceptibility is the most frequent problem. With adequate filtering and bypassing of power supply lines, it is relatively rare for an analog circuit to display anomalous behavior in conducted susceptibility tests (these must be distinguished from transient susceptibility tests, which can yield anomalous and surprising results with virtually any type of circuit).

One reason analog devices are relatively immune to conducted susceptibility problems is their relatively low slew rates. In most cases, a common operational amplifier such as the type 741 will be used. When intentional signals are of relatively low frequency, most analog circuits are immune to conducted noise because their slew rates are so modest that they do not have enough speed to respond to conducted noise.

When an operational amplifier circuit is exposed to a high frequency, especially VHF radiated field, the output voltage begins to exhibit fluctuations even when the inputs are held constant. This situation is especially true when the circuit is exposed to a modulated field, but it is also true when the circuit is exposed to a continuous wave (CW) field of sufficient magnitude (see Fig. 4.45).

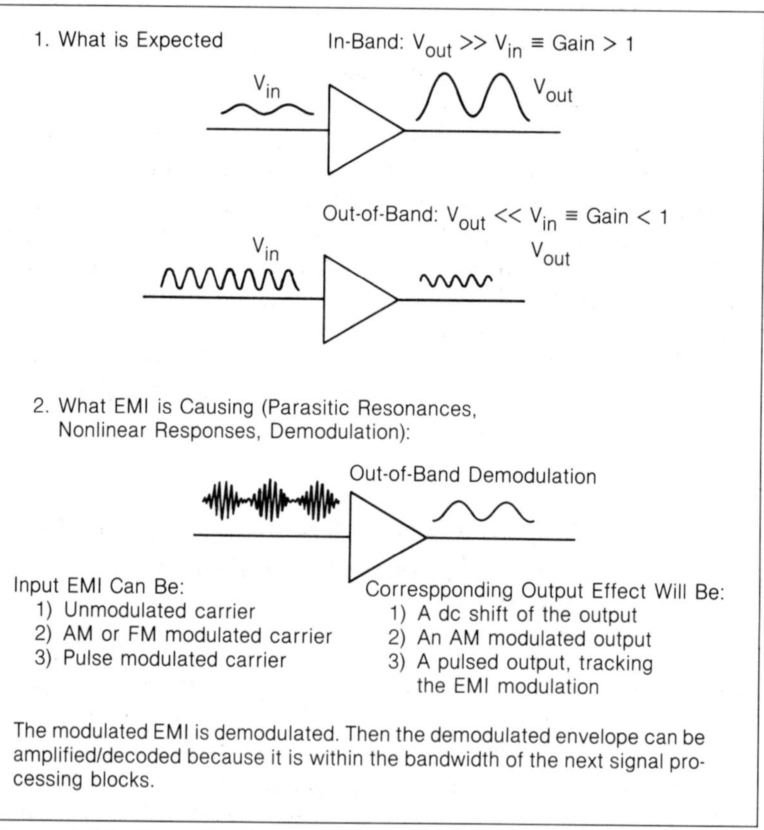

Figure 4.45—Parasitic Response of Analog Amplifiers

Changes in the output voltage will initially appear in synchronization with the modulating waveform, if one is present. In any event, the output will usually saturate in the presence of a large enough radiated field.

It has been suggested that the base-emitter junctions (or base-collector, depending on the type) act as rectifying junctions, with their responses being unmatched at certain frequencies due to imperfections in the geometries of individual devices. Thus, a strong radiated field at the proper frequency is capable of injecting artificial signals into the differential pair of the operational amplifier, thereby producing the observed behavior.

These problems are especially noticeable in preamplifier stages for two reasons. First, the preamplifier will usually be the one connected to long leads, and these leads act as antennas to RF. Second, preamplifier stages are commonly operated at high gain with a high input impedance, which in itself makes the stage more susceptible to disturbance.

Analog circuits, especially those built around operational amplifiers, are maximally susceptible to radiated EMI at frequencies between 30 MHz and 500 MHz. Figures 4.46 through 4.52, extracted from the general model for audio rectification in Section 4.1.3.1, can be used to predict the out-of band rejection of amplifier when audio rectification is accounted. It must be remembered that

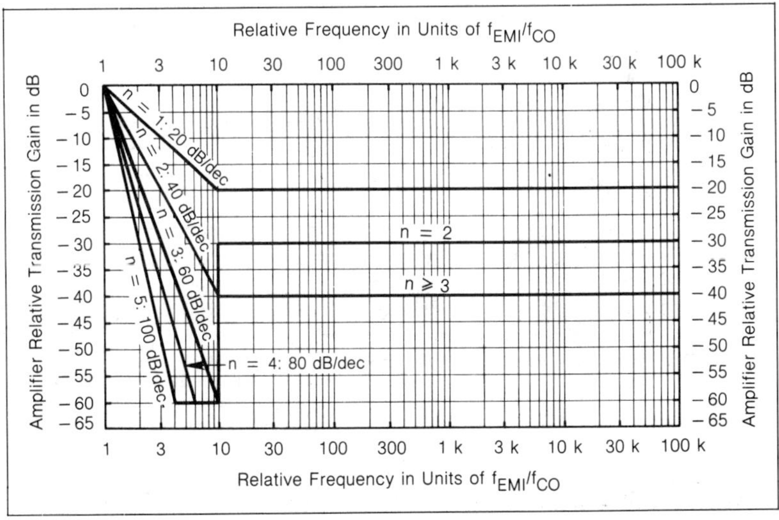

Figure 4.46—Default Model for Audio Rectification for Amplifiers with Fc > 30 MHz

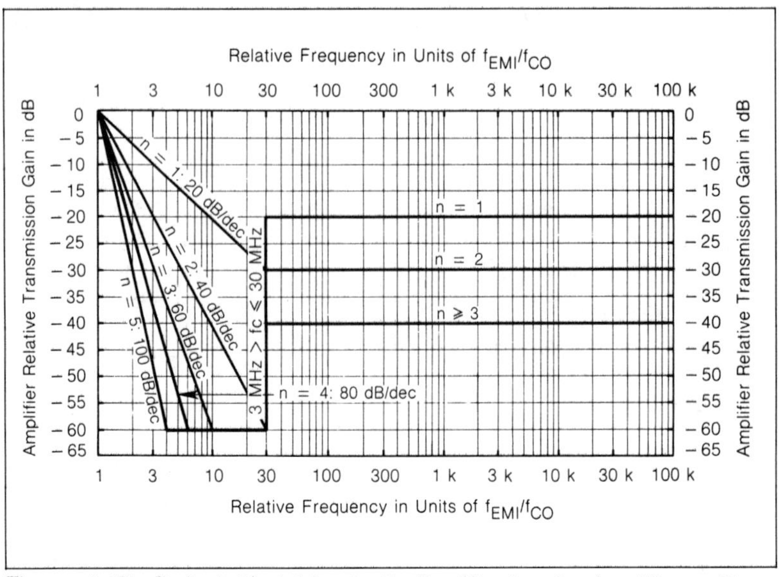

Figure 4.47—Default Model for Audio Rectification for Amplifiers with 3 MHz < Fc < 30 MHz

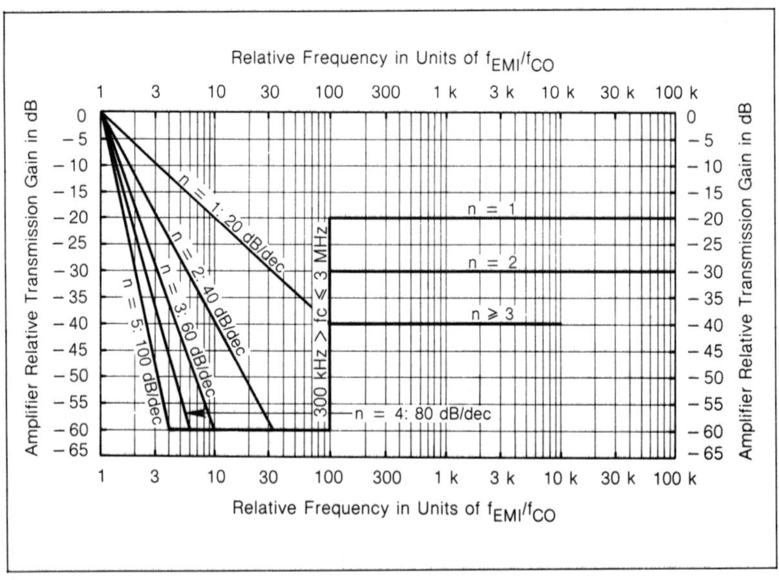

Figure 4.48—Default Model for Audio Rectification for Amplifiers with 300 kHz < Fc ≤ 3 MHz

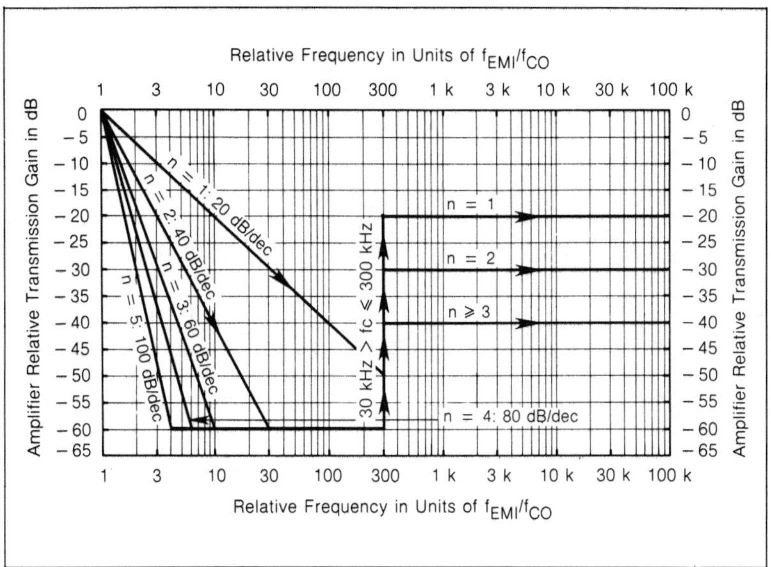

Figure 4.49—Default Model for Audio Rectification for Amplifiers with 30 kHz < Fc ≤ 300 kHz

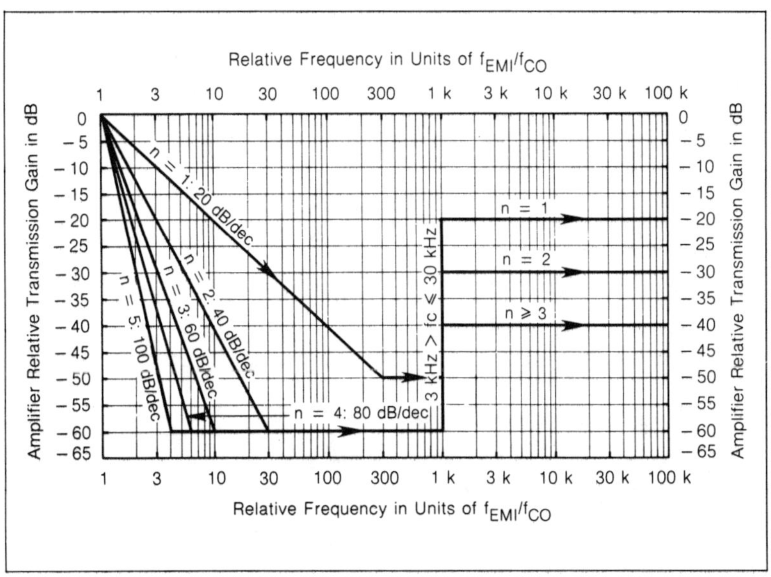

Figure 4.50—Default Model for Audio Rectification for Amplifiers with 3 kHz < Fc ≤ 30 kHz

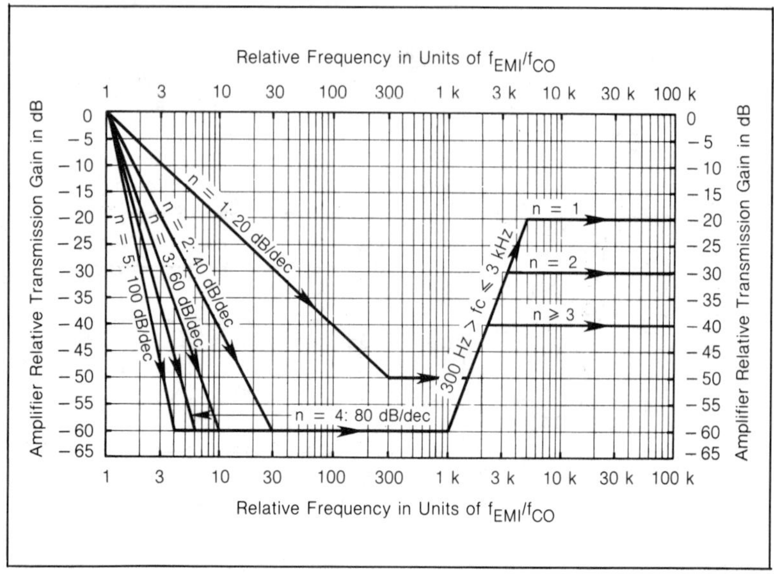

Figure 4.51—Default Model for Audio Rectification for Amplifiers with 300 Hz < Fc ⩽ 3 kHz

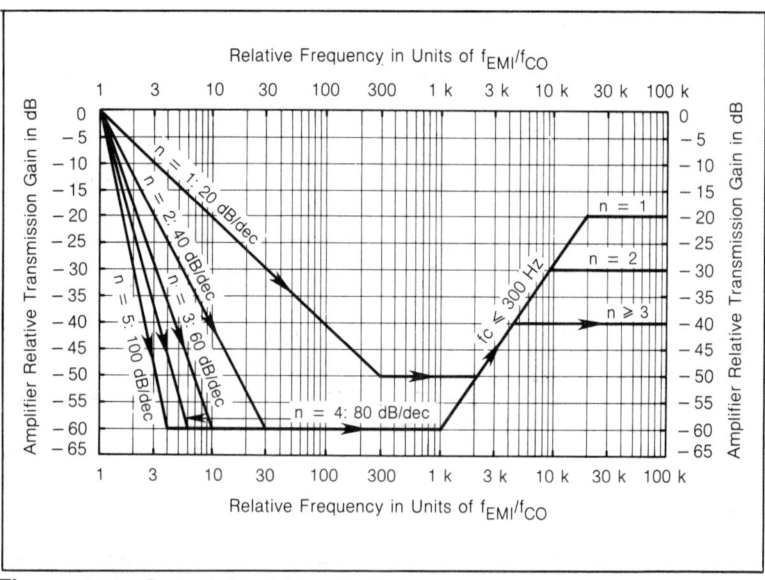

Figure 4.52—Default Model for Audio Rectification for Amplifiers with Fc ⩽ 300 Hz

for rectification to occur, the entering EMI must be at least few 100 mV, i.e., the typical barrier potential of junctions.

Besides the pure audio rectification caused by the p-n junctions in analog circuits, other spurious effects of RF exposure are:

1. **AGC Desensitization**—Many analog amplifiers have a built-in or external automatic gain control (AGC). If the EMI signal is large enough, it can drive the AGC to decrease the gain. Therefore, the useful signal is reduced (less gain) while the EMI signal is more or less amplified.

2. **RF Offset**—Rectification of an RF signal causes some shifts in polarization voltages (those provided by zener diodes or resistors bridges) which translate ultimately into output variations.

3. **Harmonic Generation**—The nonlinear nature of some components, especially when overdriven by strong RF signals, causes a distortion of the sinewave RF carrier, with generation of harmonics of each single EMI frequency.

4. **Erratic Responses due to Feedback Loop Sensitivity**—The parasitic L and C element of the feedback loop in amplifiers can go into resonance at specific frequencies, sometimes causing a negative feedback input to see an out-of-phase signal. This is equivalent to a positive feedback (no phase margin in the Nyquist criteria), i.e., parasitic oscillations excited by the EMI signal.

4.3.2.3   Erratic EMI Behavior of Cascaded Amplifiers

Figure 4.53, in parts a and b, shows the expected and not-so-expected (although quite common) behavior of an amplifier block made of several stages of amplification. The variable here is not the frequency but the amplitude of EMI stimuli, in this case a radiated field necessary to produce a given error.[9]

The maximum allowed device deviation response for both curves shown is ±10 percent. Figure 4.53a depicts a typical linear type response versus field strength. Above 45 V/m the equipment exceeds the −10 percent tolerance allowed. With the application of a programmed susceptibility test, a maximum test field of 100 V/m would be applied first. The device exceeds the −10 percent limit

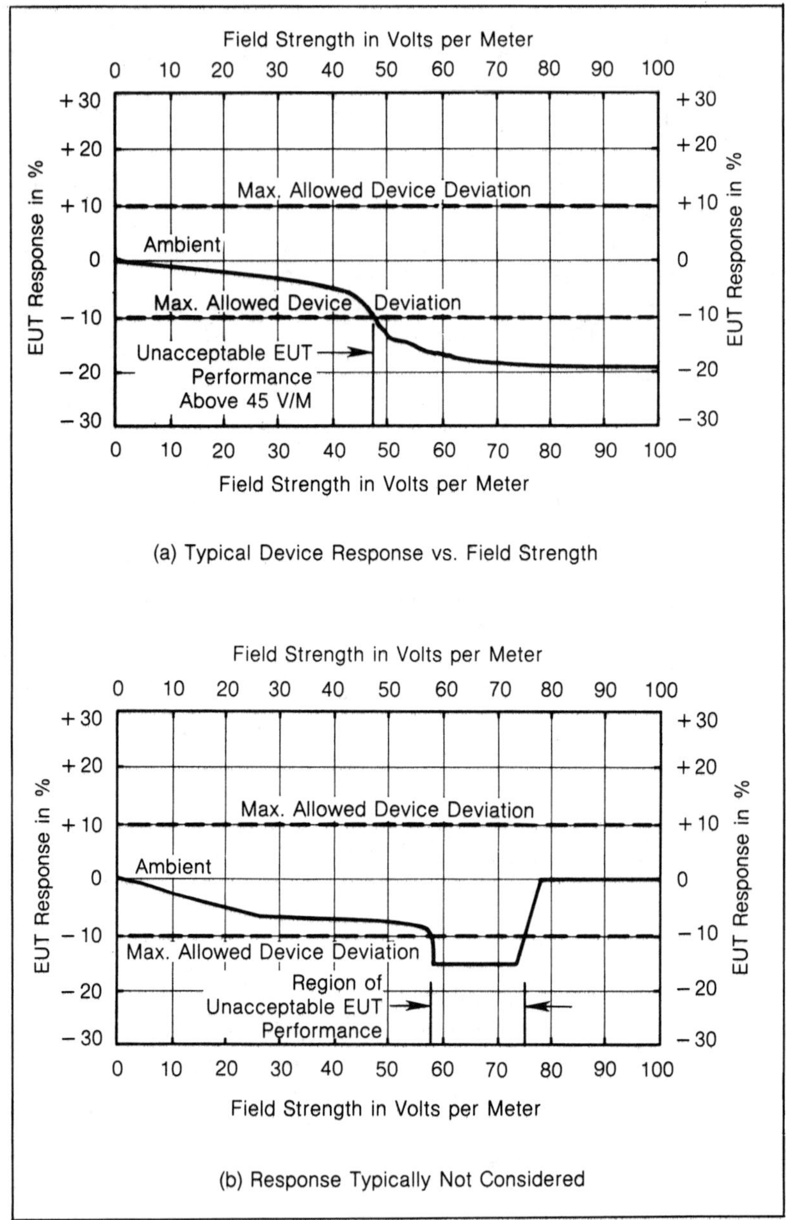

Figure 4.53—Erratic, Nonlinear EMI Response in Function of Applied Field Strength

at the 100 V/m level, and the program would then apply a 50 V/m field. Again, at 50 V/m the device exceeds the − 10 percent limit. The successive approximation routine would then continue until the maximum field strength is found where the device's response will be maintained within the ±10 percent limit.

Now consider the not-so-typical response of Fig. 4.53b. The equipment exceeds the − 10 percent limit between field strength levels from 56 V/m to 76 V/m but is within the ± 10 percent limits at 100 V/m, the maximum possible field strength level the device may encounter. This type of device response is usually not considered or is overlooked, although it is common in very complex multistage electronic control or monitoring systems.

Within such a control system, all the stages or components do not necessarily have the same threshold of susceptibility. Above a level of 76 V/m, a particular stage of the multistage system produces a response which aids or corrects the response of a separate stage or stages that causes the out-of-limit response from 56 to 76 V/m. If no previous consideration had been given to the response curve of Fig. 4.53b and a 100 V/m had been applied first, the device would have been declared "good." Its region of unacceptable performance would have been overlooked.

To properly test for this type of erratic response, the test field at each frequency must be slowly advanced from 0 at small increments up to the maximum test field level. The device's response must be checked at each increment of field strength to determine its maximum field strength or threshold of susceptibility and to maintain the device's response within the ±10 percent limits. This technique will eliminate the possibility of overlooking the region of unacceptable device response, but it will also be very time consuming.

4.3.2.4 Sensitivity to EMI Coming by the Output Port

Although the output of an amplifier having generally a much lower impedance than the input is not likely to be the predominant EMI entry port, this possibility cannot be completely ruled out. For very strong interference, such as nuclear electromagnetic pulse, lightning, powerful radio transmitter proximity, etc., an amplifier whose input has been hardened to perhaps 40 or 60 dB can still show malfunctions because of EMI entering by the "back door."

4.3.3   Measured Susceptibilities of Analog Devices

Many experiments and model validations have been conducted to measure the RF susceptibilities of digital analog devices like op amps, voltage regulators, etc. These experiments were generally made with carefully arranged test jigs to avoid parasitic effects that spoil measurements above the VHF regions. One often-mentioned experiment is the McDonnell Douglas study where the devices under test were mounted on a microwave test fixture and EMI injection and monitoring were done with coaxial fittings and strip-line leads.

The following is a summary of the essential findings of Refs. 8 and 10 through 12. Note that the amount of injected EMI is generally expressed in terms of **absorbed power**, i.e., an accurate monitoring of the forward and reflected power was made in order to determine the actual amount of EMI entering the device.

4.3.3.1   Susceptibility of Op Amps

The interference effect monitored was the amount of RF power necessary to cause an equivalent input offset of several values ranging from 50 to 200 mV (input offset can be derived by taking the output change divided by the gain).

In the case of an unmodulated RF carrier, this effect would be predominant. The "susceptibility domain" is defined as the one below and up to the permanent damage level (see Section 4.1). The results are shown in Fig. 4.54. For modulated RF carriers, some results are shown on Figs. 4.55 and 4.56, for a 0.45 V RF signal, 80 percent modulated by a 1 kHz AM signal.

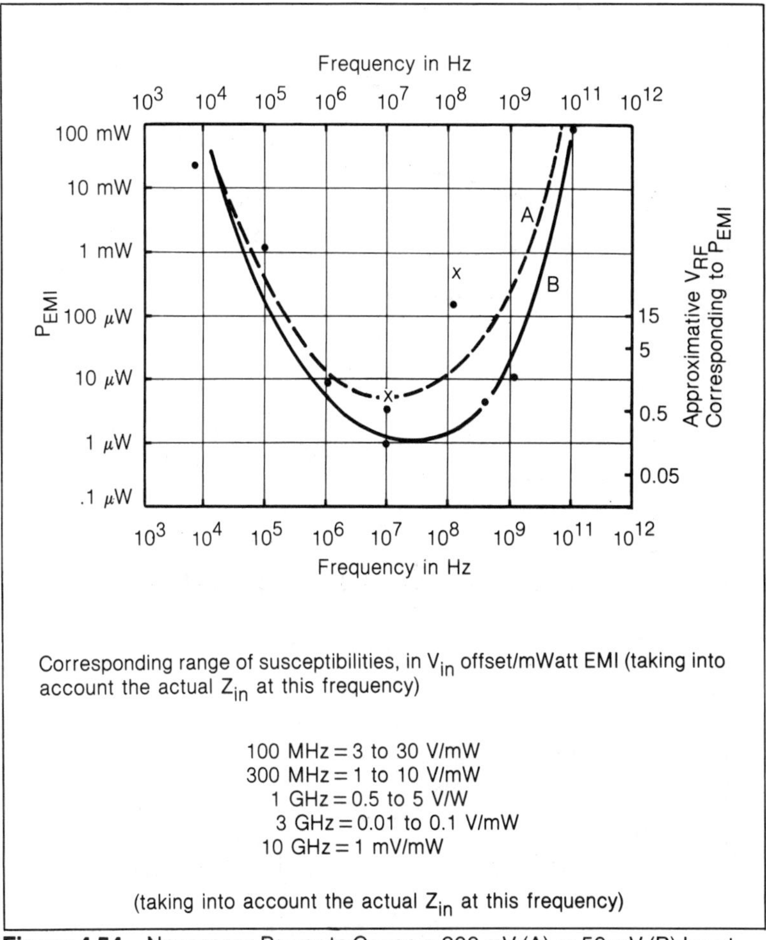

Corresponding range of susceptibilities, in V_{in} offset/mWatt EMI (taking into account the actual Z_{in} at this frequency)

100 MHz = 3 to 30 V/mW
300 MHz = 1 to 10 V/mW
1 GHz = 0.5 to 5 V/W
3 GHz = 0.01 to 0.1 V/mW
10 GHz = 1 mV/mW

(taking into account the actual Z_{in} at this frequency)

Figure 4.54—Necessary Power to Cause a 200 mV (A) or 50 mV (B) Input Offset for Several Types of Op Amps (741, 108A, 201A, 207, 0042C, 531) from Ref. 2 and 10

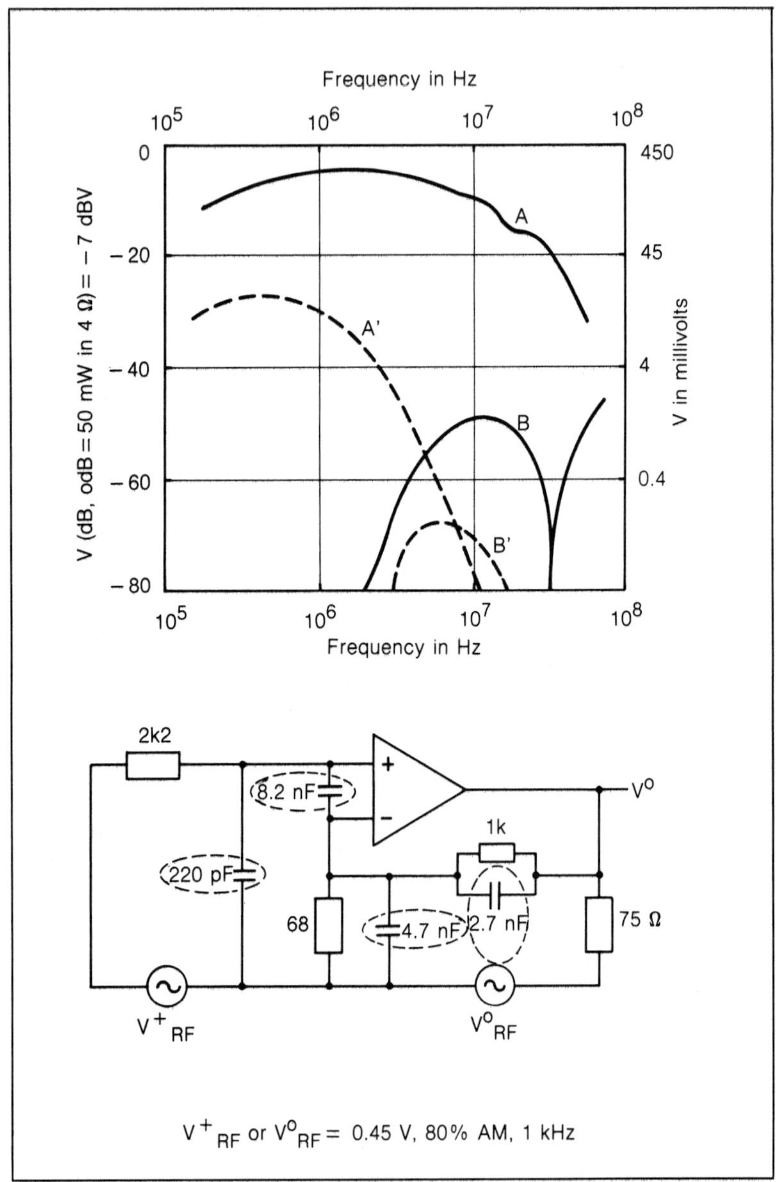

Figure 4.55—Detected 1 kHz AM from a 0.45 V Modulated EMI Applied at Input (Curve A) or Output (Curve B), measured at output V_o. The dotted lines are the improvement when the capacitors shown on the schematic are added (from Ref. 10).

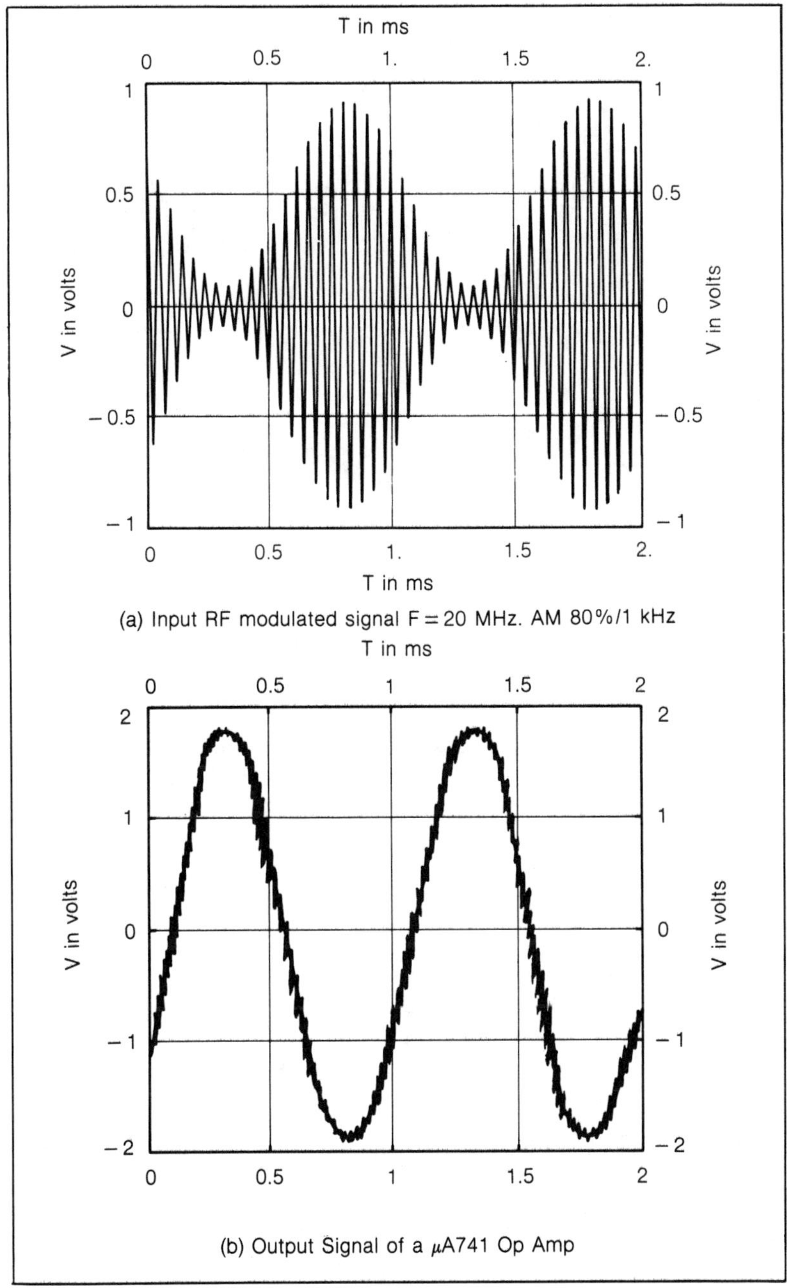

Figure 4.56—Other Experiments from Ref. 10 (continued next page)

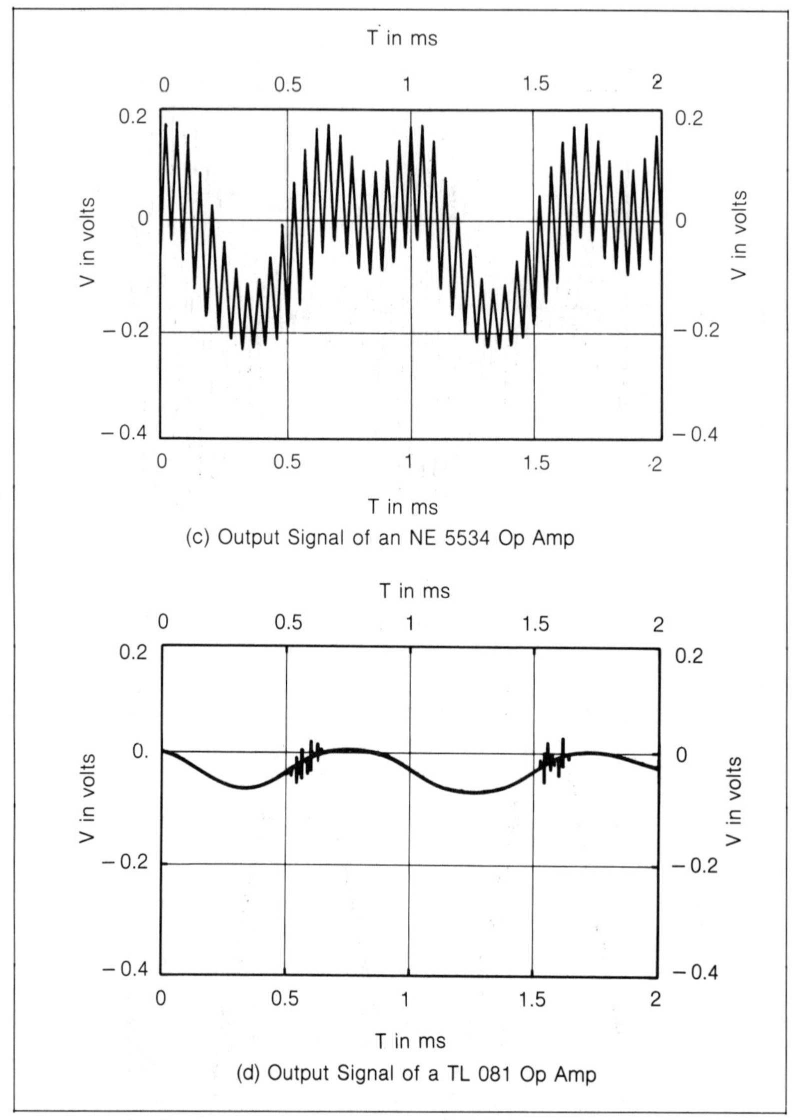

(c) Output Signal of an NE 5534 Op Amp

(d) Output Signal of a TL 081 Op Amp

Figure 4.56—(continued)

4.3.3.2 Susceptibility of Other Typical Linear Devices

Among the analog devices covered by the study of Ref. 8 several linear voltage regulators, comparators and line receivers were included. Since line transceivers belong more to digital applications, they will be covered in Section 4.4 (digital devices). For voltage regulators, tests were conducted on 3-pin regulators (309, 320, 78MO5) with 5 V regulated output and various loadings, and on 8-pin regulators (300 and 305) with 12 V regulated output.

In both cases, the criterion of unacceptable malfunction was an output voltage change greater than ± 0.25 V. RF was injected successively in each of the 3 or 8 pins, and the worst-case results (lowest power level to cause malfunction) is reported in Fig. 4.57.

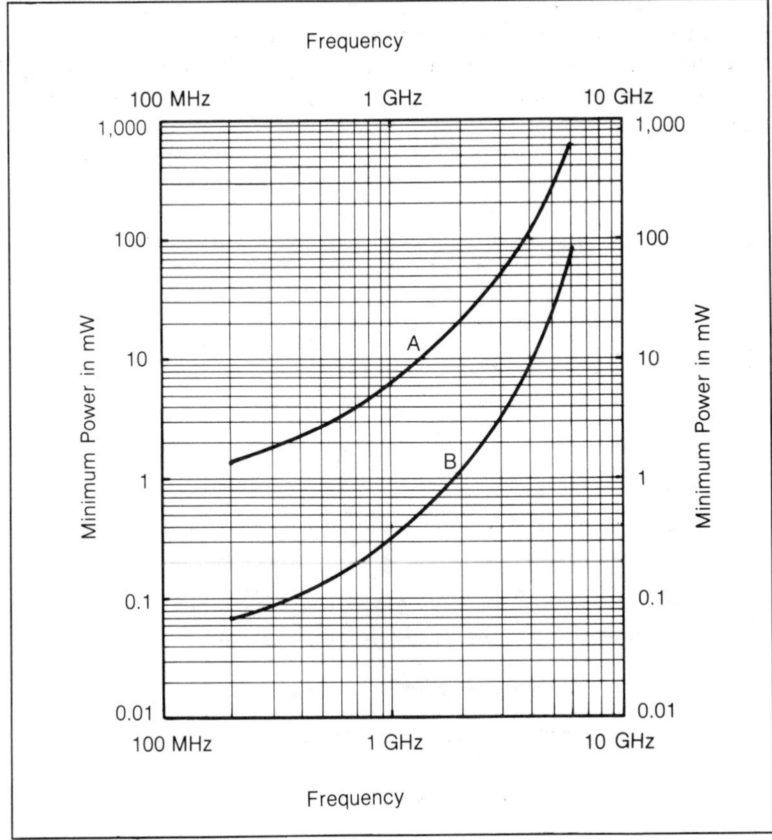

Figure 4.57—Worst-Case Susceptibility Values for 3-pin (A) and 8-pin (B) Voltage Regulators

Results indicated that 8-pin regulators were more susceptible, apparently because these devices have an error-amplifier op amp directly accessible from the "feedback" pin. The rectified RF causes a dc offset similar to that of a standard op amp, "fooling" the regular operation. Consequently, the curves show that 3-port regulators are 10 to 16 times (10 to 12 dB in power terms) less susceptible.

For comparators, tests were conducted on 306, 311, 339, 360, 710 and 760 types. Comparators are devices performing basically a one-bit analog-to-digital conversion; that is, the output switches from 0 to a given dc level when a linear input is varied a few mV around a threshold point (Fig. 4.58). One of the EMI effects is again a parasitic RF detection causing a shift in the device's switch point.

The criterion was the EMI power necessary to cause a switch-point to change by 50 mV, 200 mV and 500 mV. Here again, the worst-case susceptibility occurs in the upper VHF, where only 20 or 30 μW are enough to cause the first degree of malfunction (Fig. 4.59).

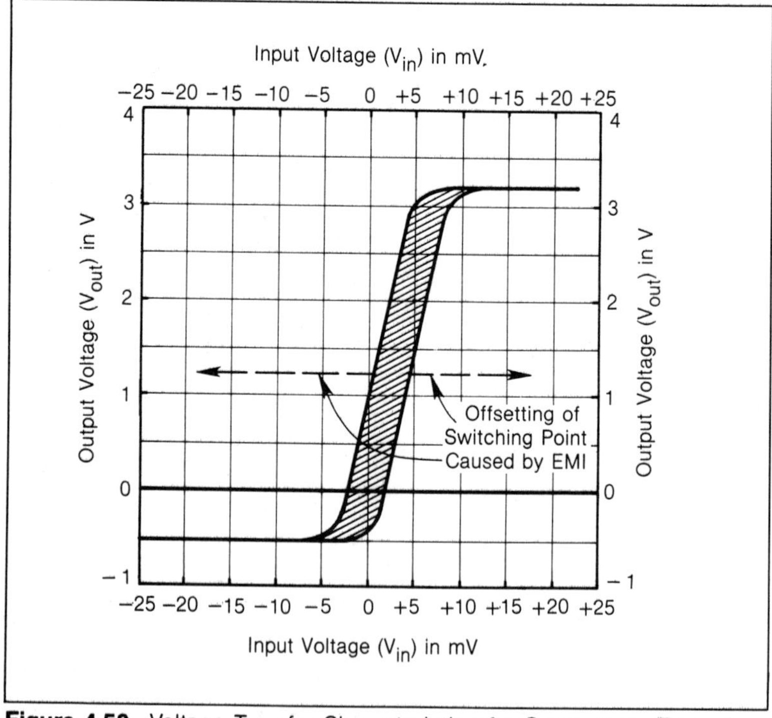

Figure 4.58—Voltage Transfer Characteristic of a Comparator (Typical)

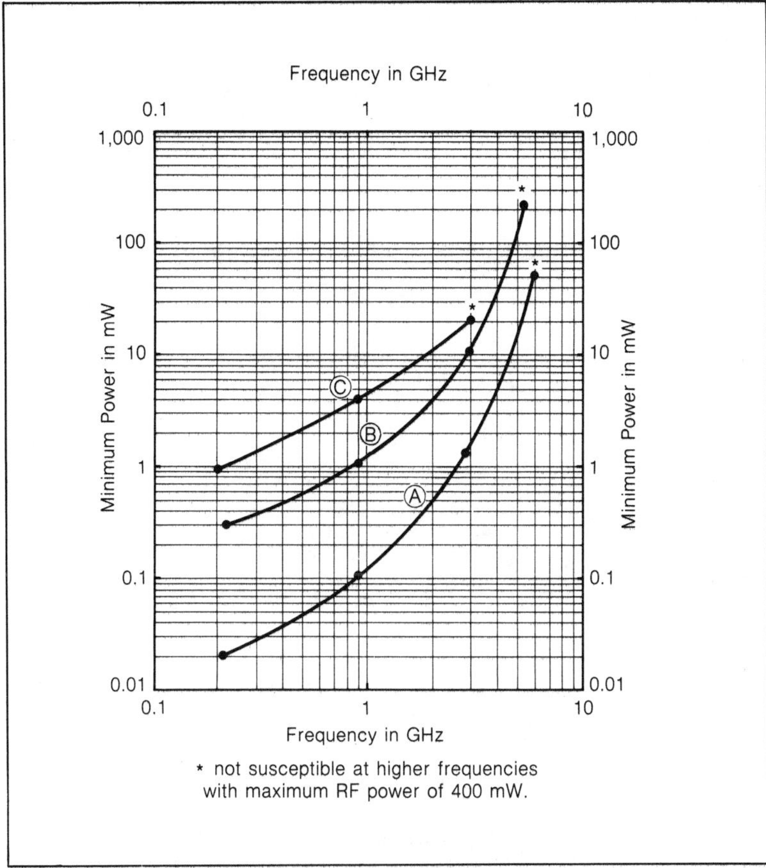

Figure 4.59—Worst-Case Susceptibility Values for Comparators, Showing the RF Power Needed to Cause a Change in Threshold Point of 50 mV (A) 200 mV (B) and 500 mV (C)

4.3.4   Other Typical Analog Devices

Isolation Amplifiers

Isolation amplifiers have been designed to provide a high degree of isolation between the input and the output. For instance, an ordinary op amp cannot tolerate a CM input voltage greater than its supply voltage. Isolation amplifiers, on the other hand, can withstand several thousand volts (up to 8 kV are obtainable) between their input and output without malfunction. Isolation amplifiers are made using two techniques: transformer coupling and optical coupling.

The transformer coupling technique uses a miniature transformer to separate the front-end amplifier from the output section as shown in Fig. 4.60. Since the input signal generally needs to be amplified

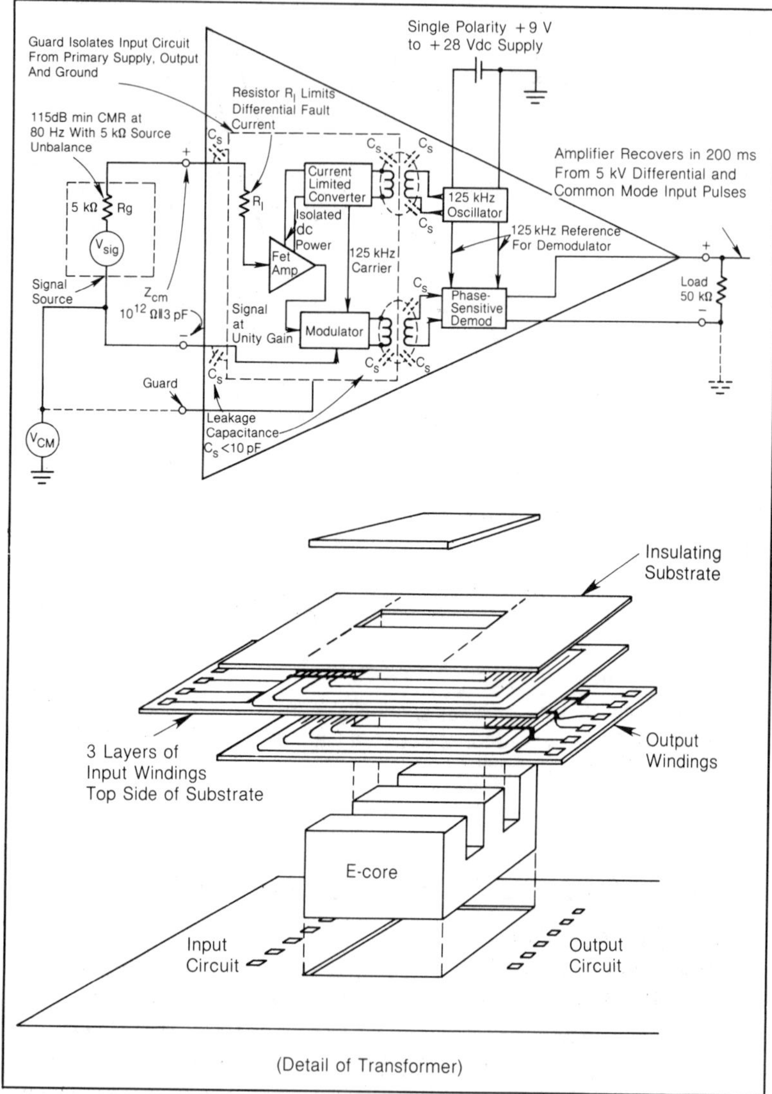

Figure 4.60—Carrier-Type Isolation Amplifier with Miniature Shielded Transformer (courtesy of Analog Devices)

or conditioned before its processing by the signal transformer, a miniature isolated power supply is also built in to provide power to the front-end FET amplifier without breaking the input-output isolation.

Such amplifiers are used in industrial process control instrumentation where high CM transient voltages may exist, and also in medical equipment where virtually no ground current should flow into the patient's body. The CMRR is equal to or better than the best conventional op amp, i.e., 100 to 120 dB up to 100 Hz. Additionally, another parameter sometimes specified is the **isolation mode** rejection, which is slightly different from the input CMRR. Both concepts are shown on Fig. 4.61.

The optical-coupling technique uses an optical barrier to provide the electrical isolation via an LED-photodetector pair. Normal optocoupled isolation amplifiers do not have a built-in isolated power supply. The input section must be supplied by an isolated power

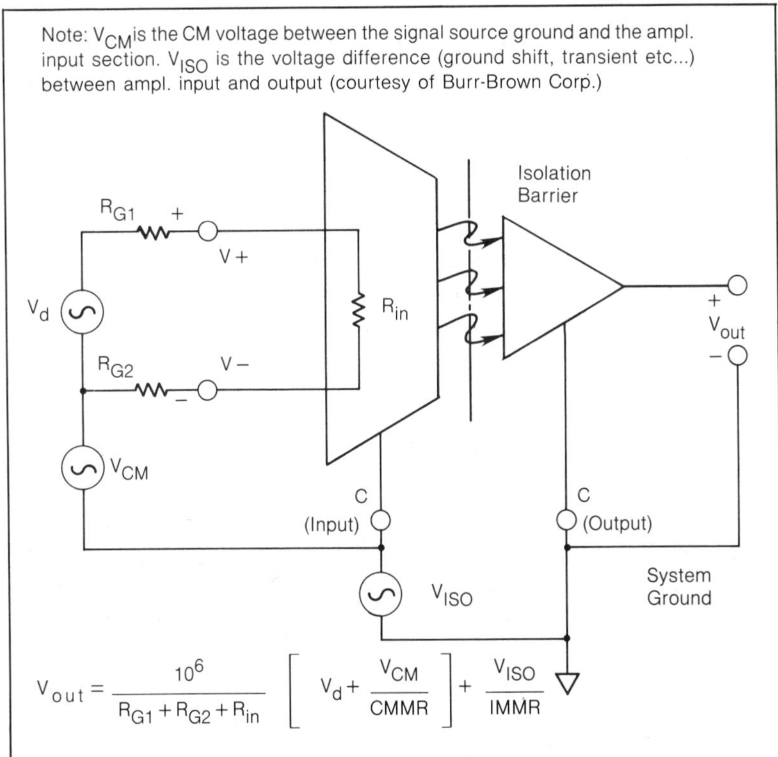

$$V_{out} = \frac{10^6}{R_{G1} + R_{G2} + R_{in}} \left[V_d + \frac{V_{CM}}{CMMR} \right] + \frac{V_{ISO}}{IMMR}$$

Figure 4.61—Optically Coupled Linear Isolation Amplifier

source, otherwise, the benefit of isolation is somewhat compromised. Typical CMRR and **isolation rejection** for such devices are, respectively, 80 and 110 dB up to 100 Hz where they start to roll off at 20 dB/decade. Above a few MHz, ground-loop immunity becomes quasi-null.

Digital-to-Analog Converters (DACs)

A DAC consists basically of an op amp delivering a stepped voltage proportional to the content of the input digital word. For a digital input of "n" bits, the output is made of 2^n steps. For instance, with an 8-bit DAC and a 12 V full output range, the ramp will consist of 256 steps of 46.8 mV each.

A typical EMI symptom would be an output not corresponding to the digital content, shifting up or down by one or several steps. The cause of the shift will invariably reside in the op amp vulnerability to a conducted EMI causing the ground reference to fluctuate (CMR problem), the Vcc to fluctuate, the voltage reference of the weighting network to fluctuate or a strong RF interference (probably of a radiated origin) causing the typical audio rectification or dc offset discussed previously.

Analog-to-Digital-Converters (ADCs)

The ADC poses a more complex EMI problem because it contains both an analog comparator (an op amp operating in the threshold mode) and a digital stage made of a successive approximation register (SAR), clock, enable/disable circuits, output buffers and a parallel-series converter. Simpler ADC ("flash") converters are made of a set of comparators (one per bit, from the least significant bit, LSB, to the most significant, MSB) feeding a digital encoder.

In any case, the merging of a linear amplifier and digital switches in the same package is a sure recipe for EMI troubles if the layout is not done properly (as will be discussed in Chapter 8, "PCBs"). A typical symptom of EMI will be a digital output word not corresponding to the linear input. To avoid the digital section jamming the analog stage, some precautions must be taken. For example, the analog common should be distinct from the digital common to avoid transient currents from digital switching to flow into the analog section causing, for instance, ground shifts. Additionally, the supply pins must be well decoupled, with the same precautions as for a simple op amp.

Voltage-to-Frequency Converters (VFCs)

A VFC is a form of ADC that produces a serial output of pulses of constant amplitude, with a frequency proportional to the input amplitude. This process is particularly well adapted to long-range transmission of analog variables via cables, fiber optics and radio. Here again, the heart of the device is an op amp configured as a comparator, coupled with a monostable multivibrator and a driver. The precautions are similar to the conventional ADC: segregation of analog supply and ground from the digital supply and ground.

Line Drivers and Receivers

Although they are based on analog amplifiers, line drivers and receivers belong generally to the digital world because their applications are now used widely for logic interfacing. In that respect, they include pulse-regenerating, hysteresis enable/disable circuits which make them more relevant to the "threshold" or two-state category of components. The line drivers and receivers will be covered further in Section 4.4.

4.3.5 Emission Aspects of Analog Devices

Unless overdriven or used outside their normal dynamic range (bandwidth and slew rate), analog circuits will have no radiated or conducted emission of a magnitude capable of disturbing other equipment or exceeding EMI emissions limits. There are, however, a few circuits which can possibly generate nonintended emissions large enough to become an EMI problem. They are found primarily in transmitter circuits, power amplifiers and motor control circuits and will be addressed in the corresponding chapters.

4.3.6 EMI Protection Measures in Analog Devices

Following is a set of protection measures which are mostly common to all EMI problems plaguing analog devices.

Bandwidth and Sensitivity Optimization

To minimize the EMI effect on analog inputs, the bandwidth should be reduced to what is just necessary to process the desired signal without objectionable distortion. One should first compute the "bandwidth of interest," f_o, of the useful signal to be process-ed and apply all the passive and active input filtering techniques to narrow the passband of the device down to f_o. Too often, op amps are seen with a 300 kHz bandwidth, used to amplify quasi-dc or audio type signals. The unused portion of this bandwidth is an open door to every possible EMI. With f_o determined, it is possible that the gain setting of the amplifier, directed by considera-tions other than EMI, results in an effective bandwidth larger than f_o.

For example, assume an amplifier with a unity-gain bandwidth of 10 MHz. If a 20 dB voltage gain is desired, a proper resistor network has been devised in the feedback circuit, and the effec-tive bandwidth is now 1 MHz. Assume that the shortest rise time of the signals to be processed is 3 μs corresponding to a 3 dB bandwidth:

$$f = \frac{1}{\pi tr} \cong 100 \text{ kHz} \qquad (4.53)$$

It is clear that our existing bandwidth is 10 times larger than necessary. If all other parameters must remain unchanged, a decoupling must be used. This will be addressed in forthcoming paragraphs.

The matter of sensitivity is interactive with gain, bandwidth and input resistance so that it is almost impossible to discuss sensitivi-ty alone. However, if the sensitivity criterion is set at S = N, a device should be selected with the highest level N compatible with the minimum signal which is expected, while preserving a certain design objective of signal/noise ratio. For instance, if the lowest statistical value of the wanted signal is 1 mV, and a S/N ratio of 20 dB is considered adequate, it is sufficient to have an input sen-sitivity N equal to 100 μV.

A lower sensitivity threshold will not bring any performance im-provement but will certainly invite more EMI problems. Since sen-sitivity is related to input resistance (see RS criteria in Section 4.1), one might deduce that high resistance inputs are preferable. This deduction is not true, as will be explained next.

Input Resistance Optimization
The choice of input resistance, notwithstanding what technology

and industry have available, has to satisfy the following constraints:
1. Highest value of threshold N (calls for large R)
2. Lowest device input resistance to decrease induced EMI voltages (calls for small R)
3. Optimization of the signal source-to-amplifier power transfer (calls for a unique value of R)

Hopefully, these three apparently contradictory requirements are convergent. Section 4.1 has shown that "optimal source resistance" is better achieved when resistances of both signal source and amplifier input are low. Lowering the input resistance will cause less voltage and power to appear at the input port of the victim device for a given EMI current. The N level will thus decrease, which seems to make the device more sensitive, but the fact is that N decreases as $\sqrt{R}$ while EMI immunity will improve as R, an overall benefit evolving as $R/\sqrt{R}$, that is, $\sqrt{R}$. Even with very low level instrumentation amplifiers where input impedances in the order of megohms are usual, the concept of the lowest possible input resistance should be considered.

Decoupling

One of the first applications of decoupling is direct decoupling of a signal input. In the previous example, the analog bandwidth had to be brought down from 1 MHz to 100 kHz. If the input resistance and signal source resistance in parallel represent a total impedance of 10 kΩ, it is necessary to RF decouple the amplifier input so that as $RC\omega = 1$ for 100 kHz. Therefore, a capacitor,

$$C = \frac{1}{2\pi \times 10^4 \ \Omega \times 10^5 \ \text{Hz}} = 150 \ \text{pF} \qquad (4.54)$$

can be put across the input.

An additional improvement can be obtained by RF decoupling the feedback circuit as well so that an additional pole is introduced in the rejection slope. Goedbloed (Ref. 10) has shown, in this case, that the decoupling capacitors in Fig. 4.62 must be such as $R_1C_1 = R_2C_2$ so that the feedback time constant and the input time constant are compatible. (See also improvement in Fig. 4.55.)

If EMI is expected at frequencies greater than 10 times f_o, much care must be taken in the selection and physical layout of these capacitors to avoid parasitic LC resonances at VHF or UHF frequencies and the resulting deterioration of the out-of-band rejection (see Sections 4.1.3.1 and 4.3.2). Beyond about 50 MHz, the only capacitors which can perform adequately are feedthrough, chip or leadless type capacitors.

The next front where decoupling has to be used to combat EMI is power supply input and returns. We have seen in Fig. 4.39 how the power supply rejection ratio (PSRR) can be improved with an analog device. With linear amplifiers having a power driver at the output, the designer must be aware of the Vcc drop caused by transient current demand. *In any driver, the current demanded by the load is drawn from the driver's power supply pin.* The corresponding Vcc drop can cause bias variations of the input stage if not decoupled. This is assimilated to a reaction of the output onto the input (parasitic feedback) which may go as far as oscillation. The solutions in this case are to provide capacitive decoupling right at the analog V_{cc} supply pin, provide a separate supply pin in the analog amplifier module which will correspond to the output driver stage

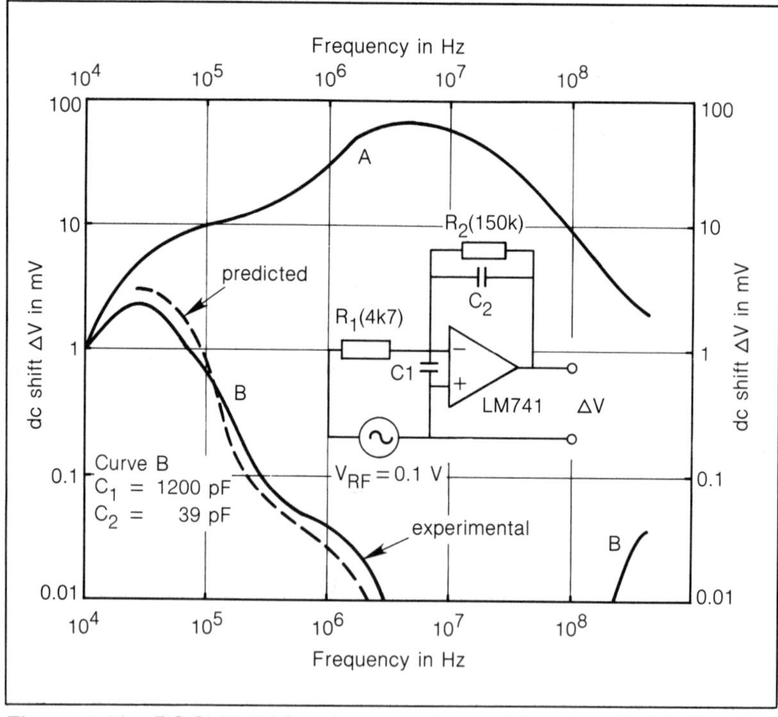

Figure 4.62—DC Shift ΔV Created by an Unmodulated 0.1 V EMI Signal without Decoupling (Curve A) and with Decoupling Capacitors as Shown (Curve B)

only and design a single point ground in the circuit where the device will be used (see Chapter 8, "PCBs").

Interstage Decoupling

For high-current stages such as a power-output stage shown in Fig. 4.63a, the collector current should be supplied from a low-impedance source. Signal currents that flow through Z_c are supplied by the charge on decoupling capacitor C_d and power supply V_{cc} in series with Z_d. If signal current flows through the source of V_{cc}, there will be a voltage drop in both the source and in the interconnecting wiring. This voltage will then affect all other circuits connected to that supply, and objectionable interaction may result. In addition, the current through the wiring may induce voltage in other wires. Therefore, a low impedance path, C_d, is provided. A series impedance Z_d is provided to effect an HF attenuation on the source side. The value of Z_d and C_d must be such that the dc or low-frequency content of I_c is provided by V_{cc}, but the high-frequency contents are supplied by C_d.

The signal current that flows through Z_L is the desired output. This current should also be returned to the emitter with a minimum of disturbance to other circuits.

If the emitter and the load are both connected to the chassis or PCB ground, the return current can provide a voltage drop in the ground bus impedance that might interfere with other circuits. A loop is also created that can act as a transformer to couple into other loops. For low-frequency circuits, where the signal system is grounded at only one point, the best return path is a twisted wire with the outgoing wire to Z_L.

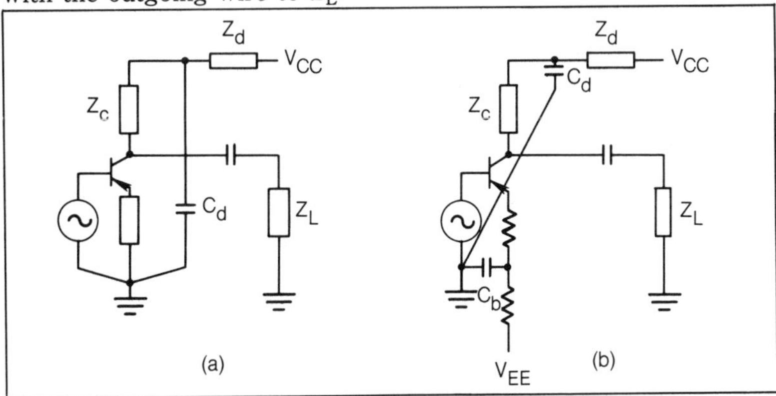

Figure 4.63—Output Stage Decoupling

At high frequencies, the capacitance of the lead connected to Z_L will cause currents to flow to the chassis along its length and to adjacent wires. This current must also return to the emitter. The solution is to provide a shield around the cable to establish a controlled capacitance. The return current for the cable capacitance then can be sent back to the emitter by connecting the shield to the emitter. The return current for the load can also be passed through the shield, particularly if either the load or the emitter can be floating.

If both emitter and load are grounded, a small percentage of the return current will pass through interchassis grounds. Twisted pairs inside shields and transformer coupling decreases this coupling problem. Figure 4.64a shows the emitter bypass method used with two-supply biasing. With the connection shown in Fig. 4.64b, the signal current flows through the emitter side of C_d so that the collector current does not pass through the emitter bypass capacitor. That connection allows disturbances on the supplied power to be coupled into the base emitter signal loop.

Interference involving tuned-output currents is minimized by connecting the resonating capacitance across the tuning coil rather than between collector and ground as shown in Fig. 4.64a. With the capacitor across the coil, current through the decoupling capacitor at resonance is $e_o/Q\omega L$, where Q is that of the tank circuit, ωL is the impedance of one arm of the tank at resonance and e_o is the output voltage. If the capacitor is connected to ground as shown in Fig. 4.64b, C_d becomes part of the series resonant circuit, and current through it is e_o/L, which is higher by a factor Q, approximating 100. Thus, for easier decoupling, the former con-

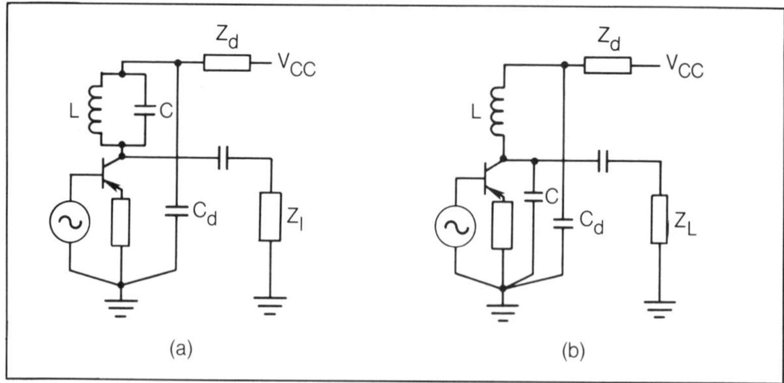

Figure 4.64—Tuned-Output Stage Decoupling

nection is preferred. In many cases, most of the current passes through the distributed capacitance to the chassis and for which C_d must handle the full tank current.

If the amplifier has signals at a frequency such that $\omega^2 = 1/L(C + C_d)$, then series resonances of Z, C and C_d as a pi-network can give an objectionable amount of voltage across the decoupling capacitor. The amount of voltage depends on Z_L and Z_d. To approximate the effect, consider the impedance of both to be infinite, then the current through C_d is the short-circuit output current of the transistor. Finite values of Z_L and Z_d reduce this current appreciably.

Emitter-follower stages should be provided with a collector decoupling network to return the collector signal current to the emitter without flowing through the collector supply. Figure 4.65 shows the collector bypassed to the emitter ground. However, if $Z_L \ll R_g$, which is a typical case when a separate emitter supply is used, the current through the chassis can be reduced by returning C_d to the point at which the load current is returned to the chassis.

Interstage coupling of a pair of transistors is shown in Fig. 4.66a. The subsequent transistor is represented by its input impedance, Z_i, and its base biasing resistors by R_1 and R_2. The function of this stage is to amplify the input signal represented by e_g and supply maximum current in Z_i and minimum current to the impedances in common with other circuits. At the same time, disturbances on the supplies or in the chassis impedances should supply minimum current to Z_i.

The emitter is shown bypassed to the input signal ground to return the base current signal directly to the driving source without going through the chassis impedance. The ground point of C_d has

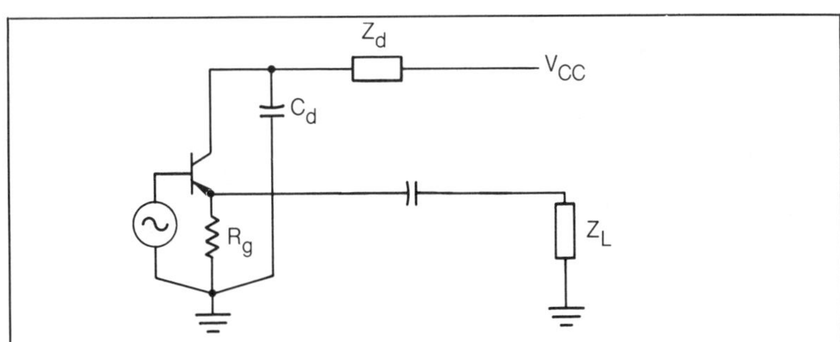

Figure 4.65—Emitter-Follower Decoupling

conflicting requirements. In the connection shown, all of the transistor current flows through the chassis so that there may be coupling to other stages. If C_d is connected back to the first-stage ground, this chassis current is reduced by the amount of current that flows through Z_c and R_1. If a current exists in the chassis due to some other source, however, the configuration of Fig. 4.65a minimizes the amplification of this undesired current by the following stages. This event is demonstrated in equivalent circuits shown in Fig. 4.66b and 4.66c.

The transistor has been replaced by its output impedance, r_o, and e_g and z_g represent the interference source causing the chassis currents. Any current that passes through z_i must pass through r_o, which is usually a high impedance. However, in Fig. 4.66c, which represents the circuit with C_d returned to the emitter ground, the current flows through R_1 and Z_c which are usually much lower impedances than r_o.

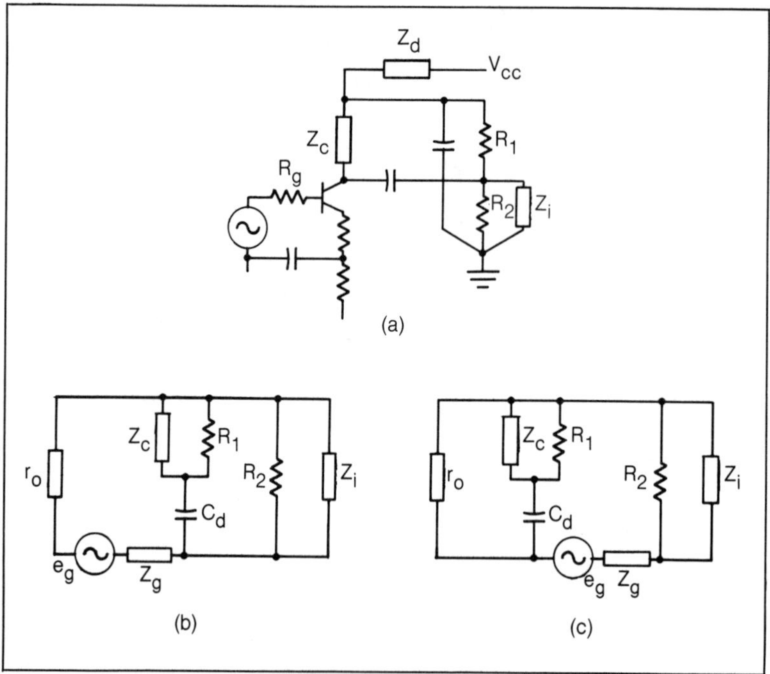

Figure 4.66—Interstage Decoupling

Shielding and Guardshields

Enclosing analog devices in a small shielded compartment can accomplish two things:

1. Prevent stray capacitance coupling between nearby noise source and amplifier input. The shield should be connected in such a way that it is referred to the amplifier common and that *no EMI current drained by the shield can flow in an active wire of the analog device.*

2. Protect the analog circuit from ambient electromagnetic fields, provided that shield openings at the wire leads' penetration are of minimum size and preferably equipped with feed-through capacitors. Integrated analog devices enclosed in metal can have a definite advantage since copper or aluminum can act as a shield.

Guardshields are used around low-level amplifiers intended to work in noisy installations where long hauls will cause common-mode voltages between the signal origin (sensor or measurement point) and the amplifier input. The guardshield is tied to the amplifier common. This point also is kept at the same potential as the source reference by connecting both of them by a cable shield. Since no current should float into this shield, the guardshield is **not** connected to the equipment chassis or main shield. The latter, in turn, can be earthed for safety without creating a ground loop, at least at low frequency. Different aspects of op amp shielding are shown in Fig. 4.67.

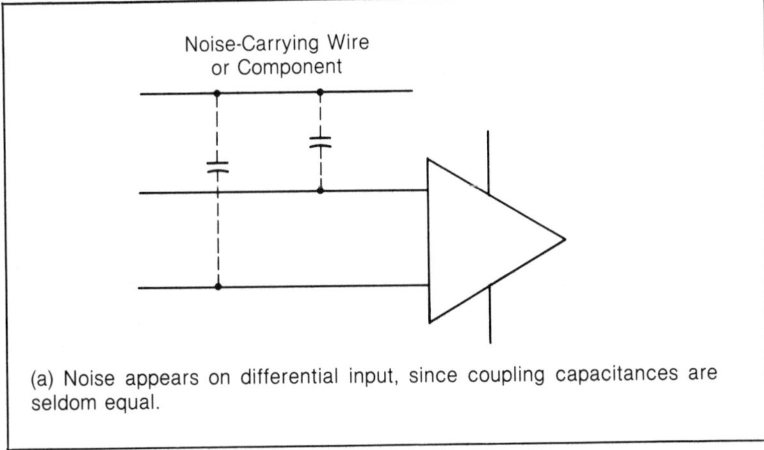

(a) Noise appears on differential input, since coupling capacitances are seldom equal.

Figure 4.67—Shielded Analog Amplifiers (continued next two pages)

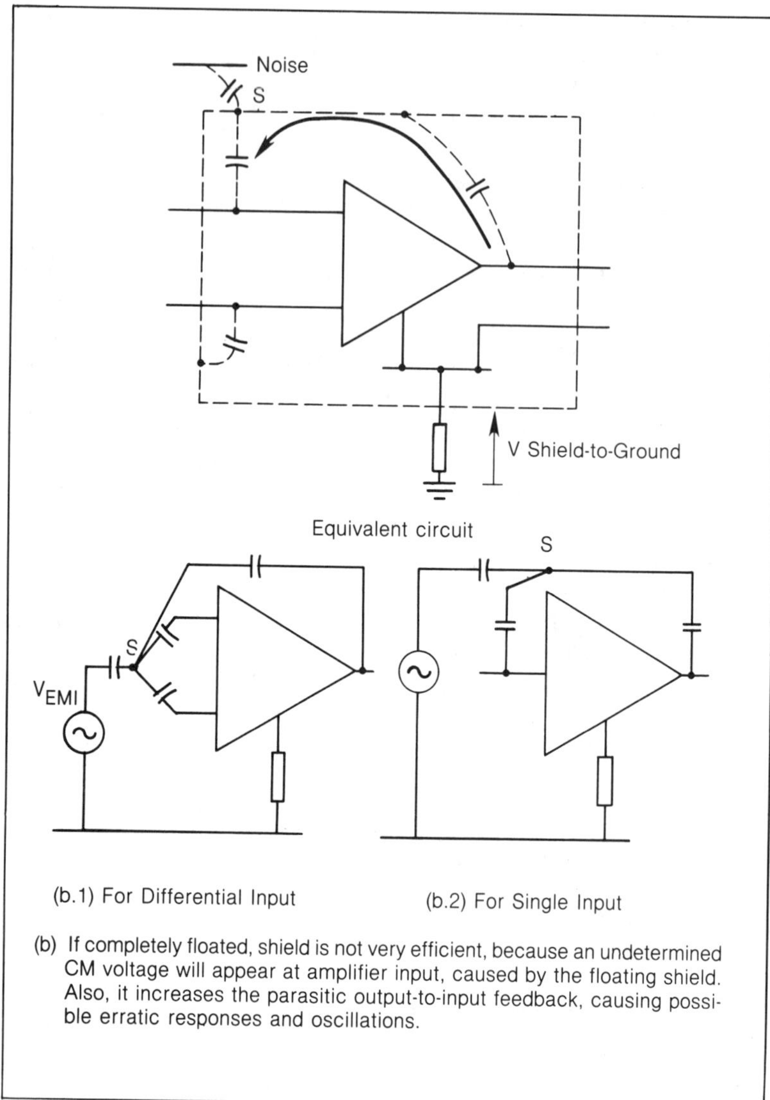

(b) If completely floated, shield is not very efficient, because an undetermined CM voltage will appear at amplifier input, caused by the floating shield. Also, it increases the parasitic output-to-input feedback, causing possible erratic responses and oscillations.

Figure 4.67—(continued)

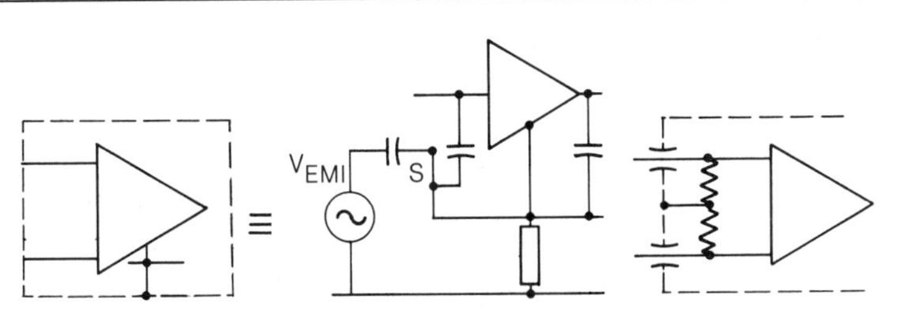

(c) With the shield tied to amplifier common, the device is now surrounded by a Faraday cage referred to its common, or to a virtual ground for floated amplifiers.

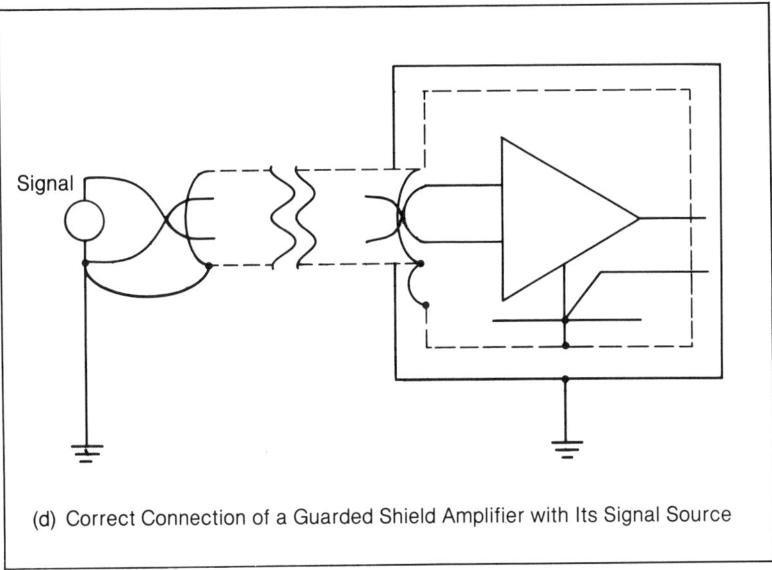

(d) Correct Connection of a Guarded Shield Amplifier with Its Signal Source

Figure 4.67—(continued)

Balanced vs. Unbalanced Input

A symmetrical (or balanced) input will always provide better immunity to common-mode EMI, provided that a decent percentage of balance can be achieved. A one percent balance over the whole useful bandwidth will provide at least 40 dB of CMR up to frequency f_o. Beyond that, the parasitic capacitance to ground of each input may spoil the CMR because they are generally unequal. Several solutions exist to balance an otherwise unbalanced (or poorly balanced) analog amplifier. Two of them are shown in Fig. 4.68.

1. Adding **precision** series resistors to a differential input. This scheme can improve the total symmetry of the signal input. Provided that R_1 and $R_2 > Z_1$ and Z_2, the percentage balance will be imposed by the precision of R_1, R_2. A serious drawback is a loss of sensitivity of the input, proportional to Z/R.

2. Adding a balanced-unbalanced transformer (BALUN) to the input.

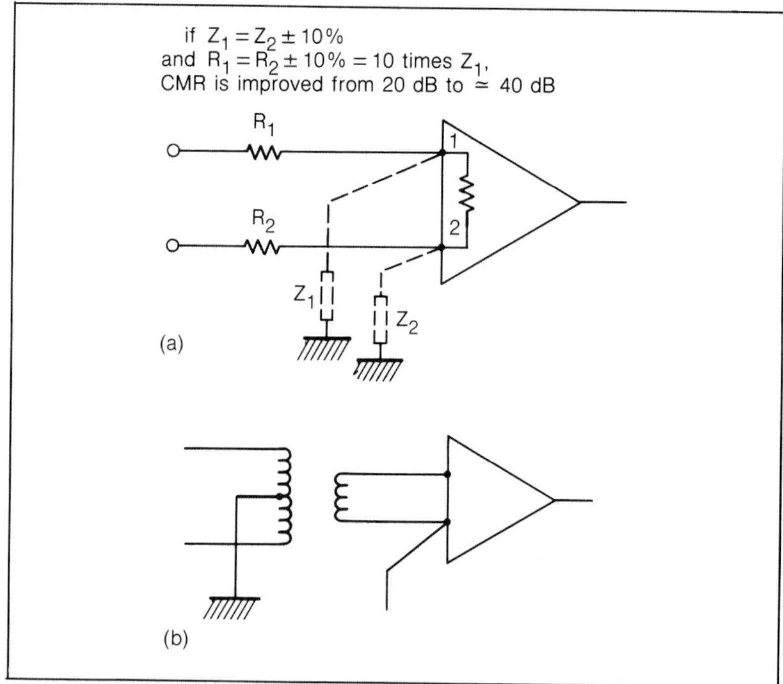

Figure 4.68—Balancing Analog Devices. (a) The CMR of a balanced input can be improved by adding series resistances. (b) An unbalanced input can interface a balanced line via a Balun transformer

Mixing Analog with Digital or Power Components

Proximity, common supply and ground paths of analog and logic should be avoided by all means. However, this is easier said than done. Many circuits require a matching of an analog and a digital device. The rules in this case are to decouple the digital device Vcc to ground so that no pulsed current from the logic flows into the analog section (including analog ground), segregate digital and analog grounds, (even in the same hybrid or monolithic IC) and decouple the analog device as well.

Prevention of Audio Rectification and Other Out-of-Band EMI

As seen previously, audio rectification typically occurs in the VHF-UHF range. Unfortunately, that is where conventional decoupling components show physical problems. To prevent a high level of RF (high level being levels above 50 or 100 mV, which start to drive pn junctions into a detection mode), specific measures can be taken. Feed-through capacitors can be used at the input, output and power supply leads. Lossy ferrite beads, acting as dissipative and low-Q inductors especially efficient above 30 MHz on power and output leads, can be added. As an alternative, ferrite-loaded sockets and substrates are now available (Fig. 4.69). The devices can be filled with lossy, absorptive material.

Galvanic Effects and Dissimilar Metals

The parasitic voltages caused by corroded contacts IR drop or dissimilar metal junctions are well known in regard to bonding and grounding problems. But they might cause analog amplifier malfunction as well. Corroded or dissimilar metal contacts are nonlinear junctions which can produce RF rectification just like a p-n junction and cause dc or modulated noise at the input of analog circuits.

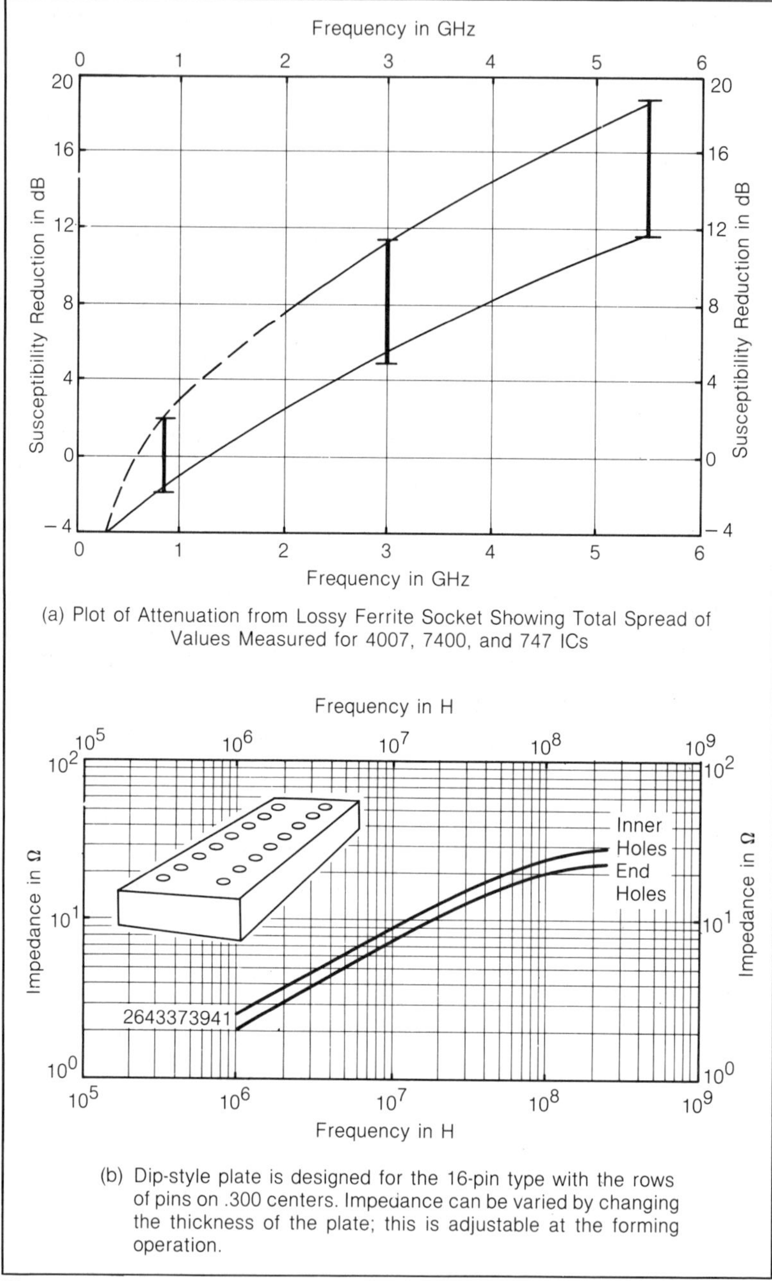

(a) Plot of Attenuation from Lossy Ferrite Socket Showing Total Spread of Values Measured for 4007, 7400, and 747 ICs

(b) Dip-style plate is designed for the 16-pin type with the rows of pins on .300 centers. Impedance can be varied by changing the thickness of the plate; this is adjustable at the forming operation.

Figure 4.69—Two Examples of Lossy Ferrite Sockets Made by Emerson & Cumming (a) and Fair-Rite (b)

4.4 Logic Devices

Logic devices exhibit completely different EMI characteristics from analog ones. They are two-state devices, and their sensitivity threshold, typically a few hundred millivolts, is not determined by the thermal noise in some input resistance but by the minimum energy level required to trigger the device from one state to the other. This threshold is translated generally into the "noise margin," ac or dc. The logic is a contrast with analog devices where sensitivities range from a few nanovolts to millivolts. They are basically less dc-sensitive, but their bandwidth is much wider (see criteria in Section 4.1). Their EMI emission is significant and is probably the biggest problem with digital electronics. They exhibit significant spectral contents in the VHF and UHF range. ECL or advance STTL logics can run at clock speeds exceeding 100 MHz, i.e., the No. 10 harmonics will be in the gigahertz range.

4.4.1 Some Logic Families

Most analog devices have relatively narrow bandwidths, i.e., less than 1 MHz. In contrast, the bandwidth of all logics is greater than 1 MHz. The popular TTL logic has a bandwidth of about 30 MHz, for example, and the high-speed logic families have more than 100 MHz. Thus, while logic tends to be more immune to narrowband EMI than analog devices, it may be more susceptible because of its enormous bandwidth occupancy, i.e, it falls within the frequency range of many sources.

Table 4.2 lists typical characteristics of various digital logic families. The bandwidth is shown as $1/\pi\tau_r$, τ_r being the rise time measured between the 0.1 to 0.9 amplitude points. Some switching-state transition currents are listed, and the table shows CMOS current to be low relative to those of the high-speed family. Several of the entries in Table 4.2 will be used in later illustrative examples to compute specific thresholds or immunities.

The speed × power product is a commonly used figure of merit obtained by multiplying the quiescent power dissipation of one gate by the minimum pulse width (transition time or Tpd) required to flip that gate. The lower the picojoule level, the better the figure for low power consumption and low EMI generation, but the more vulnerable the device. Low-power technologies with slow speed and speedy technologies with excessively high consumption will both show a high speed-power factor.

Table 4.2—Typical Characteristics of Various Logic Families.
All parameters are for one single gate.

Logic Families	Typical Output Voltage Swing (V)	Rise/Fall Time (ns)	Bandwidth (MHz) $1/\pi\pi$	Permissible Max Vcc Voltage Drop (V)	No load Power Supply Transition Current (mA)	Input Capacitance (pF)	Input Current (1) for 1 Gate Transient/Steady	Speed ×Power pJ (3)	DC (2) Noise Margin mV	Noise (4) Immunity Grade %
Emitter Coupled Logic (ECL-10k)	0.8	2/2	160	0.2	1	3	1.2/1.2 mA	50	100	12%
Emitter Coupled Logic (ECL-100k)	0.8	0.75	420	0.2		3	3/0.5 mA		100	12%
Transistor-Transistor Logic (TTL)	3.4	10	32	0.5	16	5	1.8/1.5 mA	100-150	400	12%
Low Power TTL (LP-TTL)	3.5	20/10	21		8	5	1/0.3 mA	35	400	12%
Schottky TTL Logic (STTL)	3.4	3/2.5	120	0.5	30	4	5/4 mA	60	300	9%
Low-Power Schottky (LS-TTL)	3.4	10/6	40	0.25	10	6	2/0.6 mA	20	300	9%
Advance Schottky (AS)	3.4	2	160	0.5	40	4	7/1 mA	15	300	9%
Advance Low Power Schottkty (ALS)	3.4	4	80	0.5	10	5	4.3/0.3 mA	5	300	9%
Fast (Fairchild)	3	2	160	0.5		5	8/0.5 mA	10	300	10%
Complementary Metal Oxide Logic (CMOS) 5 V or (15 V)	5 (15)	90/100 (50)	3 (6)		1 (10)	5	0.2/ <10 μ A	5-50 (0.1 to 1 MHz)	1 V (4.5 V)	20% (30%)
High Speed CMOS (5 V)	5	10	32	2	10	5	2.5/ < 10 μA	10-150 (1 to 10 MHz)	1 V	20%
GA AS (1.2v)	1	0.1	3,000			$\cong 1$	10/μA/	0.1-1	100	10%
MODFET (5)	0.6 - 0.8	0.03	10,000	0.1		0.6	12-16 μA/	0.03	100	14%

Notes:
1) Current that the driving device has to feed into each driven gate, in the worst case input status
2) dc Noise Margin = difference between V_{out} of driving gate and V_{in} required by driven gate to recognize a "1" or "0".
3) For some technologies, like CMOS and HCMOS, the speed-power product depends on the clock frequency, due to the mostly capacitive nature of their input impedance.
4) Expressed as noise Margin/voltage swing — the larger the %, the better the relative immunity.
5) New Technology, still in development stage

4.4.2 Noise Immunity

The noise margin is defined in Fig. 4.70. It is the difference between the true noise immunity of a driven device (the worst-case level above which a low input will no longer be recognized as a "zero," or the level below which a high input will not be recognized as a "one") and the V_{OL} or V_{OH} of the driving device. From Fig. 4.69:

V_{NSH} (negative noise over a high level)

$$= V_{OH\ min} - V_{inH\ min}$$

V_{NSL} (positive noise over a low level)   (4.55)

$$= V_{in\ Low\ max} - V_{OL\ max}$$

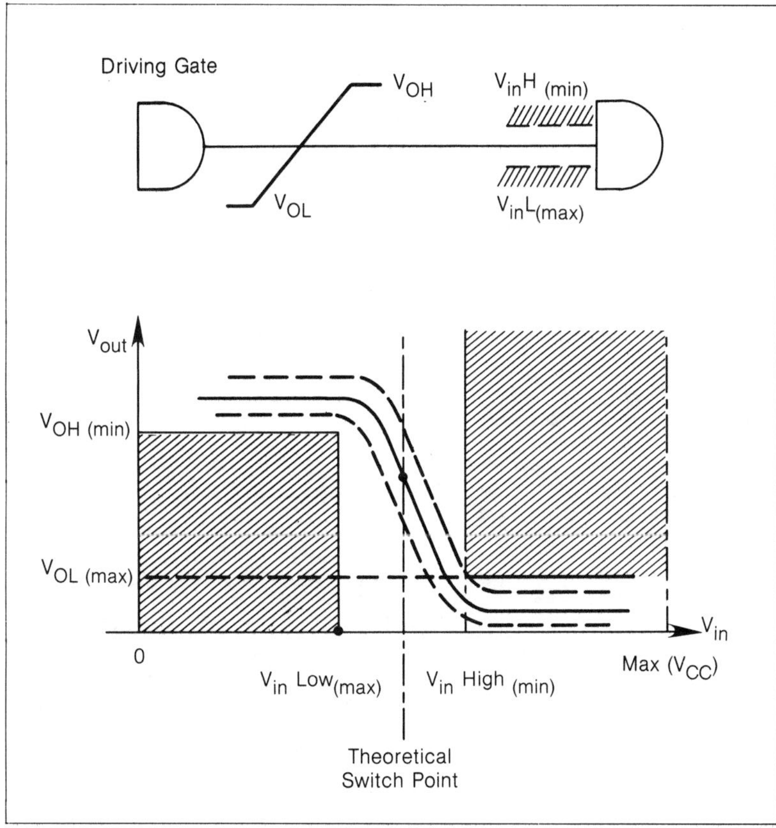

Figure 4.70—Direct Current Noise Immunity of an Inverter Gate

This margin represents the maximum "budget" allowable for noise riding over the input signal. For any input noise exceeding the margin, the device may go into the transition curve and its status is no longer guaranteed (Fig. 4.71).

The dotted line zone around the ideal transfer curve of Fig. 4.70 corresponds to temperature and V_{cc} supply drifts. Noise margin also depends on the dynamic load during transfer. Therefore, it is generally defined under specific output load, such as 30 or 50 pF.

For instance, the theoretical switch point for standard TTL is about half the voltage swing of 3 V; i.e., 1.5 V. However, the true value beyond which a low input will not be recognized is about 0.8 V worst-case (generally taken at 85°C or 125°C junction temperature). In addition, the signal delivering this logical "0" level may drift up to 0.4 V above the ground reference. This V_{OL} max

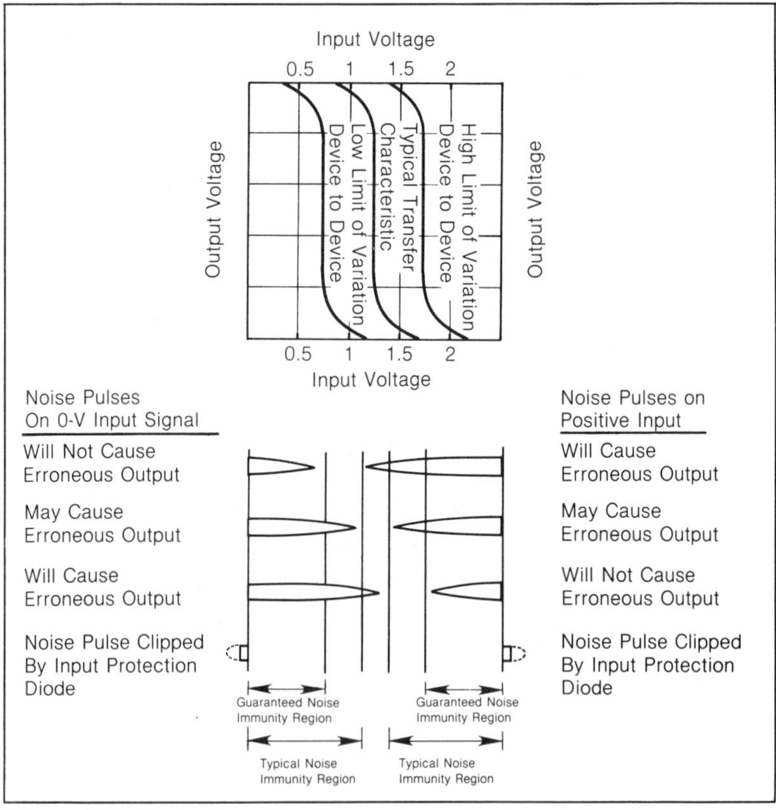

Figure 4.71—Illustration of Positive and Negative Noise Immunity (Inverter Gate) (courtesy of RCA Corp.)

is caused by the V_{CE} saturation of the output transistor in the driving device. Consequently, the margin left for noise is:

$$0.8 \text{ V} - 0.4 \text{ V} = 0.4 \text{ V (worst-case)} \qquad (4.56)$$

For TTL technology, the gate is the most susceptible to positive noise over a "low" input state. For CMOS, high and low immunities are about symmetrical.

The AC Noise Margin

For noise pulses shorter than the logic switching time, or more exactly, the logic Tpd (see Fig. 4.72 for the subtle difference between the two) noise immunity improves. In other words, to provide the minimum level of picojoules required for switching,

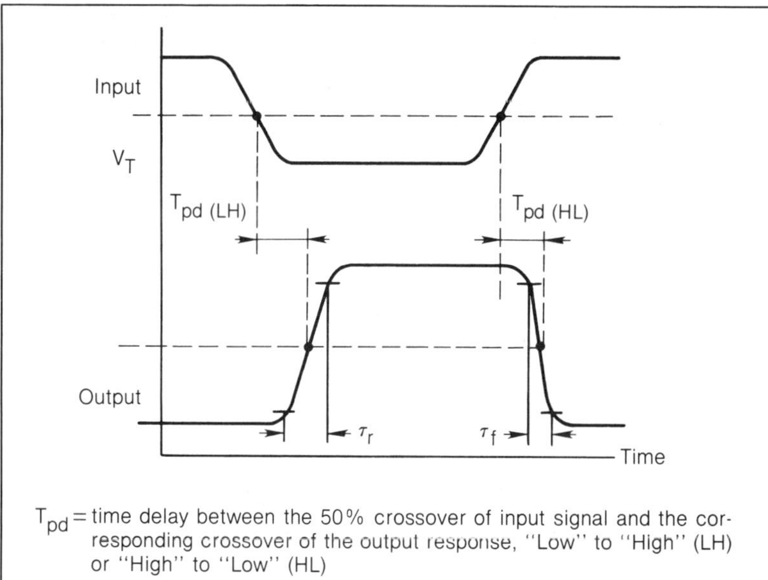

T_{pd} = time delay between the 50% crossover of input signal and the corresponding crossover of the output response, "Low" to "High" (LH) or "High" to "Low" (HL)

τ_r, τ_f = transition rise and fall times

T_{pd} is important for maximum data rate, and noise immunity aspects

τ_r, τ_f are important for crosstalk, power bus pollution and noise emission in general

Figure 4.72—Propagation Delay (T_{pd}) and Transition Time of Logic Devices (Shown for an Inverter Gate)

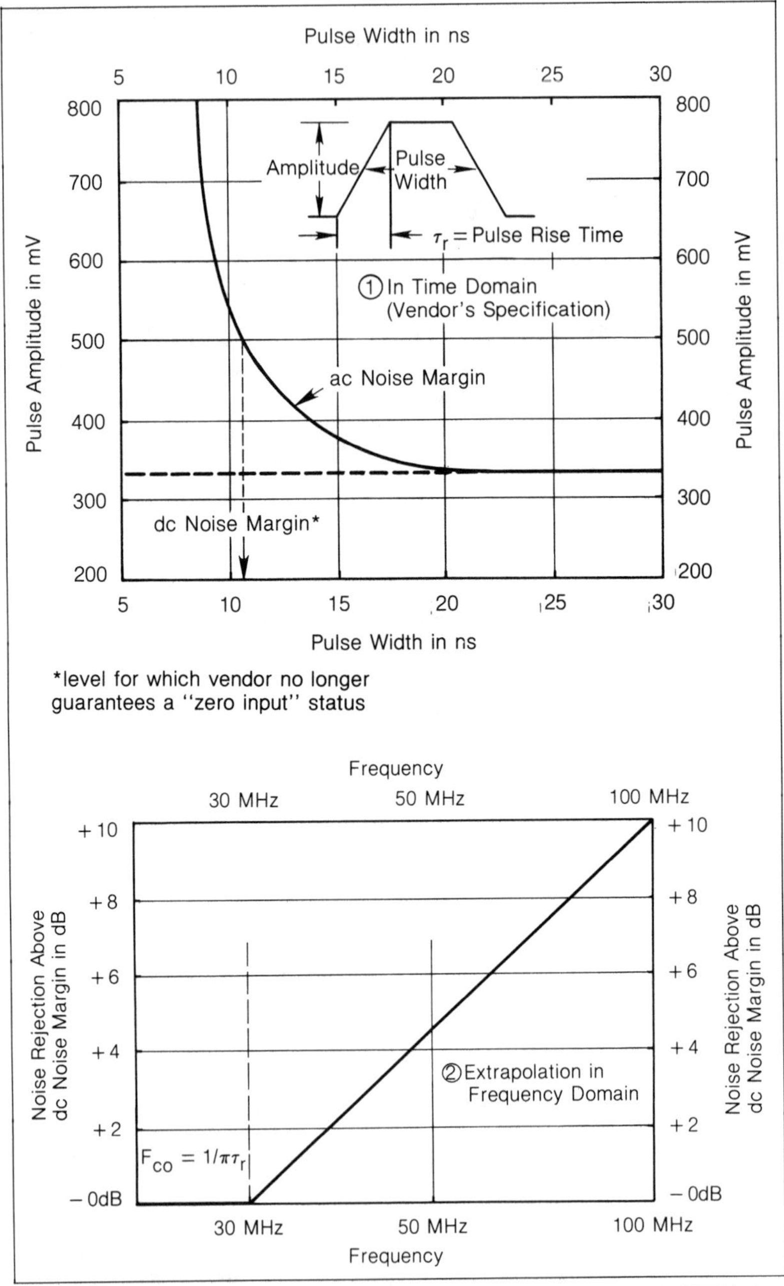

Figure 4.73—Example of Noise Rejection of a Logic Circuit (LS-TTL), in (1) Time and (2) Frequency Domains.

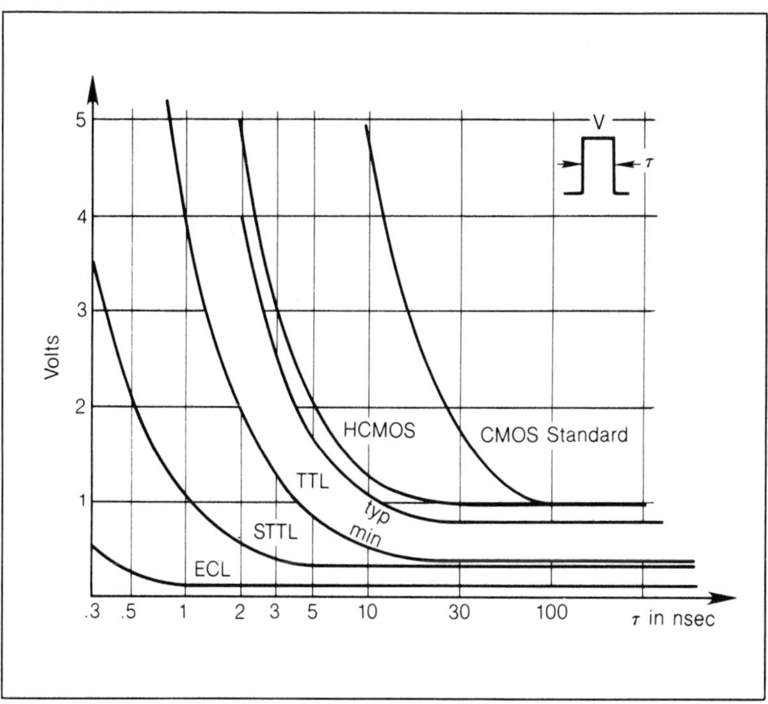

Figure 4.74—Noise Margin of Logic Families versus Pulse Width

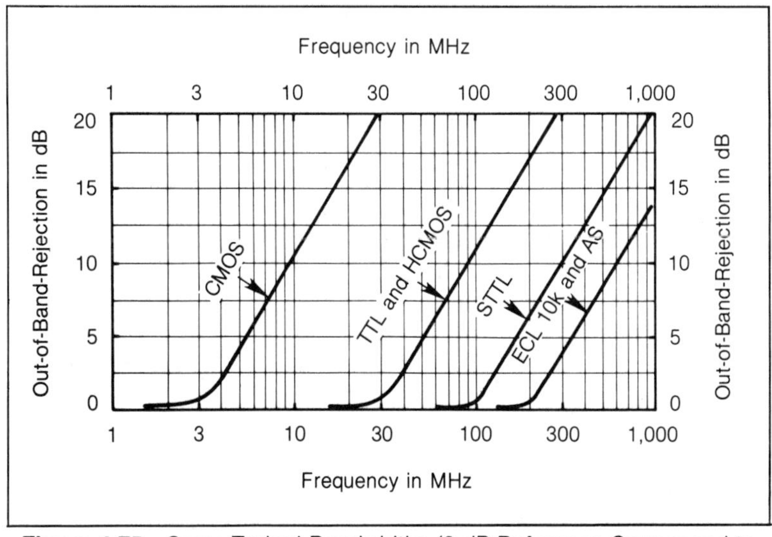

Figure 4.75—Some Typical Bandwidths (0 dB Reference Correspond to the In-Band Sensitivity)

amplitude needs to be greater when pulse duration shortens. The ac noise margin is equal to or greater than the dc noise margin.

Figure 4.73 shows a parallel between the noise rejection of a TTL gate in the time domain and in the frequency domain. For noise glitches shorter than 10 ns, the gate should not be activated; however, it will be activated if the amplitude is sufficiently high, i.e., 1 V for 5 ns pulse duration. The reciprocal of this is shown in the frequency domain where the same TTL gate is pictured as a low-pass filter. Above a certain frequency (which is shown in the column "bandwidth" of Table 4.2), it does reject the noise, but only to a certain extent (see Figs. 4.74 and 4.75).

Another aspect of ultra-short pulses is their possible repetitive nature. Figure 4.72 assumes a single isolated pulse. If instead of a single isolated pulse the noise is a burst of repetitive spikes, a capacitive storage effect may partially store and add up the effect of each pulse if their repetition period is shorter than the device's time constant. Therefore, another parameter is sometimes used to characterize noise immunity (regardless of whether it is ac or dc) of a digital device, as explained next.

Noise-Energy Immunity

Noise-energy immunity takes into account not only the voltage amplitude and width of the noise pulse, but also the impedance that the coupled noise "sees" on a line. A typical noise-energy plot looks like Fig. 4.76 according to the equation:

$$E_{N\ (joules)} = \frac{V_N^2}{R_o} \times \tau \tag{4.57}$$

where, V_N = Noise amplitude in volts to cause circuit switching
$\quad\ \tau$ = Pulse duration in seconds (average or mid-height)
$\quad R_O$ = total impedance across the line (device's input resistance paralleled with the driving device output resistance)

or,

$$E_{N(nJ)} = \frac{V_N^2}{R_o} \times \tau_{ns} \tag{4.58}$$

The shape of the curve shows that E_N reaches a minimum value, corresponding to $\tau \cong 2 \times$ Tpd. The introduction of R_o in the immunity criteria is interesting because, for a same given V_N, the technology with the highest resistance will be penalized since it

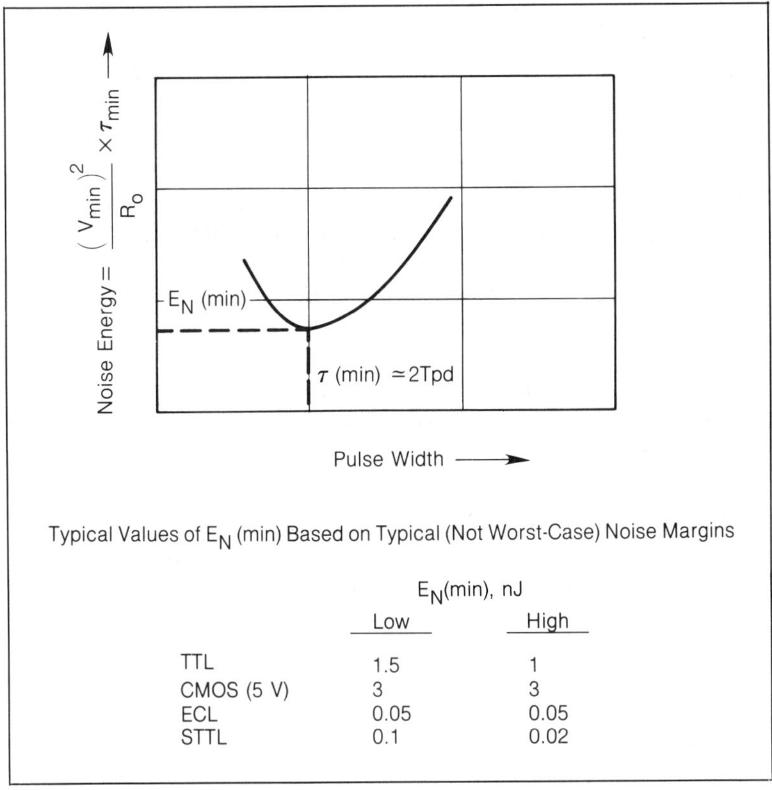

Typical Values of E_N (min) Based on Typical (Not Worst-Case) Noise Margins

	E_N(min), nJ	
	Low	High
TTL	1.5	1
CMOS (5 V)	3	3
ECL	0.05	0.05
STTL	0.1	0.02

Figure 4.76—Typical Noise-Energy Immunity (May Vary from One Manufacturer to Another)

will be more vulnerable. The lower the E_N figure, the more vulnerable the logic.

Immunity to Power Supply and Ground Noise

Although signal inputs are most vulnerable, the effect of EMI superimposed on Vcc and ground has to be taken into account. The immunity to supply voltage shifts is generally of the same order or better than the signal line noise immunity.

This immunity curve follows the same law as the signal line dc versus ac noise margin, i.e., immunity improves for shorter pulses. The measurement of power supply noise immunity requires some precautions (as shown in Fig. 4.77), otherwise, misleading results may occur. The device under test (DUT) should be fed by an input signal at its worst case tolerance, e.g., if the DUT is deemed to stay in a "low" input status, dc input signal equal to $V_{OL\ max}$ (0.4 V with TTL, for instance) will be applied. The noise pulse will be injected from a pulse generator via a coupling capacitor (C) and its amplitude monitored (A). Typically, the low impedance of the power source would prevent the pulse generator from delivering enough amplitude at point A. To resolve this, a decoupling inductance with good HF properties (ferrite) should be used (such that L/τ is much larger than τ/C).

For example, with a 1 nF coupling capacitor, an inductance of at least 1 μH will provide a satisfactory blockage (20 dB attenuation) for any pulse shorter than 10 ns, and at least 90 percent of the current will go to the DUT at node A.

Dynamic Resistances of Logic vs. EMI

Knowing the equivalent input and output resistance of a digital device is important when trying to predict or understand EMI coupling. Figure 4.78 shows that whatever the origin, the final EMI coupling into devices input looks like either (1) a high-impedance, constant-current source across the line (electric coupling) or (2) a low-impedance voltage source in series (ground shift, magnetic coupling, etc.).

In the case of electric coupling, constant-current EMI source sees a parallel combination of Z_s and Z_{in}, or, if the signal comes from

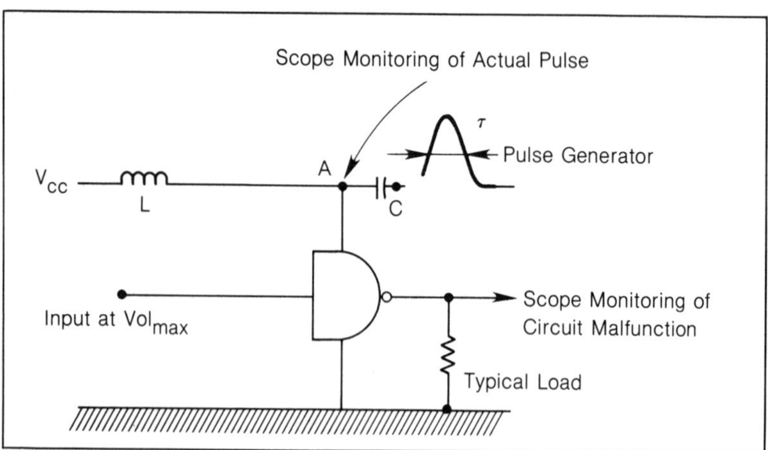

Figure 4.77—Test Setup for Power Supply Noise Immunity of a Logic Device

an electrically long path, the load is approximated by the line characteristic impedance. For example, if the logic source resistance is 30 Ω and the logic input resistance is 3,000 Ω, the load seen by an electric type of EMI coupling (stray capacitance, capacitive crosstalk, etc.) is about 30 Ω, a relatively low coupling.

The situation is reversed in the same conditions if the EMI appears as a series voltage (inductive crosstalk, ground-loop coupling, etc.). The voltage coupled at the device's input will be:

$$V_{in} = V_o \frac{Z_{in}}{Z_{in} + Z_s + Z_w} \qquad (4.59)$$

That is, in the previous situation, 99 percent of the coupled noise is applied at the input.

Another point of interest in understanding logic device impedances is the necessary matching for transmission line conditions. Unfortunately, a logic device has a strongly nonlinear V, I characteristic because of its two-state nature. Its impedance cannot be modeled by one simple resistance. Several macromodels, using piece-wise approximations, have been developed (Ref. 13). On the other hand, for EMI predictions, the following has to be considered:

For input impedances, the fully saturated or fully blocked states do not present much interest. The EMI critical conditions generally begin when a gate, for instance, is near leaving either a "low" or a "high" state.

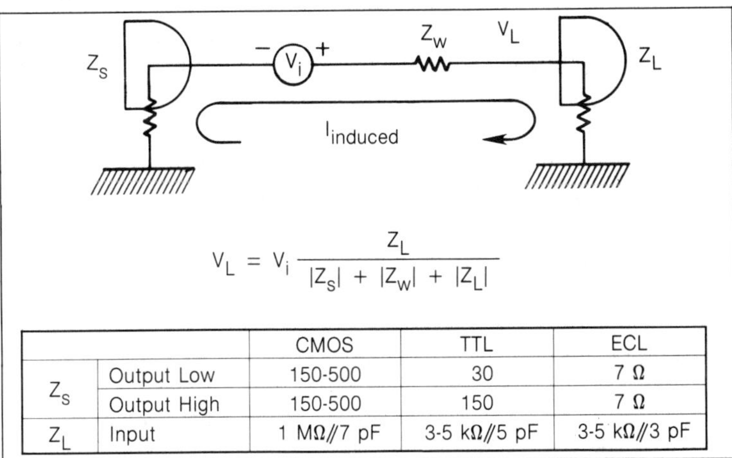

		CMOS	TTL	ECL
Z_s	Output Low	150-500	30	7 Ω
	Output High	150-500	150	7 Ω
Z_L	Input	1 MΩ//7 pF	3-5 kΩ//5 pF	3-5 kΩ//3 pF

Figure 4.78—Influence of Victim Circuit Impedance on Coupled Voltage

So the knowledge of the V_{in}, I_{in} curve during the narrow transition zone of the transfer function allows a dynamic resistance to be approximately defined during this transfer:

$$R_{in(dyn)} = \frac{\Delta V_{in}}{\Delta I_{in}} \qquad (4.60)$$

For output resistance, since the device's output generally is a more or less saturated transistor plus a limiting resistor and a clamping diode, the V_{out}, I_{out} curves can be plotted. The corresponding output resistances can be derived by:

$$R_{out\ (high)} = \frac{\Delta V_{OH}}{\Delta I_{OH}}$$

$$R_{out\ (low)} = \frac{\Delta V_{OL}}{\Delta I_{OL}} \qquad (4.61)$$

Figure 4.79 shows the equivalent resistances R_{in}, $R_{out\ (H)}$ and $R_{out\ (L)}$ for several digital families.

For instance, all digital families exhibit more than 1 kΩ of input resistance in the transition region. This input, however, is also shunted by the device's parasitic input capacitance for very short pulses. Also, from manufacturers data, output resistance in the TTL technology is significantly different in low state (10 to 30 Ω) than in high state (120 to 160 Ω).

Using device resistance and noise margins, noise energy immunity can be recalculated for dynamic conditions where the device is on the edge of its threshold (see Table 4.1).

The Consequence of EMI on Logic Performance

The most obvious consequence of EMI on logic devices is an inadvertent change of a logic state; a noise pulse will be mistaken for a "1" or a "0" and then, once reshaped by the logic block, will propagate in the form of a corrupted digital word. Some more insidious results may also occur as shown in Fig. 4.80, where a ringing noise overriding the true signal is causing a train of false switchings of the gate before it stabilizes.

The case captioned (b) is a typical ringing noise caused by improper termination of the line. This will be discussed further in Chapter 8, "PCBs".

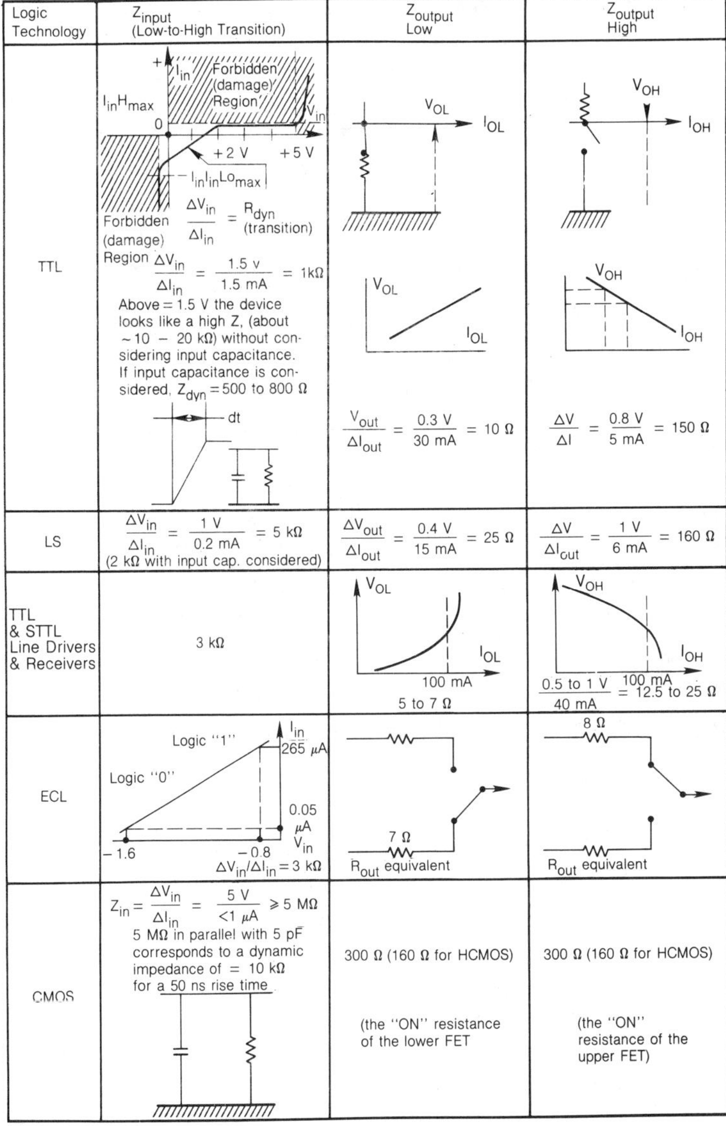

Figure 4.79—Dynamic Resistances of Digital Families

The consequence of such false triggering is that the designer has to slow down the clock speeds and use integrating or hysteresis networks to improve immunity, but at the risk of compromising circuit performance. A hysteresis circuit is shown in Fig. 4.81

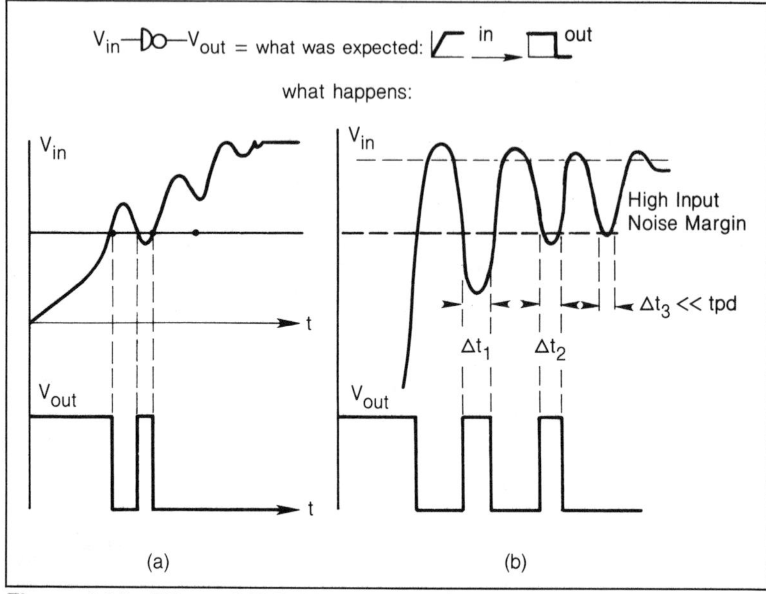

Figure 4.80—Effect of EMI Ripple over a Digital Response

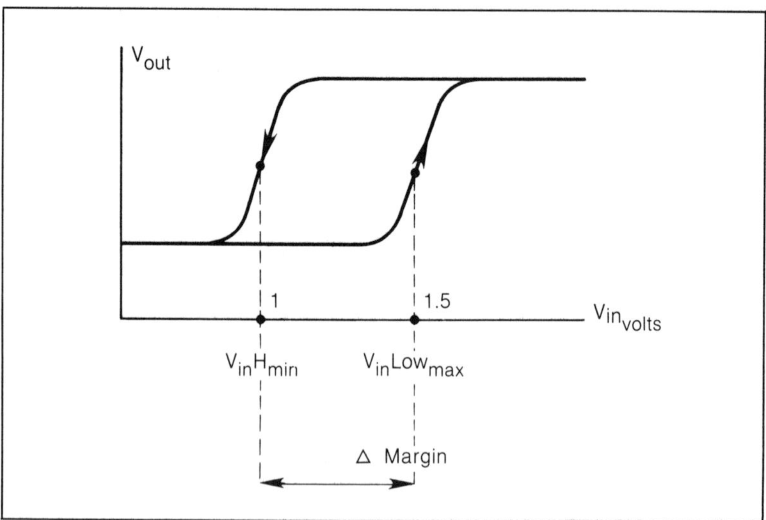

Figure 4.81—Logic Circuit with Hysteresis in its Transfer Characteristic (Trigger)

where the thresholds for an upgoing signal and a downgoing signal are different. The result is an improved noise immunity around the threshold point, but this translates into a sacrifice of the speed; that is, Tpd is increased by a value Δt that can be calculated from the edge speed of the technology:

$$\Delta t = \Delta V \times \tau_r / \text{voltage swing} \qquad (4.62)$$

For instance, if a technology having a 3.5 V swing and 5 ns transition uses an hysteresis circuit with a ΔV of 1 V in the hysteresis curve, the additional delay can be:

$$\Delta t = \frac{1 \times 5}{3.5} = 1.4 \text{ ns} \qquad (4.63)$$

Besides their differences due to technology, the vulnerability of logic circuits depends on:

1. The circuit function. Table 4.2 presents a typical gate. Other circuits (counters, triggers, drivers, decoders, etc.), although working with compatible levels, may exhibit different input impedances.
2. What the circuit is doing when a parasitic pulse appears on its input. For instance, an inverter gate will be more susceptible to a positive spike when its input is in a logical "0," and vice versa.
3. The criticality of the stressed line. The classical example is the "reset" line on a microprocessor chip which, if activated by a noise glitch, will wipe out the ongoing transaction.
4. The maximum "not-to-exceed" levels of the technology. Ultra-short transients above 30 V can permanently overstress FET and MOS chips. Negative transients can bring CMOS devices into a "latch-up" mode, requiring a power-off to return it to normal.

4.4.3 Measured Susceptibility of Logics to Out-of-Band EMI

In the McDonnell-Douglas study already mentioned in Section 4.3.3, the engineers evaluated the amount of UHF and microwave power necessary to upset some typical logic chips. The devices tested were TTL (7400 series) and CMOS (4000 series) gates and line drivers and receivers. Some of the results are shown in Figs. 4.82 through 4.84. For TTL and CMOS, based on NAND gate results, the curves show the amount of power necessary to reach the following degradation criteria:

1. A = EMI causing the output to exceed V_{OL} max (output low) or to go below $V_{OH\ min}$ (output high). This means, that in a given logic link, the available noise margin is reduced and operation becomes risky, although malfunction is not certain.

2. B = EMI causing the output to exceed the nominal threshold point, where the next driven gate will no longer recognize a low level as a "0" (0.8 V for TTL, 1.05 V for CMOS) or a high level as a "1" (2 V for TTL, 3.95 V for CMOS).

3. C = EMI causing the output to enter the zone where a change in the logic state is certain, i.e., a 100 percent error ("stuck-at" condition).

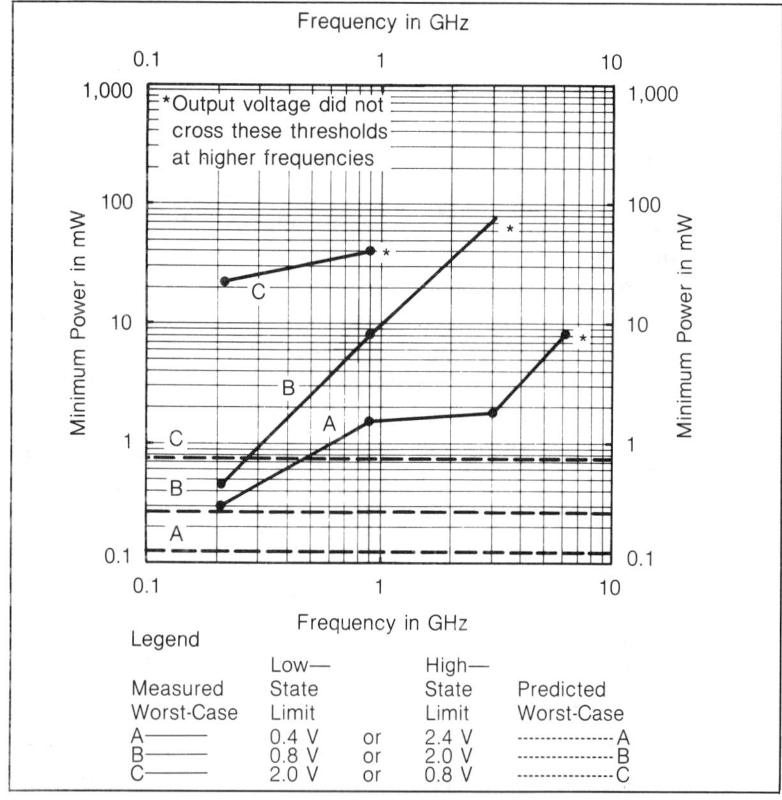

Figure 4.82—Worst-Case Susceptibility Values for TTL Devices

For all the three criteria, CMOS appears to be slightly less susceptible than TTL since it takes, at the worst frequency of 200 MHz, about 3 to 10 times more power to cause its malfunction than with TTL.

For the line drivers (type 9614, 8830, 55109 and 5110) and receivers (9615, 8820 and SN 55107A), the receivers were found to be most susceptible since some of them have a differential input threshold as low as 50 mV. What happens during EMI injection into the line receiver is a rectified voltage bias shift (similar to what occurs in analog comparators, Section 4.3.3). This causes a fluctuation of the detected position of the leading edge of the intentional digital pulse, appearing on the output as a pulse position "jitter." In a typical link this may oblige the designer to use slower data rates to prevent the percentage of jitter from becoming intolerable when the pulse position imprecision approaches the

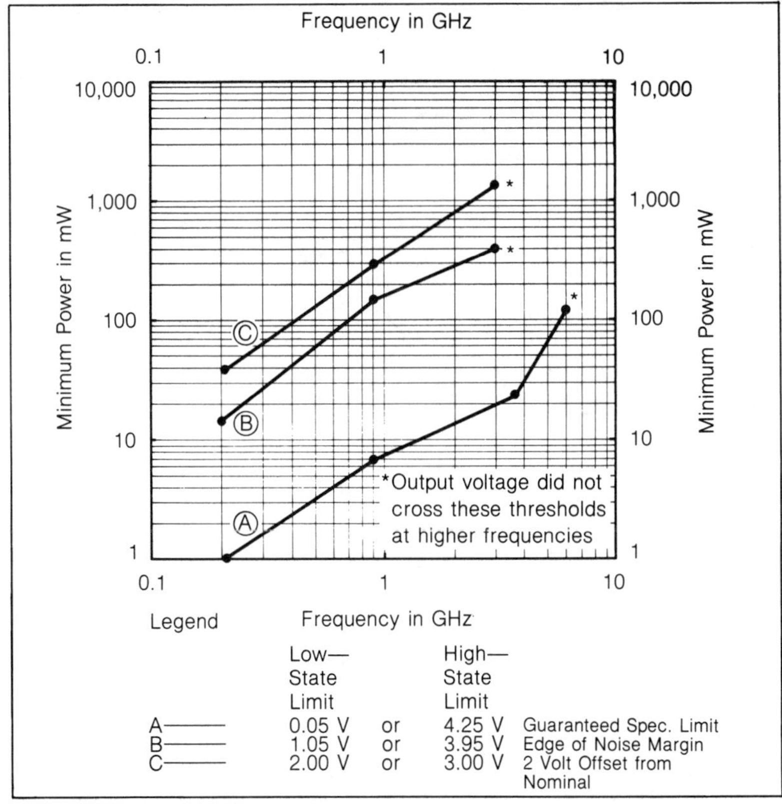

Figure 4.83—Worst-Case Susceptibility Values for CMOS Devices

nominal duration of the pulse. An extreme example is when the input offset caused by rectified EMI exceeds 1 V. The effects of EMI can totally override the intentional digital signal: the device stays "stuck at" high or low state (with unmodulated EMI) or it tracks the EMI modulation envelope. A parallel study was made by Tront (Ref. 13) where simulation and experiment shows a small decrease in the output pulse rise delay t_{pdr} and a significant increase in its fall delay t_{pdr}. In a logic message, logical "ones" will be "shortened" and "zeros" lengthened (Fig. 4.85).

EMI susceptibility measurements conducted on devices with and without integrated electrostatic discharge (ESD) protection clamps show that, in general, the presence of the protection more or less modifies the immunity to out-of-band EMI.

Although the ESD clamp performs its task by clipping isolated transients, it may cause a slight degradation of immunity to a CW

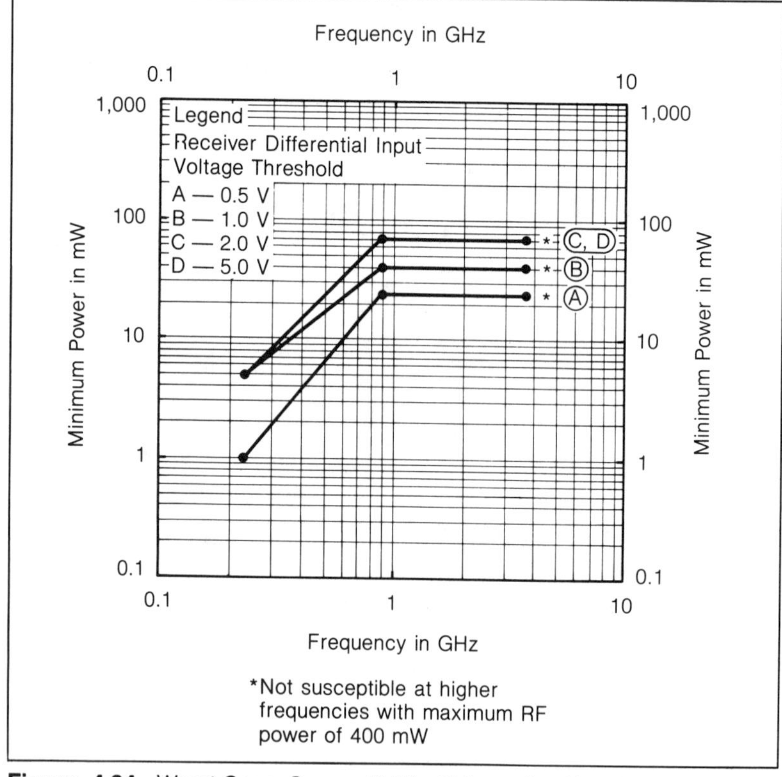

Figure 4.84—Worst-Case Susceptibility Values for Line Drivers and Receivers

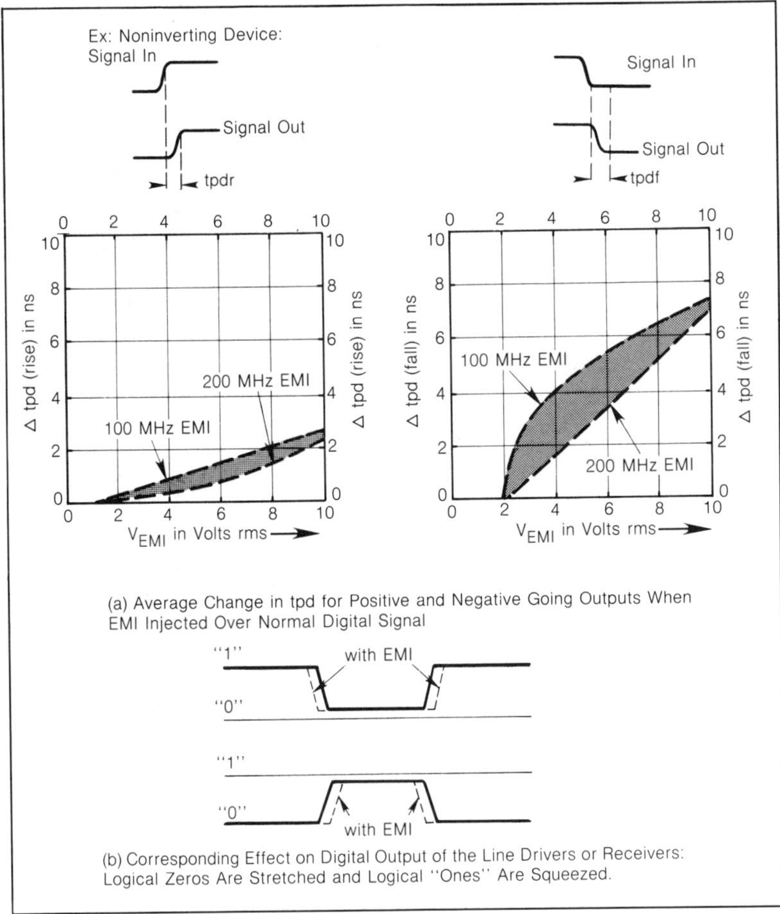

Figure 4.85—Change in Pulse Front Caused by EMI on Line Drivers/Receivers

type of interference because of another nonlinear element at the device's input. Degradation from 0 to 6 dB has been reported (Ref. 14).

4.4.4 Noise Emission

Logic devices are generally more troublesome because of their EMI emission than their susceptibility, although both aspects are reciprocal. Logic devices are primary sources of two kinds of EMI: **conducted interference** that they create on power supply bus, ground bus and signal lines, and **radiated interference** either directly from the chip or indirectly via the interconnect traces and wires acting as antennas. The relevant features of that noise generation appear in several columns of Table 4.2:

1. The shorter the transition time, the wider the occupied spectrum (Fig. 4.86) of the corresponding logic pulses. Since EMI problems increase in severity like F (for conducted EMI and crosstalk) and often F^2 (for radiated), the consequence is obvious.

2. The power supply transition current, which the device demands during its switching, can be very large and have nothing to do with the quiescent current during an established "1" or "0." In TTL, and to some extent in the fast HCMOS technologies, this inrush current is due to the partial conduction overlap of the two output transistors arranged

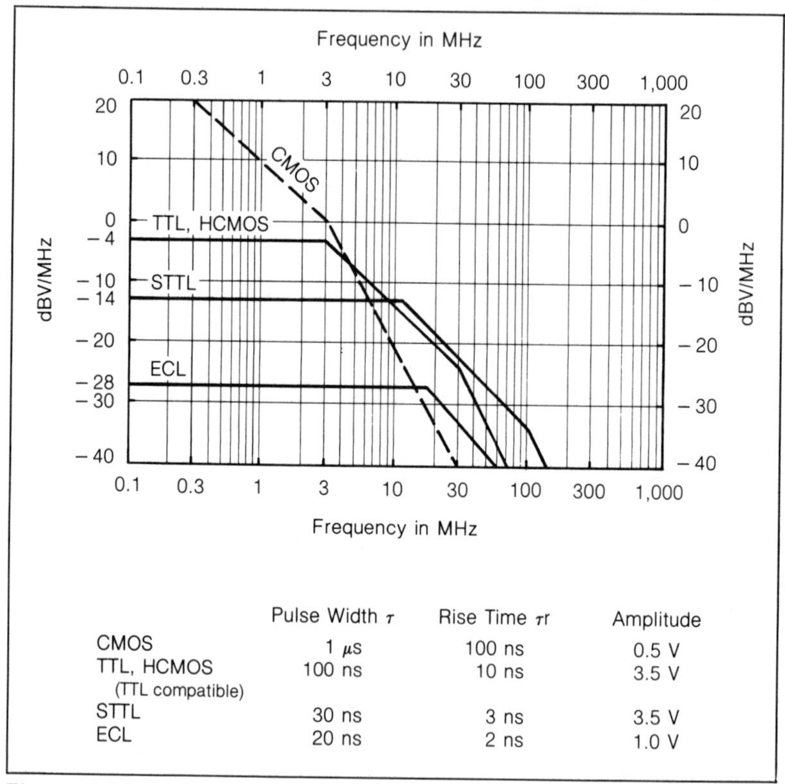

	Pulse Width τ	Rise Time τr	Amplitude
CMOS	1 μs	100 ns	0.5 V
TTL, HCMOS (TTL compatible)	100 ns	10 ns	3.5 V
STTL	30 ns	3 ns	3.5 V
ECL	20 ns	2 ns	1.0 V

Figure 4.86—Voltage spectrums of Typical Logic Devices, Assuming Typical Bit Pulse Width for Each Technology

in a "totem pole." During this overlap, the V_{cc} bus is virtually shorted to ground through two partially saturated transistors plus a limiting resistor.

More recent designs have reduced this effect by using Schottky barrier diodes (SBD) to prevent the output transistors from going into excessive saturation. However, the current peak is still significant and can pose PCB problems or even on-the-chip problems with gate arrays or other highly populated chips.

3. The voltage swing relates directly to electric radiation and capacitive crosstalk.
4. The current that the gate is forcing into (low-to-high transition) or pulling from (high-to-low) the driven gates is also larger than the quiescent current. For very short lines, this load current can be calculated by:

$$I_L = C \, \frac{\Delta_v}{\Delta_t} \qquad (4.64)$$

where C is the sum of the driven traces' capacitance to ground (0.1 to 0.3 pF per cm for single-layer boards) and the input capacitance to ground of the driven gate(s), given in Table 4.2. For instance, for a 3 ns rise time, 3 cm-long traces and a fanout of 5, the transient output current is:

$$I_L = (5 \times 3 \text{ cm} \times 0.3.10^{-12} \text{ F/cm} + 5 \times 4.10^{-12} \text{ F}) \, \frac{3.5 \text{ V}}{3 \text{ ns}}$$

$$I_L = 30 \text{ mA} \qquad (4.65)$$

This peak-current combines with the power supply transition current I_p previously mentioned in a nonsymmetric way. For low-to-high transitions, this current adds up to I_p since the gate is "feeder." For high-to-low, this current subtracts from I_p, since the gate is "sink" and the capacitive charge from the load has to discharge in the driving gate output, which appears as a short to ground (Fig. 4.87).

When the driven line is electrically long, such as when its propagation delay comes within the range of the pulse rise times (in epoxy glass, T_p is approximately 15 cm/ns, therefore transmission-line problems appear when length exceeds $\cong 5$ cm

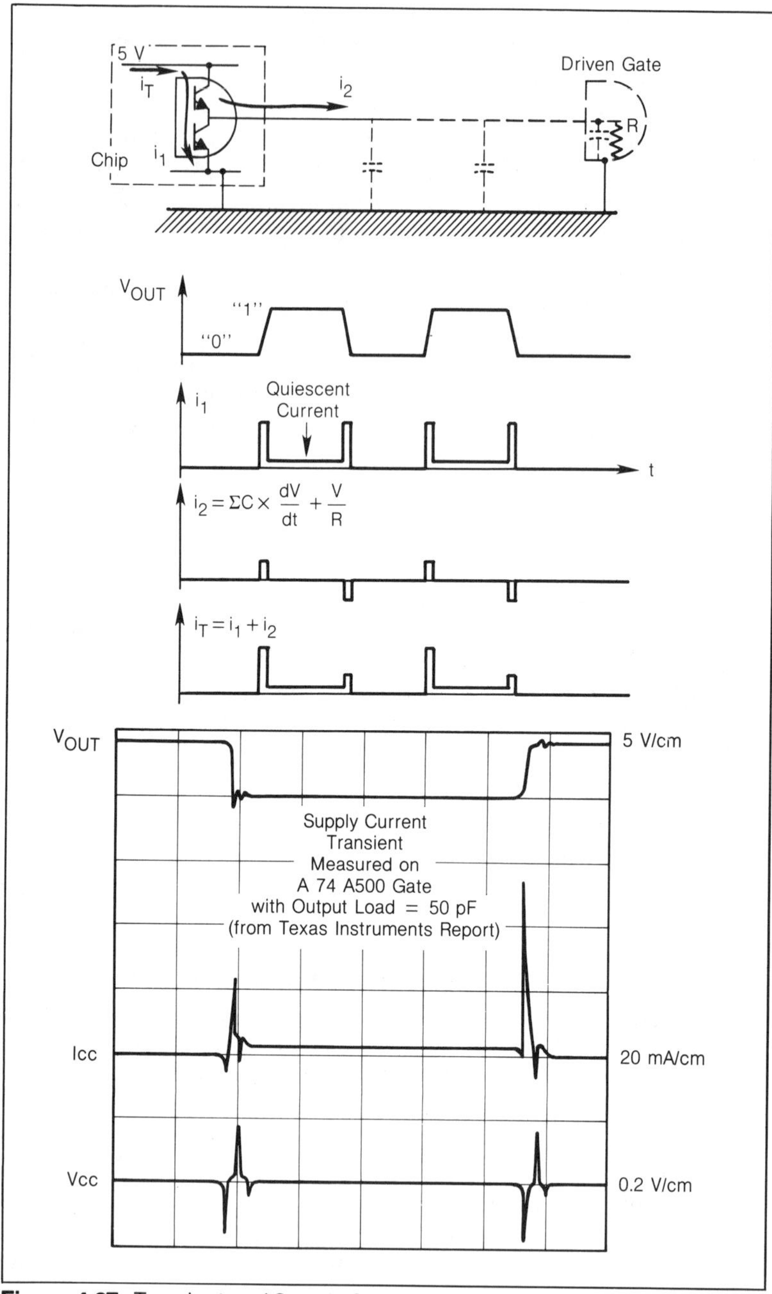

Figure 4.87—Transient and Steady Current Drawn by a Gate off the Power Bus

per ns of rise time), the output current during transition becomes equal to:

$$I_L = \frac{\Delta V}{Z_o} \qquad (4.66)$$

where Z_o is the characteristic impedance of the driven trace. For instance, if Z_o is 90 Ω , a 3 V swing will cause:

$$I_L = \frac{3}{90} \cong 33 \text{ mA} \qquad (4.67)$$

Ultimately, since very low line impedances would create very high currents, I_L is limited by the output resistance of the gate. Figures 4.88 through 4.90 show measured and computed values of conducted currents and radiated fields during the logic devices' operation. Since the basic building block of an LSI logic chip is still generally a gate, the radiation from an elementary gate mounted in a DIP container can be calculated as shown in Fig. 4.90. However, since thousands of gates can exist in a chip, the radiation from the whole chip cannot be the arithmetic sum of all gate radiation. Not all gates operate synchronously, and as the orientations of the radiating doublets in the chip are random, they can add or subtract as well.

Either some randomization is taken into account in math models or actual measurements are made of complex chips. For microprocessors (since many operations take place inside the chip with different data rates), radiated and conducted profiles contain all frequencies corresponding to internal transactions between the ALU, the registers, buffers, etc. Their frequencies are the clock and its submultiples, plus all their harmonics.

Another aspect of radiated EMI from logic chips is the wide variance between manufacturers for the same IC device. Reference 15 shows the following differences between two vintages of a 74LS flip-flop:

E-field (dBμV/m) at 3 m		
	Vendor A	**Vendor B**
30 MHz	32.8	42.8
40 MHz	27.5	33.0
50 MHz	23.0	23.0

4.127

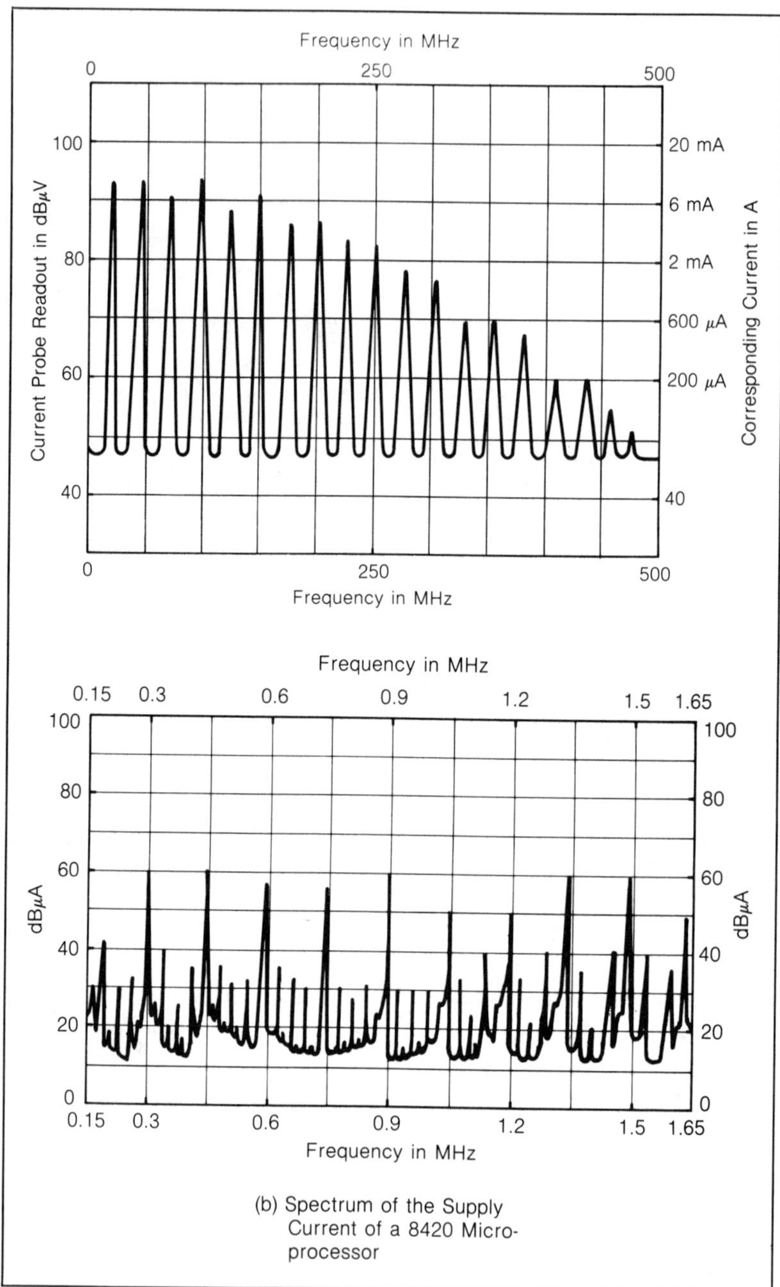

Figure 4.88—Frequency Spectrum of the Current Demanded by Some Logic Devices, and Radiation Vertically above Two Microprocessor Modules. **(continued next page)**

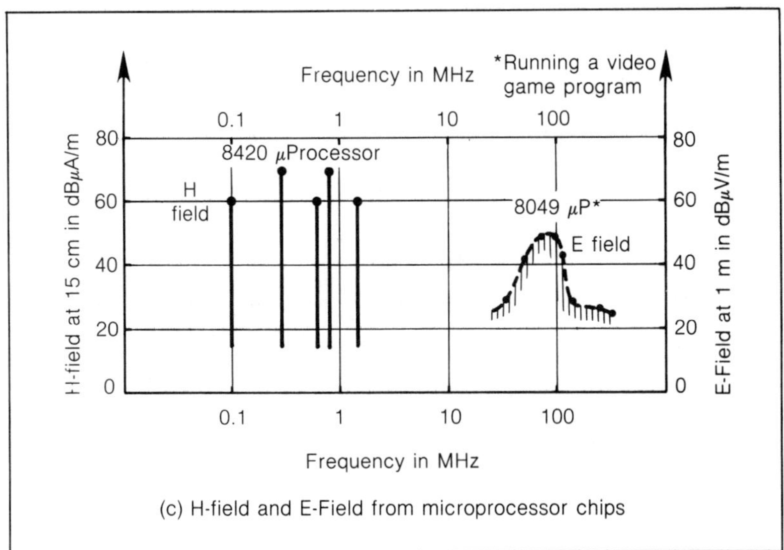

Figure 4.88—(continued)

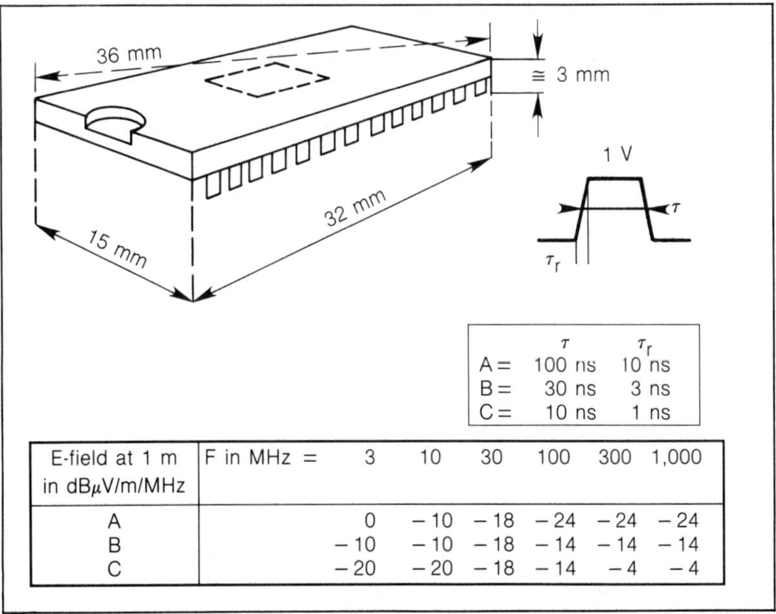

		τ	τ_r
A =		100 ns	10 ns
B =		30 ns	3 ns
C =		10 ns	1 ns

E-field at 1 m in dBμV/m/MHz	F in MHz =	3	10	30	100	300	1,000
A		0	− 10	− 18	− 24	− 24	− 24
B		− 10	− 10	− 18	− 14	− 14	− 14
C		− 20	− 20	− 18	− 14	− 4	− 4

Figure 4.89—Broadband Radiated E-field at 1 m from 1 Gate in a Typical 28-Pin Module, for 1 V Pulse and 3 Different Pulse Characteristics. (If actual amplitude is ≠ 1 V, radiated fields would vary accordingly.)

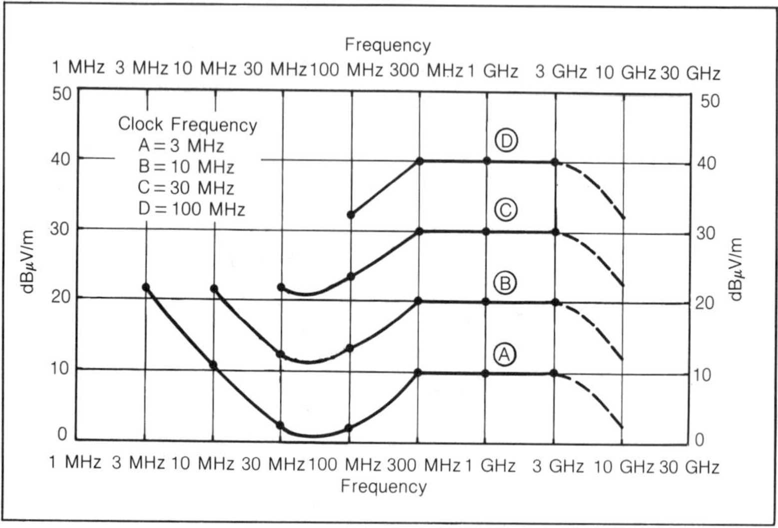

Figure 4.90—Narrowband Radiated E-field at 1 m from One Gate Array Module, 28-Pin, for Several Clock Frequencies. Assuming = V = 3.5 V $\tau_r \, \tau_l$ = 1 ns. Duty Cycle 50%. 100 Gates/Chip. 30% of Them Clocked (Synchronous)

Reducing EMI at Logic Chip Level

Success in making a chip which is functionally good with an integration of 10,000 devices in a 0.25 cm^2 die is enough of a challenge; generally, designers satisfy themselves with just avoiding self-jamming inside the chip and cannot afford too many constraints regarding susceptibility to, or generation of, outside EMI. However, several practices exist which reduce noise problems at the chip level:

1. In TTL-type outputs, use a clamping Schottky to avoid having the output transistor go into deep saturation. Overlap during transistor turn-off is reduced, which in turn reduces the excessive current demand on the supply bus.

In other studies, even an IC model from the same manufacturer shows different EMI characteristics depending on the plant of origin or the fabrication date. This is because EMI is generally not a monitored parameter in the quality control of a chip, and non-functional changes in the masks or wafer process can influence emission profiles.

2. In all technologies, improve noise immunity by a steeper V_{in}/V_{out} transfer curve. This increases the noise margin. Unfortunately, this also increases EMI emission since it translates into a faster transition.

3. Try to use a die-to-pin bonding wire layout which minimizes the radiating area for the larger current paths (i.e., power supply and I/O). For instance: Prefer device pin-out which places the ground and V_{cc} pins near the die at the center, rather than diagonally opposed. This configuration decreases the internal pin-to-die inductance (henceforth reducing noise spikes during transitions) and decreases the radiating loop area. (See Ref. 16 and Fig. 4.91.)

4. Install on-site chip decoupling by using a multilayer ceramic pellet included in the substrate (see Section 2.2, "Capacitors"). A promising technique (Ref. 17) consists of using **multilayer IC substrates** where supply voltages and zero-volt (0_V) returns are made by internal planes. Signals between I/O pads and die are sandwiched between ground planes. A multilayer alumina capacitor is formed between the 0_V and V_{cc} planes. When packaged in a surface-mount logic array, this mounting allows two voltage planes, two signal layers and several ground planes to be stacked with minimum

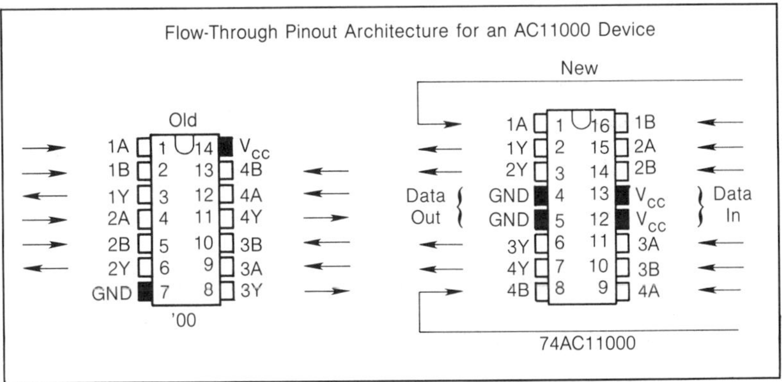

Figure 4.91—Texas Instruments' Advance CMOS family uses flow-through architecture which places the ground and V_{cc} pins in the center of the package rather than the traditional end pin location. This configuration enables designers to use simultaneous-switching while allowing the maximum performance to be obtained from the device. A flow-through architecture has also been implemented to ease board layout and save board real estate. For a typical high-speed logic transition, the noise glitches on the power distribution near the devices are reduced from about 2 V to 0.4 V.

intra-IC crosstalk, supply and ground noises. Compared to the typical 5 nH of package inductance of a standard pin-grid array, the inductance is decreased to less than 300 pH. For instance, with such a device having 224 I/O pads, 8 bus drivers can switch 190 mA each in 10 ns, with less than 50 mV of common ground noise at the chip level. The need for an external decoupling capacitor is eliminated since it is built in (Fig. 4.92).

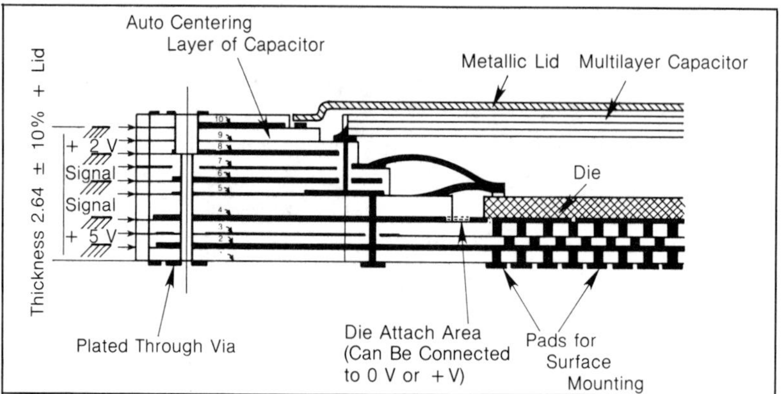

Figure 4.92—Example of Leadless IC Mounting with Integrated Decoupling Capacitor and Multilayer Power. Ground and Signals Arrangement (from C.Val., Ref. 27)

5. Use metal canned modules with a can connection available for grounding at the 0_V reference.
6. Use "quadrupole grounding," (Ref. 18) where a metal top over the chip is used to carry some supply current and form a cancelling loop with the other normal path.
7. Given a necessary transition time, use the lowest voltage swing compatible with the function.
8. Given acceptable power consumption, use the lowest input impedances.
9. Prefer leadless (surface mount) technologies.

Also, most of the solutions to reduce logic component noise belong to their mounting techniques, (described in Chapter 8, "PCBs").

4.4.5 Line Drivers and Receivers

Concerning EMI, the behavior of line drivers and receivers

resembles the comparators discussed in the analog section (4.3.3.2). There are mainly three categories of line driver-receiver pairs: unbalanced, pseudo-balanced and truly balanced (differential).

The unbalanced type (like those used for RS-232 or 423 interface) is the simplest. Signal is simply transmitted or received between a single wire and the logic ground. The main drawback of this simplicity is a vulnerability to ground noise and loop-induced voltages since any common-mode voltage will appear in series with the wanted signal (Fig. 4.93).

The true balanced scheme (also known as true differential or

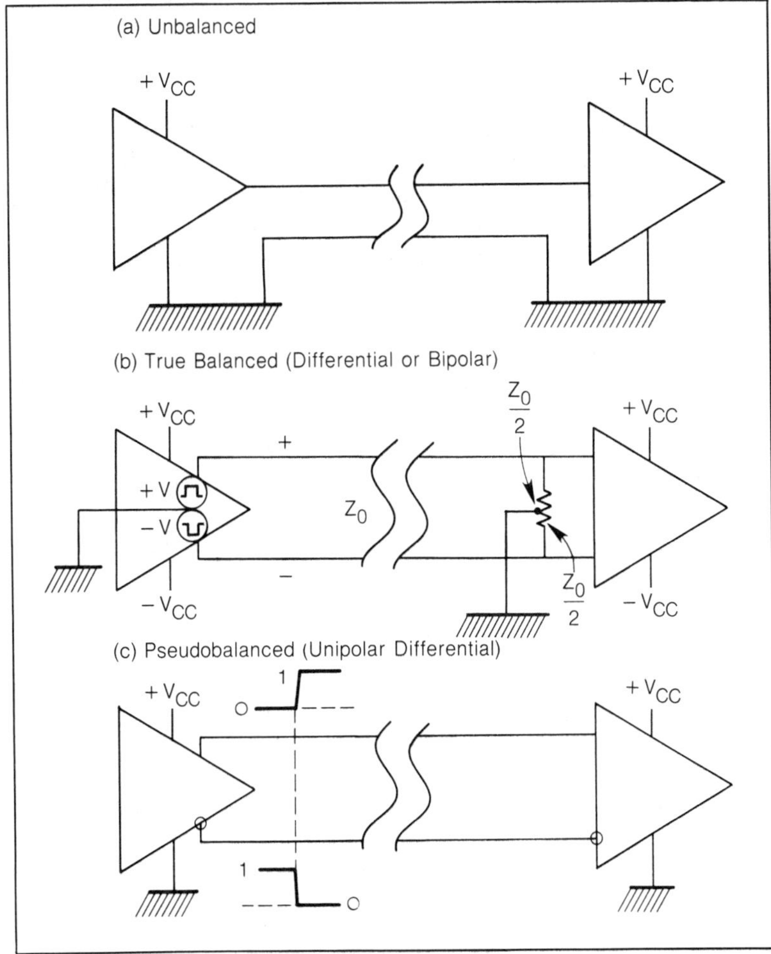

Figure 4.93—Different Types of Line Drivers/Receivers Balancing

bipolar, like those used for RS-422 or 485) consists of transmitting a signal with equal positive and negative amplitude relative to ground via a wire pair (or a twinax). Therefore, there is no need for a ground connection between the driver and receiver to carry the signal return. Receiver input only responds to differential voltage, and common-mode noise is, in theory, rejected since it causes both inputs to shift with the same polarity. In reality, rejection cannot be perfect since the common-mode noise never appears with exactly the same amplitude or phase on both receiver inputs at the same time. Also, the CM rejection can only be as good as the symmetry of the two drivers' outputs and the receiver's input impedance to a ground, as well as the wire pair unbalance (whichever is the worst). Both parameters generally deteriorate at high frequency.

For instance, if a balanced differential receiver has an input sensitivity of 100 mV and is said to withstand up to 10 V common mode, this translates into a steady CMR of $10/0.1 = 100$ times, or 40 dB. However, if two 10 V pulses show up on the line with a rise time of 10 ns and a phase difference of 2 ns, this corresponds to a differential voltage of $(1 \text{ V/ns}) \times 2 \text{ ns} = 2 \text{ V}$, with a duration of 2 ns; 2 V is 20 times larger than the sensitivity of 0.1 V, therefore, a receiver with a transition time equal to or shorter than $2 \text{ ns} \times 20 = 40 \text{ ns}$ (that is, a bandwidth greater than 8 MHz) will react to such common mode transients.

Balanced receivers are also ideal to use with a balanced terminated transmission line, where a pair of characteristic impedance Z_0 will terminate in two resistances ($R = Z_0/2$) tied to ground, which will decrease EMI vulnerability, crosstalk and ringing.

The penalty to pay with balanced drivers and receivers is the need for a double power supply, $+V_{cc}$ and $-V_{cc}$. This can be provided now by miniature watt-size isolated dc-to-dc converters mounted on the PCB.

The pseudo-balanced, or unipolar differential, is a driver-receiver pair where the driver delivers at the same time a logic pulse A and its complement $\overline{A}$. The signal is still referenced to a ground, like with an ordinary unbalanced link; however, dynamically, when one output goes up, the other goes down. The receiver tracks that scheme and it functions like an edge detector because it will respond only if one input goes up at the same time the other goes down. So, immunity to CM pulses is improved since they will rise with same polarity on both inputs.

Some typical characteristics of balanced line receivers are:

1. Input differential sensitivity =
   a. Normal types, for short distance, motherboard-to-motherboard: 500 mV (+ and −250 mV vs. ground)
   b. Sensitive type, for long hauls, party line: 20 mV (+ and −10 mV vs. ground)
2. CM voltage range: up to 30 V (for type a)
   up to 3 V (for type b)
3. Input impedance (without matching resistances): ? kΩ
4. Output impedance ≅ 5 to 20 Ω
5. Drive current capability: 0.1 A

Figure 4.94 shows typical immunity to isolated transients like lightning or other high-energy pulses.

Improvement of Driver/Receiver EMI

1. Hysteresis circuit improves immunity to short noise pulses, but penalizes speed.
2. Balancing resistors in series on input (for balanced receivers) improves symmetry, and thus CM rejection, but degrades sensitivity.
3. Clamps (Schottky or else) on the input improve immunity to high energy pulses, ESD and isolated transients. They decrease ringing (by leveling off oscillations) in case of mismatch, but can also aggravate parasitic rectification of CW interference.
4. Enable/disable port allows the gating of line input by an undisturbed strobe signal.
5. For drivers with high current capabilities, use close decoupling and a separate set of V_{cc} and GND pins for the drive section than those used for the input section. These drivers should be packaged close to the I/O connectors on PCB to minimize their induction to nearby victim circuits and the amount of external radiation (RFI).

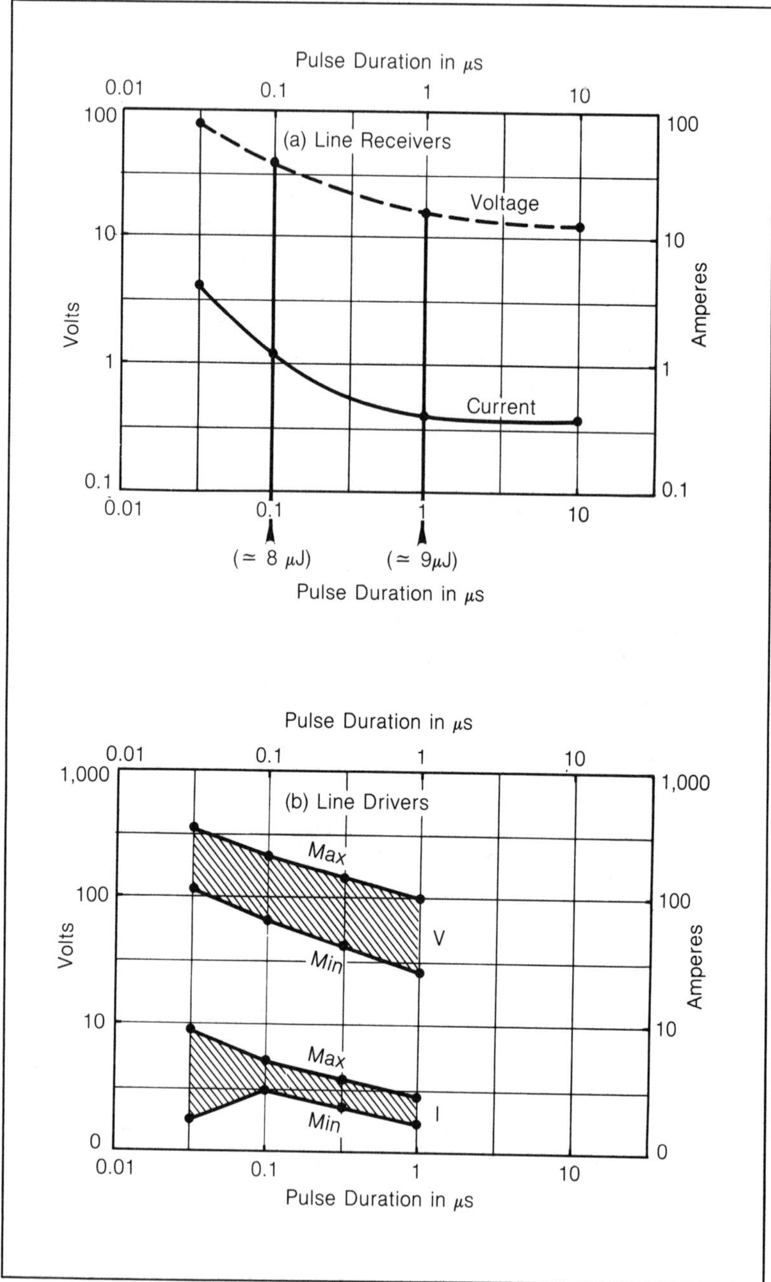

Figure 4.94—Damage Threshold to Single Transient of Typical TTL Receivers (a) and Drivers (b)

4.5 Topics on Microintegration vs. EMI

The increasing trend to ultra-integration of logic devices (and to some extent, analog as well) poses certain EMI problems inside the chip itself.

Geometric Factors

Reduced size will affect the degree of EMC because of common-mode conductive coupling and induction resulting from both magnetic and electric fields. Three induction-field cases must be distinguished: compatibility problems (1) within a module, (2) between modules and (3) between equipments.

In the conducted EMI case, coupling occurs by means of current flow through a resistance element that is common to both the source and the receptor. If the assumption for the discrete-component case is made that a form factor for microelectronics is similar to that for conventional construction, the relative resistance depends only on the relative linear dimensions involved.

For those cases where both the source and the receptor are microminiaturized, it can be expected that any change in relative sensitivity of the receptor will permit a corresponding change in the square root of relative power (p) of the source. Thus, for construction using similar materials and thicknesses, conducted EMI can be expected to be less severe in proportion to the reduction in linear dimension.

On the other hand, when the same material is used but the thickness is proportional to the linear dimensions, then no significant change can be anticipated in conducted EMI since the coupling resistance will remain constant. When this is not the case and common conductor thickness is less reduced than the length/width reduction, common impedance problems will be decreased.

For coupling on a given thin-film wafer, the reduction in common-mode path thickness may be quite severe, with a corresponding increment in coupling resistance. The relative resistance depends not only on relative linear dimensions, but also on relative sheet resistivity. For the LSI case, coupling on a given substrate may involve radically different materials from the conventional case. The high relative resistivity of the substrate may also increase the severity of the interference problem.

In intramodule compatibility, for the magnetic field, the relative source-antenna efficiency will be proportional to the relative loop areas involved or proportional to the square of a linear dimension.

The relative receptor efficiency will behave similarly. For close spacings involved within a given module, true induction-field operation may be assumed, with the field strength at the receiver being inversely proportional to the cube of the linear dimension. Intramodule compatibility problems due to magnetic field coupling can be expected to be less severe in proportion to the reduction in linear dimension, due to low currents and high impedances generally used.

For the electric-field case, the relative antenna efficiency will be proportional to the capacitive effect, as will the receptor efficiency. Electric-field effects can thus be anticipated to be more severe than in conventional construction by the inverse ratio of linear size reduction.

For side-by-side modules, or MCMs (multichips), intermodule compatibility is essentially the same as for the intramodule case. The only special consideration here is a possible change from source and receptor in separate equipments to source and receptor in adjacent modules, made possible by the volume reduction of microelectronics.

Interequipment compatibility is the problem which includes the source of EMI in one piece of equipment and a receiver of such interference in another piece of equipment, either part of the same system or of different systems. This equipment spacing is determined primarily by construction techniques. Microminiaturization will permit equipments to be closer together, with relationships similar to those developed for the intramodule case.

The following considerations and tradeoffs are to be made when dealing with intradevice EMI.

Crosstalk: Fine lines within a chip and from chip pads as well as substrate pins are prone to crosstalk, as in any circuit. The small size of an LSI device should not hide the fact that, due to scaling factor, two fine lines, 3 mm, long and separated by 0.1 mm are equivalent to two 30 cm traces separated by 10 mm.

The problem is even worse because when the length of adjacent conductors is reduced by a factor K, capacitive coupling is also reduced by K. But when spacing is reduced by the same factor K, capacitive coupling **increases** approximately like K^2. Therefore, a 1/100-scale reduction of a given parallel-wire configuration results in an aggravation of crosstalk of $100^2/100$, if the height above ground does not change significantly. The ϵ_r of ceramic and SiO_2 also aggravates capacitive crosstalk, compared to the same geometry in the air.

For instance, two fine lines 10 mm long, separated by 0.1 mm and with a trace width of 0.05 mm have a mutual capacitance of 0.2 pF. A 3 V, 1 ns pulse in one line will induce in the other (assuming 100 Ω impedance line-to-ground) a noise equal to:

$$V_n = RCdV/dt = 100 \times 0.2 \times 10^{-12} \times 3/10^{-9} \qquad (7.68)$$
$$= 60 \text{ mV}$$

Figures 4.95 and 4.96 show other illustrations of this problem at module level.

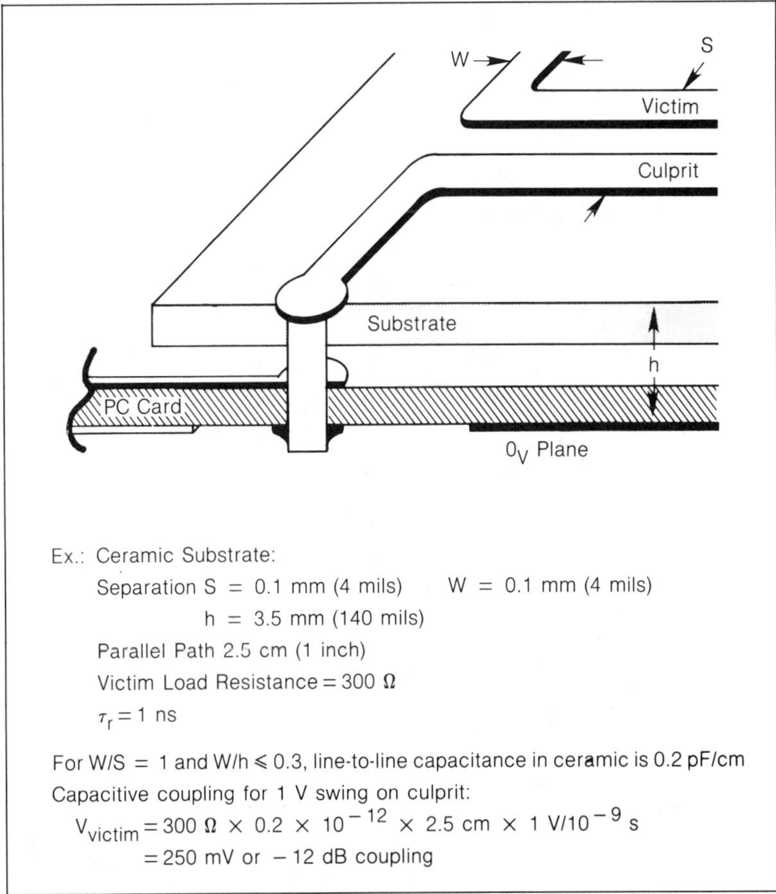

Ex.: Ceramic Substrate:

Separation S = 0.1 mm (4 mils)   W = 0.1 mm (4 mils)

h = 3.5 mm (140 mils)

Parallel Path 2.5 cm (1 inch)

Victim Load Resistance = 300 Ω

$\tau_r = 1$ ns

For W/S = 1 and W/h ≤ 0.3, line-to-line capacitance in ceramic is 0.2 pF/cm

Capacitive coupling for 1 V swing on culprit:

$V_{victim} = 300 \text{ Ω} \times 0.2 \times 10^{-12} \times 2.5 \text{ cm} \times 1 \text{ V}/10^{-9} \text{ s}$

$= 250$ mV or -12 dB coupling

Figure 4.95—Conductor-to-Conductor Crosstalk in Macro Elements (Chip Supports, IC Sockets, Hybrid Circuits Etc.)

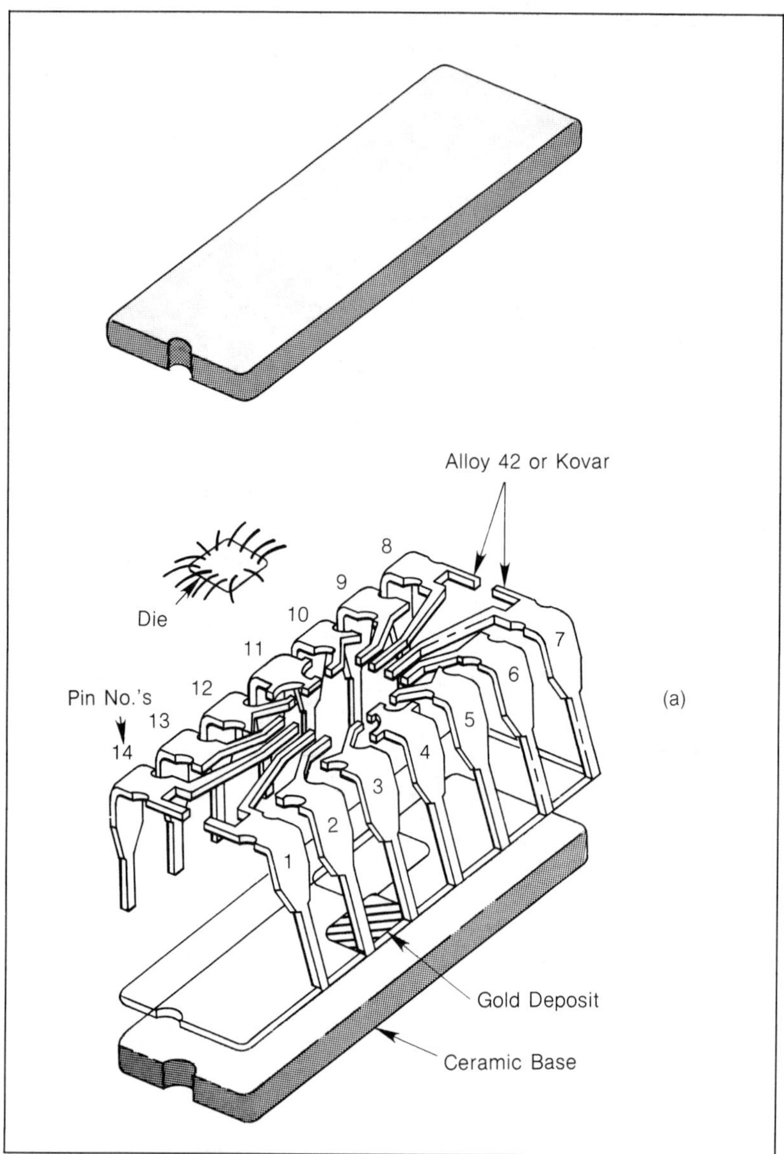

Figure 4.96—Isolation between Pins of a Standard IC Package (a). in (b) (next page) the comparison between DIP and flat pack, with die in place shows a 10-12 dB better isolation for the flat pack. In (c) without die, the DIP improves by about 6 dB, i.e. isolation due to the die and isolation of the DIP were identical. Also the opposite pins can be up to 25 dB better isolated than adjacent ones.

Figure 4.96—(continued)

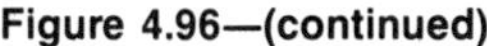

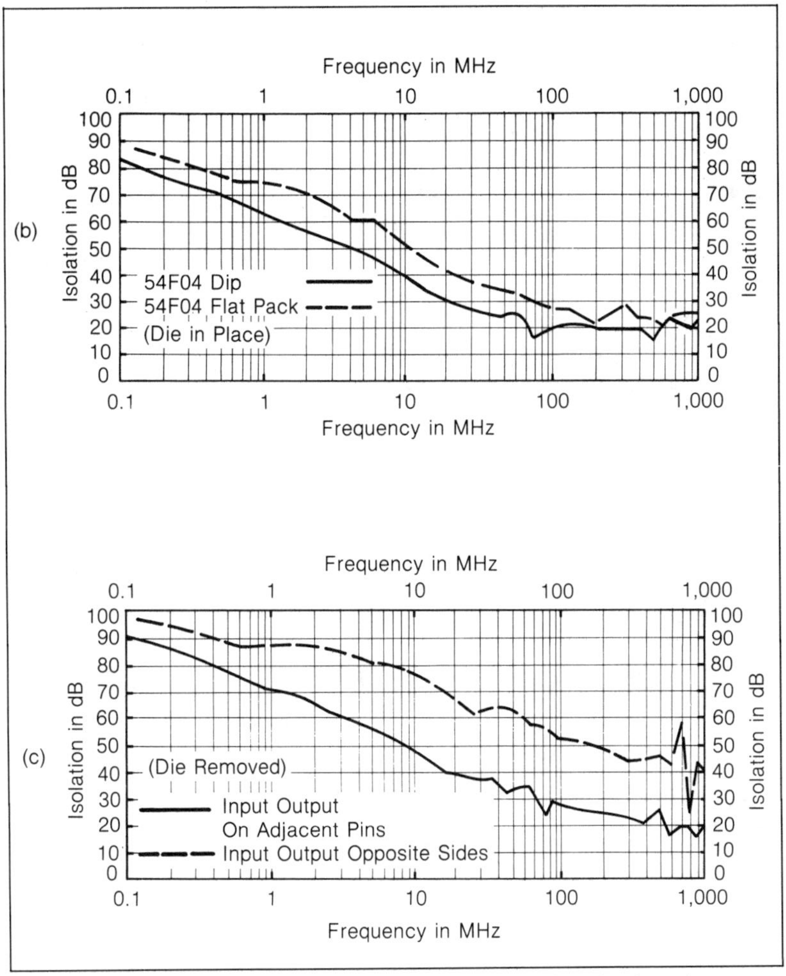

Protection devices: Integrated transient protection devices (zener clamps, etc.) are necessary for ESD and transient protection, but they may adversely affect audio rectification. Also, since they are mounted across the inputs, their parasitic capacitance (typically 0.5 to 1 pF) must be carefully controlled to avoid bandwidth degradation of the logic.

Faster rise times are used for intrachip functions than for input/output connections. Typically, a chip with 1 to 3 ns transition times for signals in and out will use 300 or 500 ps transition times inside. To allow these fast transitions and avoid a prohibitive addition of internal propagation delays, the voltage swing for inner functions is generally reduced, to approximately 1 V instead of 3 V for I/Os. In other words, the edge speed in V/ns is constant, but voltage excursion is less. Therefore, crosstalk and radiation are not aggravated. On the other hand, since signal amplitude is decreased, the EMI margin is badly reduced. Hopefully, since $V_{OL\,max}$ and $V_{OH\,min}$ are better controlled, they can be reduced to 100 or 200 mV. So, for a 1 V swing, the mid-point threshold would be 0.5 V and the typical internal noise margin* would be:

$$0.5 \text{ V} - 0.2 \text{ V} = 0.3 \text{ V} \tag{4.69}$$

So, the reserve for internal EMI (self jamming) is not shrunk in proportion to the signal amplitude.

Faster transitions inside chips means impedance matching concerns even at chip level. For a 13 to 15 cm/ns typical speed in an LSI substrate, a 300 ps signal needs to see a matched termination when line length approaches 1/2 Tp that is beyond 2.2 to 2.5 cm. Otherwise, ringing will occur.

4.6 Measured Sensitivity of Microelectronics to Radiated EMI Fields

Although EMI radiation interferes with micromodules primarily by using associated cables and circuit boards as pickup antennas, studies were also conducted to find the field intensity required to produce a malfunction in a linear IC where only the area of the module itself was exposed (Ref. 19).

A field creating an interfering signal in a device is basically an antenna modeling problem. At low frequencies, quasistatic approximation is reasonable, and the sensitivity of the circuit to E-field and H-field can be addressed separately by modeling the integrated circuit successively as a small receiving dipole and a small loop. (see Fig. 4.97, parts a and b). At higher frequencies it is often more

---

*At this point, it could be pertinent to wonder if the bandwidth of the logics does not become so large that thermal noise becomes a limiting factor. With 0.3 ns speed, the bandwidth is 1 GHz. Assuming a gate input resistance of 10 kΩ for logic with low power dissipation, the thermal noise at 85°C junction temperature is about 0.5 mV, not yet a problem. However, considering future logic trends, it could be said that for a 30 ps logic and higher input resistances, the thermal noise will reach 5 to 10 mV and become a non-negligible criteria in the device's S/N ratio.

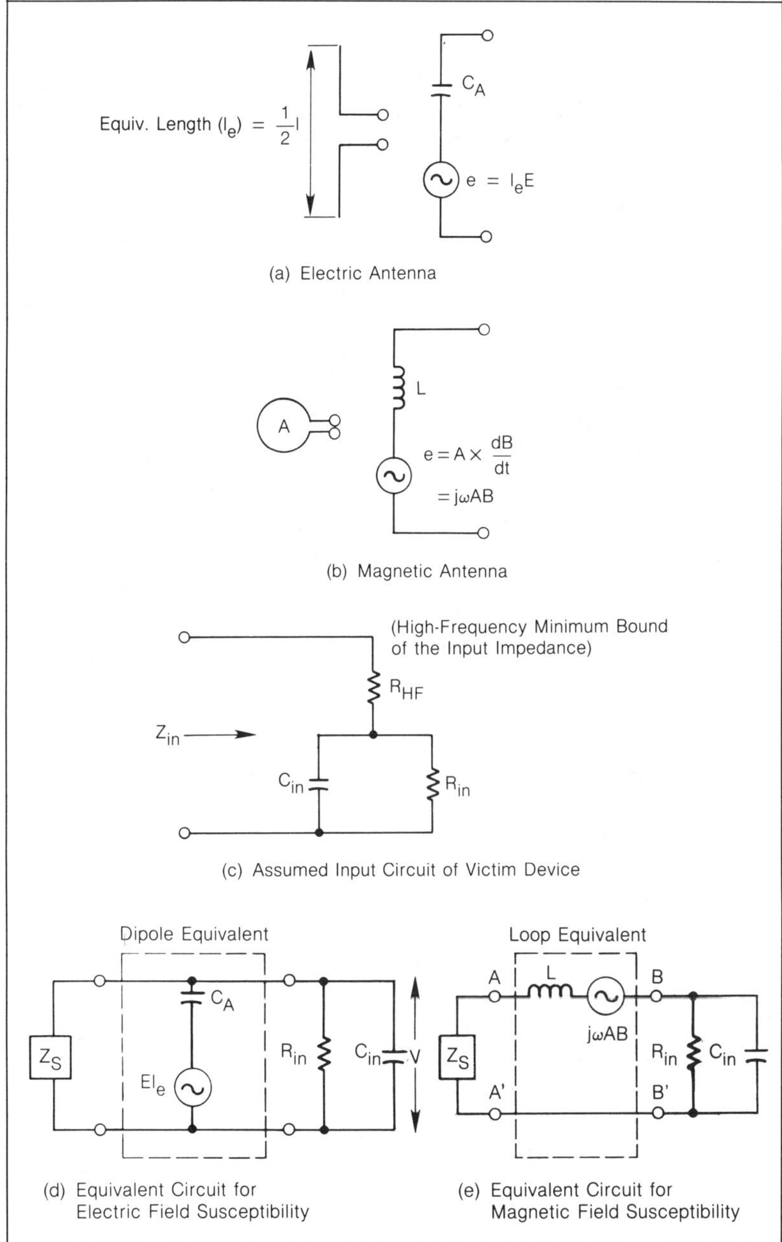

Figure 4.97—Model Used for Prediction and Analysis of Integrated Circuits Radiated Susceptibility (Ref. 30)

convenient to consider the incident EMI power density. For analog circuits, the criteria can be the one used in the definitions section (Section 4.1); that is, EMI exists when the induced signal equals the device's input noise. For digital circuits, the criteria would be an induced signal exceeding noise margin.

As long as the largest circuit in the IC module is smaller (less than one-tenth) than the wavelength, the worst-case equivalent electrical antenna can be taken as the longest pin-to-pin (or pad-to-pad) path of the module. The worst-case magnetic loop can be taken as the widest closed area formed by a sensitive line in the module and a return path. Then a model is required for the transfer of the antenna output voltage to the actual circuit input impedance. The basic input configuration is shown in Fig. 4.97c.

R_{HF} is the high-frequency asymptote for a minimum-input impedance, used only when λ_{EMI} approaches circuit dimensions (non-quasistatic case). Otherwise, for lower frequencies (essentially below 1 GHz), R_{HF} can be ignored. The circuit impedance offered to the antenna is basically a parallel RC, with values of R in the 10 kΩ to several MΩ range for linear ICs and in the 100 to 1,000 Ω range for digital ICs (since the output impedance of the driving device has to be considered in parallel with the input of the addressed gate).

Quasistatic Electric Field Pickup

An equivalent circuit for the electric field pickup can be drawn as shown in Fig. 4.97d. The high-frequency resistance R_{HF} has been ignored since it becomes significant only at frequencies where the quasistatic model is of dubious value.

Here the effect of the source impedance Z_s may be included simply by defining R'_{in} and C'_{in} as the equivalent combined parallel resistance and capacitance of both Z_s and Z_{in}. A conservative estimate may be made by taking $Z_s = \infty$, that is, a floated input (worst case for E-coupling). The voltage V may be calculated by the following equation:

$$V = El \ \frac{R'_{in} \ C_A \ j_\omega}{1 + R'_{in}(C_A + C'_{in})j\omega} \tag{4.70}$$

By setting $V = V_s$, (the voltage susceptibility level), we can calculate the magnitude of the E-field required for susceptibility.

$$E_s = \frac{V_s}{jlR'_{in}C_{A}\omega} \times [1 + R'_{in}(C_A + C'_{in})j\omega] \qquad (4.71)$$

This is sketched asymptotically in Fig. 4.98. Clearly, the antenna length, the antenna capacitance and the equivalent parallel

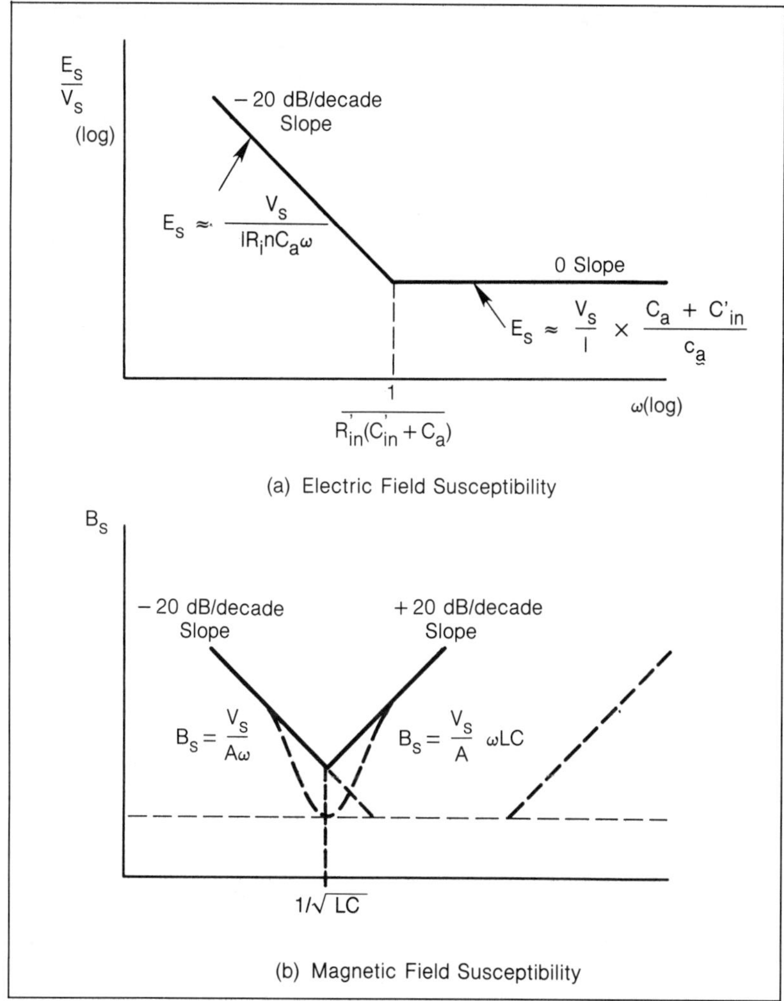

Figure 4.98—Profiles of Field Strength Required to Cause a Voltage V = V_s to Appear at Victim Device's Input (Quasistatic Hypothesis)

resistance R'_{in} are the dominant elements in the low-frequency range. In the high-frequency range, the antenna length, antenna capacitance and total circuit capacitance are the dominant parameters.

Quasistatic Magnetic Field Case

The combined equivalent circuits of the loop antenna model, the circuit input and the normal driving source are shown in Fig. 4.97e. Z_s represents the output impedance of the normal driving source. The central section of the circuit represents the pickup loop with induced voltage $BA\omega$ and inductance L. All parasitic shunt impedances and other shunt circuit elements are assumed to be included in either Z_s or the input circuit, as appropriate. The circuit input impedance, Z_{in} is represented by the $R_{in} C_{in}$ combination.

It is apparent that Z_s will affect the resultant voltage V across the input terminals. Normally, Z_s might be expected to be comparable to Z_{in} so that about half of the induced voltage would appear across each. To simplify the analysis, we assume Z_s is of negligible magnitude compared with Z_{in}; that is, all the loop voltage appears across Z_{in} (worst case for H-field). The solution of the circuit shows that:

$$V = \frac{j\omega BA}{\left[1 + \dfrac{j\omega L}{R_{in}} + (j\omega)^2 LC_{in}\right]} \tag{4.72}$$

If $V = V_s$ (the susceptibility level), we can then calculate the level of the field B_s at which the circuit is susceptible:

$$B_s = \frac{V_s}{A} \frac{\left[1 + \dfrac{j\omega L}{R_{in}} + (j\omega)^2 LC_{in}\right]}{j\omega} \tag{4.73}$$

The actual curve departs significantly from the asymptotes at $1/\sqrt{LC}$. The amount of this departure depends on the damping term in the quadratic. Nevertheless, the minimum value of B_s can be seen to be $V_s L/AR_{in}$ (for $\omega = 1/\sqrt{LC}$), regardless of the location of the high-frequency asymptote. The low-frequency asymptote is, of course, fixed. The dashed line indicates the form of response obtained when the quadratic is heavily damped.

Nonquasistatic Case

In the range of frequencies where the dimensions of the circuit, circuit board or system are comparable to an appreciable fraction of a wavelength, the quasistatic model becomes invalid. The calculation of the performance of the circuit as an antenna in this frequency range is impractical. We use the idea of effective area (A_{eff}). The maximum possible power absorbed is given by:

$$P_{max} = A_{eff} \times P_{DI} \qquad (4.74)$$

where P_{DI} is the power density of the incident field. If all of the intercepted power is assumed to be applied to the circuit input, the resulting estimated susceptibility is quite conservative. Two models may be used to determine the effective area. Assuming that the input wiring constitutes a dipole, the effective area is given by:

$$A_{eff} = 1.64\lambda^2/4\pi = 0.13\lambda^2 \text{ for the tuned dipole case}$$
$$A_{eff} = 1.5\lambda^2/4\pi = 0.12\lambda^2 \text{ for the short dipole} \qquad (4.75)$$

Because the difference between resonant and nonresonant is only 10 percent, the tuned dipole can be taken as a model without creating a gross overestimate.

A second assumption would be to think that the effective area of the complete circuit is equal to its actual area, and that all power incident on the circuit was absorbed. For a given circuit, the larger of the two effective areas should be used to assure a conservative prediction.

In this frequency range, it is assumed that any interference will be out of band, that is, beyond the bandwidth of any analog circuit and above the maximum operating clock rate of any digital circuit. Accordingly, the dominant element of the circuit input impedance is taken as R_{HF}. Since a measurement of this impedance is difficult, an estimated value must be chosen. Presently, it appears that a choice in the neighborhood of 50 to 100 Ω seems reasonable since this is of the order of the cable and stripline impedances. Since the frequencies are far out of band, interference can happen only through nonlinear mechanisms such as rectification or detection. In the case of an analog circuit, the susceptibility criterion is that the demodulated envelope of a 100 percent modulated EMI (CW) is equal to thermal noise. For the digital circuit, the susceptibility level is that level at which the rectified envelope could cause false triggering of a flip-flop. For example, if the EMI field comes from a radar, the rectified signal is a pulse train corresponding to the pulse repetition rate of the radar. This is normally in a frequency range which would be expected to affect both analog and digital circuits.

As shown in Eq. (4.71), the incident power density is multiplied by the effective area. The RF voltage is given by:

$$V^2 = R_{HF} \, P_{DI} \, A_{eff} \qquad (4.76)$$

If the susceptibility voltage level is V_s, then the maximum tolerable field strength is obtained from the power density:

$$P_{DI} = \frac{E_s^2}{377} = 377 \, H_s^2 = \frac{1}{A_{eff} \, R_{HF}} \, (V_s^2) \qquad (4.77)$$

then, letting $R_{HF} = 100 \; \Omega$ and replacing A_{eff} by $0.13\lambda^2$:

$$E_s = 1.8 \times 10^{-2} \, F_{MHz} \times V_s \qquad (4.78)$$

R_{HF} is the only significant parameter in this circuit model. Figures 4.99 and 4.100 show the susceptibility levels predicted using this concept in the microwave region. The horizontal curve (solid) is based on the use of the actual area as the effective area. The dashed line (with slope 45°) is based on the effective area of a dipole [see Eq. (4.75)].

Experimental Results

Validations were made by Showers et. al. (Ref. 19) of the model described before. Several test jigs were used, some of them replicating TO5 cans, DIPs and flat packs. For an actual IC radiated susceptibility test, all the cables and conductors (other than the IC itself and its leads) were carefully shielded.

Provided this, the exposed area was about 2 cm², with average lead length l of 1.4 cm and parasitic input capacitances of approximately 10 pF across Z_s and R_{in}. The in-band voltage sensitivity was about 10 mV for the linear IC and 0.8 V for the digital. The plot for the analog IC had many measurement points and it shows remarkably close agreement in the quasi-static zone and a coarse tracking in the "power density $\times$ A_{eff}" zone, where the model is pessimistic and overpredicting EMI by about 20 dB. The plot for

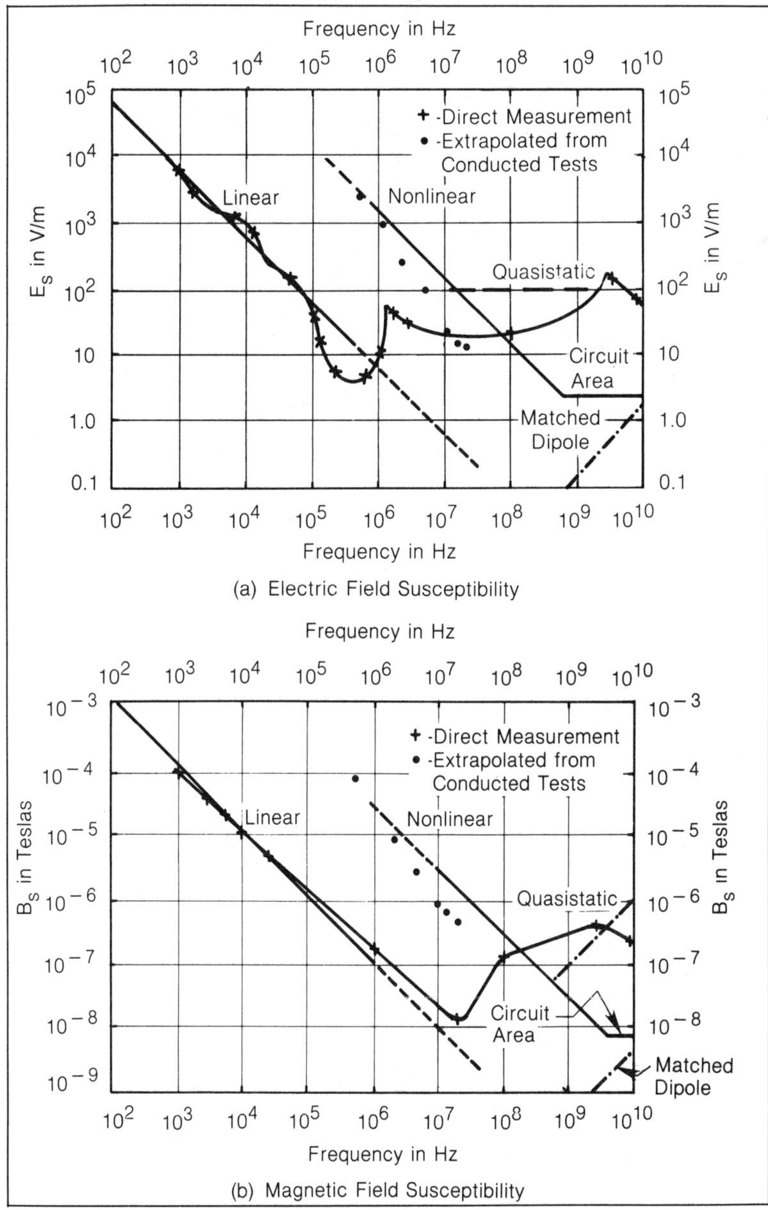

Figure 4.99—Predicted and Measured Susceptibility for a Linear I.C. (RCA CA-3001), for an In-band Susceptibility of $\cong$ 10 mV. The curve captioned "Linear" correspond to EMI just superimposed on a desired signal. "Nonlinear" correspond to audio rectification.

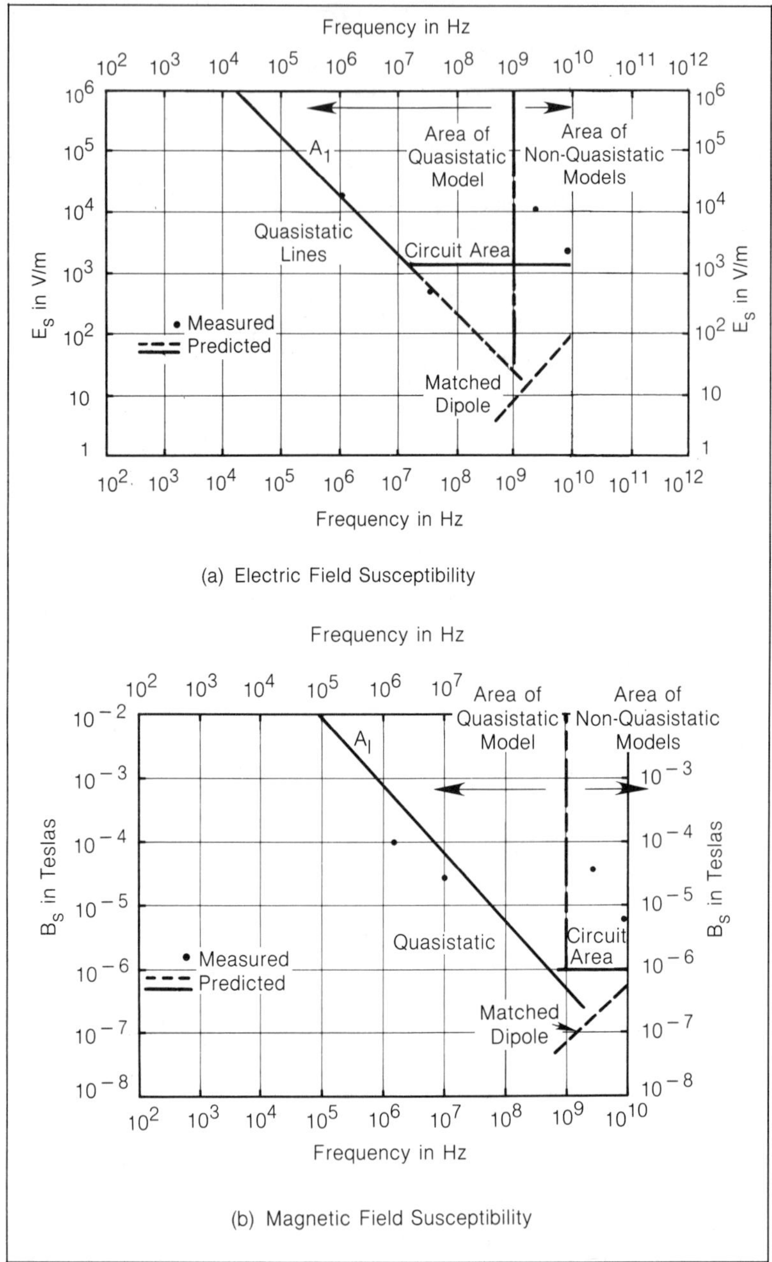

Figure 4.100—Predicted and Measured susceptibility for a Typical Digital Circuit (RCA CD-2203 flip-flop) Flat Pack, with dc Noise Immunity $\cong 0.8$ V.

the digital IC had only a few measurement points and is much coarser and more conservative in the microwave region due to the hypothesis that *all* the captured incident power is absorbed. Also note that, in the AM region for instance (around 1 MHz), it takes about 10 V/m to upset an analog amplifier and 10,000 V/m for the digital IC. By that time, any ordinary system would have long since run into terrible EMI problems due to its cables and PCB radiation pickup. However, this study gives a feel for *what remains susceptible and what it is susceptible to when all cables and circuit boards have been hardened for a severe EMI ambient.*

4.7 ESD Sensitivity of Microelectronic Devices

Direct electrostatic discharge (ESD) to an electronic component is one of the most severe threats to component reliability during the manufacturing and handling of electronic parts. It costs *millions* of dollars in dead chips and severe breeches in the percentage yield of IC manufacturing lines, not to mention the latent wounds which will later show up as unexplained failures in the field.

This problem was recognized in the late 1960s, and all the large manufacturers of electronic components have now implemented strict static control programs which comprise:
1. Static awareness for employees
2. Static-free work areas and manufacturing hardware
3. Monitoring of percentage yields and chip infant mortality to detect ESD symptoms
4. Antistatic precautions for card handling in the field

This section does not address component failure but rather the malfunctions of a complete equipment. The reader who is interested in the component failure aspect of ESD is referred to the abundant literature existing on the subject (e.g., Ref. 20). We will simply give a brief reminder of this aspect of ESD by showing a few typical examples.

In Fig. 4.101a, the module is on a workbench and is "zapped" by a person who touches it. Note that:
1. The culprit person may not even notice the ESD event if he was charged at less than 1.5 to 2 kV (threshold of feeling); therefore, no warning exists of possible damage.

2. The sad thing is that if the workbench was conductive and grounded (a good antistatic practice), this would only make things worse by decreasing the impedance of the ground path.

The moral is that antistatic precautions should be applied across the board. Nobody should handle modules without using a grounded wrist strap or, at the very least, touching a grounded structure first; the workbench should be conductive, yes, but "softly" grounded via a few hundred kilohm resistor.

Plastic-encased modules are prime victims of this type of event since the field gradient creates arcing or charge transfer from the case surface to the chip itself, then sinks to ground via the leads. Metal encapsulated modules have better chances of survival if the arc flows on the metal can and breaks the can-to-benchtop air gap.

Figure 4.101b shows the opposite situation where the module is standing with the leads facing upward. Here metallic-canned modules may be as badly damaged.

Figure 4.101c shows a more insidious scenario: the module has become charged by sliding in carousel rails or by being put in ordinary plastic bags, crates, skin packing, etc. The device can remain charged a significant amount of time until it finds an occasion to get rid of its excess (or lack of) electrons, hence receiving a self-inflicted wound.

Few reports exist of the actual current waveforms in this latter case, but given that the impedance of the device can be very low, and that the benchtop may be conductive (for the very purpose of avoiding static!), it is assumed that the peak current and the total energy can be significant. This scenario, known as the "charged device model," has been identified and discussed in several papers (Refs. 20 through 22). The solution is to always ship ESD-sensitive devices in antistatic bags, preferably the shielded type, and to have their pins shorted out in a conductive foam pad.

Remember also that all the above scenarios repeat themselves throughout the whole life of a device:

1. During wafer processing, testing and cutting
2. During chip transport, test, substrate mounting and encapsulation
3. During module handling, testing and storage
4. At the next step, i.e., PCB level (contrary to common myth, an IC mounted on a PCB *is* still vulnerable)

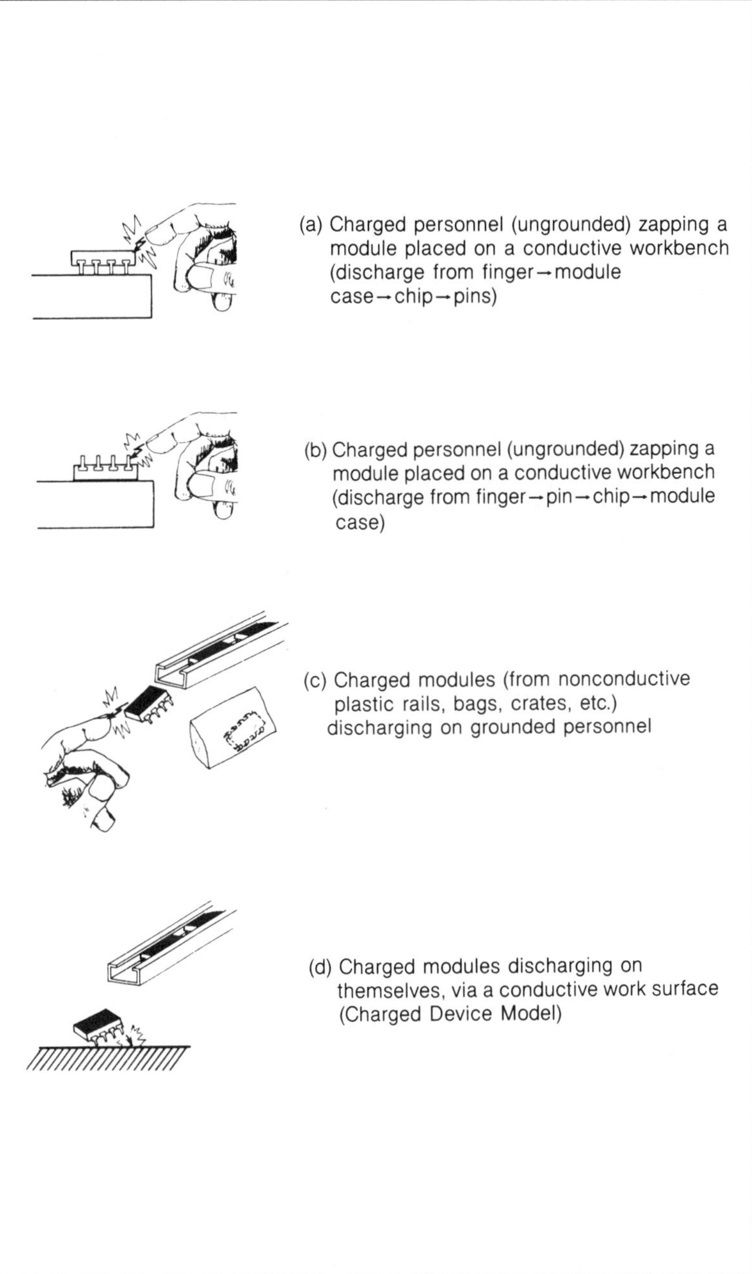

Figure 4.101—Integrated Circuit Damage During Handling

5. In the host machine itself, during maintenance, etc. Figure 4.102 shows a typical static-free work station, and Fig. 4.103 shows some elements of static protection at the factory.

What ESD level does it take to damage a component which is at rest and unpowered? The answer, of course, depends on:

1. The type of technology
2. The type of built-in protection provided by the manufacturer (assuming that the clamping diodes are themselves ESD resistant, which is not always true)
3. The dimensional rules: silicon oxide thickness, width and thickness of traces, traces and pads spacing.

ESD failures generally occur by upsetting the breakdown voltage of the SiO_2 isolation. For instance, in MOS structures with typical oxide thicknesses of 1,000 Å (10^{-7} m), the rigidity of 5 to 7 MV/cm is exceeded if a voltage of 50 to 70 V is reached between two isolated channels. Another common failure is by the joule effect where a junction or a metallized trace will deteriorate and even blow up like a fuse. In either case, the criterion is often the available pulse energy. For current MOS technology, for instance, the threshold of damaging energy is on the order of 1 μJ.

Since energy = power × time, and given that the 50 percent

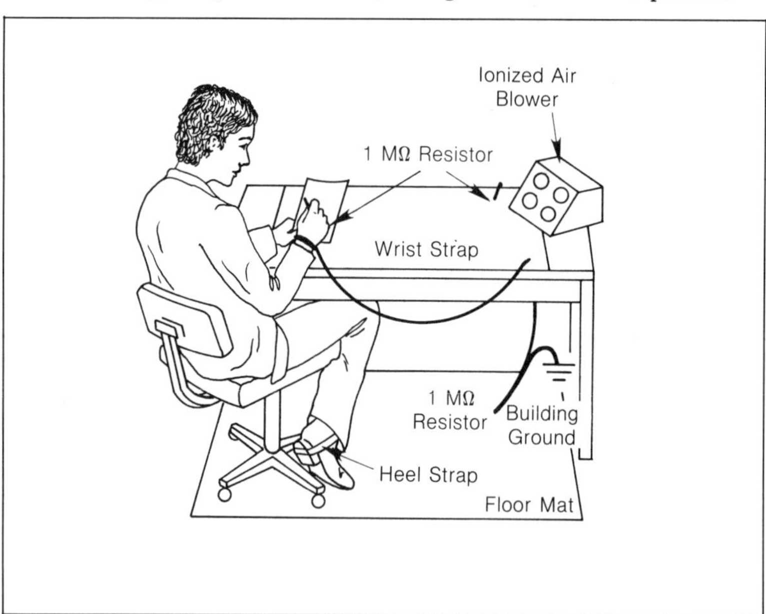

Figure 4.102—Example of a Static-Free Workstation

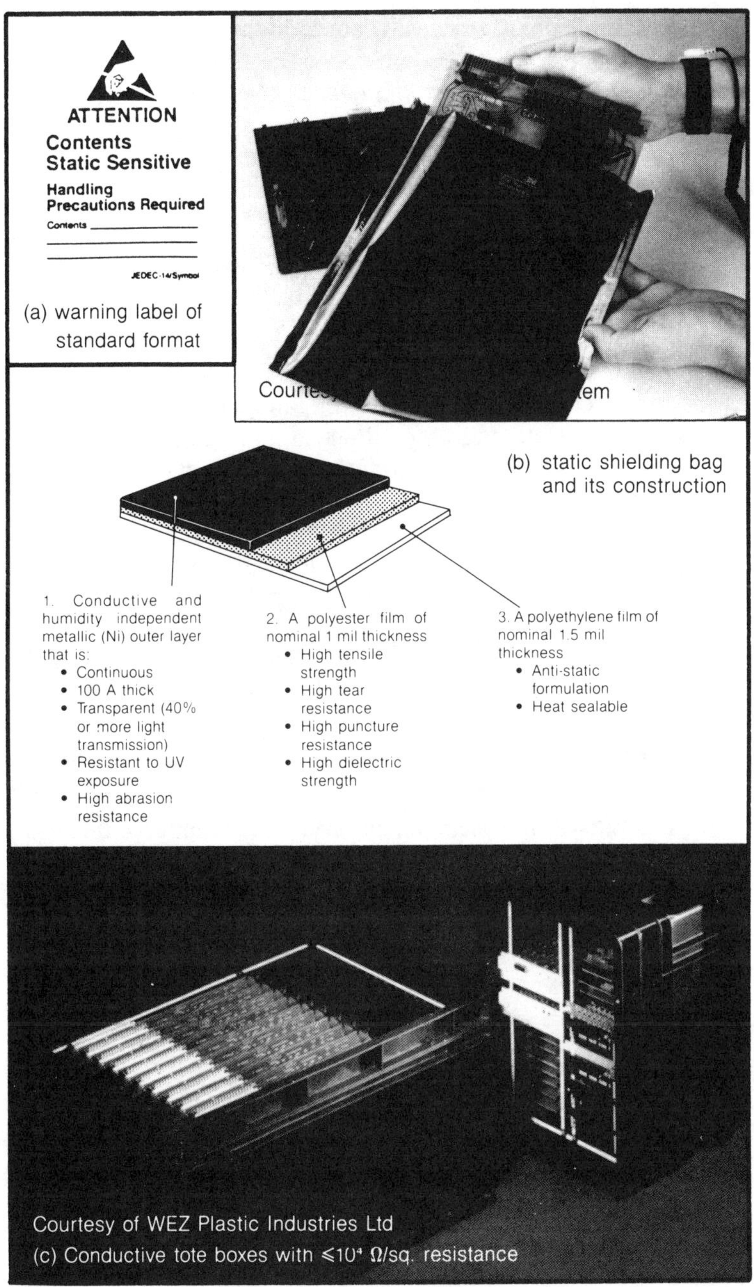

Figure 4.103—Static Protection at the Factory

pulse duration of human ESD is about 150 ns, this indicates a not-to-exceed pulse power of:

$$P = \frac{10^{-6} \text{ J}}{150 \times 10^{-9} \text{ s}} = 6.6 \text{ W} \qquad (4.79)$$

Figures 4.104 and 4.105 show some typical ESD damages in microelectronics. Although CMOS and FET devices are notorious for their fragility to ESD, other technologies can be victims too. The trend toward hyperintegration (VLSI and ULSI with 10,000 or 100,000 gates per chip and gate dielectric down to 250 Å) will only make things worse since all the spacings and thicknesses will be further reduced. It is not an overstatement to say that if ESD vulnerability is not addressed drastically, it can become a major obstacle in the challenge toward faster speeds and shorter propagation delays.

Table 4.3, often reported in ESD literature (Ref. 21), shows the susceptibility of various electronic devices to a standard human body ESD. For someone used to dealing with a certain degree of accuracy, this table may appear as some kind of joke. Which ESD level destroys a JFET: 140 V or 7,000 V? Or a CMOS: 250 V or 3,000 V? How can any protection strategy be optimized if the data are so vague?

Nobody is to blame. There was no sloppiness in the gathering

Table 4.3—ESD Susceptibility of Various Electronic Devices

Device Type	Range of ESD Susceptibility (Volts)
VMOS	30 to 1,800
MOSFET	100 to 200
GaAsFET	100 to 300
EPROM	100
JFET	140 to 7,000
SAW	150 to 500
OP AMP	190 to 2,500
CMOS	250 to 3,000
Schottky Diodes	300 to 2,500
Film Resistors (Thick, Thin)	300 to 3,000
Bipolar Transistors	380 to 7,000
ECL (PC Board Level)	500 to 1,500
SCR	680 to 1,000
Schottky TTL	1,000 to 2,500

(Courtesy of 3M Static Control Systems)

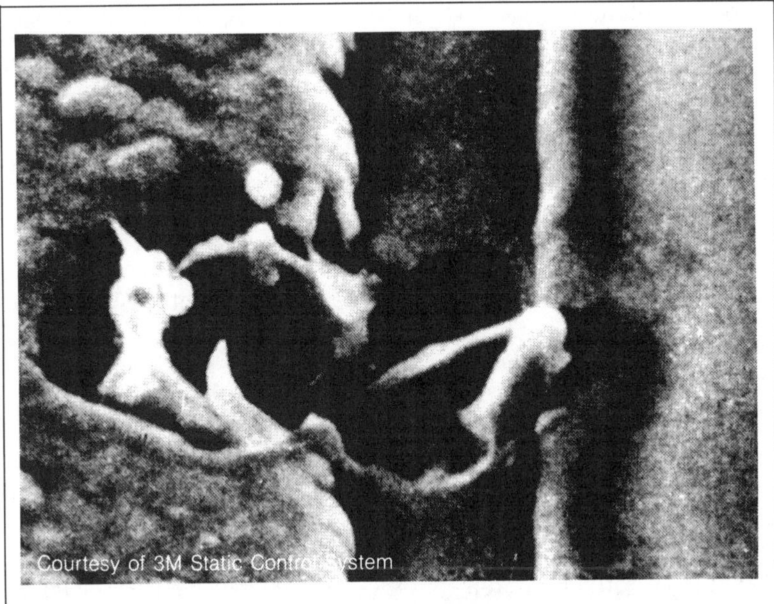

Figure 4.104—Magnified View of Integrated Circuit Damaged by ESD

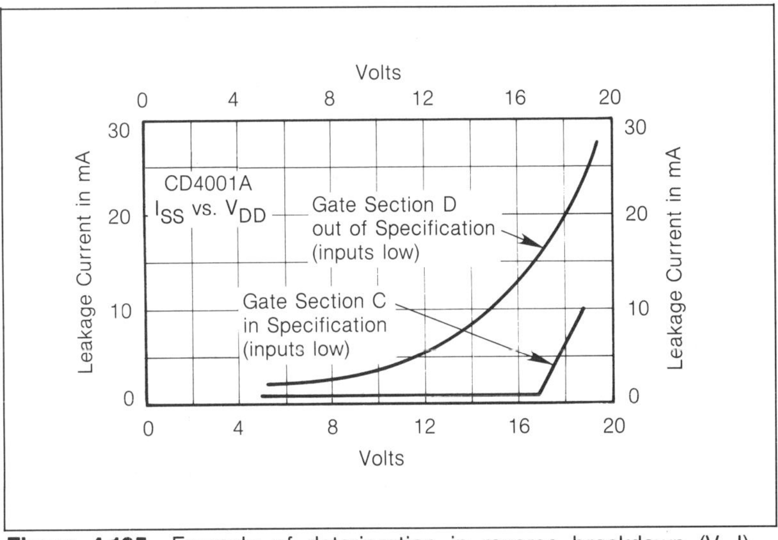

Figure 4.105—Example of deterioration in reverse breakdown (V, I) characteristics. V_{BR} is often used as an idicator of ESD overstress, either by monitoring V for a given current, or monitoring $I_{leakage}$ for a given voltage. The parameter surveyed here was the quiescent leakage current I_{SS} on a CMOS NOR gate.

or compilation of the data. Actual thresholds of ESD failure can, indeed, vary by one order of magnitude or more, based on the following:

1. Damage thresholds differ depending on how the pulse is applied: 1 pin against all other (N-1) pins shorted together? Or each pin selectively against the grounded case? Or all possible combinations?
2. Damage thresholds differ according to pulse polarity.
3. Damage thresholds of the same part number and same manufacturer vary from one vintage to another, depending on the plant of origin or the serial number of the masks used.
4. Damage thresholds are not the same for a unique test pulse as for repeated pulses.
5. Damage thresholds of identical, compatible part numbers from different manufacturers vary widely.
6. Reported levels depend on the criteria selected: was a part declared failed when it was functionally wrong? Or as soon as its dc parameters deviated from the initial specification?

Figure 4.106 from Ref. 23 shows, on a Weibull distribution graph,

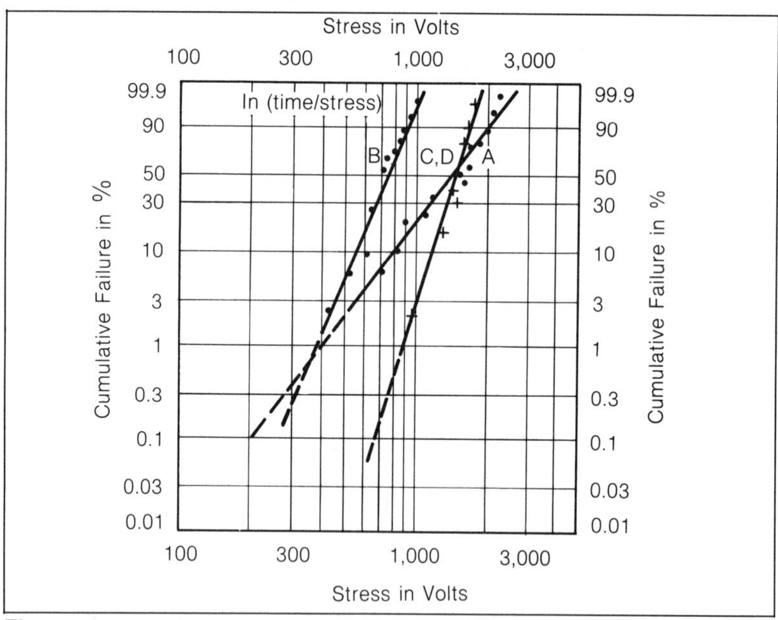

Figure 4.106—ESD Failure Levels for 4001 Inverters from Manufacturers A, B, C and D

This is a case where all the usual methods of stress tests and QC statistics have to be used to determine a reliable number which the ESD threshold of CMOS inverters from four different manufacturers. The monitored parameters included an input current > 1 μA (indicating a damaged gate isolation) or output not responding to input change, whichever failure came first.

The graph shows ratios of 3 to 1 or 4 to 1 between the best part and the worst part of the sample. Also, the deviation varies widely from one manufacturer to another.

Another example showing that the ESD damage level is never a "green light/red light" situation is shown in Fig. 4.107 from Ref. 24. The product tested was a bipolar logic gate type 54L04 (TTL low power). One hundred devices were tested for multiple pulses until a failure occurred.

The criterion for failure was the deviation of two critical parameters from their normal value. The normal value of $I_{input High}$ was 10 μA_{max}, and failure was declared if $\Delta > 20$ nA. The normal value of $V_{out low}$ was 0.3 V max; failure was declared if $\Delta > 0.05$ V. The results are shown in number of parts that did not survive to N pulses at level V.

All the parts survived after the application of up to 175 pulses at 2,000 V, then 10 died between 176 and 200 pulses. The survivors were stressed to 2,750 V, where casualties are seen as soon as the first pulse. Finally, of the 15 survivors who came victoriously through the 200 pulses at 5,000 V, 10 died at the first application of a 5,750 V electrostatic discharge. Can we then say that we have a product that can withstand ESD at 2,000 V? Or up to 5,000 V?

Pulse Level	Number of Pulses to Failure									
	1	2	3	4	8	50	75	125	175	200
5750	10									
5000	12	1			2					
4250	16				1					3
3500	9			2	1					8
2750	2		1		1	1	1	1	2	11
2000										10

Figure 4.107—Failure decision is never a "step function." Example: damage testing of a 54 L04 (TTL LowPower). One hundred devices were tested. Normal specification of $I_{input high}$ was 10 μA max, and the device failed if $\Delta > 20$ nA. Normal specification of $V_{out low}$ was 0.3 V max; and failure was declared if $\Delta > 0.05$ V.

will characterize the ESD immunity with a certain confidence level. In all these tests, an unstressed sample lot should be monitored to avoid introduction of an uncontrolled variable not related to ESD.

An interesting question arises: Would not the 15 heroes that came through 5,000 V have gone even further if they had not been inflicted at lower level 600 discharges before? Sometimes, to investigate this, fresh parts marked for identification are reintroduced into the lot to replace the victims as the test goes on.

Another related problem is one of latent failures. A part which still *appears* undamaged after an ESD test may, in fact, have its lifetime affected. There are several confirmations that parts which have been subjected to ESD (either by accidental or intentional events) become "walking-wounded" and exhibit abnormal failure rates in the field.[22] But even this is not always true. Some sample lots which were ESD-stressed have exhibited, during accelerated life tests, better lifetimes than unstressed sample lots! This self-healing phenomenon has been given several explanations, the descriptions of which would be far beyond the scope of this book.

Since 1980, the U.S. Department of Defense has issued a standard document, DoD-STD-1686, defining all the requirements of an ESD control program for electronic components and assemblies. This standard, originally intended for suppliers and subcontractors of the DoD, has become a commonly used reference for industry in general.

The DoD standard calls for:
1. Identification and tagging of ESD-sensitive items (class 1 with sensitivity $\leqslant$1,000 V and class 2 with sensitivity between 1,000 and 4,000 V). Details of the test are given in Chapter 3.
2. Built-in circuit protection at chip and card level
3. ESD-proof handling, shipping, etc.
4. QC and audits
5. Field maintenance precautions

The standard also requires that subcontractors rule out class 1 devices when a class 2 device is available which would perform identical functions.

As a complement to DoD-STD-1686, the Department of Defense has also issued *Handbook #263* which gives ESD control guidelines and details of failure mechanisms. The handbook also contains classifications of ESD protection equipment, materials, manufacturing and shipping procedures.

4.7.1 ESD Protection Integrated in Chip Design

The drastic reduction in area and thickness of the active elements occurring in recent generations of semiconductors makes each new logic family more prone to ESD damage than the former one. Technologies achieving more than 1,000 gates/mm^2 and offering speed $\times$ power products much less than 1 pJ and propagation delays inferior to 100 ps will have a channel length of 0.25 μm. This corresponds to a theoretical breakdown voltage of only 20 V!

Designers of integrated circuits, hybrids and microelectronics can generally build in a certain level of ESD hardening through layout precautions and integrated protection networks. This hardening will make the chips or modules safer to handle and will relieve the user of some of the cost and burden of hardening. The following guidelines are taken from Military DoD-Handbook 263 and from technical notes from manufacturers such as National Semiconductor and Texas Instruments.

4.7.1.1 Protection of CMOS Devices*

General. Various protection networks have been developed to protect sensitive MOS. These circuit protection networks provide limited protection against ESD. Many of the protection networks designed into MOS devices reduce the susceptibility to ESD to a maximum of 800 V. MIL-M-38510 V— ZAP test voltages for CMOS, for example, vary from 150 to 800 V. Protection circuitry of some devices is improving, and protection to 4,000 V appears to be achievable for some MOS devices. However, electrostatic potentials of tens of thousands of volts can be generated in uncontrolled environments.

The protection afforded by specific protection circuitry is limited to a maximum voltage and a minimum pulse width. ESDs beyond these limits can subject the part's constituents to damage or damage the protection circuitry constituents themselves. These too are often made of moderately or marginally sensitive parts. Damage to the protection circuitry constituents could result in degradation in part performance or make the ESD-sensitive part more susceptible to subsequent ESDs. The degradation, for example, could be

---

*This section is taken from MIL-HDBK-263

4.161

a change in speed characteristics of the parts or an increase in leakage current. Multiple ESDs at voltages below the single ESD pulse sensitivity voltage or energy level can also weaken or cause failure of the part performance or protection circuitry constituents, resulting in degradation or failure of the part. Loss of protective circuitry may not be apparent after an ESD.

In summary, protection networks reduce but do not eliminate the susceptibility of a part to ESD. This reduction in ESD sensitivity, however, results in a lower incidence of ESD part failure.

The sensitivity of the same type of ESD-sensitive part can vary from manufacturer to manufacturer. Similarly, the design and effectiveness of protection circuitry also varies from manufacturer to manufacturer.

4.7.1.2 Design Precautions

Various design techniques have been employed in reducing the susceptibility of parts and assemblies to ESD. Diffused resistors and limiting resistors provide some protection, but are limited in the amount of voltage they can handle. Zeners require greater than 5 ns to switch and may not be fast enough to protect a MOS gate. Furthermore, zener schemes, diffused resistors and limiting resistors reduce performance characteristics which in many instances are the primary consideration for which that part was designed.

4.7.1.3 Part and Hybrid Design Considerations

Some design rules to reduce ESD sensitivity for parts and hybrids are as follows:

1. MOS protection circuitry improvement techniques are increasing diode size, using diodes of both polarities, adding series resistors and utilizing a distributed network effect.
2. Avoid cross-unders beneath metal leads connected to external pins; otherwise, treat the part as electrostatic sensitive. Also, since cross-unders are diffused during the n+ (emitter) diffusion process, the oxide over the diffusion will be thinner, causing this area to have a lower dielectric breakdown. Deep n diffusions, rather than n+ diffusions, should be used for cross-unders if a deep n diffusion step is used in the fabrication process.

3. MOS protection circuits should be examined to see if the layout permits the protection diodes to be defective or blown without causing the circuit to be inoperative.

4. Distance between any contact edge and the junction should be 70 μm or greater on bipolar parts.

5. Linear IC capacitors should be paralleled by a p-n junction with sufficiently low breakdown voltage.

6. For bipolar parts, avoid designs permitting a high transient energy density to exist in a p-n junction depletion region under ESD. Use series resistance to limit ESD current or use parallel elements to divert current from critical elements. The addition of clamp diodes between a vulnerable lead and one or more power supply leads can improve ESD resistance by keeping critical junctions out of reverse breakdown. If a junction cannot be kept out of reverse breakdown, physically enlarging the junction will make it more ESD resistant by reducing the initial transient energy density in inverse proportion to its area.

7. The protection of a transistor from ESD can be improved by increasing the emitter perimeter adjacent to the base contact. Enlarging the emitter diffusion area also helps in some pulse configurations.

8. As an alternative to using clamping diodes, which consume chip area and can cause unwanted parasitic effects, a "phantom emitter" transistor can be used to improve ESD resistance. The phantom transistor incorporates a second transmitter diffusion shorted to the base contact. This creates a deliberate separation of the base contact from the normal emitter without interfering with normal transistor operation. The second emitter provides a lower breakdown path BV_{CEO} between the buried collector and the base contact.

9. Avoid pin layouts which put the critical ESD paths in corner pins which are prone to ESD.

10. Avoid metallization crossovers where possible. These crossover areas are typically separated by thin dielectric layers. Crossovers often impose a number of metallurgical requirements which are frequently incompatible. For example, once the first metallization layer (Al) is deposited, the circuit cannot be subsequently heated in excess of 550°C because the eutectic point of the Al-Si system is 575°C. Thus, the dielectric layer (SiO_2) should be deposited by a low-

temperature process such as pyrolytic deposition. This layer is prone to breakdown from ESD for two reasons:

    a. A low temperature growth of SiO_2 generally is not uniform in thickness and not free from pin holes.

    b. The dielectric layer is thin and thus the breakdown voltage is very low.

11. Avoid parasitic MOS capacitors whenever possible. Microcircuits with metallization crossing over low-resistance active regions, that is, V_{CC} over n+ isolated diffusions, are moderately sensitive to ESD. Such constructions include microcircuits with metallization paths over n+ guard rings. The n+ guard rings are used in the n-type epitaxial islands to inhibit possible inversion of the n-type semiconductor to p-type semiconductor and to reduce the leakage current. Since the final oxide layer over the n+ guard ring is relatively thin, parasitic MOS capacitors of relatively low breakdown voltage are created when a metallization path passes over this ring. These MOS capacitor structures are ESD-sensitive.

12. Caution is advised in the use of microcircuits and hybrids containing dielectrically isolated bipolar parts which are generally moderately sensitive to ESD. Failure occurs due to breakdown of the thin dielectric layers between these small geometry parts from an ESD.

4.7.2 Assembly Design Considerations

Procedures are as follows:

1. Latchup in CMOS, with the exception of analog switches, can be avoided by limiting output current. One solution is to isolate each output from its cable line with a resistor and clamp the lines to V_{DD} and V_{SS} with two high speed switching diodes. The use of long input cables poses the possibility of noise pickup. In such cases, filter networks should be used.

2. Additional protection can be obtained from MOS by adding external series resistors to each input.

3. Where practicable, an RC network consisting of a relatively large value resistor and a capacitor of at least 100 pF should be used for sensitive inputs on bipolar parts to reduce effects from ESD. However, if circuit performance dictates, two parallel diodes clamping to 0.5 V in either polarity can be used to shunt the input to ground. This reduces disturbances to the input characteristics.

4. Leads of sensitive parts mounted on PWBs should not be connected directly to connector terminals without series resistance, shunts, clamps or other protective means. Assembly designs containing ESD-sensitive items should be reviewed for incorporation of protective circuitry.

4.7.3 ESD Part Protection Networks

Manufacturers have incorporated protection circuitry on most MOS devices (see Fig. 4.108). The purpose of these protection networks is to reduce the voltage across the gate oxide below the dielectric breakdown voltage without interfering with part electrical performance. Differences in fabrication processes, design

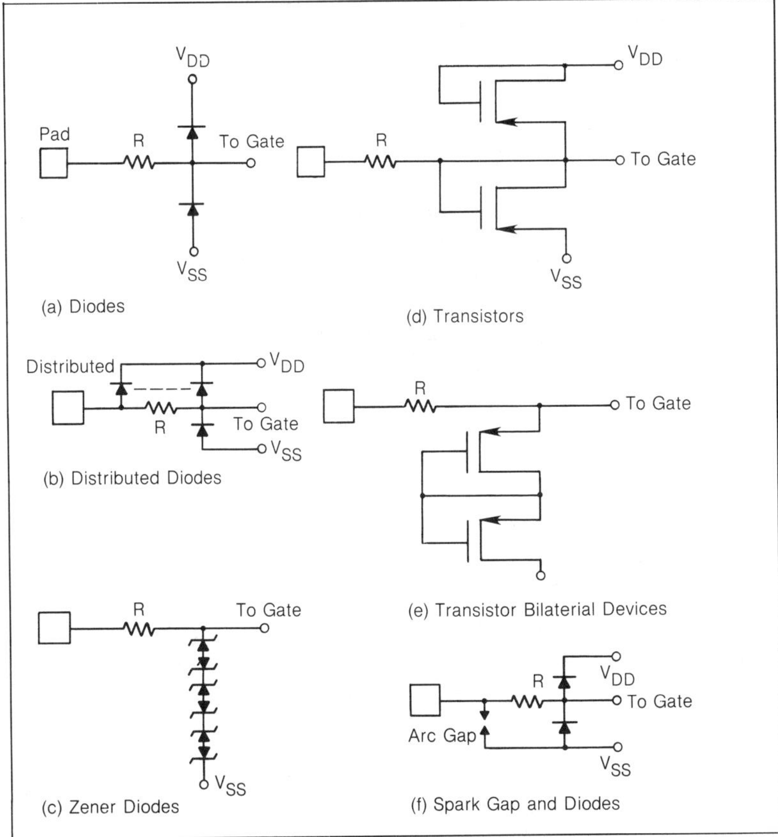

Figure 4.108—Gate Protection Networks

philosophies and circuitry have resulted in difference gate protection networks.

Standard Input Protection Networks*

In order to protect the gate oxide against moderate levels of electrostatic discharge, protective networks are provided on all National CMOS devices, as described below.

Figure 4.109 shows the standard protection circuit used on all A, B and 74C series CMOS devices. The series resistance of 200 Ω using a P^+ diffusion helps limit the current when the input is subjected to a high-voltage zap. Associated with this resistance is a distributed diode network to V_{DD} which protects against positive transients. An additional diode to V_{SS} helps to shunt negative surges by forward conduction. Development work is currently being done at National on various other input protection schemes.

Other Protective Networks

Figure 4.110 shows the modified protective network for CD4049/4050 buffer. The input diode to V_{DD} is deleted here so that level shifting can be achieved where inputs are higher than V_{DD}.

Figure 4.111 shows a transmission gate with the intrinsic diode protection. No additional series resistors are used, so the "on" resistance of the transmission gate is not affected.

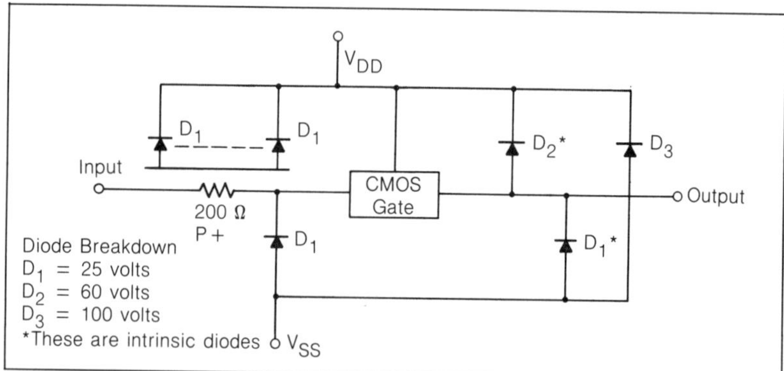

Figure 4.109—Standard Input Protection Network (courtesy of National Semiconductor)

*This section is taken from National Semiconductor, **Application Note 248.**

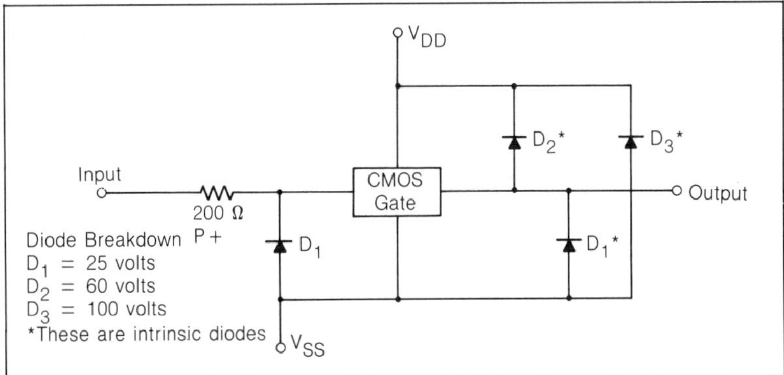

Figure 4.110—Protective Network for CD4049/50 and MM74C901/2 (courtesy of National Semiconductor)

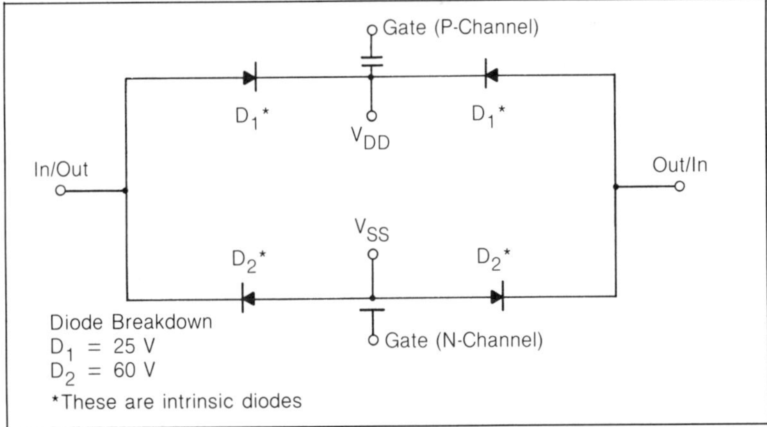

Figure 4.111—Transmission Gate with Intrinsic Diodes to Protect against Static Discharge (courtesy of National Semiconductor)

4.8 Electron Tubes

EMI problems associated with electron tubes and their circuits are similar to those with transistors, differing primarily because of power levels and a greater number of elements as well as greater spacing between elements.

Although ordinary vacuum tubes operating in the linear or switching modes have EMI characteristics similar to those of transistors, gaseous conduction tubes (such as thyratron and ignitron) are quite different. The ignitor in an ignitron can be a source of noise because it acts like a spark-producing device. When a gaseous conduction tube is conducting, the space between anode and cathode contains a plasma that can oscillate by itself because of internal instabilities.

The principal types of tube noise are:
1. Shot effect
2. Partition noise
3. Induced noise
4. Gas noise
5. Secondary emission noise
6. Flicker effect
7. Filament noise (hum)

Shot noise (also called Schottky noise or Schot noise) is due to the random fluctuations in the rate of electron emission from the cathode. When the cathode temperature is the current-flow limiting factor, the shot-noise component is:

$$i_{sn}^2 = 2\ eI_bB \qquad (4.80)$$

where, i_{sn} = rms shot-noise current in A
$\quad$ e = electron charge = 1.6×10^{-19} C
$\quad I_b$ = average plate current in A
$\quad$ B = bandwidth in Hz

When the plate current is limited by the space charge, many of the fluctuations in the plate current are reduced due to the smoothing effect of the reservoir of electrons in the virtual cathode set up by the space charge. In this case, the following approximations can be used. For diodes:

$$i_{sn}^2 = 4\ K\ (0.64)\ T_c\ gB \qquad (4.81)$$

and for negative-grid triodes:

$$i_{sn}^2 = \frac{4\ K\ (0.64)\ T_c\ g_mB}{\delta} \qquad (4.82)$$

where, K = Boltzmann's constant = 1.38×10^{-23}
$\qquad$ W/°Kelvin/Hz
$\quad T_c$ = cathode temperature in °K

g = diode-plate conductance in mhos
g_m = triode transconductance in mhos
δ = tube parameter (between 0.5 and 1)
B = bandwidth in Hz

The shot-effect noise of a triode can be expressed by considering a resistance which, if applied to the driving grid of a noiseless tube as a source of thermal noise, would produce the same anode-current noise component as is actually present. For triode tubes, the equivalent noise resistance R_{eq} is equal to $2.5/g_m$. In amplifiers, the noise voltage generated by R_{eq} is considered to be applied in series with the grid as shown in Fig. 4.112:

$$e^2 = 4\,KT_c\,R_{eq}\,B \qquad (4.83)$$

where, e = rms noise voltage
R_{eq} = equivalent grid-noise resistance

Partition noise occurs in multicollector tubes and is due to fluctuations in the division of current between the electrodes.

The noise of a negative-grid pentode amplifier is represented by the equivalent grid resistance approximated by:

$$R_{eq} = \frac{I_b}{I_b + I_{c2}}\left(\frac{2.5}{g_m} + \frac{20\,I_{c2}}{g^2_m}\right) \qquad (4.84)$$

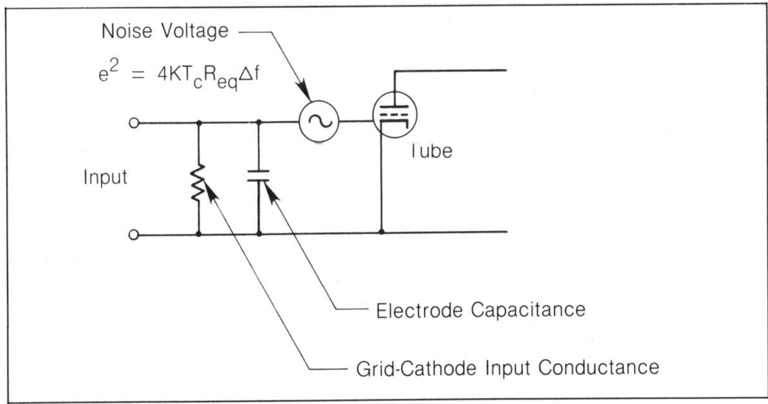

Figure 4.112—Equivalent Circuit Representing Plate-Current Noise by a Generator in Series with a Tube Grid

where, I_b = dc plate current in A
I_{c2} = dc screen current in A
g_m = transconductance in mhos

The values of R_{eq} of pentodes are usually 3 to 10 times as great as those of comparable negative-grid triodes. Electron-wave tubes, such as traveling-wave tubes, also exhibit partition noise.

At VHF (over 30 MHz), fluctuations in the number of electrons passing a negative grid induce noise currents. Increasing with frequency, the noise currents are introduced into the input circuitry by electronic conductance.

Gas noise is produced by erratic motion of gas molecules in gas or vacuum tubes. Ionization by collision produces noise when ionized gas atoms or molecules liberate bursts of electrons when they strike the cathode.

Secondary emission noise is caused by fluctuations in the rate of the production of secondary electrons. Flicker noise, more common in oxide-coated cathodes, varies as 1/F and is caused by a low-frequency variation in cathode activity. Some EMI problems involving electron tubes are listed in Table 4.4.

Table 4.4—EMI Problems, Causes and Corrections in Electron Tubes
(continued next page)

Problem	Cause	Corrections
Shot effect	Random fluctuations in the rate of emission of electrons from the cathode	Proper design
Partition noise	Fluctuation in the division of current between electrodes	If important, use tubes with fewer elements
Induced noise	At VHF, fluctuations in the number of electrons passing a negative control grid induces noise currents	Use tube with better grid geometry
Gas noise	Ionization by collision of ionized gas striking the cathode, liberating bursts of electrons	Proper design or use of other electronic components

Table 4.4—(continued)

Problem	Cause	Corrections
Flicker noise	Inversely proportional to frequency: due to a low-frequency variation in cathode activity; more common in oxide-coated cathodes	Avoid oxide-coated cathodes
Microphonic effect (vibration of socket with or without sound emanation)	Shock or vibration changes element spacing and damped mechanical oscillation causes changes in plate current, usually setting up vibration through the tube socket or by means of sound waves	Reduce by use of ruggedized tubes
Hum	Use of ac for filament and heater-type tubes	Use dc, use humbucking circuits, use balanced filament supply, bias cathodes negative with respect to filaments
Other tube noise	Leakage from the grid to another electrode, particularly a positive electrode; improper contacts (particularly with low level signals)	Tube replacement or ensure mechanically and electrically good contacts
Large number of unwanted frequencies at amplifier output	Caused by possible feeding in of unwanted frequencies compounded with operation on the non-linear portion of the characteristic curve	Proper design and shielding of the grid circuit
Interfering signals in an RF field	Pickup of RF signals through tube envelopes	Use tube shields
Affected by nuclear radiation	Change in tube characteristics	Use ceramic tubes
Noise current in catcher of klystron amplifier	Electron beam passing through the catcher causes noise current in the shunt impedance of the catcher	Proper design for minimum effect
Transient flashover	Secondary emission from grid, aggravated by impurities. Grid becomes a temporary "cathode."	Tube replacement, quality monitoring by tube manufacturer

4.9 Power Electronics Semiconductors

Power electronics semiconductor EMI considerations are different from those for small-signal or logic devices. Generally, since they handle large currents or voltages, the main problem is their EMI generation; in some cases, like thyristors, their susceptibility can be a concern in the sense that false triggering can turn on a powerful machine or process and become damaging or hazardous.

4.9.1 Power Diodes and Transistors

As semiconductors, power diodes and transistors share many dynamic characteristics. Diodes act as both sources of EMI (rectifier diodes) and EMI suppressors (clamping diodes). EMI suppressors will be discussed in subsequent chapters on relays, switches and transient protectors.

Under conditions of forward bias, a solid-state semiconductor stores a certain amount of charge in the form of minority current carriers in the depletion region. If the diode is then reverse-biased, it conducts heavily in the reverse direction until all of the stored charge has been removed. The resulting conditions are presented in Fig. 4.113. The duration, amplitude and configuration of the recovery time pulse (also called switching time or period) is a function of diode characteristics and circuit parameters (Fig. 4.114). These current spikes generate a broad spectrum.

Rectification involves switching from conduction to cutoff repetitively. This causes high dI/dt values dependent upon the input frequency, minority carrier storage in the diode and the circuit characteristics. The interference effect can be minimized by one or more of the following measures:

1. Placing a bypass capacitor in parallel with each rectifier diode
2. Placing a resistor in series with each rectifier diode
3. Placing ferrites on diode leads
4. Placing an RF bypass capacitor to ground from one or both sides of each rectifier diode
5. Using diode with smooth recovery
6. Operating the rectifier diodes well below their rated current capability

The high dI/dt during recovery spikes has several severe consequences:

1. The peak current creates I × R drops in the circuit using the diode.

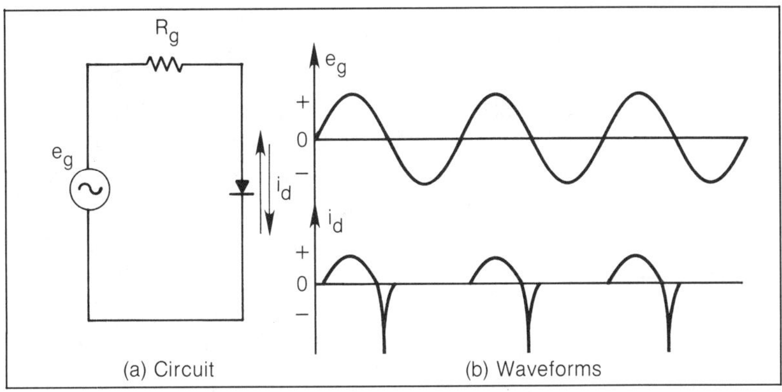

Figure 4.113—Diode Recovery Periods and Spikes

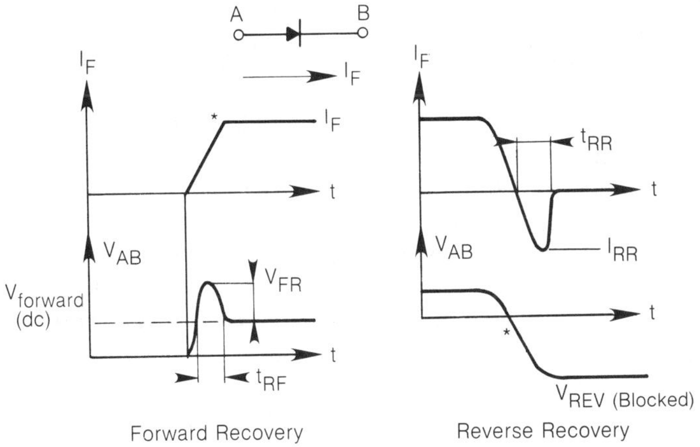

Forward Recovery

Reverse Recovery

*Note: I_F and V_{AB} are shown as ideal. In fact, V_{FR} causes in turn I_F oscillation and I_{RR} causes V_{REV} oscillations.

	Schottky	PN Fast Recov. (Gold Diffused)	Ultra-Fast Recovery
Rev. Rec. Time	10 ns	100-200 ns	25 ns
Recovery Waveform	Soft	Med	Abrupt
Example of Current and Voltage Overshoot for a 50 A Switching with a dI/dt of 400 A/µs	3 A, − 22 V	12 A, + 40 V	3 A, − 75 V

Figure 4.114—Forward and Reverse Recovery with Power Diodes

2. The LdI/dt creates an inductive transient in the same circuit.

3. The sudden current change creates a changing field around the diode circuit which can induce noise in nearby loops.

To decrease these effects, the area of the branching circuit should be kept minimal. Also, diodes should not be mounted on a steel bracket or on any magnetic metal which would create an increased inductance (L) around the diode body.

On the other hand, the highest spectrum of diode recovery spikes can be attenuated by lossy ferrite rings around the diode leads (see ferrites, Chapter 5). Some power diode manufacturers have even incorporated these ferrite rings into the diode construction. Ferrites only exhibit inductive and resistive drop at high frequencies. This fix shows no losses in the useful current bandwidth but some losses at the undesirable frequencies.

The ripple filter that normally follows a rectifier should not be relied upon to filter out the transient emissions. The usual large-value capacitors used for ripple filters exhibit too much series inductance to function effectively as RF interference filters. The filter capacitors should be shunted with a second capacitor having a value of two or more orders of magnitude of less capacitance.

Generally, in a rectifying circuit with low source impedance like the secondary of a power supply regulator, the RF filtering should be at least an LC, with the inductance looking toward the diode.

Diodes are also used to switch at a particular voltage level. The switching action results in high values of dI/dt. Diodes are used as limiters or clippers to cut off input waveforms at a certain level. The greater the amount of limiting, the greater the chance that spurious frequencies will occur due to the steepening waveform. These switching actions can result in EMI interaction with other components distributed or actual impedances and signal discontinuities.

Several design considerations that minimize switching EMI transients are:

1. Operating at the lowest possible voltages and currents
2. Anticipating diode-to-diode variation of characteristics
3. Using the lowest possible switching rate or rise time and amplitude
4. Selecting diodes with high working and peak inverse voltages
5. Using diodes with a slow recovery time (inherent with larger current ratings)

RF voltage can change the bias of a diode, resulting in improper switching, distortion or output. All diodes are subject to reverse breakdown if they are exposed to RF voltages greater than their reverse breakdown voltages. Low-power devices (generally those rated 25 mW or less) and small-junction devices (such as point contact diodes) that operate in the vicinity of a strong RF field can absorb sufficient radiated energy to be degraded or burned out. Large junction diodes have a large junction capacitance, of the order of 10 to 15 pF, which will pass high frequencies. If this energy added to the normal energy exceeds the thermal limit of the device, damage can occur. Therefore, diodes subjected to RF fields should be shielded.

When used as an amplifier, a tunnel diode may couple with related circuitry inductance or capacitance to produce parasitic oscillations, usually about 1 MHz. The oscillations should be suppressed by circuit design methods. All zener diodes generate shot and 1/F noise, but the noise level is higher in alloy zeners than in zeners made by a diffusion process.

Generally, noise increases with an increase in current, but the noise may occur at one point on the zener curve and not on others. This noise is called "spotty." Most commercial zener diodes exhibit noise levels from 1 μV to 1 mV.

Diodes with mechanical imperfections may generate noise when physically agitated. Such diodes may not cause trouble if used in a vibration-free environment.

Another cause of EMI with diodes is their unavoidable parasitic capacitance; that is:
1. Junction capacitance, which more or less bypasses the diode above a certain frequency such that the diode is a perfect dc blocker but a potential RF path for interference
2. Diode-to-mounting hardware (generally a heat sink) capacitance which causes C dV/dt leakage currents to chassis and grounds

The latter problem is difficult to eliminate because it depends on the footprint of the diode body and eventually the ϵ_r of the insulating washer. Its value can be tens of picofarads (see Fig. 4.115).

For power transistors, since they are more or less used as switches in power supply regulators, power controls (motors, relays, actuators) and electronic ignitions, the EMI aspects are the same as for diodes.

Modern power transistors, bipolars or MOSFET's are characterized by their fast switching times of 100, 50 or even

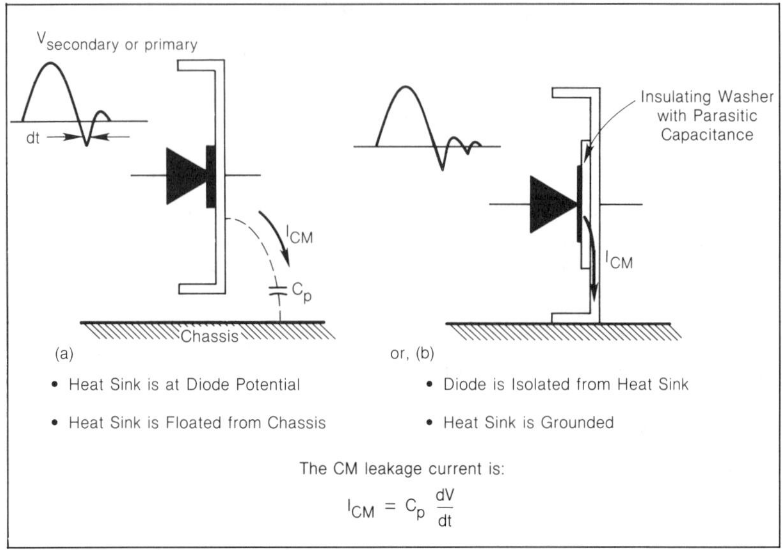

$$I_{CM} = C_p \frac{dV}{dt}$$

Figure 4.115—Contribution of Primary or Secondary Diodes to Common-Mode Current

30 ns. When the transistor is switching heavy currents in low-voltage applications, (such as V/I that is much smaller than 377 Ω), the EMI radiated is mostly an H-field mode.

In a switched-mode power supply, when the transistor is switching line voltages like rectified 115 Vac (160 V peak) or 230 Vac (320 V peak), it is radiating both in E-field mode when blocked and H-field mode when saturated. Therefore, as for diodes, the transistor must be mounted to minimize the area of radiating elements. With TO3 transistors, one could expect that the metal can acts as a shield, but since it is usually at collector voltage, the can acts as the "antenna" when switching to V_{ce} max. With a plastic can like TO220, the whole device's area is an active radiator.

The parasitic capacitances which influence EMI in a power transistor are: (1) Collector-to-base and (2) Collector-to-heat-sink, or more generally, to the resting surface of the transistor.

Collector-to-base capacitance can represent around 10 pF for power transistors in the 10 to 100 W range. It causes the switching command pulses to appear on the source side (collector, in the case of an npn) and also the output noise to couple back to the collector side via the conductive base-emitter junction.

Collector-to-heat-sink capacitance can represent 10 to 100 pF (Fig. 4.116) and cause leakage currents during switching of large voltages. For instance, if a bulk dc voltage of 500 V is turned on and off in 100 ns with a 30 pf parasitic capacitance, since the voltage source has generally one point to ground (neutral), a peak leakage current will flow into the chassis of the equipment equal to:

$$I = 30 \times 10^{-12} \times 500/10^{-7} = 0.15 \text{ A} \qquad (4.85)$$

The ways to minimize EMI from power transistors are very similar to those for diodes:

1. Use an RF bypass across the collector-emitter, presenting a low impedance outside the useful bandwidth of the transistor. These RF "snubbers" consist of plastic or ceramic capacitors in the 0.1 to 1 μF range in series with a few ohms carbon resistor. This snubber must be designed with extremely short leads to avoid parasitic LC ringing which would make the problem worse, since the circuit is under-damped (small resistance). Provided this precaution, the capacitor is a short

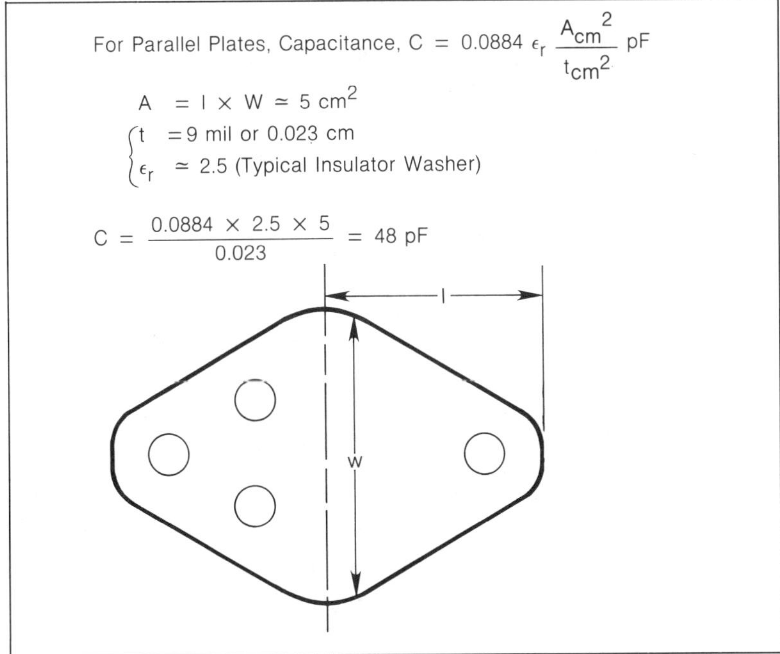

Figure 4.116—To-5, Transistor-to-Heat-Sink Common-Mode Capacitance

for the higher harmonics of V_{ce}, which dissipate in heat within the noninductive resistor.

The use of a snubber also has the advantage that some of the switching losses will be dissipated in the resistor instead of being dissipated in the transistor. No matter where these losses are, they penalize the efficiency, but letting them occur in the transistor will also increase the junction temperature. This in turn creates an increase in V_{ce} (or $R_{DS"on"}$ with FETs) which increases conduction losses, and an increase in the failure rate since, according to the Arrhenius model, when temperature increases from Tj_1 to Tj_2, failure rate increases also $(Tj_2/Tj_1)^6$.

2. Add (or integrate in the device) a clamping device across the collector-emitter (or drain-source) leads.

3. Use a transistor with the smoothest transition time compatible with the function. In power-switching applications where the designer tries to avoid a class A operation and tries to bring the transistor as quickly as possible from blocking into saturation and vice versa, a long transition time is in conflict with the low dissipation requirements. However, there is a compromise to make between power efficiency and EMI. If a certain amount of power has to be switched at 50 kHz, for instance, the power represented by the sum of the harmonics beyond harmonic #30 (that is, 1.5 MHz) is only a small percentage of the total power. The 1.5 MHz frequency corresponds to a rise time of 200 ns; any transition faster than this will not greatly improve the efficiency but will certainly create EMI. The cost, volume, etc., of additional components required to deal with this EMI might outweigh the small percentage of better efficiency. In fact, these components **may very well absorb in losses** the very same percentage of improvement which was realized from the faster transition. Often one selects the fastest transistor available, then tries to slow it down by decoupling to reduce noise!

4. Mount the transistor to heat sink via an insulating washer which incorporates a Faraday shield, connected to emitter, to decouple leakage current back into the transistor loop instead of letting it sink to chassis (Fig. 4.117).

5. Use an isolated collector scheme (Fig. 4.118) which reduces the collector-to-chassis capacitance by 10 times.

6. For high-power switching transistors in the kilowatt range,

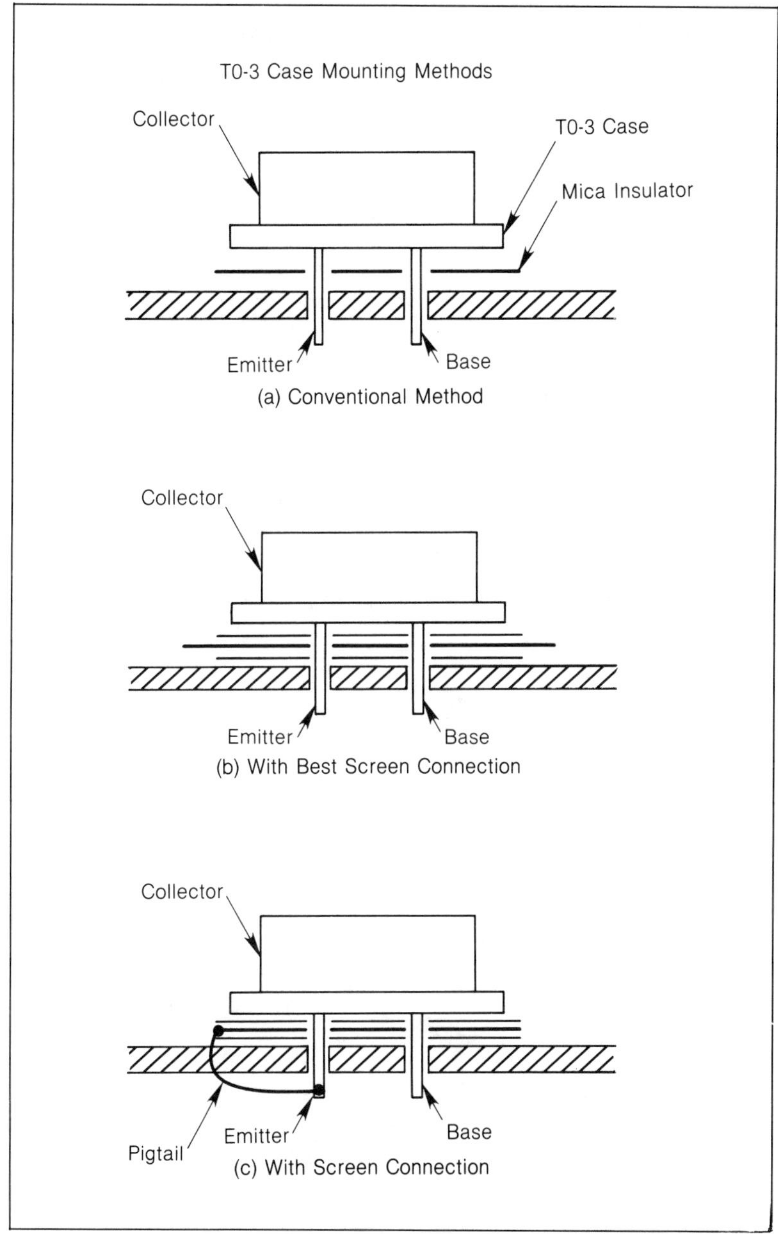

Figure 4.117—Neutralization of Transistor (or Diode) Leakage Capacitance by a Faraday Shield. The Bottom Drawing Is a Shielded Insulating Washer Made by Bergquist Corp. (continued next page)

Thickness (total)	0.4826 ± 0.0508 mm (0.019 ± 0.002")
Shield Thickness	0.0381 mm (0.0015")
Approx. Thermal Resistance (TO-3)	0.85-1.0°C/W
Minimum Breakdown Voltage between Device and Copper	5,000 V
Capacitance at 1,000 Hz and 5 V (TO-3)	50.5 pF
Dissipation Factor at 1,000 Hz and 5 V (TO-3) Power Factor	0.0156
Dielectric Constant at 1,000 Hz and 5 V	4.50
Continuous Use Temp. °C	− 60 to + 200

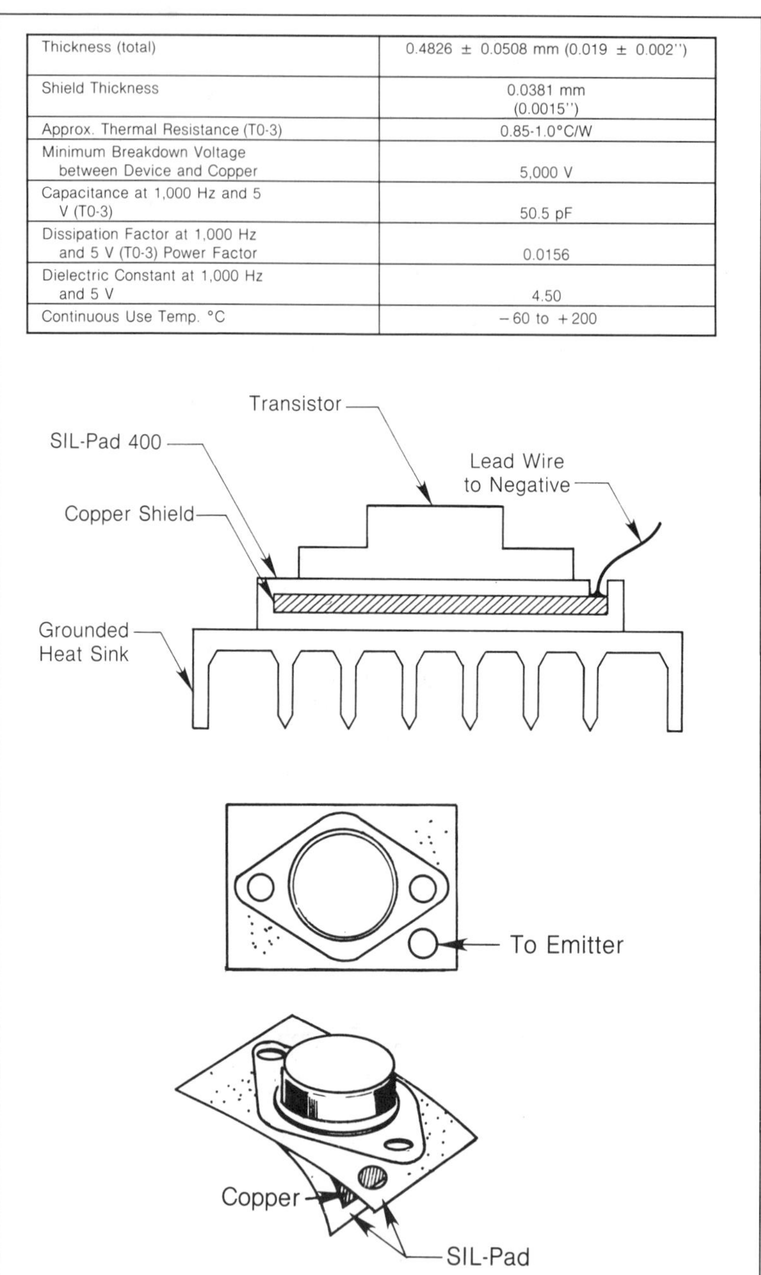

Figure 4.117—(continued)

GENERAL SEMICONDUCTOR INDUSTRIES, INC.

Isolated collector High Speed, High Power Switching Transistor. The isolated collector package configuration has been designed for use in switching applications where EMI (Electromagnetic Interference) must be controlled. Combining economy of the industry standard TO-3 power package and the versatility of an isolated collector design this family of transistors addresses numerous problems encountered in high voltage switching circuits.

Reduction of transistor collector to case capacitance can decrease conducted interference by 20 to 30 dB. Also, ground loop currents are reduced to an extremely low level. Costly assembly operations are eliminated by removal of insulating hardware. Fewer piece parts must be purchased and the "hi-pot" testing is eliminated. This design also provides increased reliability. Transistor case and heatsink nonuniformities no longer can break through insulating surfaces or washers. Lower operating junction temperatures result from the reduced junction to heatsink thermal resistance.

These transistors have been specifically designed and engineered for high speed, high voltage switching applications where the designer is concerned with optimizing power conversion efficiency, maintaining EMI at acceptable levels and providing a high degree of reliability. This series utilizes General Semiconductor Industries unique C2R© process. A manufacturing technology that provides surface stabilization for high voltage operation and enhances long term reliability.

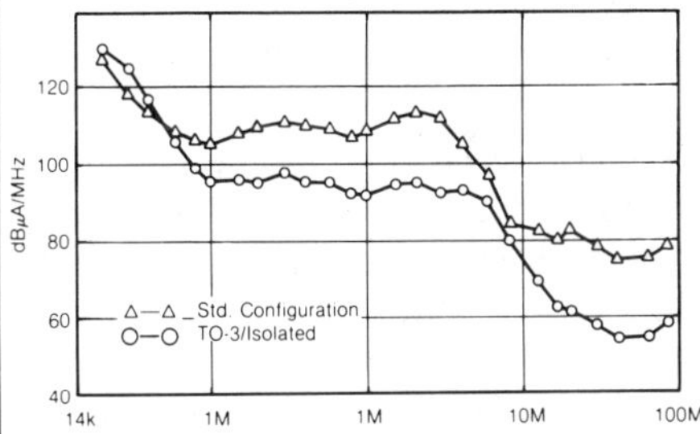

Characteristics

- Capacitance Collector to Case
  -Reduced from 250pf to $\cong$ 8pf
- Junction to Heatsink Thermal Resistance
  -Reduced 0.2 to 0.5° C/W
- Isolation Voltage Collector to Case
  ->1500V (pin limited)
- Isolation Resistance
  ->$10^{12}\Omega$

Figure 4.118—Another Way of Reducing Power Transistor Leakage by Using Isolated Collector Package (courtesy of General Semiconductor Industries, Inc.)

use several devices in parallel to reduce the parasitic inductance effect. For instance, in a modern 500 A power transistor package capable of 1,600 A/μs switching speed, the original layout had a parasitic inductance of 100 nH, causing a voltage transient LdI/dt of 160 V. Paralleling five transistors and using flat tabs has reduced the inductance to 20 nH, causing only a 30 V spike.

4.9.2 Thyristors, Triacs and Gate Turn-Off Switches (GTO)

The thyratron-like latching characteristics of the thyristor make it ideal for eliminating interference due to contact opening. Reverse-blocking triode thyristors, commonly called silicon-controlled rectifiers (SCRs), and bidirectional triode thyristors, usually referred to as triacs, are the most common types used in switching and power control.

An SCR can turn off only when the ac current through it naturally reaches zero in the process of reversing polarity, regardless of the load power factor. Because it requires no separate and sophisticated turn-off circuitry, the SCR does not interrupt current abruptly like mechanical contacts. Instead, it opens the circuit as soon as current reaches zero after gate drive has been removed. Circuit disturbances are minimum when the current is interrupted at this instant. Turn-on logic can be developed from the opposite half-cycle. The turn-on gate control signal can be applied while the SCR is reverse-biased, but it will not begin to conduct until it begins to be forward-biased.

Because an SCR can conduct in only one direction, full-wave operation calls for two SCRs in reverse parallel with gate control circuitry to be arranged accordingly. Alternatively, a triac, which effectively functions like two SCRs in reverse parallel, can be used.

Thyristors are not entirely free of EMI effects, however. At turn-on time, a voltage spike is produced as forward bias voltage passes through the forward-voltage breakover point. At turn-off time, stored carrier charges produce a current spike until a depletion region is established. Furthermore, a thyristor gate element is susceptible to EMI of high dV/dt. Capacitive charging currents can cause the gate to turn on the thyristor even though the magnitude of the gate voltage does not reach the rated triggering level.

Because thyristors are used to control electrical power or to control electrical machinery where safety is important, it is essential that the thyristors are always under the control of the gate signals so that spurious triggering cannot occur. Thyristors often operate directly from the mains supply and are therefore subject to mains-borne voltage transients. Since the magnitude of these transients is difficult to predict accurately, it may be necessary to provide filtering or clamping to ensure that the thyristor is not triggered by voltages exceeding the forward breakover value.

As with semiconductor power diodes, three reverse voltage ratings are specified for a thyristor: continuous, repetitive and nonrepetitive values. Because thyristors may have to remain non-conducting in the forward direction, three off-state forward voltage ratings are also given. The transient energy that can be absorbed by the thyristor is limited by the maximum junction temperature; therefore, the duration of the repetitive and nonrepetitive transients must be short compared with one-half cycle of the mains supply. Sometimes a maximum duration is specified for the transient.

The relationship between the voltage ratings and a typical mains supply waveform is shown in Fig. 4.119. For clarity, the crest working reverse and off-state voltages are shown equal to the peak sinusoidal component of the supply. However, in applications where thyristors operate directly from the mains supply, it is considered good practice to choose a device in which these two ratings are twice the peak sinusoidal supply voltage.

Two other voltage ratings are usually given in published data.

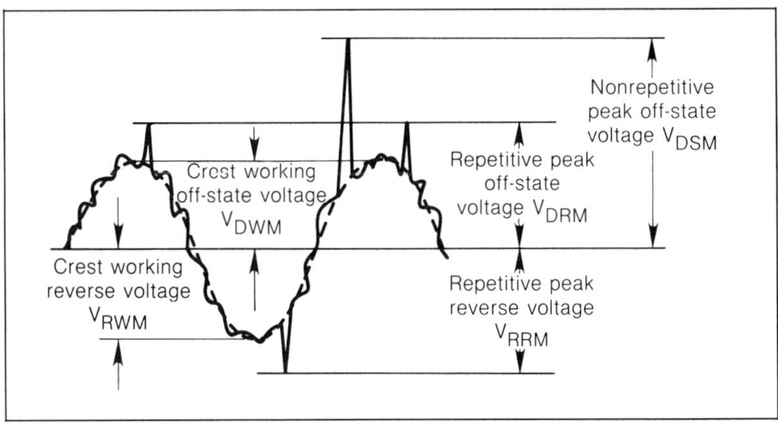

Figure 4.119—Relationship between Thyristor Voltage Ratings and Typical Mains Supply Waveform with Transients

These are the on-state voltage V_T, the forward voltage drop across the thyristor when conducting a specified current and the dV/dt rating. This second rating specifies the maximum rate of rise of the forward off-state, anode-to-cathode voltage that will not trigger the thyristor into conduction. Because of the capacitance of the depletion layer associated with the reverse-biased junction, a rapidly changing voltage across the nonconducting thyristor will induce a current across this junction. The magnitude of this current is:

$$I = C_d \frac{dV}{dt} \tag{4.86}$$

where C_d is the capacitance of the depletion layer. If this current exceeds the latching current, the device will turn on.

This unwanted triggering, called **rate effect**, can be prevented by:

1. Using thyristors with a high dV/dt rating. Typical dV/dt of a 35 A SCR is 200 V/μs at 85°C junction, but values up to 1 kV/μs are obtainable. The dV/dt rating decreases when junction temperature increases.
2. Connecting RC snubbers or voltage clamping devices across the device to limit the anode-cathode dV/dt.

A possible adverse effect of RC decoupling across the SCR is to degrade its off-state isolation against high frequency noise. EMI can now by-pass a turned-off static switch. The typical turn-on time of a power SCR or triac is about 1 to 10 μs.

Another power control device is called the GTO (gate turn-off switch). It combines the high rated voltage and current of the SCR with the easy gate drive and fast switching of the transistor since it can be turned on with a positive gate pulse, stays "on" when the gate drive is removed and then turns off with a negative gate pulse. Intrinsic turn-on/turn-off time for GTO is less than 0.5 μs.

4.9.2.1 EMI Aspects of SCRs to Power Control

The SCR and related bidirectional solid state switching devices are widely used in power control. They are smaller and lighter than equivalent variable transformers and result in less heat dissipation and power loss than rheostats, but produce high levels of EMI.

The ac power control case is reviewed here. Generation of interference in phase-control applications is analyzed using Fourier techniques, and measured EMI data are presented. An alternate technique, zero-crossover switching, is then analyzed, and measured levels are compared. Techniques of reducing trigger circuit susceptibility are reviewed, and the application of zero-crossover techniques in proportional control is discussed (Ref. 25).

Also considered is a practical design problem, the self-contained, low-interference light dimmer, which includes the effects of visual persistence and beating between a free-running oscillator and line frequency. Related low-EMI, SCR-control applications, including zero-crossover static-control switches, are discussed.

4.9.2.2 SCR Phase-Control Switching

In phase-control switching in ac power control, SCRs are triggered into conduction at a specified phase angle during each half cycle, as shown in Fig. 4.120. Back-to-back SCRs or a bidirectional switch, may be used to attain this **full-wave** control. Alternatively, one SCR or a rectifier bridge and SCR may be used to provide a dc biased waveform, but the analysis is similar. First, a square-wave truncated sinusoid, shown in Fig. 4.121. is reviewed. The development of the Fourier components proceeds as follows:

The basic square wave with the Fourier transform is:

$$\frac{2 \sin \omega T}{\omega} \tag{4.87}$$

It is shifted to the right and left by an amount D and added to yield:

$$\frac{2 \sin \omega T}{\omega} e^{-j\omega D} + \frac{2 \sin \omega T}{\omega} e^{+j\omega D} = \frac{4 \sin \omega T}{\omega} \cos \omega D \tag{4.88}$$

It is then shifted to the right by an amount D-T to give:

$$\frac{4 \sin \omega T}{\omega} \cos \omega D \times e^{-j(D-T)\omega} \tag{4.89}$$

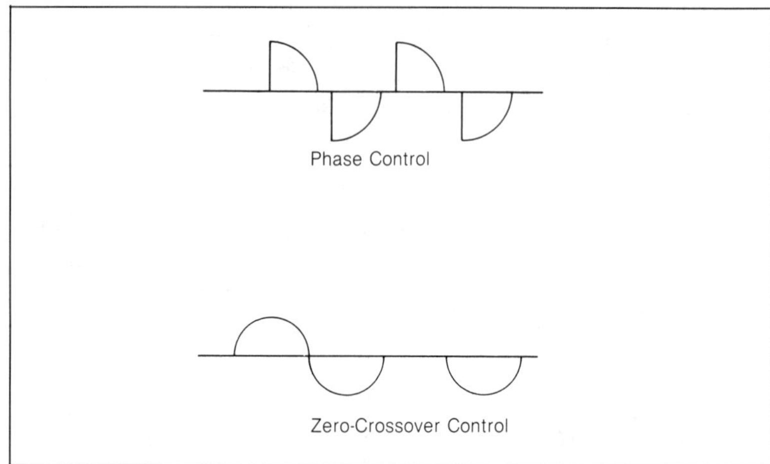

Figure 4.120—Load Voltage Waveforms

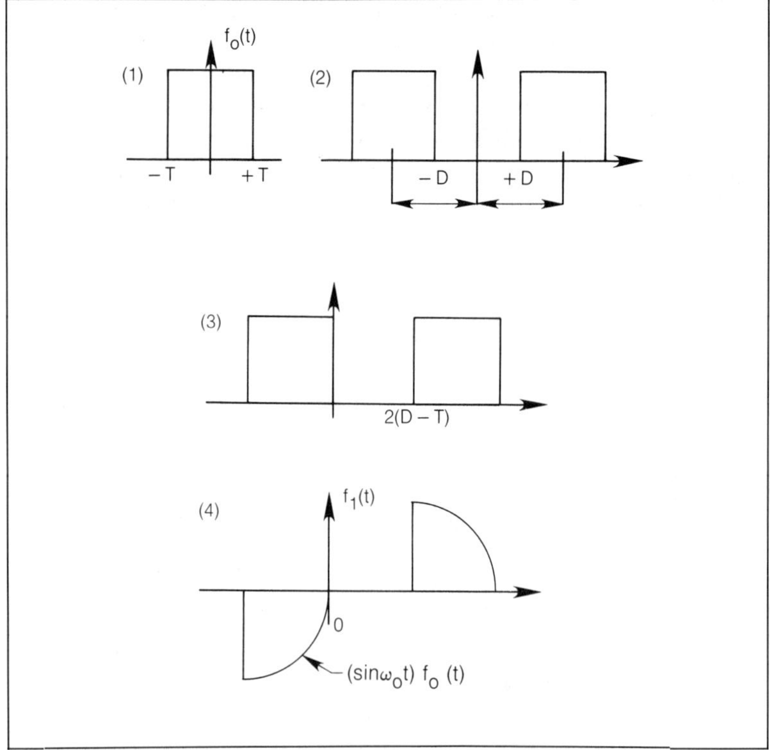

Figure 4.121—Square-Wave-Truncated Sinusoid

4.186

These square waves are then multiplied by the basic power sinusoid:

$$-j \ \frac{2 \sin (\omega - \omega_0) \, T}{\omega - \omega_0} \ \cos \left[(\omega - \omega_0) \, D \right] e^{-j \, (D-T) \, (\omega - \omega_0)}$$

(4.90)

$$- \ \frac{2 \sin (\omega + \omega_0) \, T}{\omega + \omega_0} \ \cos \left[(\omega + \omega_0) \, D \right] e^{-j \, (D-T) \, (\omega + \omega_0)}$$

The magnitude of the harmonic coefficients is:

$$\frac{2}{T_0} \ | F(j\omega_0 n) | ,$$

(4.91)

$$\text{for } n = 1,2,3...$$

A phase shift of $2 \, \omega_0 (D\text{-}T)$ is introduced between the two terms of Eq. (4.90) by the exponentials, and the phase angle of firing affects the magnitude of the sum. For the case of $D = 2T = T_0/4$, the half-power condition, the components are:

$$\frac{1}{\pi} \, [F(j\omega_0)]$$

$$\begin{aligned}
\text{which } = \ & 1 \text{ at } 3 \ \omega_0 \\
& 1/3 \text{ at } 5 \ \omega_0 \\
& 1/3 \text{ at } 7 \ \omega_0 \\
& 1/5 \text{ at } 9 \ \omega_0 \\
& 1/5 \text{ at } 11 \ \omega_0
\end{aligned}$$

(4.92)

The amplitude of a square-wave, truncated sinusoid generally falls off at a rate proportional to $1/\omega$, or at 20 dB/decade. This is the rate of decrease of the peaks of the basic $(\sin x)/x$ form. The turn-on of a real SCR may be better modeled as an exponential than as the abrupt discontinuity in a square wave. An exponential pulse train, shown in Fig. 4.122 is used to truncate the basic power sinusoid. The tail of the pulse distorts the turn-off slightly, but permits a simplified expression. For $T \ll T$, this retains the main features of the square-wave truncated sinusoid, but introduces a new break frequency at $f = 1/2 \, \pi T$, where T is the time constant of the SCR turn-on process.

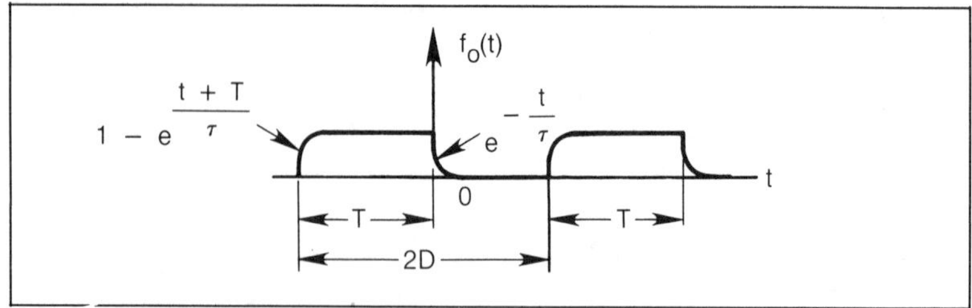

Figure 4.122—Expotential Pulse Train

Typical 10 percent to 90 percent rise times (t_r) for low current SCRs are on the order of 1 μs for switching 100 V and, for a given SCR, are inversely related to the voltage switched. The time constant T is $(1/2.2)t_r$.

For high-current SCRs, the rate of rise of the current must be held to values typically in the tens of A/μs to avoid device damage. This is accomplished by inductive or current-limiting components in the associated circuitry and provides a circuitry limited rate of rise rather than a device-limited rate of rise.

4.9.2.3 SCR Zero-Crossover Switching

Fourier analysis of a zero-crossover switched sine wave proceeds in the same manner as that for the square-wave truncated sinusoid. Eq. (4.90). is valid for this case if:

$$2T = T_o, \quad 2D = \frac{(K-1)}{2} T_o \qquad (4.93)$$

where two out of every k half-cycles are allowed to reach the load as shown in Fig. 4.123. This allows calculation of the one-of-i and two-of-i half-cycles past the condition where i = 1, 2, 3 . . . Since the phase shift terms are now integer multiples of the power frequency, the two terms add to produce a sequence proportional to:

$$\frac{1}{n-1} - \frac{1}{n+1} = \frac{2}{n^2-1} \qquad (4.94)$$

in which the smoothed envelope of the harmonics decreases in

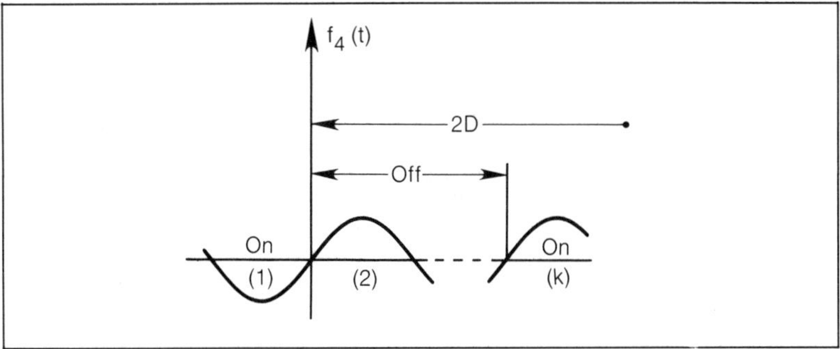

Figure 4.123—Zero-Crossover Analysis

amplitude at a rate proportional to $1/\omega^2$ or at 40 dB/decade.

In this case, ω is set equal to the harmonics of $2\omega_o/k$, and is not necessarily an integer. A half-wave, rectified sine wave with a period of 2π has an amplitude of:

$$\frac{1}{\pi} \ (1 + \frac{\pi}{2} \ \sin t + 2/3 \ \sin 2t - 2/15 \ \sin 4t \ . \ . \ .) \ (4.95)$$

Figure 4.124 shows the familiar unijunction transistor oscillator, phase-control technique. Varying the resistance varies the phase of the output waveform to the SCR gate-pulse transformer. Figure 4.125 is a simplified block diagram of the zero-crossing circuit. The **disabling mean** controls the percentage of half-cycles absent in driving the trigger circuit. Both circuits are powered by the same ac line voltage as that furnished to the SCR and load.

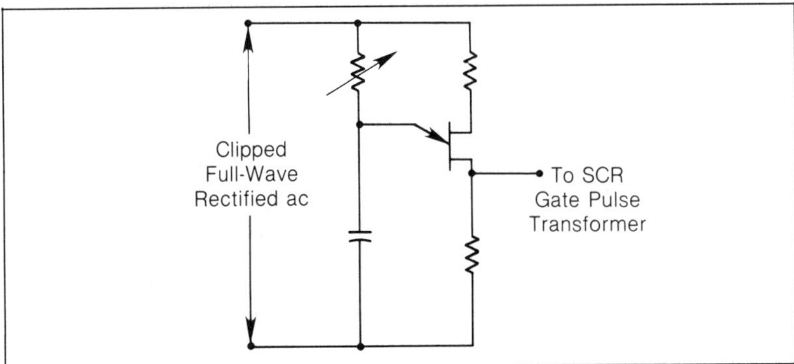

Figure 4.124—Unijunction Transistor Oscillator Trigger Circuit for Phase Control

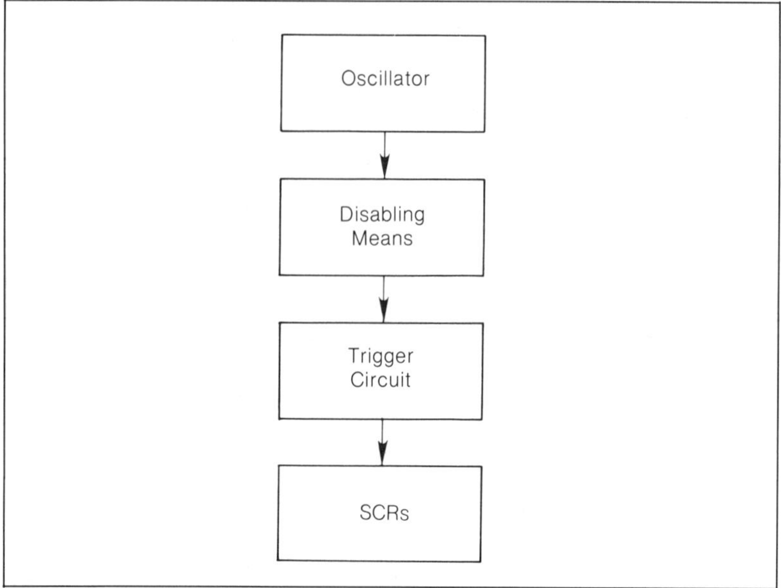

Figure 4.125—Block Diagram of Zero-Crossover Circuit

4.9.2.4 SCR Filtering

In phase control applications, filtering is often an integral part of SCR circuit design. Device dI/dt ratings for high-frequency devices are given for the SCR shunted by an **RC snubber circuit**. Protection of the device from false triggering due to dV/dt or rapid rise of blocking voltage is required. The most important filtering function, however, is control of current rise time during device turn-on. This limits conducted and radiated electromagnetic interference and protects the device from dI/dt failure.

SCR manufacturers recommend two principal methods of filtering. A single inductor in series with the load-SCR combination may be used with a parallel capacitance as in Fig. 4.126, parts a and b, if an RF ground is available to terminate the capacitance. Otherwise, the unbalanced filter creates RF currents in distributed wiring ground capacity. Then a two-inductor, balanced filter can be used as in Fig. 4.126c. It is often advantageous to use a split-inductor filter with limited inductance and high saturation-current levels.

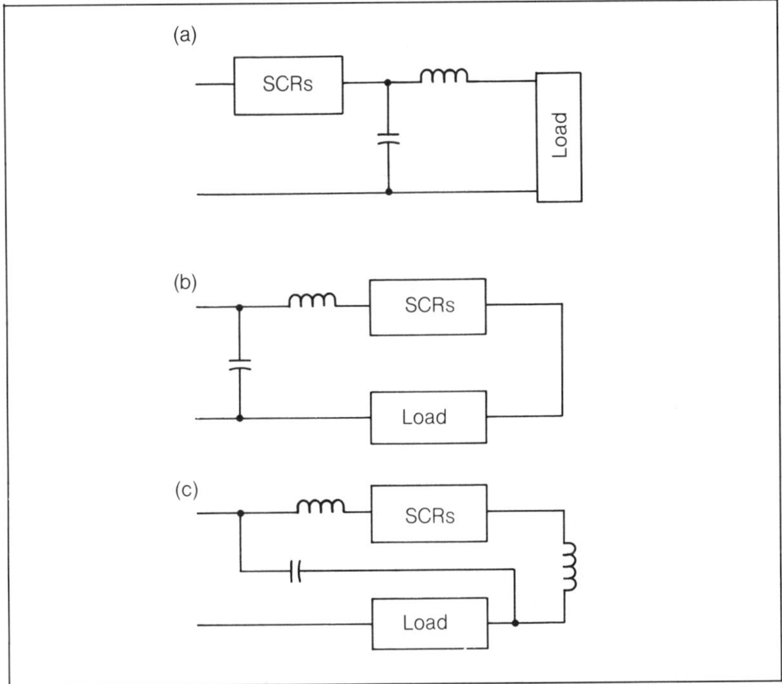

Figure 4.126—SCR Filtering

4.9.2.5 Trigger-Circuit Susceptibility

Care must be taken to avoid SCR trigger-circuit susceptibility to either line voltage EMI or to interference in the gate circuit of the controlled SCRs. Good design practice in regulated supplies for the trigger circuit, with the use of RF filtering, minimizes the effects of power-line fluctuations and transients. Protection of the gate circuit is more difficult because of the effects of the suppression devices which may adversely affect timing or firing characteristics. For example, if a trigger circuit is coupled by a pulse transformer, the HF components of the trigger-pulse waveform carries a significant portion of the pulse energy. Rapid application of trigger voltage is desirable to reduce switching stresses on the device; thus, filtering may not be a practical answer. A diode bridge as shown in Fig. 4.127 prevents any reverse interaction without reducing trigger effectiveness. Finally, opto-isolators can be used to immunize the gate command.

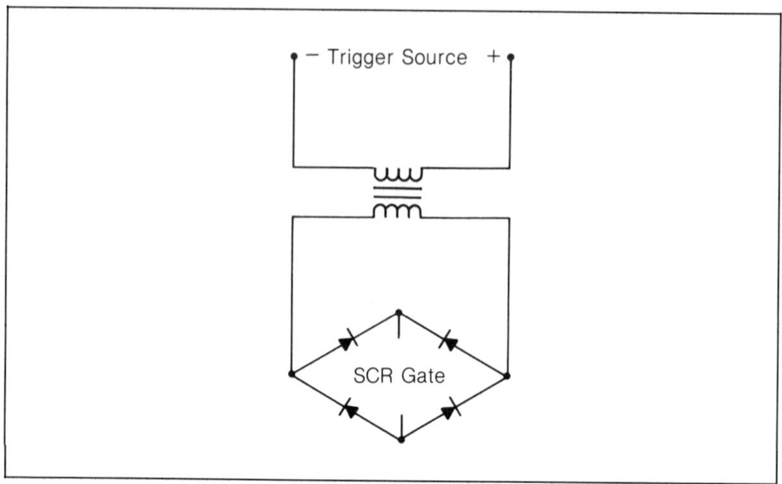

Figure 4.127—Trigger Circuit Isolation

4.9.2.6 Application of the Zero-Crossover Technique

The zero-crossover technique is appropriate in many applications requiring proportional control. Here, the term **proportional control** is used in the general sense to describe any requirement for application of continuous or nearly continuous increments of power to a load rather than in the special sense of proportional feedback, although the latter is included. Typical applications are: light dimming, heater control for industrial processes, aircraft anti-icing or comfort heating, motor speed control and control of electro-chemical processes. When the need for low weight and low-power loss coincide with an environment which is sensitive to EMI, the zero-crossover provides a viable design alternative.

One serious limitation exists. The load must exhibit sufficient damping to avoid intolerable ripple in the output function. In light-dimming this requirement can be met with 400 Hz power but not with 60 Hz power. The reason is that visual persistence integrates half-cycle pulse trains, provided their repetition rate is at least 50Hz. With 50 or 60 Hz mains, if the dimming requires that more than two consecutive half-waves are deleted, the flickering of the light becomes noticeable. In heating applications, the load thermal inertia is usually sufficient to smooth over variations in average input power over a period of several half-cycles of the power waveform. Motor mechanical load systems represent a more critical case. Again, 400 Hz power is preferable, although control is feasible in many 60 Hz applications.

The zero-crossover technique is also used successfully in solid-state relays based on GTOs. A zero-voltage detector is used to turn the device on, so it effectively conducts as soon as the line current reaches the latching current. Then, a zero-current detector is used to turn the switch off when the current is null, so the LdI/dt transient is minimum. Figure 4.128 shows the improvement in EMI signature due to the zero-crossover technique. Solid-state relays are discussed in Section 6.2.4 of this book.

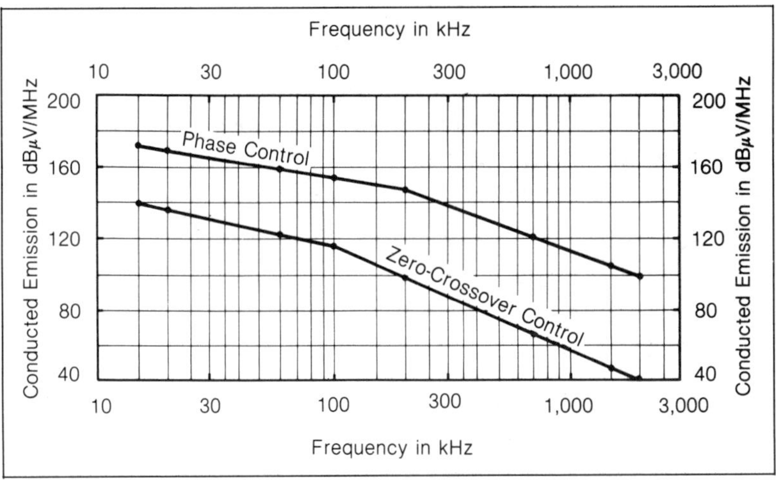

Figure 4.128—Comparison of Conducted Emissions in 400 Hz, 115 Vac Controlled 1.2 kW load. The average improvement is 40 dB. Improvement degrades when power factor departs from 0°, unless zero-crossover is performed on current.

4.10 References

1. Hopkins, P. and Carvey, D., "Spread Spectrum Interference Considerations," Zurich EMC Symposium, 1985.
2. Feher, K., *Digital Modulation Techniques,* (Gainesville, Virginia: Interference Control Technologies, Inc., 1977).
3. White, Donald R.J., *A Handbook on Electrical Filters, Synthesis, Design and Applications,* (Gainesville, Virginia: Intereference Control Technologies, Inc., 1983).

4. Larson, C. and Roe, J.M., "Modified Ebers Moll Model for RFI Analysis," IEEE International EMC Symposium, 1978.

5. Amadori, Pubielli and Richardson, "Microwave Interference in Bipolar Transistors," (*IEEE Transactions on EMC*, November, 1975).

6. Ott, H.W., *Noise Reduction Techniques*, (New York: John Wiley & Sons, 1976).

7. Morrison, R., *Instrumentation Fundamentals*, (New York: John Wiley & Sons, 1984).

8. *IC Electromagnetic Susceptibility Handbook*, (McDonnell Douglas Astronautics Co., Report MDC E 1929, 1978).

9. Howard, M., "Instrumentation and Test Method for an Automated Susceptibility System," (*EMC Technology*, July, 1983).

10. Goedbloed, J., "Increasing RFI Immunity of Amplifiers with Negative Feedback," Zurich EMC Symposium, 1983.

11. Whalen, J., "RF Pulse Susceptibility of Analog Circuits to RFI," (*IEEE Transactions on EMC*, November, 1975).

12. Elliott, M., "Susceptibility of UHF Transistors," (*IEEE Transactions on EMC*, November, 1982).

13. Tront, J., "Comparison of RFI Susceptibility of Several IC Drivers/Receivers," Zurich EMC Symposium, 1985.

14. Kennealy, D. and Head, G., "EMC Noise Susceptibility of ESD Protected MOS Devices."

15. Erwin, V. and Fisher, K., "Radiated EMI of Multiple IC Sourcing," IEEE EMC Symposium, 1985.

16. Texas Instruments, (*FYI Journal*, August, 1986).

17. Val, C., "High-Performance Surface-Mounting VHSIC Packages," 4th International Microelectronics Conference, Kobe, Japan, 1986.

18. Danker, B., "New Measures to Decrease Radiation from PCBs," Zurich EMC Symposium, 1985.

19. Showers, R.M., "A General Model for IC Susceptibility Prediction," IEEE EMC Symposium, 1978.

20. Mardiguian, M., *Electrostatic Discharge*, (Gainesville, Virginia: Interference Control Technologies, Inc., 1986).

21. Huntsman, J.; Yenni, D.; and Mueller, G., "Fundamental Requirements for Static Protective Containers," (*3M Technical Report J-SFREQ-505/11 on ESD Sensitivity*), Nepcon/West Conference, Anaheim, California, 1980.

22. Richman, P., *Electrostatic Discharge Protection Test Handbook*, (KeyTek Corp., 1983).

23. Allan, A., *Noise Immunity Comparison of CMOS vs. Bipolar Logic,* (Motorola, Application Note AN-707).
24. Bossard and Unger, "ESD Damage from Charged IC Pins," EOS Symposium, San Diego, 1980.
25. Mathias, D.W., "Interference in SCR Power Control," *IEEE EMC Symposium Record,* Vol. IEEE 69C3-EMC, June, 1969, pp. 29-34.

4.11 Bibliography

1. *AD 202 and AD 213 Isolation Amp Application Notes,* (Analog Devices).
2. *Application Note No. 15,* (Precision Monolithics, 1979).
3. Azoulay, B., "Susceptibility to RF Demodulation of Linear Integrated Circuits," IEEE Symposium, Tokyo, 1984.
4. Camiggio, S. (ITALTEL), "Macromodeling for Integrated Circuits," Zurich EMC Symposium, 1983.
5. Chase, E.W., "ESD Susceptibility of Thin Film Resistors," EOS Symposium, Orlando, Florida, 1982.
6. Comite Consultatif International de Radiocommunications (CCIR), Geneva, various reports and recommendations from Study Groups 4 and 9.
7. King, M., *Report on EMI Susceptibility,* (Cornell Dublier, 1983).
8. Rode, R., "EMC Aspects of Digital Transmission," Zurich EMC Symposium, 1973.
9. Shah, A., "Influence of Transient Disturbances on Digital Electronics," Montreaux EMC Symposium, 1985.
10. Sherwin, Jr., *Application Note No. 15,* (National Semiconductor, 1974).
11. Whalen, J., "Statistics of RFI Demodulation in Op Amp Circuits," IEEE International EMC Symposium, 1985.
12. White, Donald, R.J. and Mardiguian, M., *EMI Control Methodology and Procedures,* (Gainesville, Virginia: Interference Control Technologies, Inc., 1985).

Chapter 5

Transformers and Magnetic Coupling Components

Transformers and magnetic coupling components are devices which transfer energy from one circuit into another by using magnetic flux. There are transformers in which EMI is a side effect, and transformers in which the main purpose is to reduce EMI. Both aspects are addressed in this chapter. However, for more details on the application of transformers to power supplies, the reader is referred to Volume 4 of this EMI series, *Filters and Power Conditioning*. On the other hand, inductors have been covered in Chapter 2 and will not be addressed here.

5.1 Power Transformers

Transformers used in power applications include:
1. Transformers for ac/dc or dc/dc converters, low frequency (50/60 or 400 Hz) or high frequency (15 to 300 kHz) for switched-mode regulators
2. Transformers for power-line regulation (automatic tap changers, ferroresonant regulators, etc.)
3. Pure isolation transformers with a 1:1 ratio for mains isolation and EMI protection

Although the term "isolation transformer" is generally used for the third category, the two other kinds also are isolation transformers. In fact, any transformer (except an autotransformer), is an isolating device since it provides a galvanic isolation between the primary and the secondary; however, this isolation which is quasi-infinite at dc can decrease rapidly to 0 dB if the transformer is not specially constructed.

EMI problems related with all transformers include primary-to-secondary leakage capacitance, magnetic-field leakage and voltage distortion, i.e., generation of harmonics.

5.1.1 Problems with Primary-to-Secondary Isolation: Faraday Shields

Any ordinary transformer exhibits a parasitic capacitance between primary and secondary windings. This capacitance can vary from 20 to 30 pF for a few VA miniature transformers up to 1 to 3 nF for a few hundred VA model. In the latter, for instance, if a dc insulation resistance of 100 MΩ exists, the 1 nF capacitance will start to bypass it above 1.5 Hz.

The parasitic capacitance causes unwanted primary-to-secondary couplings of two kinds: common-mode and differential-mode. Isolation transformers are used to **break ground loops**, i.e., to increase the impedance of ground loops, hopefully to levels where common-mode current and differential-mode voltages will not be troublesome. Figure 5.1 shows the capacitance between primary and secondary windings of an isolation transformer.

The figure also shows an equivalent circuit in which the undesired common-mode voltage, e_n, is shown pushing common-mode current through the ground loop.

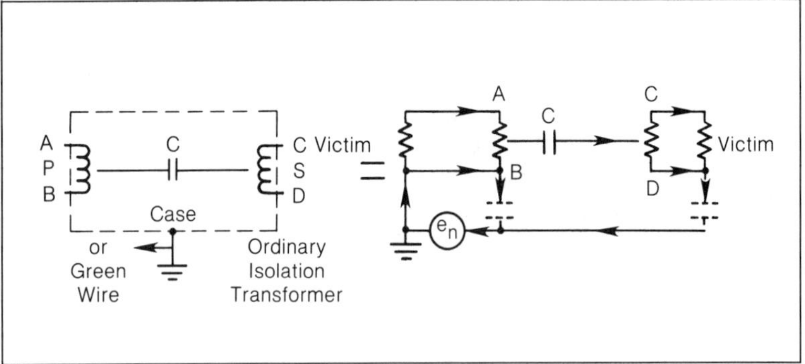

Figure 5.1—Isolation Transformer Showing Common-Mode Noise Originating in Primary and Coupling into Secondary

At low frequencies the capacitive reactance between the transformer primary and secondary is so high that there is no significant threat from common-mode current. At high frequencies, however, the potential EMI problem becomes significant. For example, a typical 1 kW isolation transformer has input-output capacitance which produces only a few dB reduction to a 1 μs transient from the primary to the secondary. Thus, unshielded (no Faraday shield) isolation transformers may perform well at low frequencies only. Other measures such as Faraday-shielded isolation transformers should be considered for use to protect against common-mode problems at intermediate and high frequencies.

Figure 5.2 shows a differential-mode (sometimes called "normal" or "transverse") voltage appearing across the primary and being transferred to the secondary via the parasitic capacitance. It should seem perfectly normal for a variable signal across the primary to show up on the secondary, since a transformer is supposed to function this way. However, above 10 to 30 times the design frequency for the intentional mode of operation, the magnetic coupling vanishes and the line-to-line voltage is no longer transferred to the secondary. This is shown in Fig. 5.3. as a DM rejection of the transformer. Then, at high frequencies, the rejection deteriorates again due to capacitance C_p.

To combat undesired emissions coupling through isolation transformers, Faraday shields are necessary. Figure 5.4 shows that the primary-to-secondary capacitance has been divided between the primary to shield and the shield to secondary. The connection to the shield is brought out externally to the transformer on an insulated standoff terminal to permit the user to connect the shield in some purposeful manner. The Faraday shield is connected to the reference for the common-mode voltage, i.e., to ground.

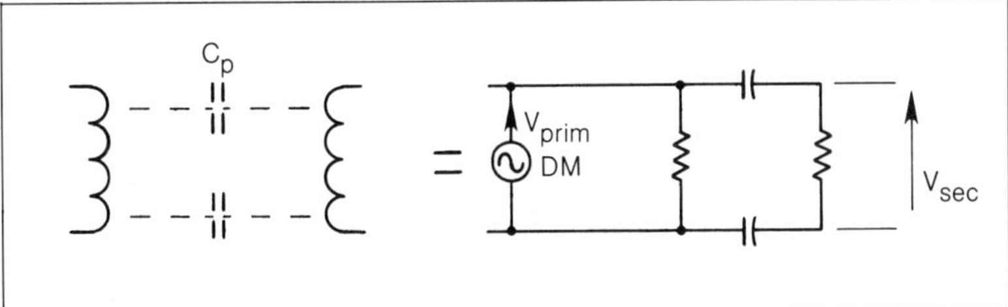

Figure 5.2—Isolation Transformer Showing Differential-Mode Noise from the Primary Coupling into Secondary (Beyond Normal Magnetic Operation of Transformer)

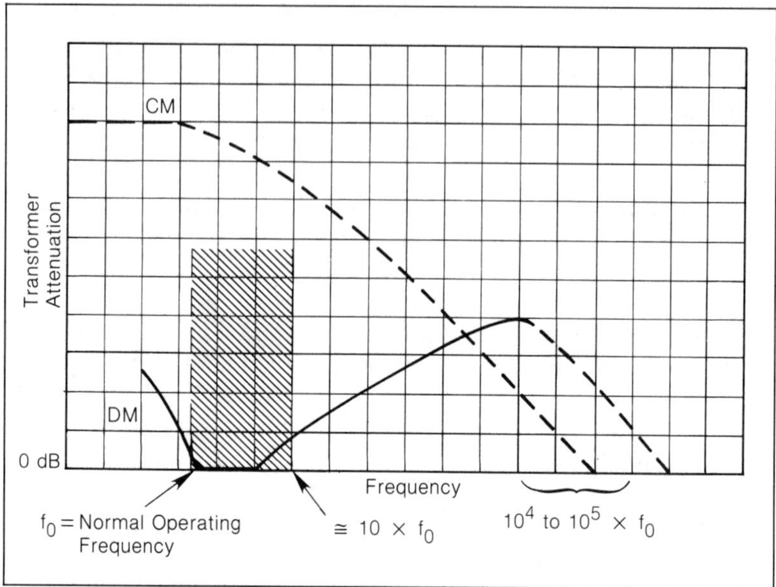

Figure 5.3—CM and DM Attenuation in Ordinary Transformers.

The schematic drawing on the right in Fig. 5.4 shows that a finite grounding impedance Z comes from the resistance and inductance of the ground strap or bus. Thus, the improvement offered by the shield by voltage-dividing action is the ratio of the shunt impedance Z to the series capacitive reactance. The common-mode suppression will be substantial at low frequencies but cannot be infinite because there always exists a dc insulation resistance of 10 or 100 $M\Omega$ between the primary and secondary. Then, CM rejection degrades with an increase in frequency like a high-pass filter.

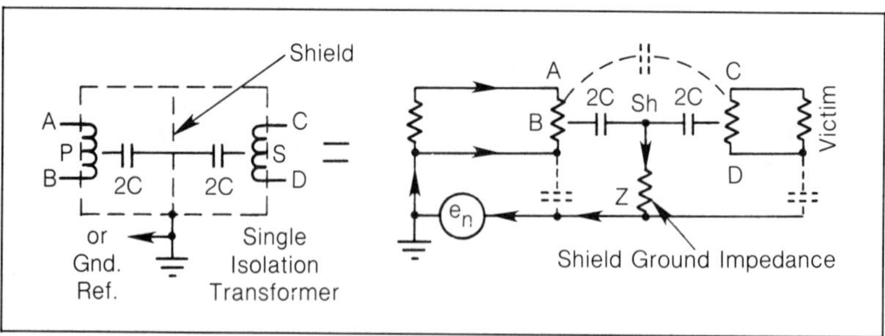

Figure 5.4—Faraday-Shielded Isolation Transformer Showing Common-Mode Noise Originating in Primary and Bypassed before Secondary

Figure 5.5 shows quantitative values of CM and DM rejection of a typical single-shield isolation transformer (residual primary to secondary capacitance < 1 pF). For contrast, an ordinary, unshielded transformer is also shown. The performance of a good Faraday shield can be spoiled if it is grounded via a conductor that is too impedant. Performances shown are with a ground bus shorter than 10 cm. Should the ground bus have been 1 m long, performances would degrade by about 20 dB.

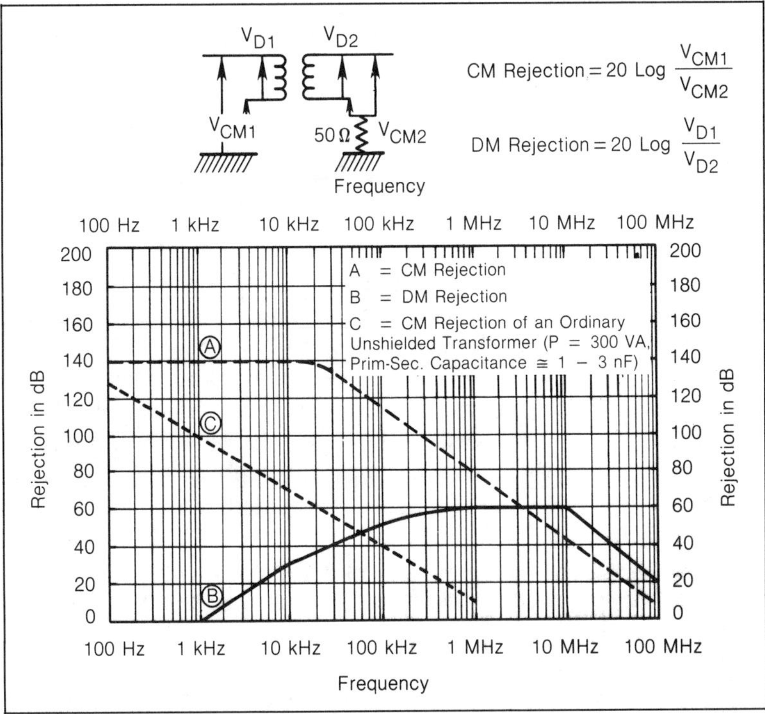

Figure 5.5—CM and DM rejection of typical single shield isolation transformer. For triple shield figure, parasitic capacitances < .001 pF are typically 40-60 dB better above 100 kHz.

For differential-mode coupling, Fig. 5.6 is the corollary to Fig. 5.2 in which the primary-to-secondary capacitance is divided into two parts by the Faraday shield between the primary and secondary. Since the differential mode appears across the primary, it is necessary to short circuit the primary at high frequencies in order to suppress the differential-mode coupling. To accomplish this, the shield should be connected to the center tap of the primary. If no center tap exists, then the shield should be connected to the low-voltage side of the transformer. The arrangement shown will be transparent at the 50/60/400 Hz power mains frequencies but will offer substantial attenuation at high frequencies.

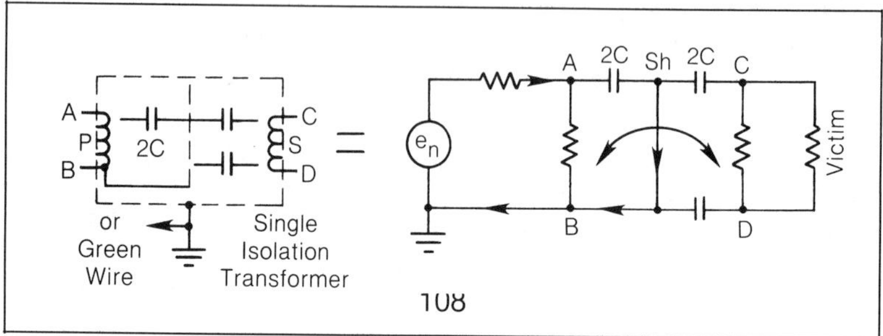

Figure 5.6—Faraday-Shielded Isolation Transformer Showing Differential-Mode Noise Originating in Primary and Bypassed before Secondary

The preceding figures indicate that it is necessary to use Faraday shields in isolation transformers to suppress conducted EMI. However, no single shield can be optimized to simultaneously suppress both common and differential modes. If both modes are present simultaneously, it may become necessary to use a **double** Faraday shielded isolation transformer. This is shown in Fig. 5.7 where the shield facing the primary is used to suppress common-mode noise, as determined by the respective grounding of both shields in Figs. 5.4 and 5.6.

The equivalent circuit is shown at the right in which the differential mode is the push-pull currents (I_d) across the primary, and the common-mode currents (I_c) are the push-push, i.e., in-phase across the primary. The remainder of the schematic shows the voltage-dividing action from the primary and secondary to their respective shields and from shield to shield.

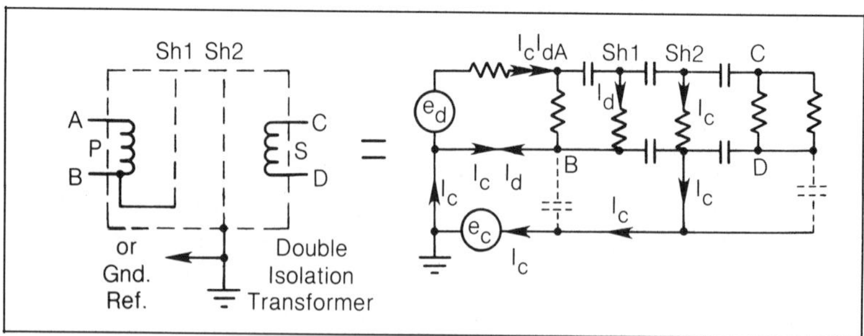

Figure 5.7—Double Faraday-Shielded Isolation Transformer in which Both Common and Differential-Mode Rejection Is Effected

Figure 5.8 is a continuation of the preceding figures. It represents a triple-shielded isolation transformer and its grounding under severe conducted electromagnetic ambients on the power bus. Triple-shielded transformers are developed for severe cases where ultra-isolation is needed. The residual primary-to-secondary capacitance can be as low as 0.01 pF or even 0.001 pF. To fully gain from this high isolation, the input and output leads must be carefully separated, with the cleaned side enclosed in a conduit and the shield grounded with an extremely low-impedance bond. The physical realization of a Faraday shield is depicted in Fig. 5.9.

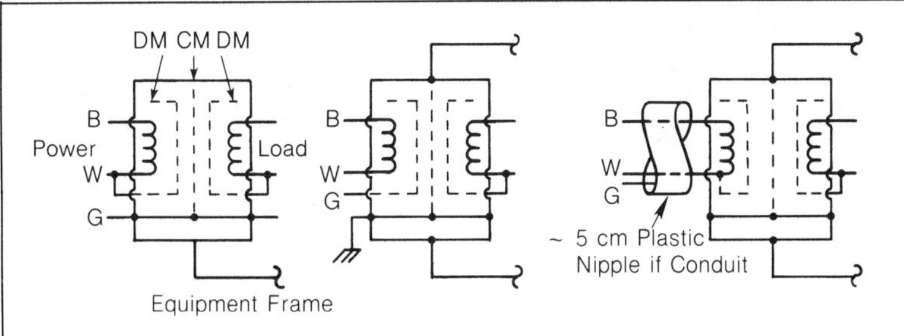

Figure 5.8—Triple Faraday-Shielded Isolation Transformer and Its Possible Grounding. Although desirable for a maximum isolation ground noise, the break of the metal conduit on the right-hand example may be in conflict with safety rules.

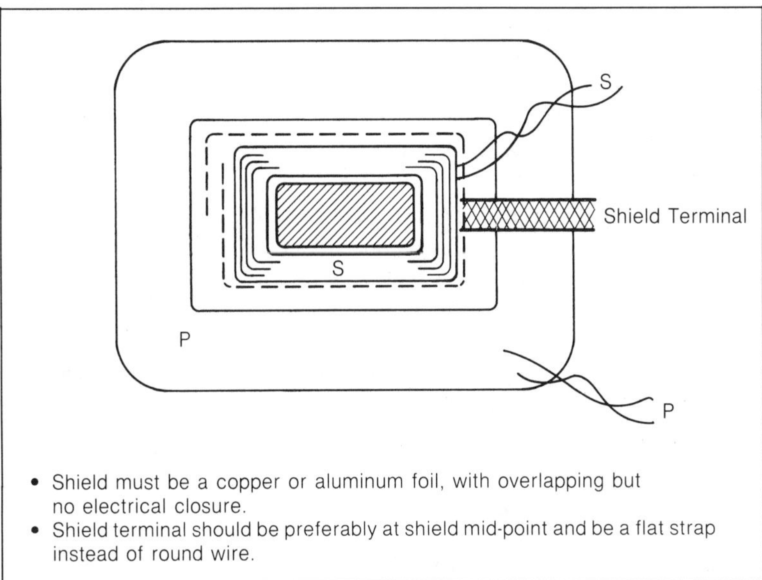

Figure 5.9—Physical Realization of a Faraday Shield

The definition of CM and DM rejection is not always clear in transformer specifications. Figure 5.10 shows the different configurations used to measure rejection. Configuration A is the one recommended by several industry associations such as ECMA or CBEMA. It uses a 10 nF capacitor to ground and the CM rejection is defined as $20 \log V_{cm} / V_2$. The 10 nF represents a typical capacitance to chassis of large electronic assemblies; therefore, the test measures the ratio of 10 nF/C_p. A transformer with a parasitic capacitance C_p of 10 pF would show a 60 dB rejection forever. Such information can be easily misused since the CMR found this way is flat with frequency. So the method shown in configuration B is often preferred, where the CM voltage is measured across a 50 Ω resistance to ground. In this case, the test actually sees the CM current, and the CMR decreases with frequency. Figure 5.11 clearly shows the different results between methods A and B when measured on a good quality shielded isolation transformer.

Figure 5.10, configuration C, is used to measure the DM rejection and requires a well balanced setup, otherwise CM couplings will combine with DM. The method shown in configuration D is sometimes used as a definition of CM rejection. However, it is a mode-conversion measurement where the lack of symmetry in the windings creates a secondary DM voltage when a CM signal is injected on the primary. The specific name of this parameter is "longitudinal balance," which should not be confused with the situations in configurations A and B.

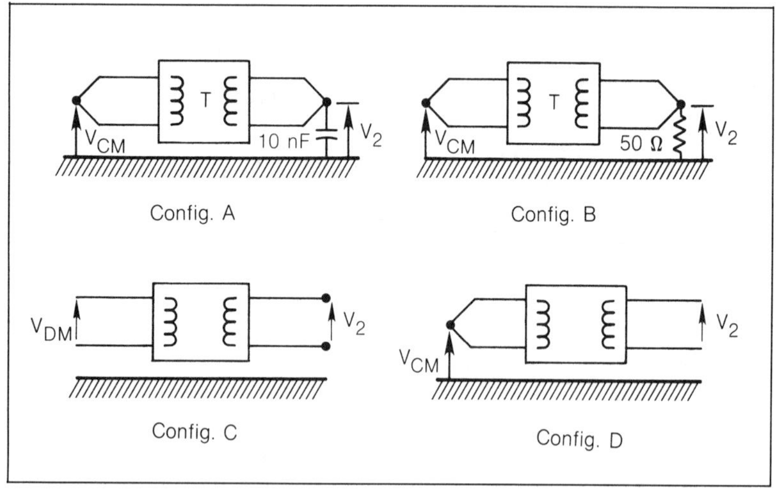

Figure 5.10—Different Configurations upon which the EMI Rejection of a Transformer Is Defined

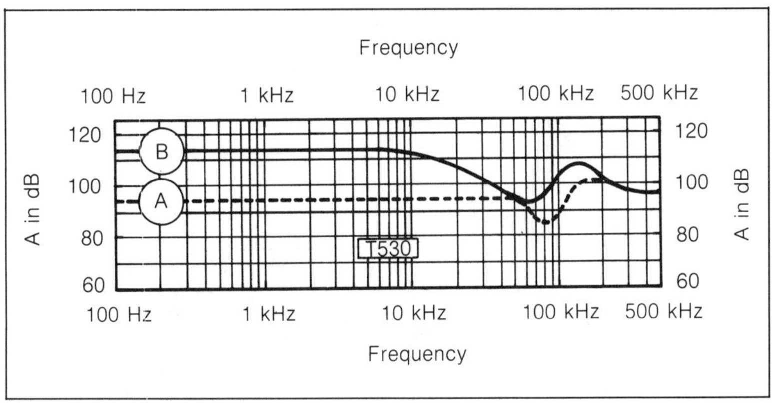

Figure 5.11—Difference between CM Rejection when Measured According to Method (a) or (b)

Figure 5.12 illustrates the calculation of parasitic capacitances. Methods other than a simple Faraday shield can be used to reduce C_p (Fig. 5.13). These are:

1. Use a toroid magnetic circuit.
2. Use a "C" core with primary and secondary on different legs.
3. Use an "E" core with primary on top of secondary.
4. Increase the dielectric thickness.
5. Use special winding patterns (see Section 2.3, "Inductors").

Unfortunately, all of these methods are in the opposite direction of what is required for low magnetic leakages.

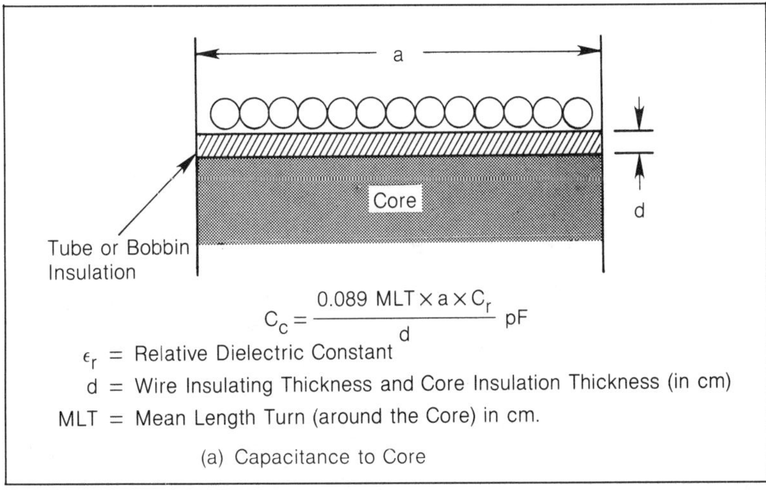

$$C_c = \frac{0.089 \; MLT \times a \times C_r}{d} \; pF$$

ϵ_r = Relative Dielectric Constant

d = Wire Insulating Thickness and Core Insulation Thickness (in cm)

MLT = Mean Length Turn (around the Core) in cm.

(a) Capacitance to Core

Figure 5.12—Calculation of Parasitic Capacitances (a) between Inner Bobbin and Core and (b) between Bobbins or Layers (continued next page)

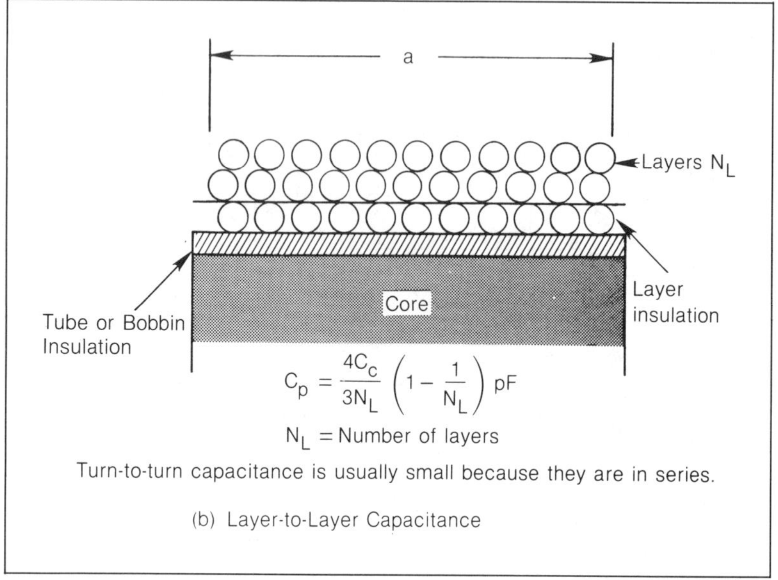

Figure 5.12—(continued)

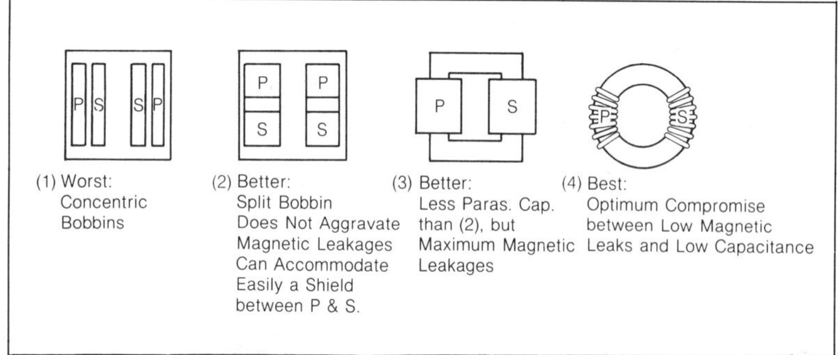

Figure 5.13—Reducing Transformer Capacitive Coupling by Core/Bobbin Configuration

5.1.2 Problems with Transformer Magnetic Leakage

The leakage flux of a transformer corresponds to the flux lines which are closing via the air instead of the magnetic circuit. To get the most V × A/cm³, the transformer designer tends to use the maximum induction compatible with his core section close to the knee of the magnetization curve.

During peak values of current, the relative permeability, μ_r, can decrease and cause more flux to go into the air instead of the magnetic core. Furthermore, with ac mains transformers, μ_r of laminates decreases rapidly above 1 kHz. In other words, a line current with high harmonic contents will cause more magnetic leakage. In an ordinary transformer of a few hundred VA, this leakage field easily represents several tens of gauss at few centimeters from the transformer. Hopefully, when the distance from the coils exceeds approximately the coil radius r, this field decreases as $1/r^3$.

Leakage flux is also greater with gapped chokes and gapped transformers such as the ferroresonant type. Transformer magnetic leakage can be a serious problem for nearby sensitive circuits including magnetic heads and their amplifiers, audio and instrumentation amplifiers and electron-beam deflection in TV sets and CRT equipment in general. Magnetic leakages can be reduced by:

1. A core material having a progressive saturation (soft knee) instead of abrupt (square cycle)
2. In low-voltage, high-current windings, using flat conductors instead of round ones
3. A magnetic core with closed or semiclosed shapes
4. A magnetic shield

The last of these can simply be a "shading" ring. The shading ring is a shorted loop made of a copper band and centered about the core winding. The optimal width is about 0.5 times the winding height; larger widths do not improve the attenuation and can be detrimental.

The shading ring works by the loop-induced current creating a field opposing the leakage field. The neutralization cannot be perfect, but in the maximum leakage azimuths, as in A and C on Fig. 5.14A, the field is reduced by 6 to 10 dB. The copper band must be thick enough to avoid excessive heating by the Joule effect. The ultimate solution is an integral outer shield made of iron or high-permeability material.

Another method used to reduce magnetic coupling between transformers and chokes or between transformers and circuit wiring is to mount them at right angles. If the magnetic fields of each are confined to a plane, then the coupling will be proportional to the cosine of the angles between their planes and will achieve a maximum practical decoupling of about 20 dB. To depend entirely on this practice may be risky, however, since fields are generally not confined in a plane and may vary with further packaging alterations.

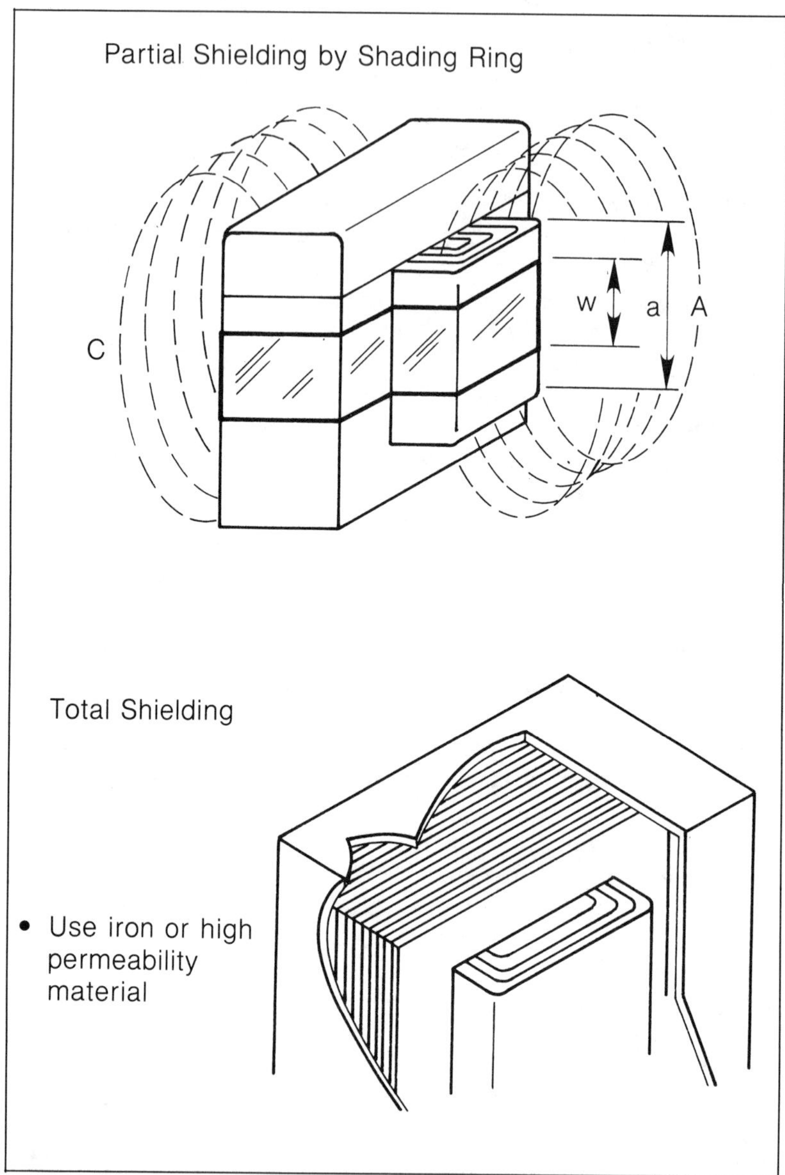

Figure 5.14A—Reducing Magnetic Leakages by Shielding (continued next page). For a shading ring, empirical data indicates that the optimal ratio is w/a = 0.5.

Magnetic leakage may be reduced by proper core selection (Fig. 5.14B). Note that core shapes without an abrupt change in field lines (toroids) and closed or semiclosed shapes are preferable. Also, the core section and material must be checked against saturation during peak load currents.

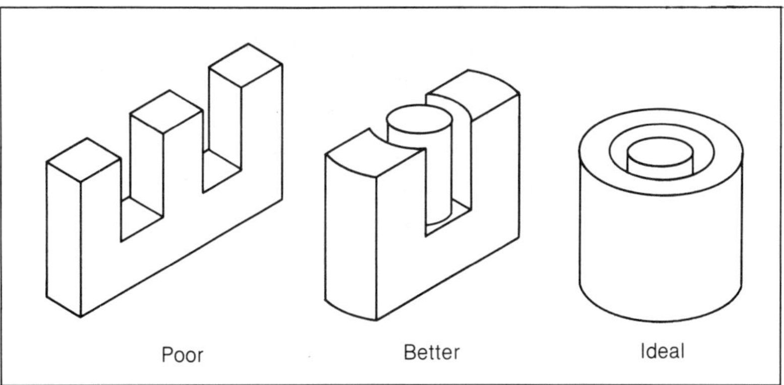

Figure 5.14B—Reducing Transformer Magnetic Leakages by Core Configuration

5.1.3 Harmonic Distortion and Disturbances Caused by Transformers

Transformers can create a more or less pronounced distortion of the secondary output waveform, even for a perfectly sinusoidal input. This distortion appears mainly as odd harmonics (3, 5, 7, etc.) of the fundamental frequency. Figure 5.15 also shows two worst-case situations. In 5.15a the turn-on of a transformer creates a peak primary current if it occurs at V_{prim} max. This in turn creates a secondary voltage disturbance. In 5.16b the MdI/dt when switch opens at I_{prim} max (Zero-crossover of V_{prim}) creates a secondary voltage disturbance.

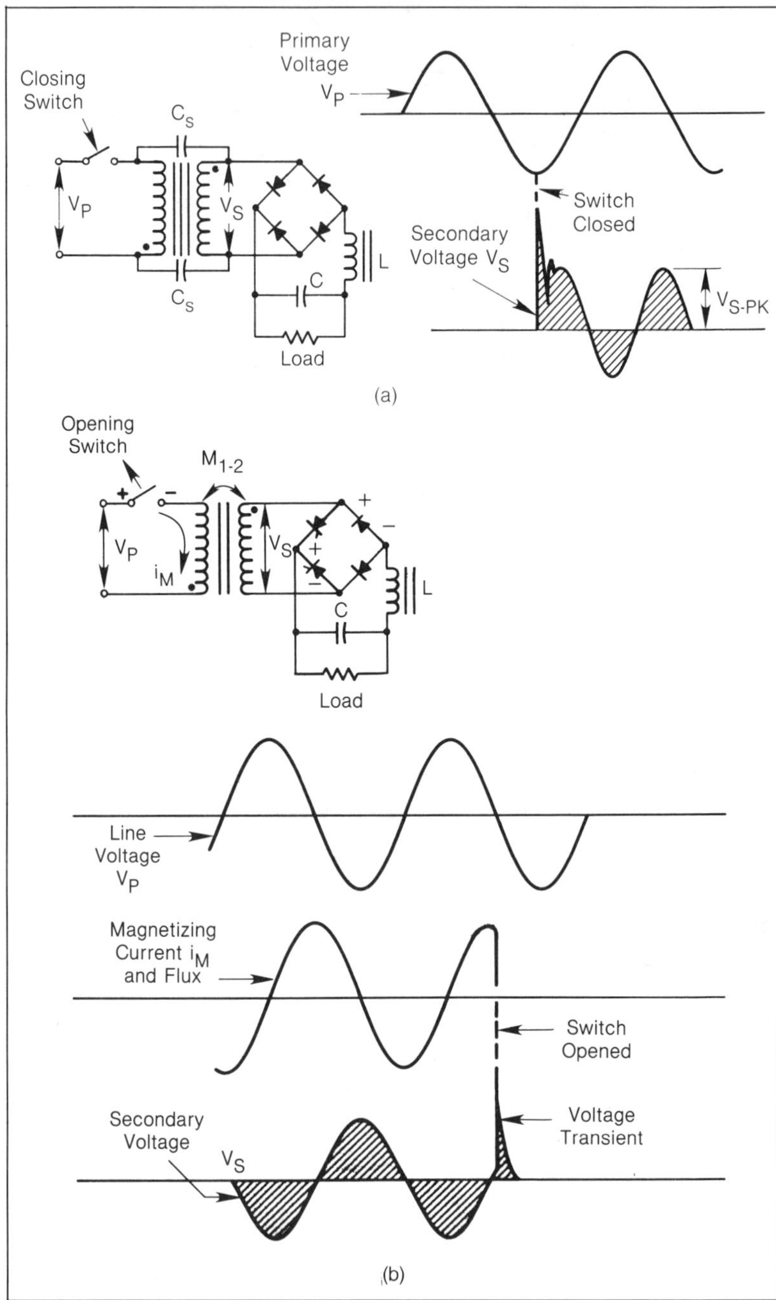

Figure 5.15—Disturbances Caused in the Secondary when Primary Current (a) Is Established and (b) Is Interrupted

5.2 Signal and Pulse Transformers

Signal and pulse transformers (see Fig. 5.16) are used in a wide range of applications. They are used from audio to video frequencies for breaking a ground loop and creating a galvanic isolation, balancing a line, triggering an SCR gate, creating an ac coupling while blocking a dc component, impedance matching and balance-to-unbalance conversion (balun).

The features of these transformers regarding EMI are their parasitic capacitance and bandwidth. The parasitic capacitance obeys the same rules as for the larger transformers of Section 5.1. Audio and miniature pulse transformers are also available with a Faraday shield for better CM rejection.

In pulse transformers, the bandwidth is sometimes expressed by the longest pulse width which can be processed without excessive droop (for instance, -3 dB) and the fastest rise time that can be processed without excessive distortion. For example, a pulse transformer with $\tau = 1$ μs and $\tau_r = 3$ ns has a 3 dB bandwidth of approximately 0.15 to 100 MHz. Table 5.1 shows some typical features of signal and pulse transformers.

Table 5.1—Signal and Pulse Transformer Features

Type of Signal Transformer	Bandwidth Lower Bound Upper Bound	CMR	Parasitic Capacitance
Video, Wideband	20 Hz-25 MHz	120 dB @ 60 Hz 30 dB @ 1 MHz	
Signal Pulse Transformer: (Dual-in-Line Package)	100 kHz-150 MHz		15 pF (Unshielded) 3 pF (Shielded)
Audio Transformer: (Voice and Lowspeed Telecom)	300-3,300 Hz Harmonic Distortion < 0.1%	60 dB (In Band)	

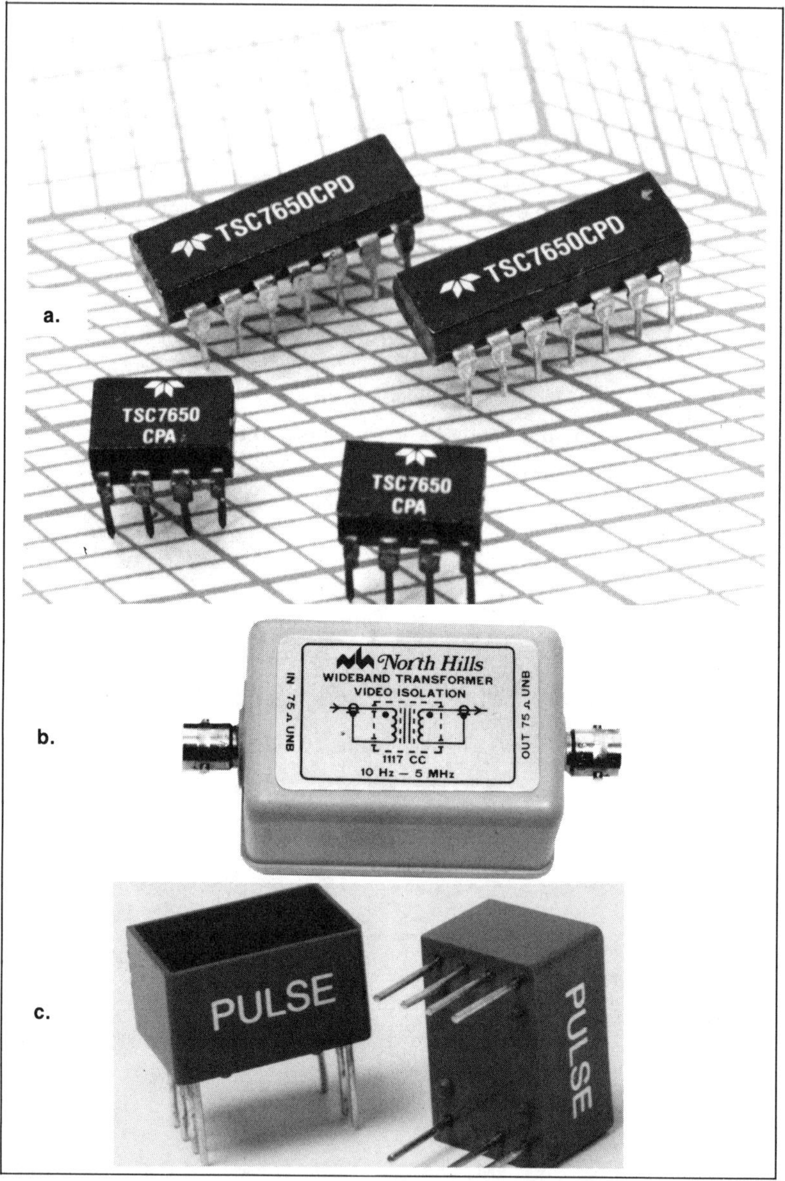

Figure 5.16—Example of Commercial Types of Isolation Signal Transformers; Type a) is for audio applications (.2 to 4 kHz); Type b) is for fast clocks and video applications (10 Hz to 5 MHz). They provide common mode isolation from 40 dB to 60 dB, i.e., only 1% or .1% of the common mode noise appears on the Output Terminals. Types c) are miniature, PCB mount transformers with shield options and upper frequency up to 150 MHz

5.3 Ferrites and Baluns

Ferrite beads are hollow cores or toroids which can be slipped over a wire and behave as a lossy inductance with one or few turns. Although they are popular among EMC specialists as quick "last chance" fixes, they can be incorporated into original designs and achieve remarkable EMI reduction provided their theory is properly understood.

In contrast to ferrites used for coupling between circuits in microwave applications (couplers, circulators, etc.) where low loss and best efficiency are the aims, EMI ferrites are made of lossy materials having a good magnetic permeability (preferably with μ_r being flat over a wide frequency span and with typical values of 300 to 3,000). They also display a resistance of 10 to a few hundred ohms. Although ferrite beads are generally thought of as inductors, they are in fact **transformers**, where the wire being filtered is the primary (one or a few turns) and the secondary is made by the eddy currents in the bead, creating losses by Joule effect. Therefore, the impedance of the ferrite is more like:

$$Z = \sqrt{R^2 + L^2\omega^2} \qquad (5.1)$$

The L term depends of the relative permeability μ_r, which makes generally few hundreds for a typical ferrite. It can be demonstrated that if L_o was the self-inductance of the segment of conductor prior to the ferrite mounting, the increase in inductance caused by the ferrite can be predicted from:

$$\frac{L}{L_o} = \mu_r \qquad (5.2)$$

An example is a 1 cm long ferrite, with $\mu_r = 300$ slipped over 1 m of wire. Since the inductance for an ordinary wire is about 1 μH/m, the initial value of L_o for 1 cm was:

$$L_o = 1 \ \mu H \times 10^{-2} \ m = 10 \ nH \qquad (5.3)$$

Once the ferrite is installed, L becomes:

$$L = 10 \ nH \times 300 = 3 \ \mu H \qquad (5.4)$$

Since, on the other hand, the initial inductance of the 1 m wire was 1 μH, the increase in L caused by the ferrite is 3 μH/1 μH, i.e., three times.

Due to the generally small size of ferrite beads, they can easily saturate for the normal current and become inefficient against the EMI current. The amount of current a bead can handle without significant decrease of μ_r is given by the manufacturer. It is related to:

$$\ln (r_2/r_1), \qquad (5.5)$$

where r_2 and r_1 are the bead's outside and inside diameters. Therefore, beads with proportionally small holes will behave better.

The permeability is also affected by frequency. Some beads are optimized to work below 10 MHz and others are suitable from 10 to 100 or even 1,000 MHz.

Figure 5.17 shows the resistive and inductive parts of a typical bead impedance. Because of their low Q, beads are especially efficient to damp high-frequency contents of switching transients and parasitic resonances. Because they add series insertion loss, they are an inexpensive way to create EMI losses without affecting dc or low-frequency intentional currents. If a circuit or cable is exposed to a high-frequency EMI coupling, the beads will prevent the circulation of induced currents. Conversely, if a signal source contains undesired spurious noise, ferrites will prevent these currents from propagating and making the wire a radiating antenna.

To gainfully use ferrite beads, it must be understood that they work by series insertion loss. Therefore, the attenuation provided by the ferrite will be:

$$A_{(dB)} = 20 \log \frac{V_o \text{ without ferrite}}{V_o \text{ with ferrite}}$$

$$= 20 \log \frac{\dfrac{Z_L}{Z_g + Z_W + Z_L}}{\dfrac{Z_L}{Z_g + Z_W + Z_L + Z_B}} \qquad (5.6)$$

$$= 20 \log \frac{Z_g + Z_W + Z_L + Z_B}{Z_g + Z_W + Z_L}$$

where,

$$A_{(dB)} = \text{attenuation in dB}$$
$$Z_g, Z_L = \text{circuit source and load impedances,}$$
$$Z_W = \text{wire impedance}$$
$$Z_B = \text{ferrite bead impedance}$$

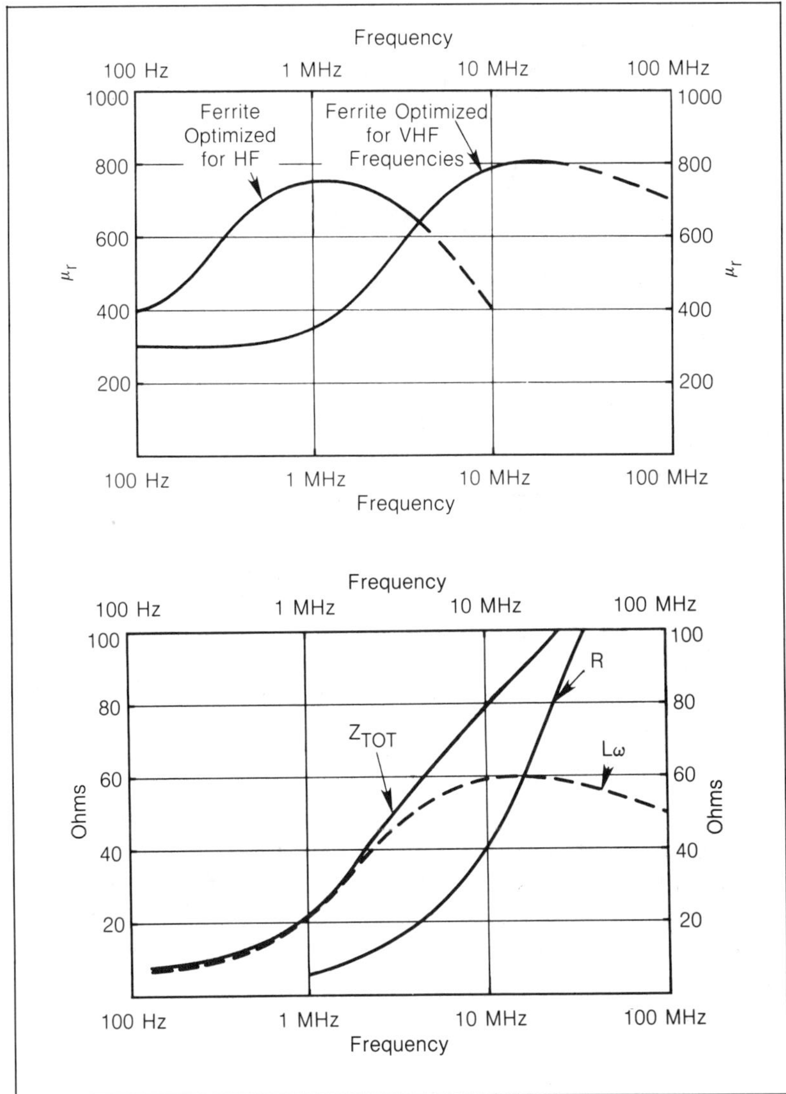

Figure 5.17—Typical Values of μ_r and Combined Impedance Z_{TOT} for an EMI Suppressing Ferrite

This equation produces two revelations:
1. Ferrites will not work efficiently in high-impedance circuits. Although significant progress has been made by manufacturers in the 1980s, the best ferrites today achieve values of Z_B in the 300 to 600 Ω range, above 50 MHz. Neglecting the wire impedance, the best attenuation they can provide in a 100 Ω/100 Ω configuration is:

$$A_{(dB)} = 20 \log \frac{100 + 100 + 600}{100 + 100} = 12 \text{ dB} \qquad (5.7)$$

Conversely, ferrites will be extremely efficient in low-impedance circuits like power distribution, power supplies or radio-type circuits where impedances are 75 or 50 Ω.
2. If the wire impedance itself is significant, the ferrite performance may be disappointing because of the presence of Z_w in the equation.

An extremely useful application of ferrite is in the blockage of common-mode currents. As explained in Section 2.3.2 with inductors, if the two wires of a signal pair are threaded in the bead, the ferrite will affect only the undesired EMI currents and will have no effect on the intentional differential-mode current. The same is true when a ferrite is slipped **over** a coaxial cable.

The limitations of ferrites, besides their limited impedance are:
1. When bead length approaches $\lambda/4$, the bead becomes inefficient.
2. The end-to-end parasitic capacitance of the ferrite (typically 1 to 3 pF) may bypass its resistance above a certain frequency and cause its attenuation to collapse.
3. Beyond about 1,500 to 2,000 G, saturation occurs and efficiency decreases.
4. When slipped over multipair cables, they may increase inductive crosstalk between adjacent pairs.

The saturation problem can be controlled by checking the flux density $B = \phi/S$. Figure 5.18 shows that ferrite bead flux ϕ is given by $B \times L (r_2 - r_1)$. However, if B is too large the core will saturate and μ_r will decrease.

Illustrative Example:
A ferrite has the following dimensions:

$$\text{length L} = 1 \text{ cm}$$
$$\text{outside radius } r_2 = 0.5 \text{ cm}$$
$$\text{inside radius } r_1 = 0.2 \text{ cm}$$

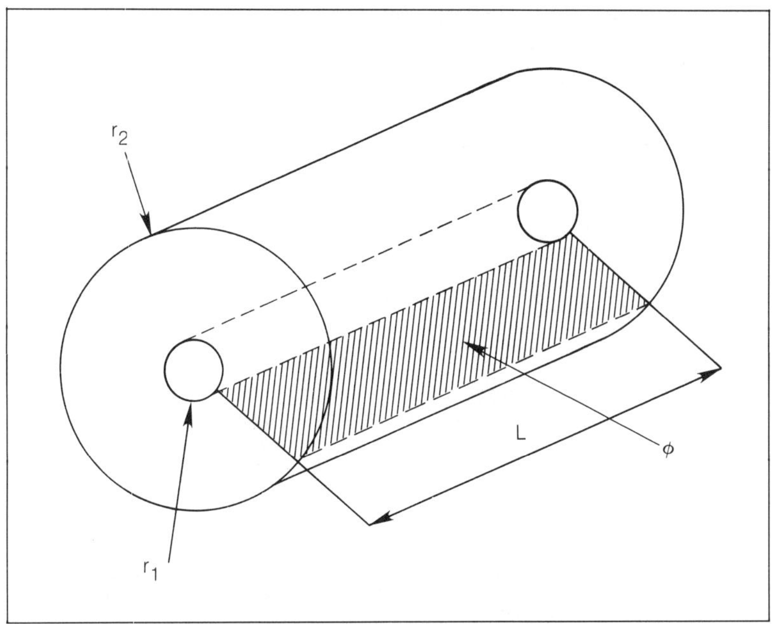

Figure 5.18—Ferrite Bead Flux ϕ, Given by Flux Density B $\times$ L(r_1 $-$ r_2)

According to manufacturer's data, the equivalent inductance L_B is 10^{-6} H, provided that B does not exceed 1,500 G. What is the maximum current this bead can handle without losing efficiency?

$$I = \phi/L_B = \phi/10^{-6} \tag{5.8}$$

Also,

$$\phi = BS = B \times L(r_2 - r_1) = B \times 1.10^{-2}\,(0.5 - 0.2)\,10^{-2}$$
$$\phi = B \times 0.3 \times 10^{-4}$$

Since,

$$B\ max = 1{,}500\ G = 1.5 \times 10^{-1}\ T \tag{5.9}$$
$$\phi\ max = 1.5 \times 10^{-1} \times 0.3 \times 10^{-4} = 0.45 \times 10^{-5}$$
$$I\ max = \phi\ max/10^{-6} = 4.5\ A$$

If the attenuation with a ferrite bead is not sufficient, this can be improved in several ways. One method is to make more than one turn of the wire in the bead hole, using two or three turns. However, this may rapidly bring the ferrite into saturation. Also, the turn-to-turn capacitance may ruin the inductance improvement. Putting several beads back to back is another method. However, this will create a cascade of parasitic capacitances. If a few beads do not work, it is doubtful that multiplying their number to reach several tens of beads will ever work. The preferred method is to use a multihole bead (Fig. 5.19) to provide a good magnetic circuit around each turn and optimize the r_2/r_1 ratio.

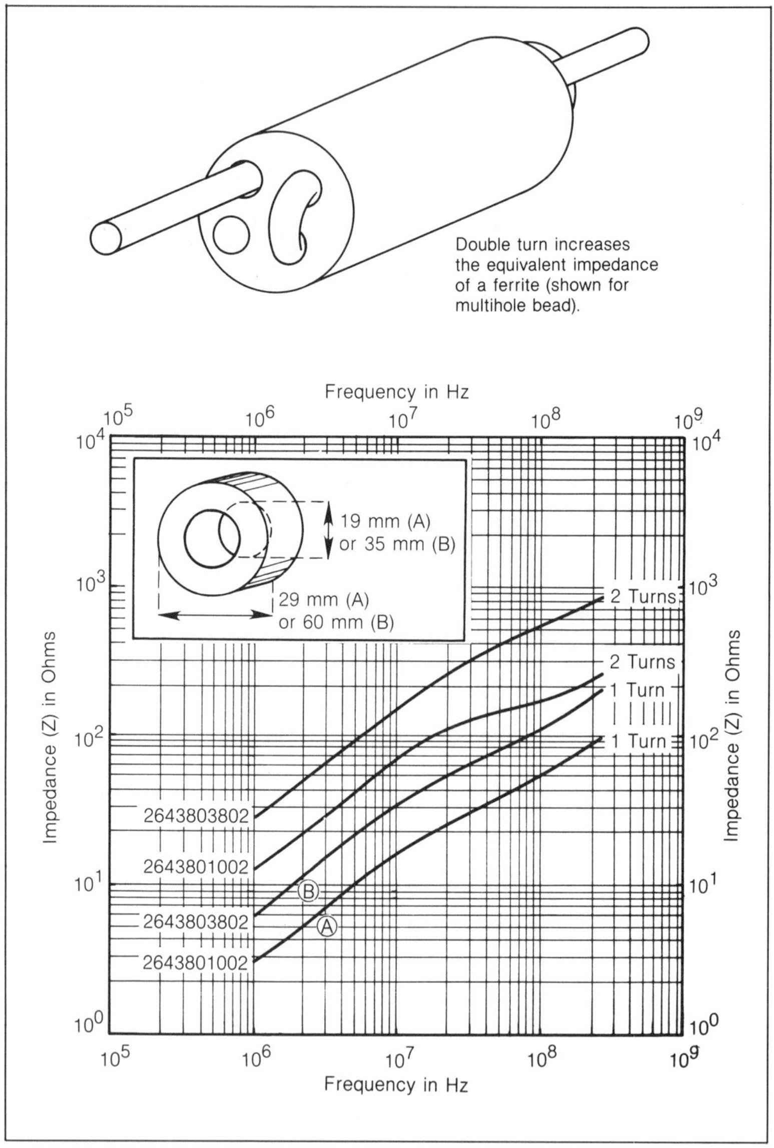

Figure 5.19—Impedance Increase by Making Several Turns in Ferrite Window (Courtesy of Fair-Rite Co.)

Figures 5.20 to 5.25 show the shapes and performances of the principal types of lossy ferrites available.

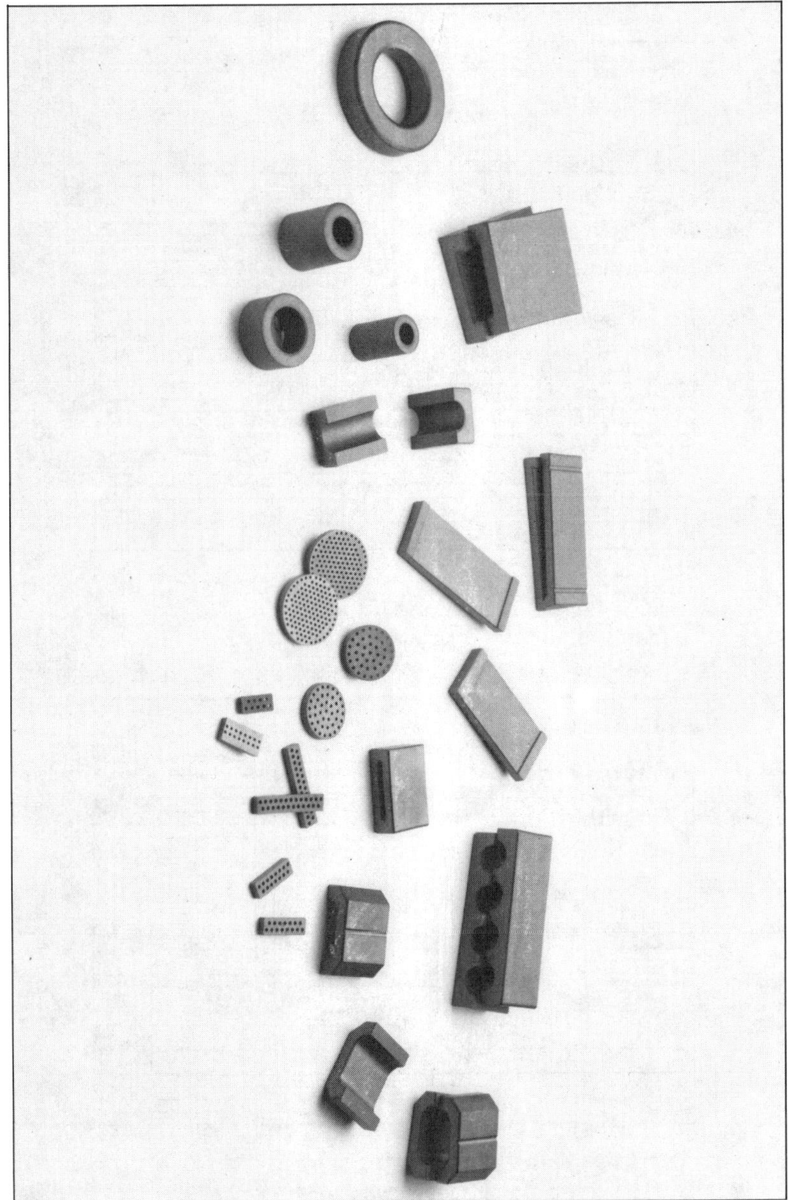

Figure 5.20— Various Ferrite Beads Available (Photograph by Fair-Rite Co.)

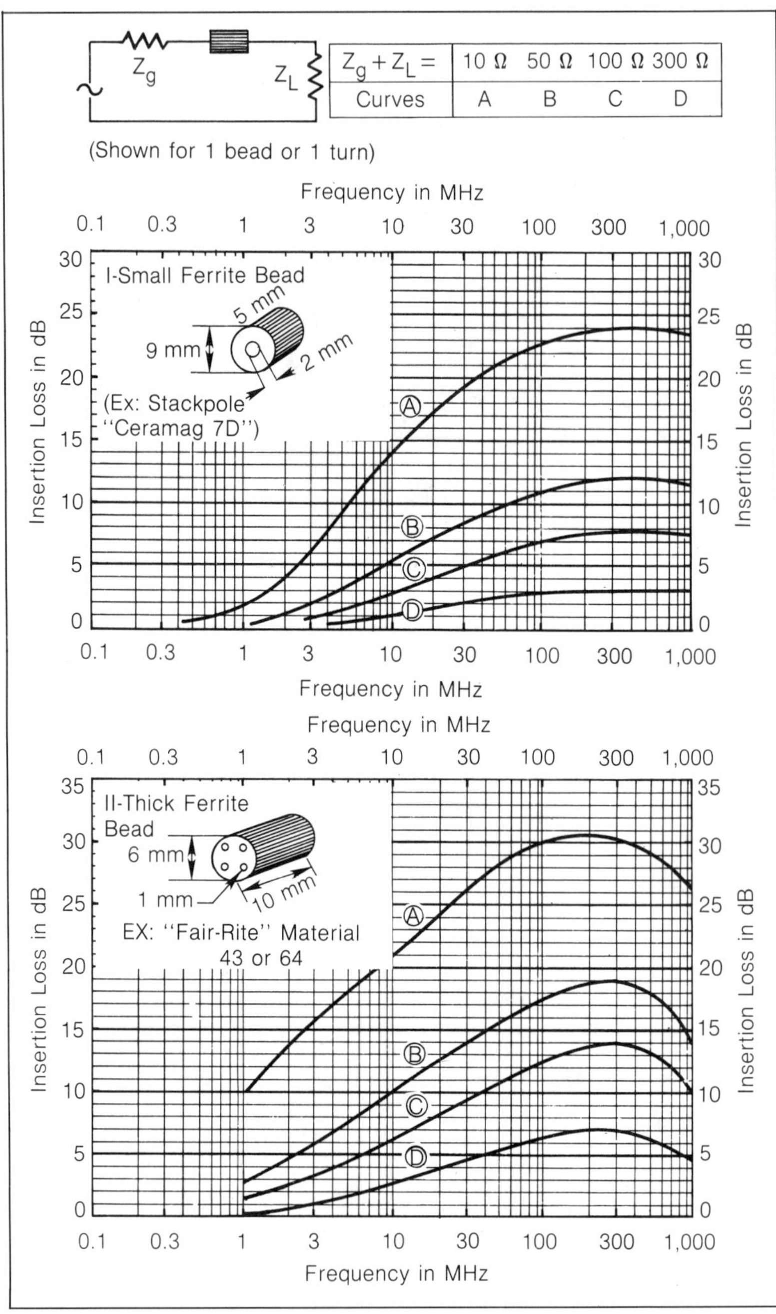

Figure 5.21—Insertion Loss of Small Ferrite Beads: (1) Small Ferrite, (II) Thick Multi-Hole Ferrite

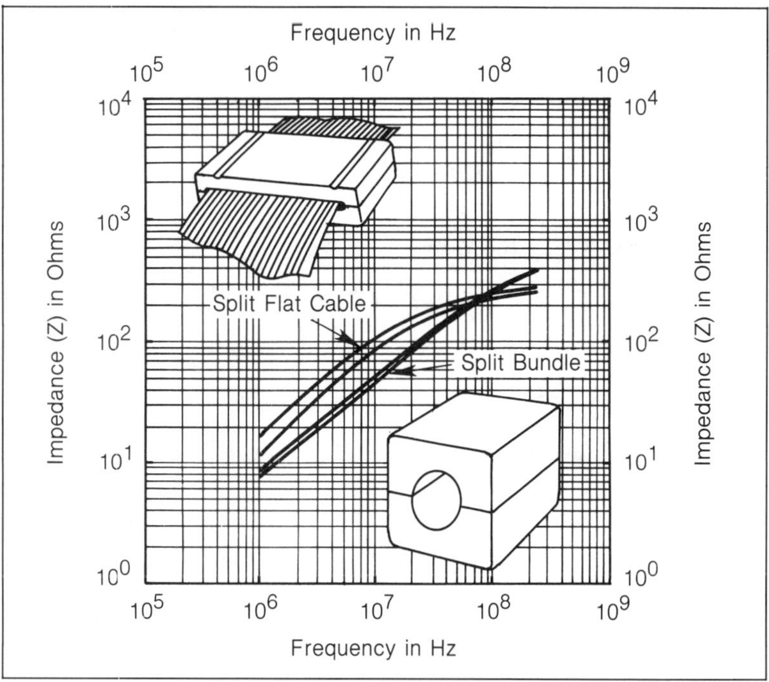

Figure 5.22—Impedance vs. Frequency for the Split Flat Ribbon Cable Bead and the Split Bundle Cable Bead

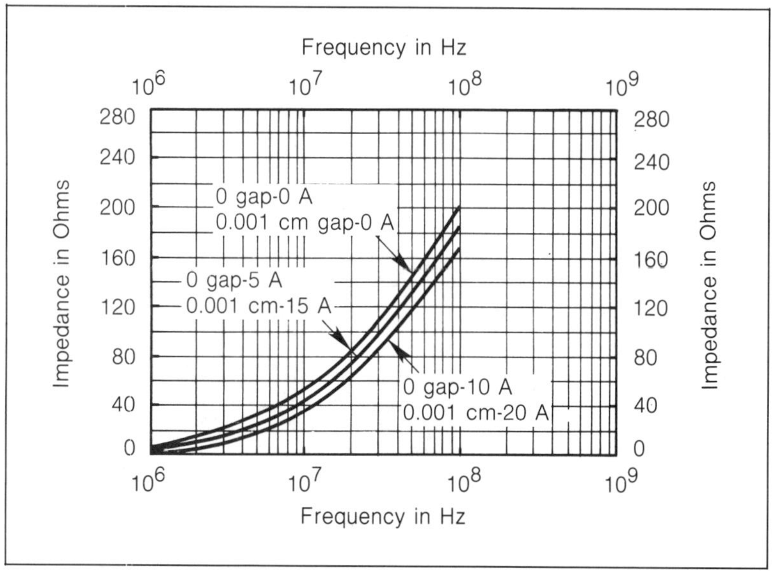

Figure 5.23—Impedance vs. Frequency as a Function of Direct Current in Amperes (A) and Gap for a Split Flat Ribbon Cable Connector Bead

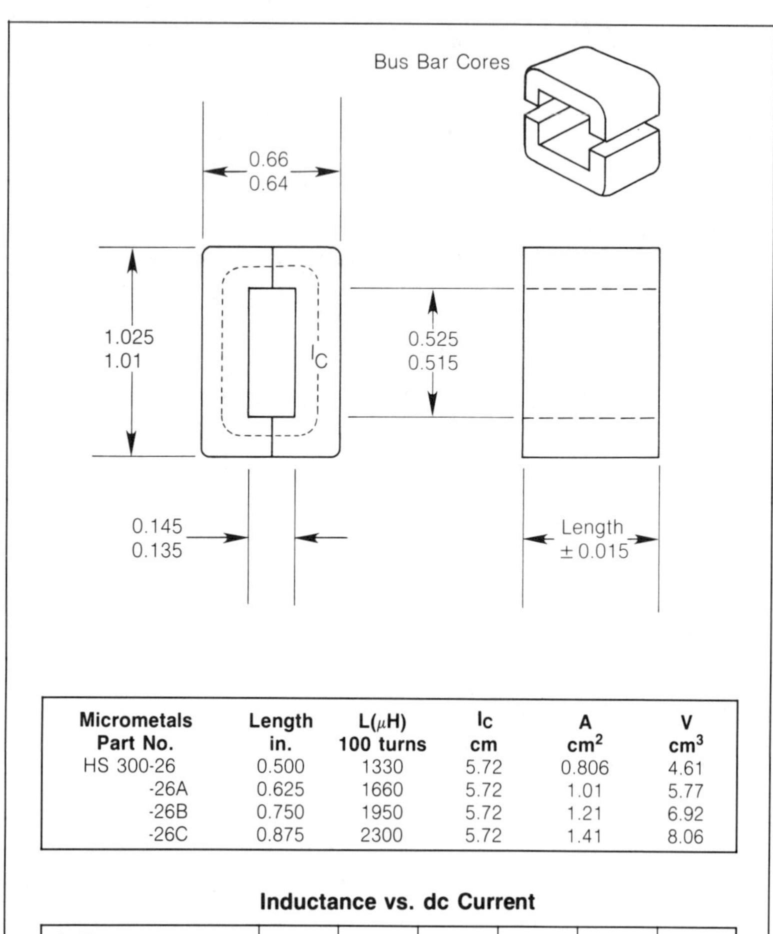

Micrometals Part No.	Length in.	$L(\mu H)$ 100 turns	I_c cm	A cm^2	V cm^3
HS 300-26	0.500	1330	5.72	0.806	4.61
-26A	0.625	1660	5.72	1.01	5.77
-26B	0.750	1950	5.72	1.21	6.92
-26C	0.875	2300	5.72	1.41	8.06

Inductance vs. dc Current

dc Current (amperes)		50	100	150	200	250	300
Part No.	Ripple	INDUCTANCE (nH)					
H5300-26	1%	125	104	87	72	63	55
H5300-26	10%	148	139	126	110	97	86
H5300-26A	1%	156	130	109	90	78	68
H5300-26A	10%	184	173	157	138	121	107
H5300-26B	1%	183	153	126	106	92	80
H5300-26B	10%	216	204	184	162	142	126
H5300-26C	1%	216	181	151	125	108	94
H5300-26C	10%	255	240	217	191	168	148

Figure 5.24—Special Flat Ferrites for Power Distribution Bus (Courtesy of Micro Metals Inc.)

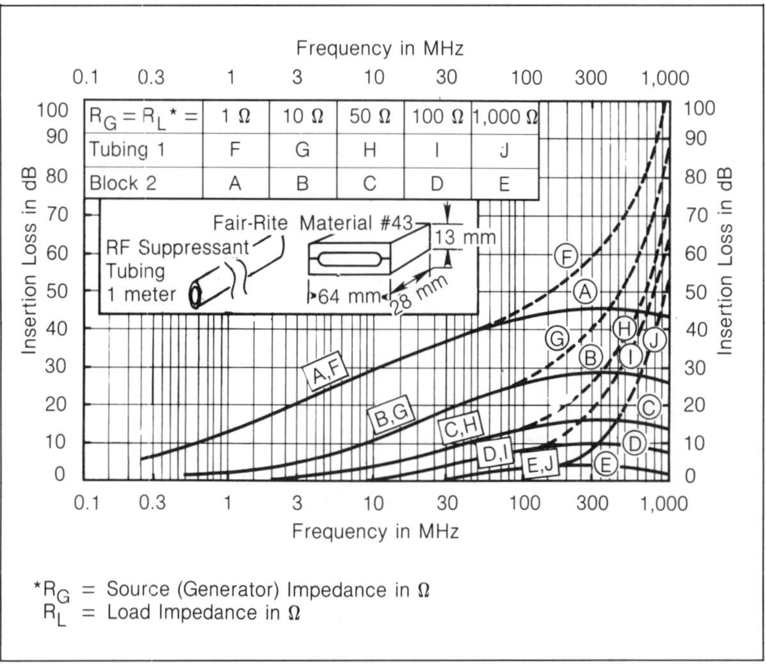

Figure 5.25—Insertion Loss of Large Ferrite Suppressors

In every case, a small analog signal is generated in the vicinity of noise sources like capstan motor and drive motor magnetic fields, head positioning electromagnetic actuators and various low-frequency and transient magnetic fields. For this reason, magnetic heads are always carefully shielded with high-permeability material.

Additionally, the output wires have to be tightly twisted and shielded and have the shortest path possible up to the magnetic head amplifier. In most of the EMI problems associated with magnetic recording, this wiring is the coupling path. For instance, in a modern high-speed digital disk unit, bandwidths up to 30 MHz are used. This is because to meet the high transfer rate requirement, high bit density and fast tape or disk rotational speeds are needed. As a result, the wavelength (or period) of the magnetized spots on a record is very short, in the range of a few micrometers, corresponding to less electromagnetic energy. A weak signal (1 mV) and a large bandwidth invite EMI problems since the amplifier will have a low threshold of sensitivity. (See "susceptibility scoring" in Chapter 4).

Inspired from the previous considerations, an interesting concept developed in the late 1960s is the ferrite-loaded wire (also called "lossy" wire) and tubing.[1,2] In these wires a conductor is coated with a flexible compound made of ferrite plus a binder, such that the lumped elements are replaced by a continuously distributed insertion loss. Because of the flexibility requirement, the ferrite content of the jacket provides a permeability of a few tens. In a 50 Ω system, little to no attenuation exists below about 5 MHz (Fig. 5.26). By eliminating the impedance discontinuities that a cascade of beads would create, lossy wires have less tendency to radiate. They share with the beads the enormous advantage of not depending on grounding or bonding techniques. Also, their distributed impedance and their lossy nature means that they can work with extremely mismatched source and load resistances without exhibiting ringing and other mismatch problems. Furthermore, the ferrite grains and their binder have an ϵ_r which can be rather large, such that a lossy line with a **predetermined** characteristic impedance can be built, since:

$$Z_{o \text{ (lossy line)}} = 377 \sqrt{\frac{\mu_r}{\epsilon_r}} \qquad (5.10)$$

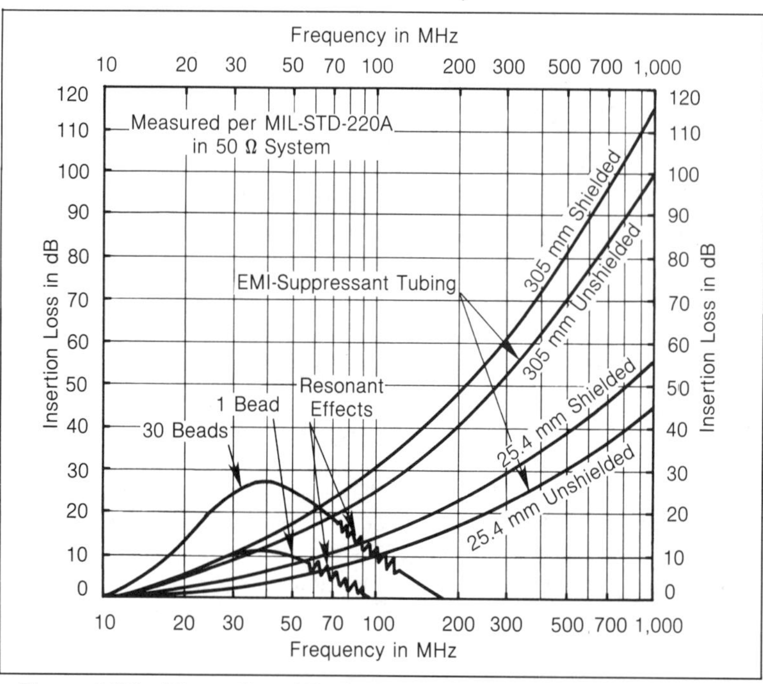

Figure 5.26—Attenuation of Ferrite-Loaded Tubing vs. Single Beads. (The beads shown were not optimized for high frequency resistance.)

5.4 Magnetic Heads

Magnetic read or write heads are used in audio, digital and video recording systems. Although these applications vary widely in speed, bandwidth and magnetic support (tape, disk, cartridge, credit card, etc.) the basic mechanisms of the magnetic pickup are the same.

The vulnerability of a magnetic head can be understood by knowing that the data on the magnetic support create a small field change when they are passing in front of the head gap. The corresponding voltage induced in the head winding is generally very small, on the order of millivolts*. Until it is amplified, this signal is especially vulnerable. Making it more vulnerable is the fact that it has to travel over some length of wire before being amplified. The corresponding amplifier (or preamplifier) must have a bandwidth corresponding to the application. For example:

1. For audio applications: 30-30,000 Hz
2. For low-speed digital data (floppy disks with 10 to 100 kBps transfer rate, 1m/s writing speed): 300 kHz
3. For high-speed digital data (hard disks, writing speeds in the 3-30 m/s): up to 30 MHz
4. For video recording: 8-15 MHz.

In a digital data disk (floppy or Winchester), the most critical situation occurs when the head is reading the tracks closest to the spindle, since this is where the background noise due to the drive motor magnetic field is the largest. For the same reason, power supplies (50/60 Hz or switch-mode) must be either kept away from magnetic head wiring or isolated by a ferrous shield.

5.5 Magnetic Bubble Memories

Although magnetic bubble memories are less favored now than when they were introduced in the mid-70s, they are still used as primary nonvolatile data storage by a few of the largest computer manufacturers. Their EMC aspects are twofold: self-jamming and susceptibility. A deeper study of their EMC features can be found in Ref. 3.

The basic bubble memory consists of a chip hosting the magnetic domains, permanent magnets and two orthogonal field coils. The

---

*For instance, a typical audio head with a 10 kHz bandwidth has a readout of 0.3 mV, with a S/N ratio in the 50 dB range.

self-jamming may occur because of interaction between incidental signals generated in the other memory circuits (coil drive and other module support functions) and the low level of detection (a few millivolts) corresponding to the presence or absence of a bubble in the detection path. The primary causes of parasitic coupling into the detection circuitry are $d\phi/dt$ noise induced in detector loops and dV/dt (capacitive) coupling between field coils and chip carrier traces.

The first type of coupling (magnetic) is reduced by careful placement of the sense loops to minimize the intercepted flux. Reducing the rate of change of the flux (lowering the magnitude or smoothing the rise and fall times) to the minimum which is necessary and making a tradeoff between square-wave and sine-wave drive are additional ways of reducing magnetic induction. Finally, performing the bubble detection in a time window where the flux is at its maximum or minimum value (when its derivative is null) is also beneficial.

The capacitive coupling is due to the coil voltage transitions, having frequency components as high as 10 to 20 MHz, "crosstalking" to the sensor circuitry. This can be reduced by a judicious routing of the chip-carrier traces, a low dielectric constant of the material between coils and sensor and a slower dV/dt. Hopefully the techniques to reduce magnetic and capacitive self-jamming are not contradictory.

Finally, crosstalk within the chip carrier itself can become a concern due to the low levels and wide bandwidth used. The solutions are similar to those described with logic circuits in Chapter 4. For instance, sense conductors and ancillary circuitry should not run parallel over extensive lengths, and the designer should seek right angle routes.

Figure 5.27 shows an example of the signal and noise in a bubble memory. It is clear that the "bubble-present" signal at 5 mV is only 3 mV above the background noise, therefore the S/N ratio (or more accurately the S + N/N) is 2.7, i.e. 8.5 dB. To improve noise segregation, a "detector-pair" (active and dummy) approach is used, where only differential signals are amplified, which eliminates a large amount of magnetic, capacitive and ground coupled noise.

The susceptibility to external EMI results from the offset of the biasing fields by external sources. As a protective measure, bubble memories are generally put in a magnetic shield. The requirements are that the internal bias field shall not show a noticeable change when the module is placed in a dc or low-frequency magnetic flux density of 40 to 50 Oe. As far as being interference sources, the relatively low frequency range (less than 1 MHz), the low dissipated energy, the small conductor size and the mandatory use of a shield make the magnetic bubble an extremely low source of EMI.

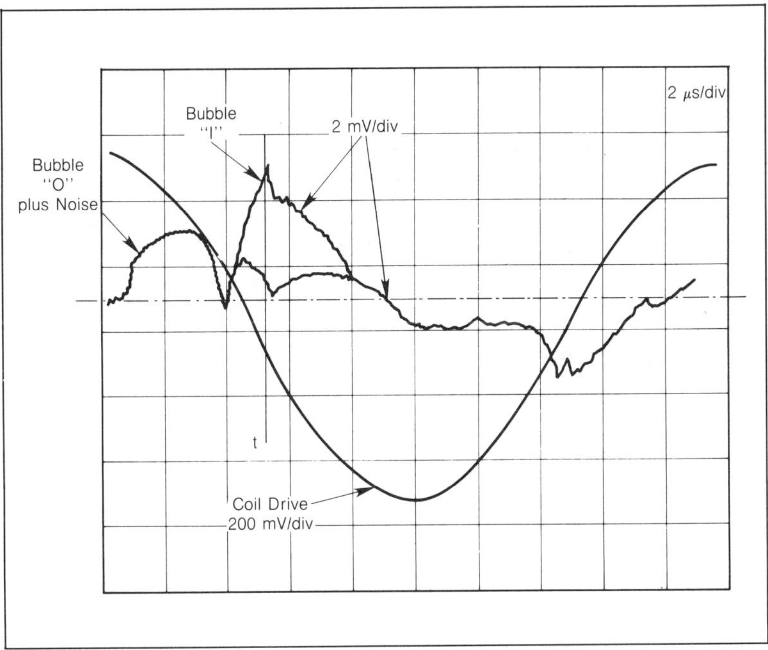

Figure 5.27—Superimposed One and Zero Waveforms and Coil Drive of a Magnetic Bubble Memory Chip

5.6 References

1. Mayer, F., "RFI Suppression Components," (*IEEE Transactions on EMC,* May, 1976).
2. Mayer, F., "Absorptive Low-Pass Cables," (*IEEE Transactions on EMC,* February, 1986).
3. Malack, J., "EMC Concerns in Magnetic Bubble Memories," *Proceedings of the 1979 IEEE EMC Symposium.*

5.7 Bibliography

1. Cowdell, R., "Don't Experiment with Ferrite Beads," (*Electronic Design,* June, 1976).
2. Hnatek, E.R., *Design of Solid State Power Supplies,* (Van Nostrand, 1981).
3. Nakauchi, E., "Techniques for Controlling Common Mode Emissions," *Proceedings of the 1982 IEEE/EMC Symposium,* Santa Clara, California.
4. Parker, C., "Prayer Beads Solve EMI Problems," (*EMC Technology,* April, 1985).
5. Pulse Transformers Catalog No. 811, Pulse Engineering Division of Varian Transformers.
6. White, Donald R.J. and Mardiguian, M., *EMI Control Methodology and Procedures,* (Gainesville, Virginia: Interference Control Technologies, Inc., 1985).

Chapter 6

Electromechanical Devices

6.1 Electromechanical Switches

Switches produce high-frequency EMI through the combined and interactive processes of arcing, bouncing and load circuit oscillations. Every time a switch closes a load circuit, a step function of voltage is applied to an RLC network. Similarly, when the switch breaks, a step function change of current is applied.

Mechanical contacts exhibit contact bouncing, sliding or rocking when closing. Random opening and closing of contacts chops the current, generating HF oscillations and their harmonic components. For inductive circuits, interruptions can lead to high induced-voltage transients, contact arcing, dielectric breakdown and associated phenomena. All cause EMI problems.

6.1.1 Glow and Arc Phenomena

The making or breaking of an electrical circuit by a mechanical switch is usually accompanied by an arc at the switch contacts. Arcing during normal operation of a switch occurs because a highly ionized gas is substituted for a part of the metallic circuit as the switch contacts move apart. The arc is extinguished when energy stored in the circuit is dissipated (including oscillation with distributed reactance) once the switch contacts exceed the arc-cover distance.

An arc is a phenomenon that dissipates energy that is either supplied to or stored in an electrical circuit. The arcing phenomenon causes deterioration of the contact surfaces where the arc is formed and may destroy the contacts if continued. The arcing phenomenon is also a prime source of interference having a broad frequency spectrum.

A glow discharge is a self-sustained mechanism occurring between two electrodes when the voltage gradient is sufficient to ionize the gas channel (Townsend discharge). Depending on the exact shape of the facing electrodes, it takes 10 to 30 kV/cm to start this discharge. The glow is then sustained by an electron avalanche in the channel. Once started, and with little influence of contact spacing, about 300 V are sufficient to maintain this glow, with a current capability in the order of a milliampere (a current-limited source with a lesser rating would not maintain the discharge.

A glow discharge will unavoidably evolve to an arc discharge if the glow voltage is maintained for a sufficient duration (> 10 ns), the contacts do not significantly separate, and a current capability of at least few hundred mA exists. With mechanical switches, the contact gap is small and glow/arcing fields thresholds are reached easily.

Due to microscopic irregularities on the contact surface, electrons emanate from these hot spots (behaving as cathode). The local current density is very high and creates microscopic melting because of the local temperature rising to a few thousand degrees Kelvin. Within a few nanoseconds the arc becomes a metal-vapor arc bridging the two contacts. If the switch is opening, this arc will extinguish and restrike as long as the voltage and the current are above the minimum values to be sustained. If the switch is closing, the arc will disappear when the two contacts touch.

In addition to its generating noise, arcing wears out the contact. Unfortunately, arcing is almost unavoidable since it is part of the closure/opening mechanism. Ac circuits do not cause as much damage since the arc will extinguish at zero crossover. And since polarity reverses every half-period, each contact is an anode and cathode alternatively.

For the better lifetime of contacts, contact material is generally classified in low-current (sub-ampere) and high-current application. Low-current contacts are gold alloys which become "wet," i.e., show low contact resistance for very low currents. But if they carry intentionally or inadvertently large currents (EMI), they will be permanently altered. Conversely, high-current contacts cannot carry small currents because they are not large enough to "wet" them.

Because of this nonlinear aspect of contact resistance, parasitic rectification can cause EMI, especially if the contacts have been abused, corroded or contaminated (see "audio rectification," in Section 4.1.3).

Contacts must also be rated as a function of the load they will commute:

1. For resistive loads, the current rating is simply V/R, provided that the value of R is considered at the lowest temperature.

2. For lamps, the cold filament current at turn-on can be 10 times the normal operating current, so the switch should have a derating factor of 10.

3. For capacitive loads, the current could theoretically reach a peak value of CdV/dt unless it is limited by the source internal resistance, whichever is less. It must be remembered that many motors and transformers have large parasitic capacitances.

4. For motors at turn-on, both the winding capacitance and the torque requirement can cause an inrush current of 5 to 10 times the steady value.

5. For inductive loads, the associated LdI/dt inductive transient can cause overvoltage and consecutive contact fatigue by arcing.

In summary, it is important to remember that capacitive loads create high inrush currents at turn-on, and inductive loads create overvoltages at turn-off. In some applications, mercury switches are used to minimize arcing and contact pitting.

6.1.2 Dynamic Study of Switching

Figure 6.1a shows an "ideal" switching condition. Let V be 1 V and R be 1 Ω. When S closes, I_s changes instantly from zero to 1 A and load voltage from 0 to 1 V. These step function changes would have the classical Fourier transform aspect with a progressive decrease of harmonic amplitudes up to infinity. The modeling of switching EMI would be simple; however, no circuit can exist without capacitances and inductances. So, Fig. 6.1b shows a more realistic view with the wiring having its parasitic self-inductance and capacitance L_1 and C_1, and the load having its resistance R_2, self-inductance L_2 and capacitance C_2.

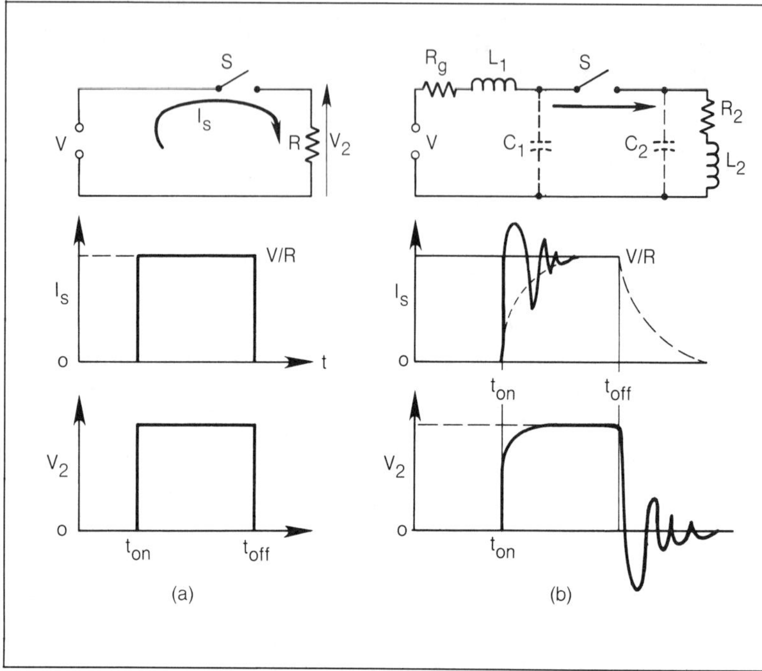

Figure 6.1—Ideal (a) vs. Actual (b) Switching Waveforms

At the opening of the switch, the overvoltage across the load would be:

$$V_1 = -LdI/dt$$

if only inductance were considered. But the solution of the integro-differential equation for R, L, C shows that V_1 cannot exceed:

$$V_{1\ max} = -I\sqrt{L_2/C_2} \tag{6.1}$$

The circuit will "ring" with decaying oscillation at a resonant frequency:

$$f_2 = \frac{1}{2\pi\sqrt{L_2 C_2}} \tag{6.2}$$

For example, assume that a dc relay has the following circuit characteristics:

V coil = 5 V
I coil = 0.1 A (for 50 Ω coil resistance)
L_2 (coil) = 100 mH
C_2 (coil) = 1,000 pF
L_1 (coil drive circuit) = 1 μH
C_1 (coil drive circuit) = 10 pF

Because I is interrupted in R, L_2, C_2, the voltage on the load side can reach:

$$V_{2\ max} = 0.1\sqrt{\frac{0.1}{10^{-9}}} = 1,000\ V$$

with a ringing frequency of:

$$f_2 = \frac{1}{2\pi\sqrt{0.1 \times 10^{-9}}} = 1.7\ kHz \tag{6.3}$$

Since the 50 Ω coil resistance is much less than the 10,000 Ω represented by the characteristic impedance the $\sqrt{L/C}$ of the coil, the circuit is underdamped and the oscillations will last for many periods.

Similarly, since a 0.1 A current is also interrupted on the driving line, an overvoltage will exist which can reach:

$$V_{1\ max} = 0.1\sqrt{\frac{10^{-6}}{10^{-11}}} = 33\ V \tag{6.4}$$

Since both voltages are opposite and the switch is still bridged by the arc, the larger of the two will prevail to predict circuit overstress.

Three items are worth noting:

1. Although large values of I and L could result in kilovolt levels of overvoltage, this value can be clamped by installation breakdowns.
2. A value of R much larger than L/C would create an over-damped situation with no ringing. However, R is dictated by the functional aspects of the circuit and cannot be selected based on a switch protection criterion.
3. A line of characteristic impedance Z_0 terminating on a load R = Z_0 cannot exhibit overvoltage greater than the line voltage V when switched off. Thus, no more than 2 × V can be impressed across the switch.

When the switch is closed, the dV/dt charges the load capacitance C_2 across the source and line impedances (mainly R_g

and L_1 since C_1 is assumed to be charged-up). The peak current may rise very rapidly to a high peak (in theory V/R_g, with a slope L_1/R_g) and then decay to its steady state value (through some oscillations) with a frequency of:

$$f_1 = \frac{1}{2\pi\sqrt{L_1 C_2}} = \frac{1}{2\pi\sqrt{10^{-6} \times 10^{-9}}} = 5 \text{ MHz} \qquad (6.5)$$

The dynamic aspect of switches opening and closing must be now confronted with the arcing parameters of the switch. Ott has carefully studied the actual contact voltages under real conditions in order to design contact protection measures.[1]

Figure 6.2 is a reminder that an actual circuit is seldom a "two-

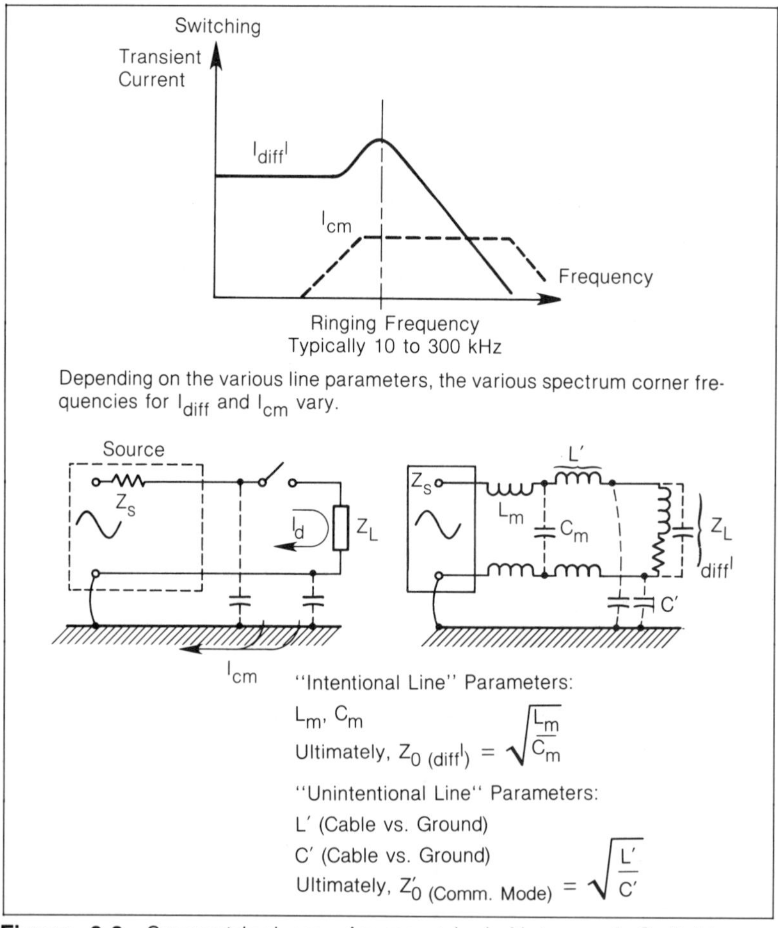

Figure 6.2—Symmetrical vs. Asymmetrical Nature of Switching Transients

wire" condition but that common-mode components, with their associated currents, exist.

Figure 6.3 illustrates the combination of load effects and contact arcing. Looking simply at the voltage across the contacts, notice that although the theoretical value could be the V_{max} calculated previously, there is a clamping effect because of the arcing phenomena with multiple restrikes. This should not create the impression that transient voltage will be limited to about 300 V in the load circuit. The voltage shown is the one across the switch, the rest of V_{max} is still distributed along all the circuit impedances and mainly across the load. Therefore, circuit and contact protection cannot be dissociated, since both the switch and the load are part of the problem.

Typical values measured during the switching of some common loads are displayed on Table 6.1 and Figs. 6.4 and 6.5.

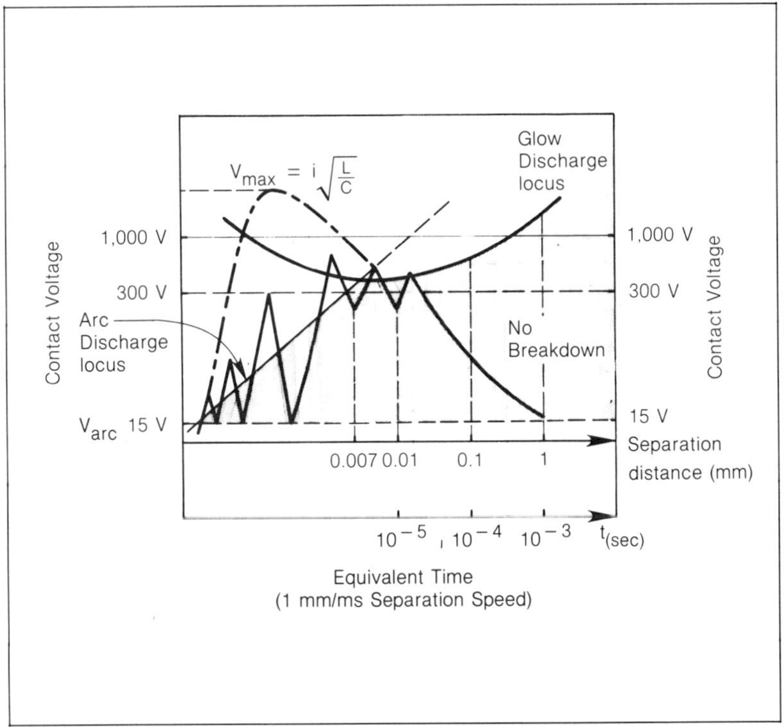

Figure 6.3—Combination of Contact Arcing and Load Effect

Table 6.1—Measured Values of Small Inductive Load Transients

	Switch 200 mA Turn-on/ Turn-Off	Relay Coil 140 mA Turn-on/ Turn-Off
Peak-to-Peak Transient	200 V/450 V	200 V/600 V
Ringing Frequency	150 kHz/150 kHz	87 kHz/150 kHz
Bounces	6	6
Time to Steady	700 μs	700 μs/1 ms

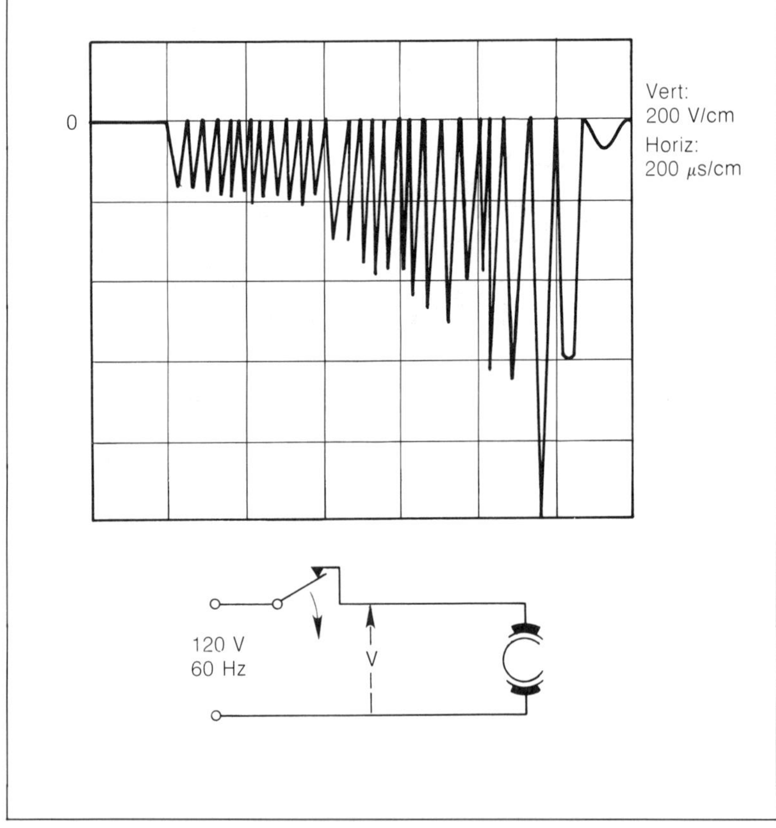

Figure 6.4—Voltage Across the Load During Switch Opening of a Small Timer Motor (Stator Inductance 6.8 H, Resistance 1,450 Ω, Parasitic Winding Capacitance 80 pF). After the Early "Showering" Arc Effect, the Highest Arc Breakdown Recorded Is 1020 V.[16]

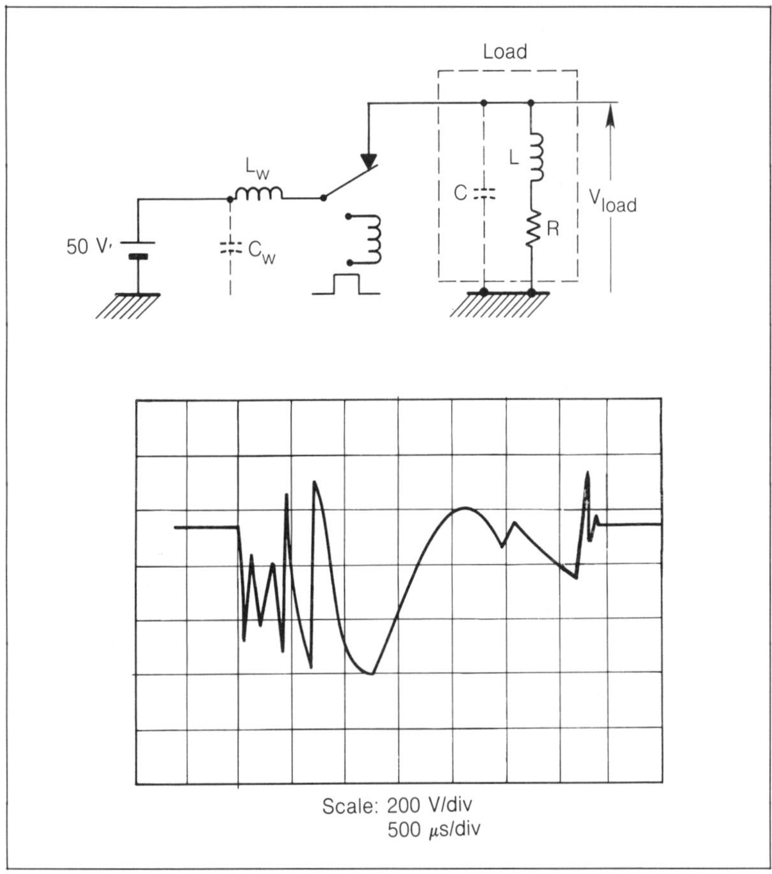

Figure 6.5—Load Voltage at Turn-Off of a Small Timer Motor. Circuit characteristics L_w = 5 μH; C_w = 14 pF. Motor = L = 6.8 H; R = 1,450 Ω C = 4,800 pF. Although the Steady-State Current was Only 34.45 mA, Ringing Voltages of 300 to 600 V are Visible. The Peak Current in the Last Arc was Estimated at About 20 A.

6.1.3 Contact Protection at the Switch Itself

The two basic rules for contact survival are to keep the voltage/distance relationship in the safe zone of Fig. 6.3 and to keep the initial rate of rise of contact voltage below the value necessary to produce an arc. A value of 1 V/μs is generally satisfactory.

Three simple, commonly used contact protection schemes are shown in Fig. 6.6. The first one, captioned "a", is the most intuitive solution: to absorb the dV/dt of the contact opening, a capacitor is put across the contact, such as during the opening phase, the transient current passes through the capacitor instead of the switch.[2]

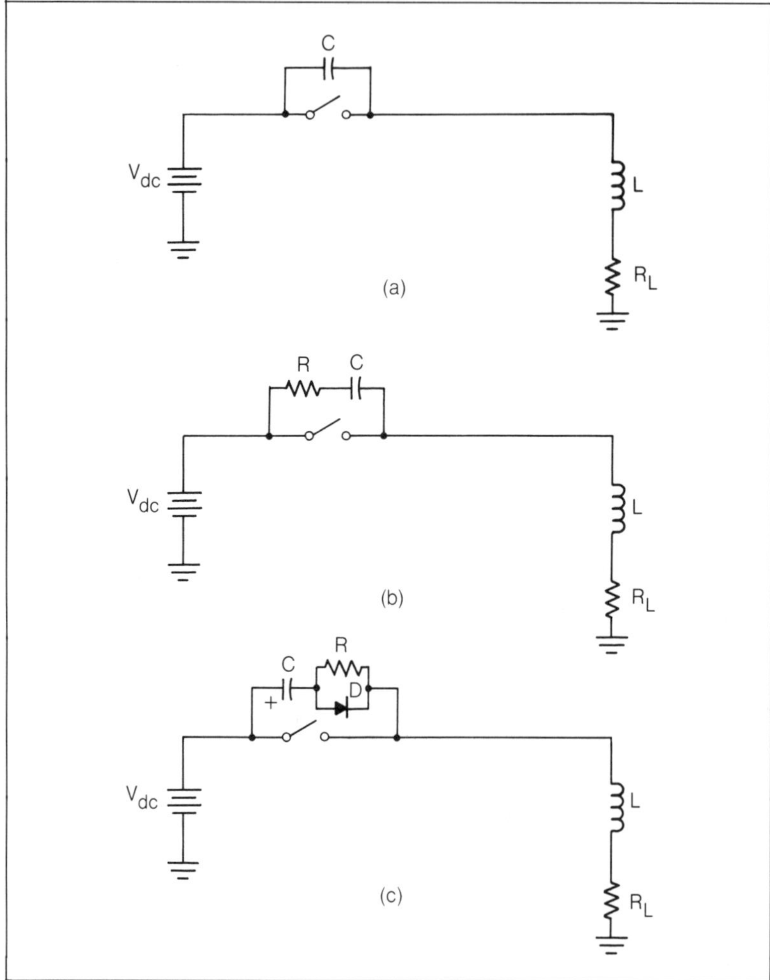

Figure 6.6—Contact Protection Networks Used Across Switch Contacts

Unfortunately, once the switch is set open, since the supply voltage is still there, the capacitor stays charged to the full V_{dc} value. At the next closure, this capacitor will release its charge across the quasishort circuit of the closing switch, causing both a possible contact wear and a radiated EMI. Therefore, the use of a single capacitor is generally not advisable. The preferred solution is shown under caption b of Fig. 6.6, where the capacitor discharge current is limited by a series resistance R. The value of R is a trade-off between: (1) the "break" condition which requires the lesser possible resistance to avoid voltage build-up and ensuing arc and (2) the "make" condition which requires a larger resistance, equal to the load resistance, to avoid a too large discharge current.

At the beginning of the break, the voltage across the opening contacts is equal to $I_n \times R$, the nominal current being interrupted times the added resistance. Therefore a safe value is to choose R equal to the load resistance R_L, such as whatever the LdI/dt voltage created by the self-inductance of the circuit being opened, the voltage $R \times I_n$ will not exceed the power supply voltage for which the switch was rated.

For switch closure, the minimum value of R is set to keep the current below the minimum arcing current I_m of the contact. Typical values of I_m are 0.4 A for gold and silver contacts and 0.8 A for palladium (Ref. 2). So the window for selecting R is as follows:

$$\frac{V_{dc}}{I_m} < R < R_L \tag{6.6}$$

This double condition, though fairly simple, could be self-conflicting. Cases will exist where the load resistance R_L is smaller than V_{dc}/I_m, the "no-glow" value. Let's take for instance a supply voltage of 28 V and a load resistance of 30 Ω. For common gold or silver contacts, I_m is 0.4 A, therefore R should be both larger than $28_V/0.4$ A, i.e., 70 Ω, and smaller than 30 Ω. The larger value should prevail (70 Ω) but in the case of inductive load, the switch should be rated to withstand the opening voltage of:

$$70 \ \Omega \times I_L$$

$$\text{or,} \qquad 70 \ \Omega \times \frac{28 \ V}{30 \ \Omega} = 65 \ V \tag{6.7}$$

which is generally not a problem.

Capacitor C is also chosen from two considerations. First the rate of rise of contact voltage must stay below 1 V/μs to avoid arcing. Using the relationship $I_L dt = CdV$, we can derive that:

$$\frac{I_L}{C} < \frac{1}{10^{-6}} \tag{6.8}$$

That is, C must be at least 1 μF per ampere of load current. The second condition is that the crest voltage across the contacts must remain below the glow discharge threshold. This voltage can be approximated from an ideal parallel L, C circuit:

$$V \text{ peak} = I_L \sqrt{\frac{L}{C}} \qquad (6.9)$$

I_L is the current of the inductive load being switched off. To avoid a glow discharge, Fig. 6.3 shows that V_{peak} must not exceed 300 V. So:

$$I_L \sqrt{\frac{L}{C}} < 300 \text{ V}$$

or approximately:

$$C_{\mu F} > L_{mH} \times 10^{-2} \times I_L{}^2 \qquad (6.10)$$

Finally, one may want to avoid the R, L, C network formed by the load and the R, C protection set to become a ringing circuit, which may add a requirement that:

$$R_{TOT} \text{ (load + contact protection)} > 2 \sqrt{\frac{L}{C}} \qquad (6.11)$$

To summarize, the set of conditions of the RC protection network with an inductive load is:

$$\frac{V_{DC}}{0.4^*} < R < R_L \qquad (\text{*0.8 in case of palladium})$$

$$C_{\mu F} > I_L \text{ amp}$$

$$C_{\mu F} > L_{mH} \times 10^{-2} \times (I_L)^2$$

$$(R_L + R) > 2 \sqrt{\frac{L}{C}} \qquad (\text{overdamping condition})$$

When someone does not want to optimize the RC selection, the following rough values are acceptable. For 6 V supply, and currents below 1 A, a 15 Ω, 1 μF set is generally satisfactory up to 100 mH of load inductance. For 28 V supply and currents below 1 A, a 68 Ω, 1 μF set is generally satisfactory up to 100 mH of load inductance.

Although it is the most commonly used, we have seen before that there are situations where a simple RC network cannot keep the contacts from arcing, mainly when some of the conflicting requirements on R cannot be met. If R is made low enough to keep $R \times I_L$ below 300 V at "break," it may cause glowing or arcing at "make," and vice versa. In this case, the third circuit of Fig. 6.6 can be used.

At the "make" time, the capacitor C discharges through R and glowing/arcing is prevented (like with the previous solution). But as contact breaks, the capacitor current flows through the directly biased diode, such that the voltage across the contacts is simply the forward drop of the diode.

Since there are no longer conflicting requirements on R, this third network provides an ideal solution, but it takes three components, with their ensuing failure rates, and being polarized it works only with dc circuits. These concepts work similarly for the protection of solid-state switches (see "snubbers" in Chapter 4).

Varistors and other transient protectors could be used for across-the-contact protection, but they are limited because the varistor must have a clamping voltage superior to the maximum supply voltage. For instance, with a 120 Vac supply, the varistor must be at least a 170 V model. This might be too high for satisfying the no-arc requirement of Fig. 6.3. Nevertheless, the varistor would still reduce a transient of a few hundred volts down to its clamp value. A parallel combination of a varistor with an RC network is also feasible. For supply voltages less than 20 V, varistors can be used across contacts with no restriction.

In all the above measures for contact protection, keep in mind that while contacts are bypassed, the transient still travels across the whole circuit where EMI is generated. A full protection scheme must also consider switching transient reduction on the load side.

One very important EMI point also has to be checked when putting an RC or semiconductor suppressor across electromechanical contacts: not atypical with EMI, fixing one problem can create another. The electrical isolation provided by an open electromechanical contact is normally infinite. When an RC circuit is put across the contact, the open-contact isolation is severely altered, especially at a high frequency. For example, if an RC network of $0.1\ \mu\text{F} + 10\ \Omega$ is put across a set of contacts, the open contact isolation will drop to about 10 Ω above 100 kHz: if a high isolation was expected, this can create an unexpected EMI path.

6.1.4 Reduction of Switching Transients with Inductive Loads

Although the contacts will be protected by a network across either the switch or the load, the latter is preferred if the load is accessible. A device across the load will protect the load **and** the switch, reduce EMI right where the cause of the transient is (allowing minimum path length for EMI) and not affect the isolation between the switch contacts when they are open.

Table 6.2 shows seven suppression techniques used for inductive loads like relays and contactor coils, solenoids and motors. Case 6.2a uses a parallel resistance to provide a closing path to the LdI/dt transient. The circuit is simple and works with ac or dc. R is chosen to equal the coil resistance R_L, which limits the transient to twice the supply voltage. The drawback is that it wastes continuous power as long as the load is energized. Also, for relays, there is a slight effect on armature release time since the new dropout time constant is $L/(R + R_L)$.

In Table 6.2, case b, the RC circuits limit the voltage surge to:

$$V_0 = I_{dc} \sqrt{L/C_T} \qquad (6.12)$$

where, $$C_T = C\ \text{coil} + C$$

provided the relay resistance R_L is negligible. The capacitor C is usually chosen to be 0.1 to 1 μF with a voltage rating of approximately 10 times the maximum dc input voltage.

The use of a capacitor alone would result in a large charging current during relay energizing which may damage the switch contacts or cause transient-noise current surges. Thus, the current-limiting resistor, R, is necessary:

$$R = V_r/I_{max} \qquad (6.13)$$

where the maximum current, I_{max}, should be limited to 10 times the normal coil operating current, I_{coil}:

$$I_{max} = 10\ I_{coil} = 10\ V_r/R_L$$

therefore: $$R = \frac{V_r}{10\ V_r/R_L} = 0.1\ R_L \qquad (6.14)$$

This circuit affects contact opening and closing times only slightly.

Table 6.2—Various Techniques for Suppressing Switching Transients with Inductive Loads (Relays, etc.)

Types of Inductive Suppression	Voltage Input	Relay Contacts		Remarks
		Closing	Dropout	
(a) Resistance Damping	ac or dc	No Effect	Function of Resistance	Increase in power consumption. Resistance should be as low as practicable. Observe power rating E^2/R and heat dissipation.
(b) Capacitance Suppression	dc	Slight Effect	Slight Effect	Need series resistance of a few ohms. Capacitance value around .01 to 1 μF. Capacitance rated 10 times input voltage.
(c) RC Suppression	dc	Slight Effect	Function of Resistance	Combination of a and b above.
(d) Diode Suppression	dc Only	No Effect	Slight Effect	Polarity critical, diode put in backward or nonconductive direction. PIV should be higher than any transient voltage plus safety factor. Series resistance of a few ohms might be needed to increase inductance life.
(e) Back-to-Back Diode Suppression	ac and dc	No Effect	No Effect	Avalance voltage should be above input voltage. Power dissipation should be sufficient for transient current. Cost of device is much greater than any of the above devices.
(f)	ac and dc	No Effect	No Effect	V_{clamp} must be above input voltage + 10 percent.
(g)	ac and dc	No Effect	No Effect	Doubles the coil volume.

The series-parallel combination of capacitance and resistance shown in Table 6.2, case c, is a combination of the above two circuits and offers no particular advantage over either circuit. The circuit of Table 6.2, case d results in a polarity-sensitive relay. When the actuating switch is closed, the diode is an open circuit since it is back-biased. Thus, the series resistor is out of the circuit. However, when the switch is opened, the relay coil develops a reverse voltage, $V_r = -L\,dI/dt$, and current flows through the diode. The voltage is limited to the forward diode voltage drop, and the voltage drops across resistor R. The peak inverse voltage rating of the diode should be higher than the maximum applied input voltage or any transient voltages, and should include a sufficient safety factor. With the addition of a zener diode in series, the inductor current will decay faster and the circuit will become protected against polarity reversal. The transient voltage in this case will reach the zener voltage plus the supply voltage.

Configuration 6.2d cannot be used with ac circuits. In this case either a varistor (6.2f), or two back-to-back zener diodes (6.2e) are used as in Configuration e. Solution 6.2g is interesting though seldom used: the coil is wound, "bifilar." In other words, a second coil is made concentric to the primary one. During flux change, the second coil absorbs the inductive energy and dissipates it into R.

6.1.5 Selection of a Protection Device

Based on the steady load current, the following guidelines have been established:[2]
1. Noninductive loads drawing less than the arcing current, in general, require no contact protection.
2. Inductive loads drawing less than the arcing current should have an RC network or a diode for protection.
3. Inductive loads drawing greater than the arcing current should have an RCD network or a diode for protection.
4. Noninductive loads drawing greater than the arcing current should use the RCD network. If voltage is less than 300 V, capacitor C does not have to meet the criteria of $C > (I_0/300)^2\,L$.

One example of contact protection design is a 100 mH/50 Ω relay coil, with 5 Vdc supply controlled by a silver contact switch. The dc coil current is 5 V/50 Ω = 0.1 A. This current is less than the 400 mA arcing for silver, so condition No. 2 of the previous guidelines applies, and a simple RC or a diode can be used. To keep the voltage gradient below 1 V/μs, the capacitance must satisfy:

$$Idt = C\, dV \text{ with } dV/dt \leqslant 10^6 \tag{6.15}$$

so,
$$C \geqslant I \times 10^{-6} \text{ or } C \geqslant 10^{-7} = 0.1\ \mu F$$

To maintain the voltage below the critical 300 V at contact opening, C must also meet:

$$C \geqslant \left(\frac{I_0}{300} \right)^2 L$$

$$C \geqslant \left(\frac{10^{-1}}{300} \right)^2 \times 0.1 = 0.011\ \mu F \tag{6.16}$$

The first requirement prevails. Finally, R is chosen such as:

$$\frac{V_{dc}}{I_{arc}} < R < R_L$$

or,
$$5/0.4 < R < 50 \tag{6.17}$$

A value of R between 12.5 Ω and 50 Ω and a 0.1 μF capacitor across the load or the contact will provide adequate protection.

6.1.6 Other Types of Switches

Although the general mechanisms described in previous sections of this chapter apply to any switch, from a few milliamps to thousands of ampere loads, there are a few switches which exhibit some specific EMI features. On the low-power side, many control panels and keyboards use membrane switches. The contact-type membrane switches use a deformable pressure pad, like a dome, under which a printed circuit pattern comes in contact with a fixed mating circuit (see Fig. 6.7). Since these switches are used to command small power drivers or relays, the switched currents are low. The contact resistance can be from a few ohms to few hundred ohms which resolves the problem of arcing and bouncing.

On the other hand, these switch arrays are sensitive to ESD (Ref. 3) and EMI because of the exposed area of the layout. Shielded versions of membrane switches are available. The capacitive type of membrane switch does not switch any current and is therefore even more bounce-free and arc-free. In turn, the switch's sensitivity to ESD is significant (see Fig. 6.8). Finally, the Hall-effect switches are EMI-free since they work simply by detecting the proximity of a magnetic plunger.

At the other end of switches, the large devices used in power substations and switchyards to control high-voltage lines create a gigantic version of the RLC switching model that was explained in Section 6.1. Additionally, the huge dV/dt and dI/dt associated with the switch's transfer create large electric and magnetic fields which can be a threat to electronic equipment in the facility itself.

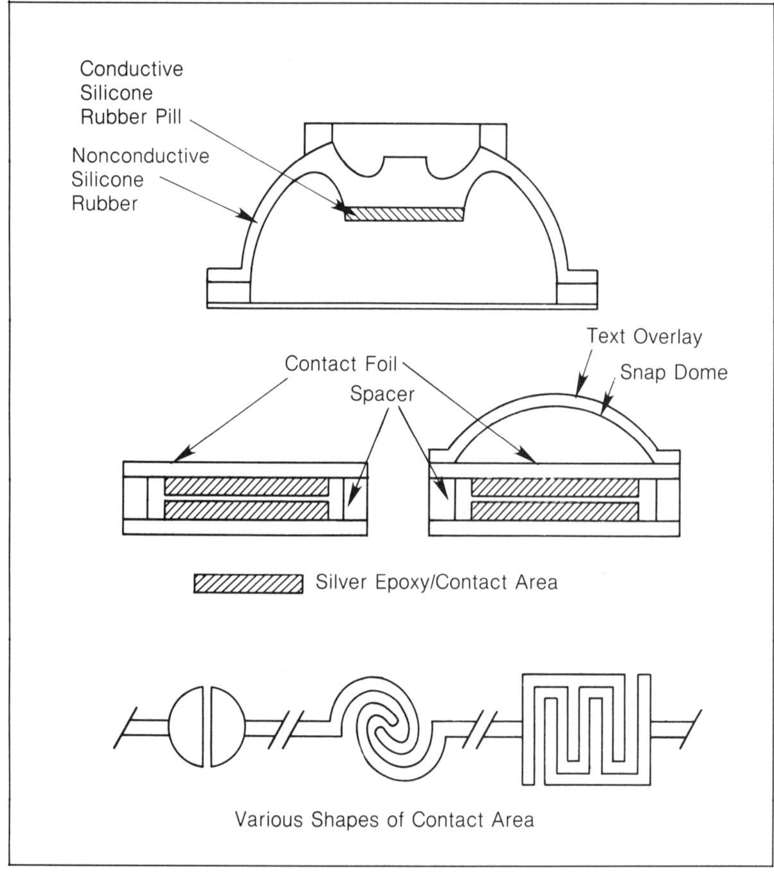

Figure 6.7—Membrane Switches Using Deformable Conductive Rubber

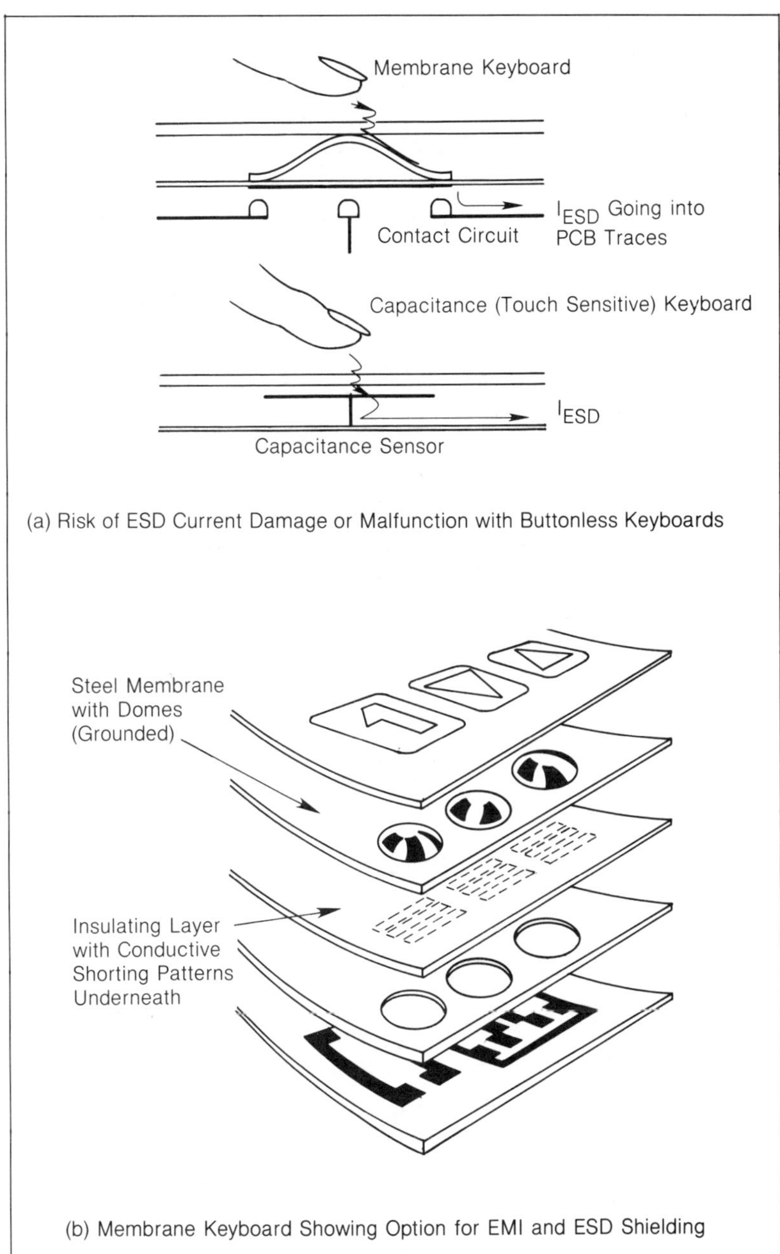

Figure 6.8—ESD Problems with Buttonless Keyboards

B.D. Russell has described peak fields of 4 kV/m and 0.6 A/m with rise times in the 100 ns range (Fig. 6.9). The survey was made using active field probes such that the antenna output is a replica of the field waveform in the time domain. Unfortunately, the study did not mention the exact measuring distance, but it seems the sensors were within 3 to 5 m of the source, and the measuring equipment was sheltered in a shielded room placed on a truck.

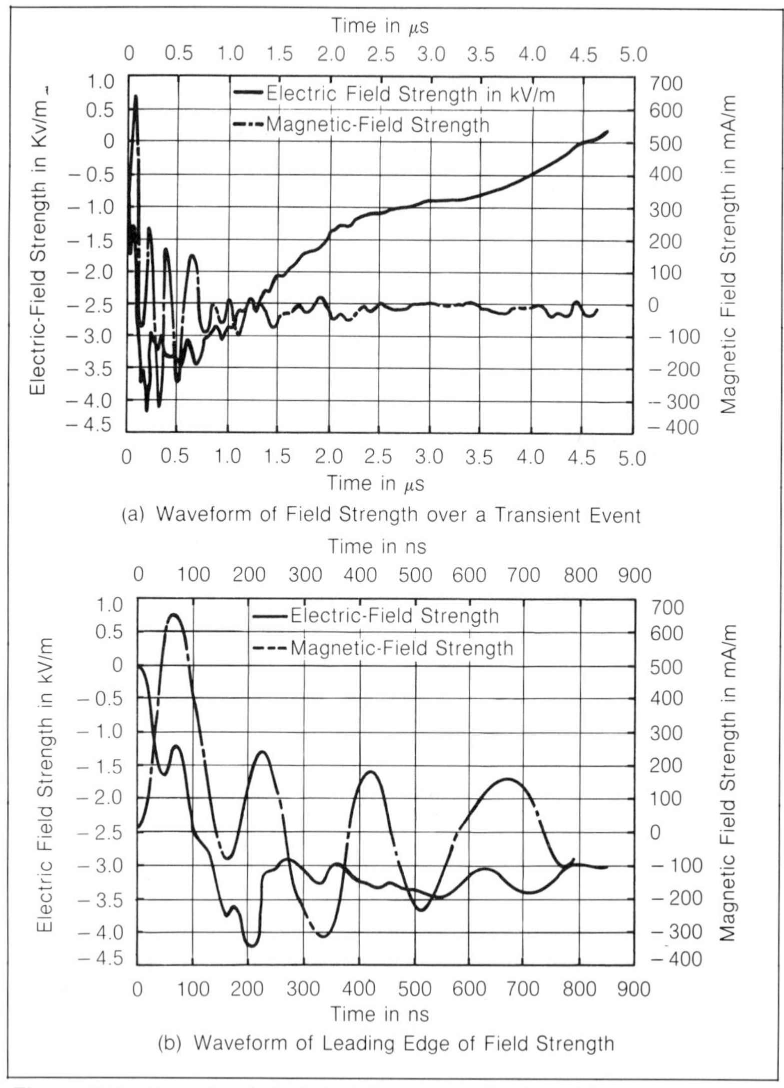

Figure 6.9—Sample of Radiated Transients Produced by Operation of a Switch on a 345 kV Bus (Courtesy of *EMC Technology*)

6.2 Relays and Contactors

A relay is a device that permits one or more circuits to be remotely switched by an independent control circuit. Several types of relays exist: electromagnetic, saturable reactor, bimetallic or reed, solid-state, photosensor and others. Many designs are available, depending on the switching to be performed, the power source available, the number and type of contacts and the cost.

Although an electromechanical relay seems like an anachronism in the modern electronic world, there are many applications where such a relay is unmatched by solid-state devices. Its primary advantages are:

1. Low contact resistance (10 or 100 times less than the "on" resistance of a transistor or SCR)
2. High isolation between open contacts
3. High isolation between control and power circuits (equalled only by optoelectronic solid state switches)
4. Excellent noise immunity

The most common relay is the electromagnetic solenoid type. EMI problems occur in both the actuator and the contact circuits. The electromagnetic solenoid has a large inductance because of the large number of turns and iron mass in the core and armature. When the coil circuit current is interrupted (de-energized), the collapse of the magnetic field generates a voltage equal to $-L \, (dI/dt)$.

This reverse potential can reach from 10 to 20 times the supply voltage in the order of a microsecond and then decay at a rate determined by the inductance, distributed capacitance and resistance of the armature winding circuit. The high-amplitude voltage surge has a steep wavefront that can cause arcing at the point of interruption along with broadband components capable of conducting and radiating interference to other circuits. EMI effects of an ac relay vary because the voltage and current are always changing in magnitude, producing results according to the momentary state at the time of switching.

Relay arcing can occur when contacts are first opened, continuing until contact spacing is too large to maintain the arc (depending on the surrounding atmosphere and the applied voltage). The making and breaking of a contact arc when opening a circuit is proportional to the instantaneous supply voltage, the circuit inductance and the rate at which the contacts physically separate. EMI may be worse for contact closing than for contact opening. Closed contacts can open or vary in contact resistance due to shock, acceleration or vibration, causing arcing or at least circuit current changes. If required, contact arc suppression is used.

All of the theory described in Section 6.1 for switches in general is directly applicable to relays, but in addition to the arcing and RLC oscillation in both the main circuit and the coil-drive circuit, relays have problems of their own. Some typical EMI parameters of a high-isolation relay (OMRON Corp.'s G4Y) for VHF/UHF applications are shown in Figs. 6.10 through 6.14[5] and are discussed next.

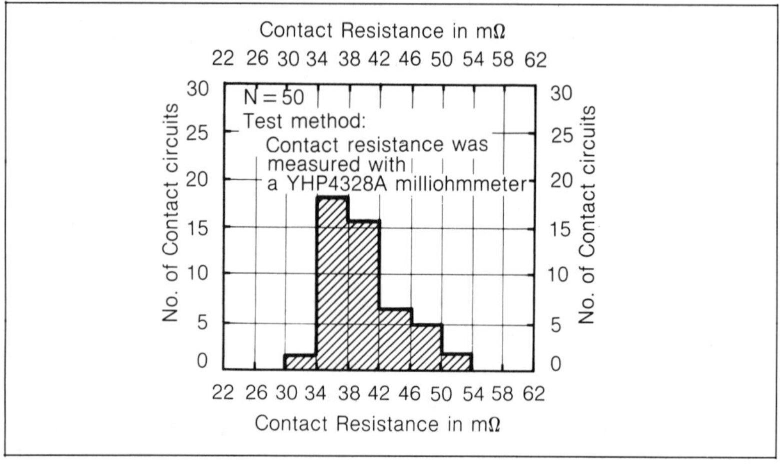

Figure 6.10—Distribution of Contact Resistance (Normally Open Contact)

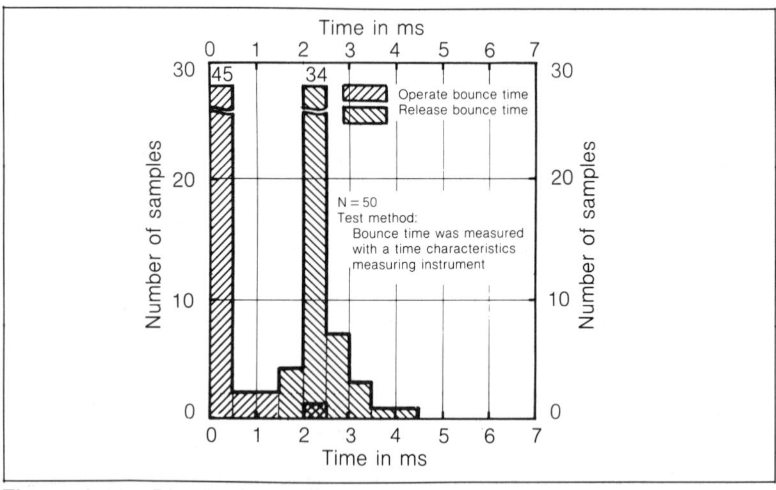

Figure 6.11—Distribution of Bounce Time

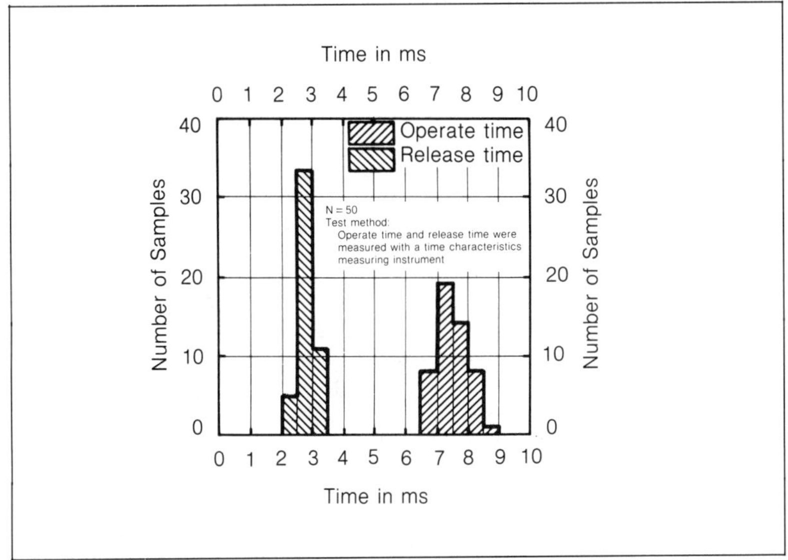

Figure 6.12—Distribution of Operate and Release Times

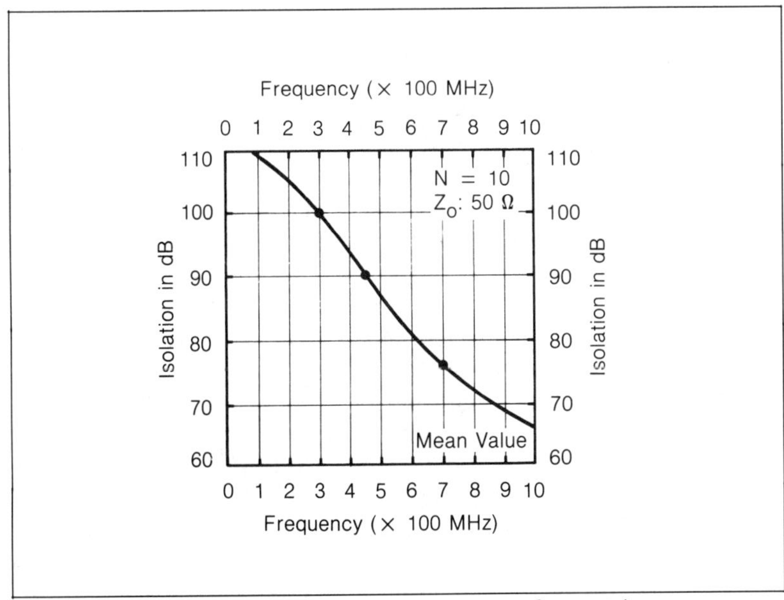

Figure 6.13—Isolation Characteristics (between Contacts)

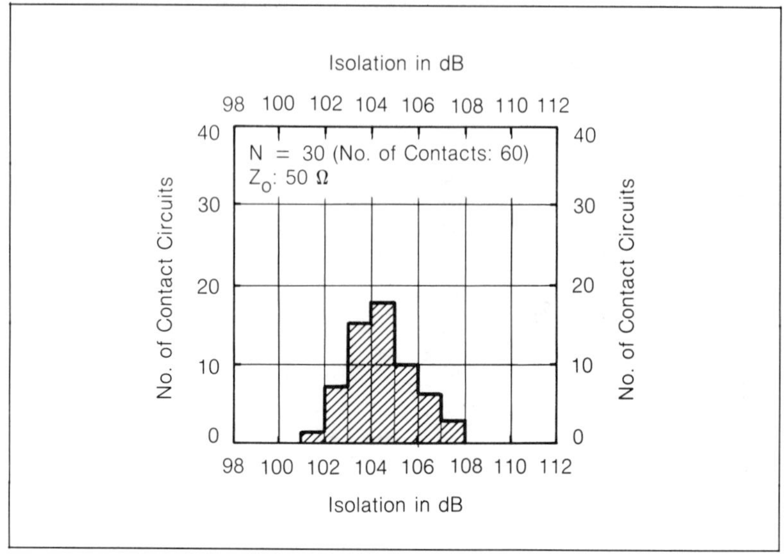

Figure 6.14—Distribution of Isolation at 250 MHz

6.2.1 Contact Bouncing

Since the armature has mass, this causes bouncing at the closing and the opening. Bounce on opening is small but (depending on kinetic energy) when contacts close, they bounce and form a closure followed by one or more openings and reclosures. In small relays, a single bounce may occur 10 to 50 μs after the initial closure; final closure may take 10 or more μs. In large relays, the bounce may be repeated several times at intervals of a few milliseconds (Fig. 6.15). Eventually bouncing may also occur during steady operation if the relay is submitted to vibration.

Bouncing can be reduced by reducing armature mass and increasing armature restraining spring force and magnetic attraction. This will also improve the operate-and-release speed but it will increase the coil energy consumption.

When used in digital systems, contact bouncing and erratic opening cause data errors. Polar or latching relays are therefore often used in such systems. A polar relay, once actuated in a given direction, will remain latched on its internal permanent magnet until its state is reversed by current in the opposite direction. Also, a "debouncer" network can be incorporated in the logic circuit. Special relays such as mercury-wetted and vacuum relays are used when minimum arcing and bouncing are required. Mercury-type switches are excellent because they are contamination-free and exhibit little contact bounce, but their use is limited to stationary system applications with respect to the horizontal plane. Vacuum switches, in which the contacts are enclosed in a vacuum envelope and operated by an external solenoid, are useful for aircraft applications.

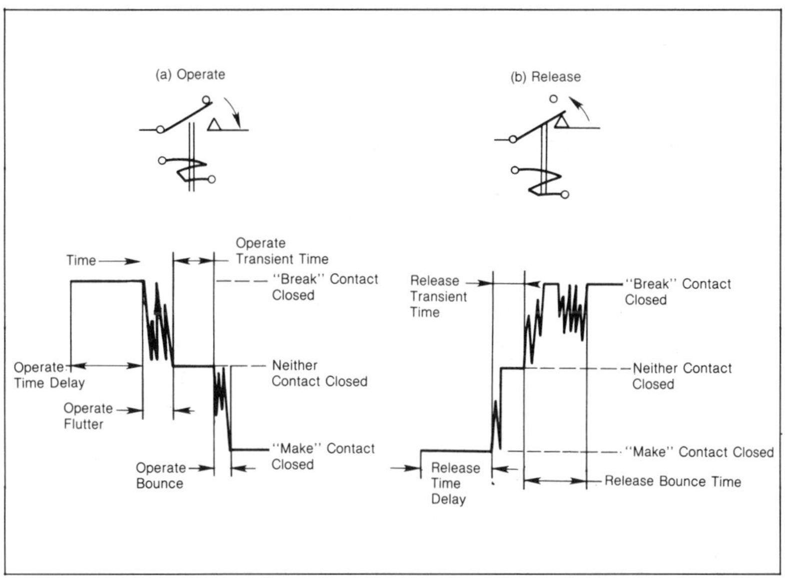

Figure 6.15—Transient Condition in a Relay

6.2.2 Field Radiation and Pickup

Relay coil leads and main leads can pick up ambient fields. The coil can induce into the main leads the noise picked up by the control leads. Under strong RF ambient, the coil and its suppressing diode and capacitor can act as an efficient pickup device, and the rectified current can wrongly activate the relay. EMI suppression is especially required for relays located in areas having susceptible circuits or equipment. Leads must be isolated, twisted or shielded to avoid magnetic and electric coupling. Filters should be used on conductors as needed at points of entry into an enclosure. Table 6.3 summarizes some of the common EMI problems with relays as well as their solutions.

The relay should be magnetically shielded where magnetic fields from the relay coil may be a potential source of EMI to nearby circuits or wiring. Shielding techniques described for chokes (Chapter 2) and transformers (Chapter 5) also apply.

Table 6.3—EMI Problems, Causes and Correction in Relays

Problem	Cause	Correction
Generation and susceptibility to propagated EMI	Relay circuit characteristics	Twist, shield and filter power leads and enclosing relay in metal
Simultaneous problems in coil and contact circuits	Movement of the armature causes coil magnetic circuit changes and varying contact dI/dt and dV/dt	Proper relay circuitry design, including filtering
Relay contacts generate EMI when they should be quiescent	Vibration	Proper relay design or contact circuit suppression
Noise, misoperation or permanent damage in semiconductors	Inductive circuit causes large voltages	Components must be properly rated and/or dI/dt and dV/dt must be diverted or suppressed
Low-frequency interference in the signal circuitry	Low-frequency spurious voltages	Single-point grounding and twisted leads
High-frequency interference in the signal circuitry	High-frequency spurious voltages	Multiple-point grounding

6.2.3 Crosstalk in Relays

In signal relays, especially those used in telecommunications and video, crosstalk between two or more circuits switched by the relay can occur due to the close proximity of the blades and contact leads. For instance, if care has been taken to guarantee 60 dB of isolation between two circuits up to 50 MHz, the weak link could be the relay having 60 dB isolation or less. Remember that if the whole circuit less the relay has 60 dB of crosstalk, adding a relay with just 60 dB will already degrade the performance by 6 dB since the two leakages will combine. The new isolation will be only 54 dB. Figure 6.16 shows examples of isolation (crosstalk) performances of relays for VHF and UHF. Crosstalk is due to not only the parasitic capacitance between contacts, which can vary from 0.1 to 3 pF, but also contact-to-frame and contact-to-coil coupling.

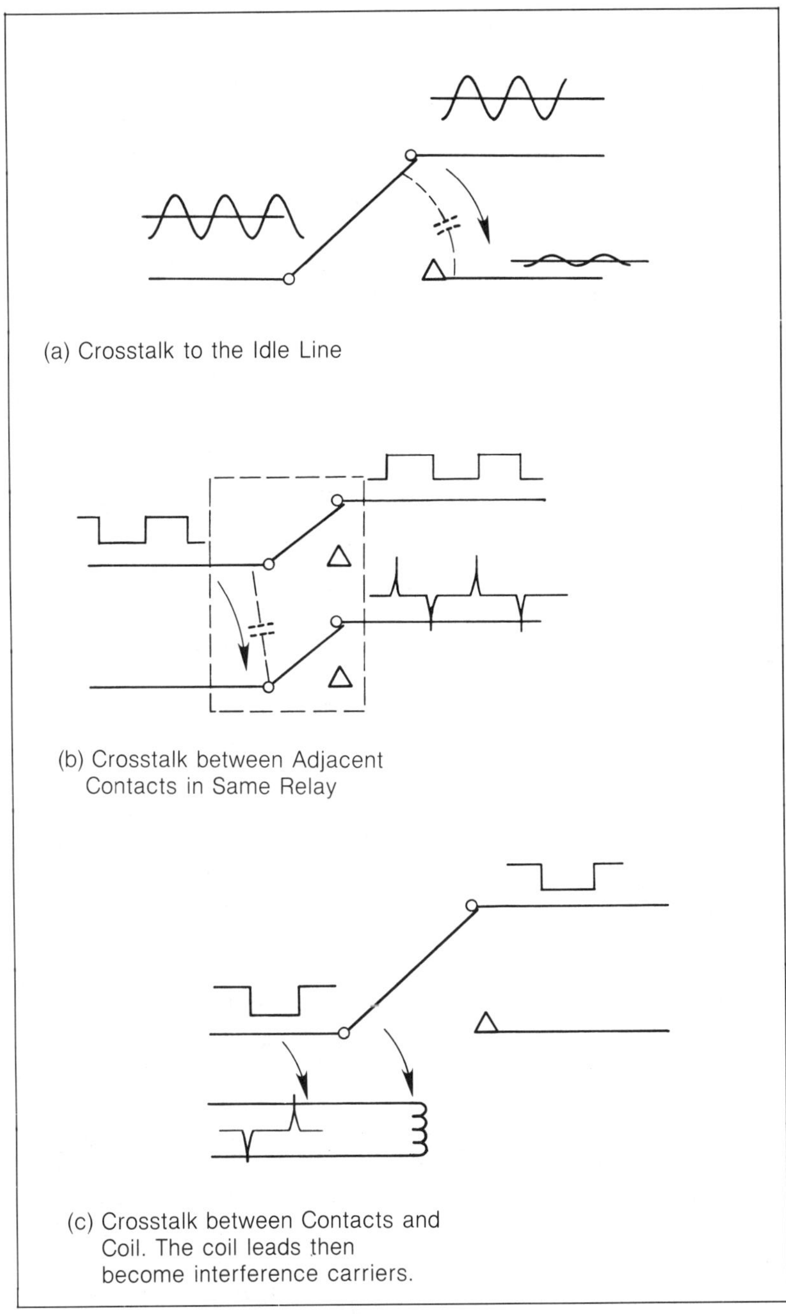

Figure 6.16—Illustration of Possible Crosstalk in Relays **(continued next page)**

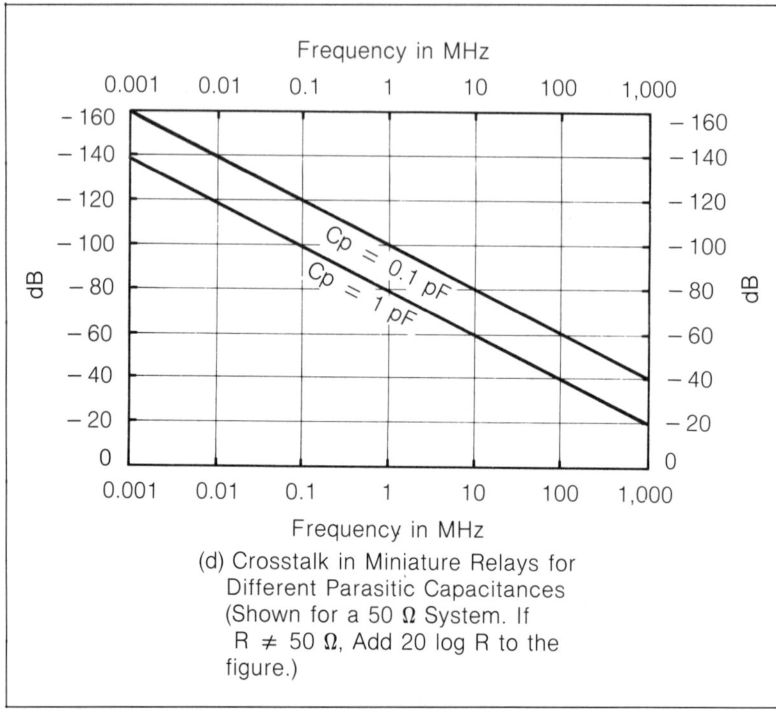

(d) Crosstalk in Miniature Relays for Different Parasitic Capacitances (Shown for a 50 Ω System. If R ≠ 50 Ω, Add 20 log R to the figure.)

Figure 6.16—(continued)

Crosstalk between two or more switched-wire circuits at a relay signal terminal is another EMI problem. While a single relay should not be used to switch both high- and low-level circuits due to crosstalk problems, applications exist in which a single relay may be used to select one or more circuits to patch to a single input circuit. One example is coaxial relays with BNC-type connectors in which it is unusual for isolation to be greater than 35 dB at 1 GHz or 50 dB above 100 MHz. Most coaxial relays of the reed type exhibit poorer performance. Larger, heavier and costlier type-N coaxial relays of the non-reed type are available to give better isolation performance.

Relays with miniature Faraday shields between contacts have also been developed. For relays with metal cans (the TO5 can for instance) the contact-to-contact capacitance may be larger due to the presence of the can: each contact has a capacitance to the can, which can act as a "jumper." The metal can should be connected to the signal ground.

A somewhat less expensive and more reliable technique is shown in Fig. 6.17 in which several inexpensive coaxial relays are combined to protect isolation requirements. Here, three two-pole relays replace one four-pole relay. The isolation between any of the four coaxial lines due to relay contact coupling is twice the number of dB obtained from any one of the single relays. When connected back-to-back to form a 0-20-40-60 dB, remote-controlled RF attenuator, Fig. 6.18 is the result. Up to 100 dB of attenuation has been realized at 1 GHz using 20 SPDT relays in which individual relay-circuit isolation is about 30 dB. VSWR padding or impedance-mismatch protection is also preferred over the more expensive multipole relays for other than the 0 dB position.

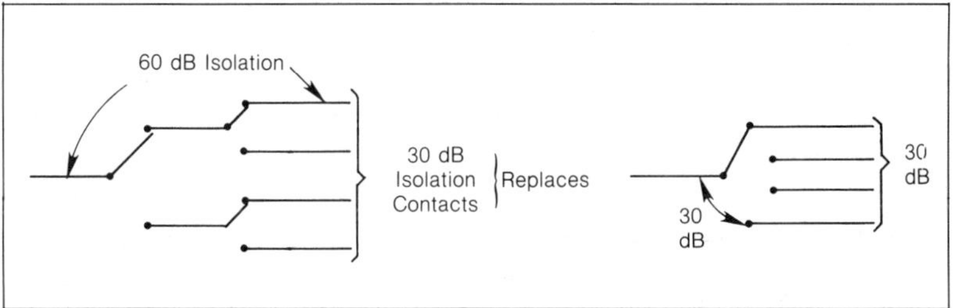

Figure 6.17—Improvement of Coaxial Relay Circuit Isolation (Crosstalk) by Combination of SPDT Relays

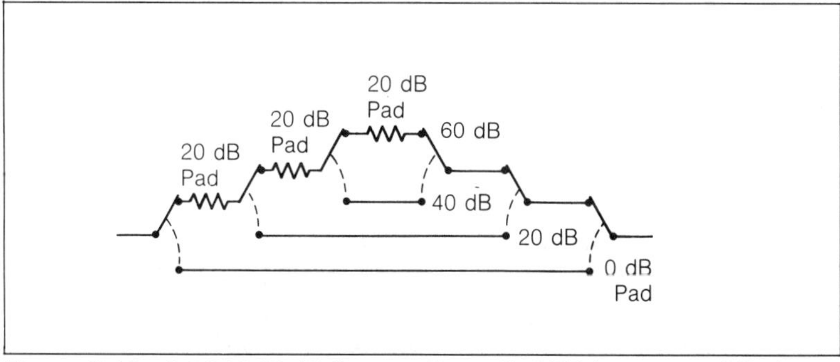

Figure 6.18—0-20-40-60 dB Remote-Controlled RF Attenuator Offering Substantial Isolation over Equivalent Two SP4T Relays

6.2.4 Solid-State Relays

Solid-state relays (SSRs) are widely used today after being introduced in the 1960s. Just as transistors and ICs are preferred by designers, solid-state relays are used because of their small size, reliability and low power dissipation. Since they have no moving mechanical parts, arcing is inherently prevented. This has the added advantage of permitting switching in explosive atmospheres.

Many solid-state ac relays are designed so that the relay contacts are closed at the time of zero-axis crossing of the ac supply voltage (called zero-voltage switching). Here the transient is significantly reduced. Additionally, many solid-state ac relays have provisions for turning off at a zero-current crossing of the ac line (called zero-current switching). This eliminates the $-L\, dI/dt$ arcing problems associated with de-energizing inductive loads.

On the other hand, SSRs can exhibit false operation in the presence of line transients. Line transients can cause an incidental "closure" of an open SSR for two major reasons.

The first reason is that if the line voltage transients exceed the maximum contact voltage, the relay contacts will close until the current waveform passes through zero, or for a maximum of one-half an ac cycle, and then reopen. While such transients will not damage the relay, back-to-back zener diodes can be used across the contacts for further EMI suppression. If a dV/dt transient across the contacts exceeds a certain rate, the contacts will close for one-half of an ac cycle. Reclosing due to inductive load switching is avoided by an internal RC network across the contacts. However, in this case, like in the case of electromagnetic relays, the presence of an RC network across the contact will spoil the high-frequency isolation between open contacts.

The second reason is that the dV/dt, or merely the peak voltage induced by an EMI ambient, may trigger the pilot circuit via the signal leads, the power leads or even the relay heat sink. Table 6.4 shows the results of a NEMA Standard Noise Immunity test of SSRs.

A sample of four SSRs was intensively tested for EMI emission and susceptibility. The results are shown in Figs. 6.11 and 6.12 for emissions.[4] In Fig. 6.19, relays were activated at a 0.5 pps rate.[6] For both Figs. 6.19 and 6.20, measurements were made per MIL-STD-461 methods with a 10 μF bypass. In Fig. 6.20, relay A is optically coupled, relay B is transformer coupled. Both A and B activated at a 0.5 pps rate. The susceptibility findings were as follows:

1. *Power-line injected EMI, modulated RF:* some relays were found susceptible to 0.5 to 1 V at a few frequencies between 0.4 and 40 MHz.

---

*An exception to this is the growing tendency to integrate some amplifying electronics or an A/D converter inside the sensor package. It becomes a high-level active device, similar to what is discussed in Chapter 4 of this book.

2. *Control-line injected EMI, modulated RF:* for transformer-coupled and optically coupled devices, susceptibility was 0.5 to 1 V at a few frequencies between 50 kHz and 40 MHz. One type which has a transformer coupling plus an optocoupler input was found susceptible to 2 V RF and 7 V transient injection.

Table 6.4—Results of NEMA ICS 2-230 Electrical Noise Immunity Test (Courtesy of *EMC Technology*)

Manufacturers	Relay	Peak-to-Peak Baseplate	Voltage Case
A	10 A, 120 Vac	>10,000	>10,000
A	25 A, 240 Vac	>10,000	>10,000
B	10 A, 120 Vac	≤10,000	≤10,000
B	25 A, 240 Vac	<1,000	<1,000
C	10 A, 120 Vac	≤6,000	≤3,800
C	25 A, 240 Vac	≤8,400	≤2,600
D	10 A, 120 Vac	≤3,300	≤4,700
D	25 A, 240 Vac	≤4,600	≤6,100
E	10 A, 120 Vac	<800	<800
E	25 A, 240 Vac	1,600	900
F	10 A, 120 Vac	<1,000	<1,000
F	25 A, 240 Vac	<1,000	<1,000
G	25 A, 240 Vac	1,000	500
H	10 A, 120 Vac	<800	<800

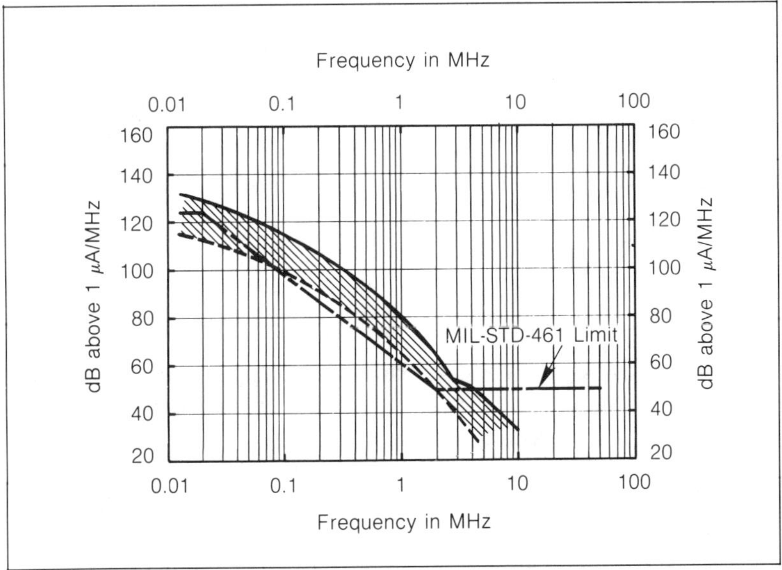

Figure 6.19—Envelope of Line-Conducted EMI Current for Four Types of 115 V SSRs, Driving a 75 Ω Wirewound Resistor

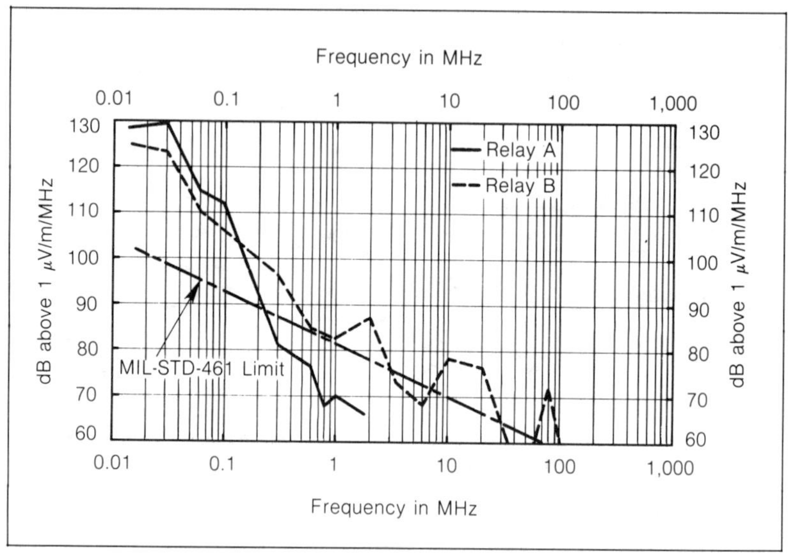

Figure 6.20—Radiated EMI Field of Two Types of SSRs Driving a 115 V, 1.5 A Induction Motor

Typical contact resistance (ratios of voltage drop to current through contacts) of a 10 A, solid-state relay is about 100 mΩ and rises with a decrease in load current. The off-state leakage is a few mA measured at a dc voltage equal to the rms voltage for which the output is designed. The maximum load current is generally rated at room temperature and smoothly decreases with an increase in temperature until it is completely derated to zero at about 100°C.

6.3 Sensors

The family of electrical "sensors" or "transducers" corresponds to devices which capture a specific physical parameter like pressure, strain, position, velocity, temperature, acceleration, etc., and translate the parameter into an electrical signal which is proportional to the magnitude of the parameter being sensed. This electrical signal, generally very weak, can then be amplified or coded by nearby or remote electronics.

There are too many types of sensors to list here, but their principal EMI features will be addressed. Generally, sensors share the following characteristics:

1. The output analog signal is low,* ranging from few microvolts or less up to few hundred millivolts. Most of the time, it is in the millivolt range. Therefore, in terms of EMI, this signal is vulnerable.

2. The internal (or source) resistance of sensors is generally high; that is, the next amplifier which will process the sensor signal must have a high input resistance too, which makes the link more vulnerable.

3. Whenever the sensor generates a voltage or simply changes a resistance, capacitance, etc., the signal current which tracks the physical parameter is low, so the amplifier will be a high-gain type. This reinforces the second statement.

4. The bandwidth is generally low since the speed of change of the physical parameters is usually slow in terms of electronics parlance; that is a few Hz to a few kHz. However, growing requirements for more accuracy and better resolution tend to call for faster sampling rates in A/D conversion. This results in a greater bandwidth.

5. Sensors are often used in harsh environments of all kinds. EMI is one of them, and others include temperature, flammable or hazardous fluids or gas, etc., where no human operator can be. Therefore error-free sensor output is a priority.

6. Some sensors are mounted in places where no dedicated grounding, voltage reference or other such commodities exist. Furthermore, although in good EMC practice low-level signals should be carried over a balanced link for better ground loop immunity, some sensors are by nature unbalanced.

If the transducer is a passive element, there is no rectification of a would-be EMI signal at the transducer level. Thus, the EMI current flows in the sensor, the capacitances to ground and finally the amplifying electronics.

The behavior of the amplifying electronics in the presence of the EMI current will dictate whether the interference is tolerable or not. Filtering, band-pass limiting, twisting wires and shielding are the classical techniques in this case. However, due to the criticality of ground loops, there is always the problem of where to tie the shield. This is explained in detail in *Shielding,* the third volume in this EMC series, as well as in Refs. 7 and 8.

The following sections will present typical sensor examples with their parameters relevant to EMI.

6.3.1 Strain Gauges

Strain gauges measure the variation of resistance in a Wheatstone bridge under mechanical strain. The gauge bridge is generally excited in the few-volts range, 10 V for example. The gauge factor relates the resistance variation to the length variation ΔL of the gauge, which itself corresponds to the applied stress.

$$\text{gauge factor} = \frac{\Delta R/R}{\Delta L/L} \qquad (6.18)$$

A full-scale range corresponds to a $\Delta R/R$ of about 1 percent, a typical full-scale signal being 10 mV. Therefore, if a minimum discernable output of 1 percent of the full scale is expected, any EMI noise above 0.1 mV can produce a reading error.

Gauges are cemented to the structure they are sensing. The gauge-to-mounting surface capacitance can be as large as 100 pF, which causes a prohibitive ground loop in the presence of ambient EMI. For instance, if a 1 V, 10 kHz common-mode EMI exists between the receiving amp and the gauge, this can cause a 3 μA current to flow. In a gauge with 100 Ω resistance, this creates an error voltage of 0.3 mV. This can be reduced to some extent by trying to balance the parasitic capacitance to ground of each gauge in the Wheatstone bridge.

6.3.2 Thermocouples

Thermocouples utilize the natural EMF of two dissimilar metals and the variation of this voltage when temperature changes. Typical output is in the 10 to 80 μV/°C.

To obtain a fast tracking of the temperature change, the thermocouple junction is usually bonded to the measured surface, which forces the circuit to be grounded at the measurement point. Since thermocouple wire is resistive and two different metals are used, an unbalance in resistance results until the leads are converted to regular copper wires, usually at the temperature reference junction.

For all these reasons, the circuit can be very sensitive to induced noise. On the other hand, the time constant is large (0.1 to 1 s) and the source impedance is low, so the input of the reading amplifier can be strongly decoupled without affecting the bandwidth.

6.3.3 Pressure Transducers

Many types of pressure sensors exist as can be seen on Table 6.4. Some devices, like the piezo-resistive bridge or thin film, have outputs in the range of a few millivolts per volt of excitation. That is, for 10 V of excitation, their full-scale output would be in the 10 to 100 mV range. In a pressure sensor with 0 to 1,000 millibar pressure range, if 10 millibar accuracy is required the corresponding electrical threshold would be about 0.1 to 1 mV, which then can be compared to an EMI threat.

Other types of pressure sensors use a flexible diaphragm capacitance which varies according to the applied pressure. This

Table 6.5—Comparison of Pressure Transducers
(Courtesy of *EDN*)

Sensor	Excitation	Output Level	Percent Accuracy	Pressure Range (psi)	Frequency Response
Capacitance	ac/dc Special	High Level and Frequency Bridge	1 to 0.05	0.01 to 20.0	0 to > 100 Hz
Strain Gauge Unbonded	ac/dc Regulated 10 V/ac/dc	Low Level 4 mV/V	0.25	0.5 to 10 k	0 Hz to > 5 kHz
Bonded Foil	10 V ac/dc	Low Level 3 mV/V	0.25	5 to 10 k	0 Hz to > 5 kHz
Thin Film	10 V ac/dc	3 mV/V	0.25	15 to 10 k	0 Hz to > 5 kHz
Diffused Semiconductor	10 V-28 V dc	Medium Level 3 to 20 mV/V	0.25	15 to 5 k	0 Hz to > 5 kHz
Bonded-Bar Semiconductor	10 V ac/dc	Medium Level 3 to 20 mV/V	0.25	5 to 10 k	0 Hz to > 5 kHz
Piezoelectric	dc Amp and Self Generating ac	Medium Level with Amplifier	1	0.1 to 10 k	1 Hz to > 100 kHz
Potentiometer	ac/dc Regulated	High Level	1	5 to 10 k	0 Hz to > 50 Hz
Differential Transformer	ac Special	High Level (5 V) with Phase Demod/ Bridge	0.5	30 to 10 k	> 100 Hz

NOTE: *Linear Compensated Range

varying capacitance is part of an oscillator whose frequency changes with pressure. In some designs, the changing frequency is mixed with a reference frequency and the two combine in a demodulator delivering a few volts signal. Figure 6.21 is an example of the EMI sensitivity of such a device.

Some manufacturers thoroughly test their sensor EMI characteristics.

For example, the KAVLICO sensor in Fig. 6.22 exhibits the following EMI sensitivity:

1. Susceptibility to conducted transient: $\geq$ 50 V/ms
2. Susceptibility to radiated fields: $\geq$ 200 V/m from 10 kHz to 1 GHz, for a maximum 10 percent change in output

These devices usually have internal oscillators working in the 10 kHz to 10 MHz range, so be careful of the possible EMI created in nearby circuits by the oscillator. Generally, all these pressure sensors are fairly small, and the biggest EMI concern is not the coupling to or from the device itself, but EMI coupling to the interconnecting leads.

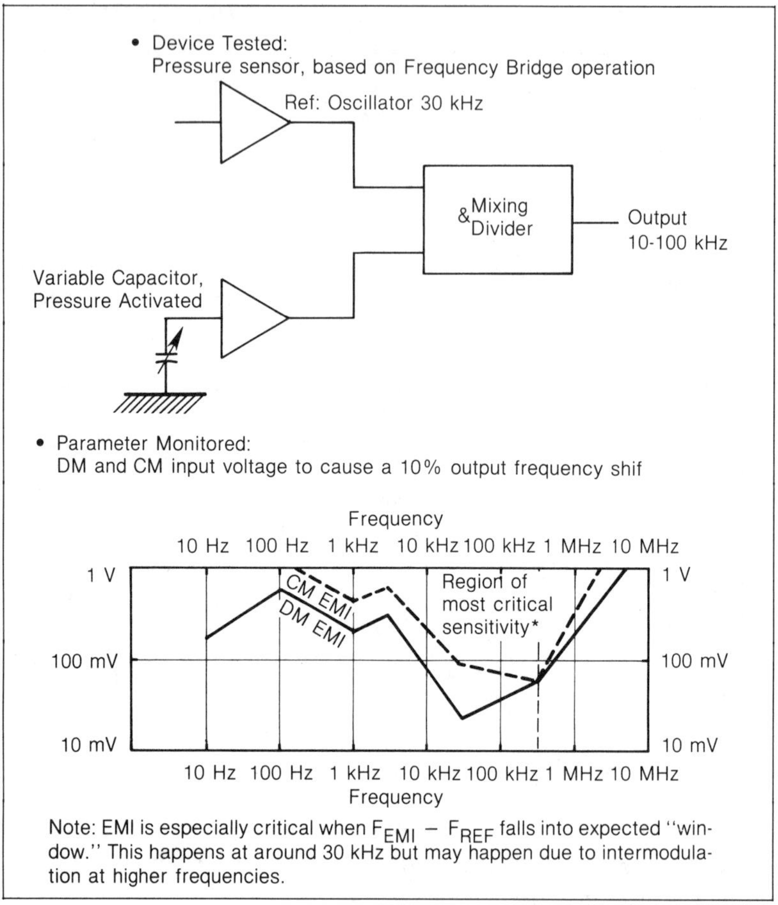
• Device Tested:
Pressure sensor, based on Frequency Bridge operation
Ref: Oscillator 30 kHz
& Mixing Divider
Output 10-100 kHz
Variable Capacitor, Pressure Activated
• Parameter Monitored:
DM and CM input voltage to cause a 10% output frequency shif
Frequency
10 Hz 100 Hz 1 kHz 10 kHz 100 kHz 1 MHz 10 MHz
1 V
100 mV
10 mV
CM EMI
DM EMI
Region of most critical sensitivity*
Note: EMI is especially critical when $F_{EMI} - F_{REF}$ falls into expected "window." This happens at around 30 kHz but may happen due to intermodulation at higher frequencies.

Figure 6.21—Example of Measured EMI Response in an Analog Sensor

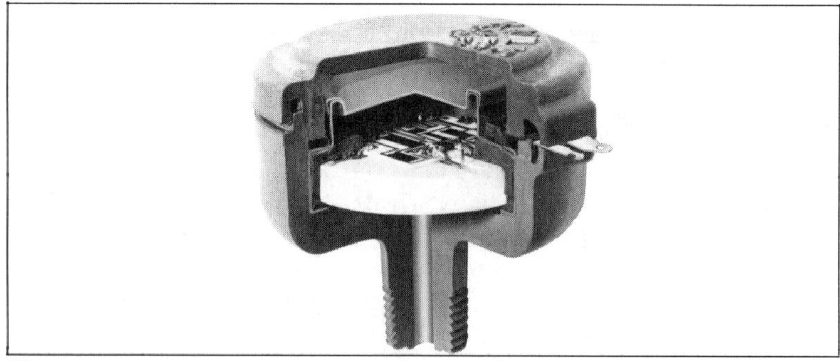

Figure 6.22—Detail of a Pressure Sensor (Courtesy of KAVLICO Corp.)

6.3.4 Position Sensors and Transducers

Many electromechanical devices detect, transmit or reproduce a linear or rotational motion. The simplest devices are the potentiometer type, generally used in rugged applications where modest accuracy is needed. Their EMI problems are limited to those described in Section 2.1 ("Resistors"), as noise generation from the wiper and noise pickup from the resistive track or winding. If the position sensor is shielded, it may be necessary to provide a bonding (by a wiping contact or otherwise) from the shaft to the shielding case.

When greater accuracy and a contact-free mechanism are needed, synchros and resolvers are used. These devices have been the workhorses of ships and aircraft position repeater systems. Even with the development of digital encoding, synchros and resolvers are still widely used in robotics, intelligent motion and automated manufacturing. In such devices a pair of matched generators and motors is interconnected such that for a rotation $\ominus$ of the generator shaft, the motor shaft on the receiving side will rotate by the same angle, with an angular precision which can be better than 10 minutes.

Since, to get a good time response and minimum overrun, the rotors of both machines have a minimum inertia, the system can be vulnerable to ambient magnetic fields. Low-frequency magnetic fields (50, 60 or 400 Hz) can cause the receiver to flutter around its resting position. This is especially critical when synchros and resolvers work in combination with servomotors.

Servomotors generally have high current demand, and the synchro in the feedback loop which controls the servoaction can be disturbed by the pulsed magnetic fields, causing the whole system to oscillate. All the same, the transmitter side can generate magnetic fields disturbing other synchros installed nearby. In both cases, the solution is a careful shielding and twisting of all the leads in and out.

Since synchros are traditionally thought of as high-impedance, audio-frequency devices, great expectations are sometimes put on the effect of a simple cable shield. However, remember that a synchro is rather a transformer with a varying azimuth between primary and secondary; consequently, its leakage inductance is high. On the other hand, the stator impedance is low, in the tens of ohms range, because when at equilibrium this impedance looks like the shorted rotor impedance transformed toward the primary.[8] Therefore, it may be that capacitive coupling is not the biggest EMI threat, but magnetic coupling is. Thus, twisted pairs (or twisted triads) and magnetic permeable shields will be more effective.

Careful separation from power wiring will avoid magnetic coupling to the synchro-resolver system. Also, due to their low impedance, stator leads can be heavily filtered against high-frequency EMI without affecting their response.

A common tendency in military vehicles and aircraft is to use the chassis or skin in place of one of the three synchro wires in order to save weight. This is a bad practice because it increases the size of the pickup loop and inserts large common impedance voltages at the power mains frequency or its harmonics in series with this stator lead.

Figure 6.23 shows an example of an EMI error budget in a synchro. It must be noted that some of the errors will be ignored if the synchro detector (usually digital) is sampling the reference voltage; capacitively coupled EMI will be ignored if they are in quadrature with the excitation voltage. The synchro shown is typical of aircraft applications. The sensitivity around null position is 210 mV/degree (i.e., 12 V $\times$ sin 1°). Therefore, to preserve a 0.1° (6 minutes) accuracy, any EMI signal must be $\ll$ 21 mV.

The installation shown has three potential problems: (a) crosstalk from power line, (b) common impedance noise due to the "savings" of the S_2 lead using the fuselage skin and (c) unbalance of S_1 and S_3 voltages introduced by the 1 μF decoupling capacitor tolerances.

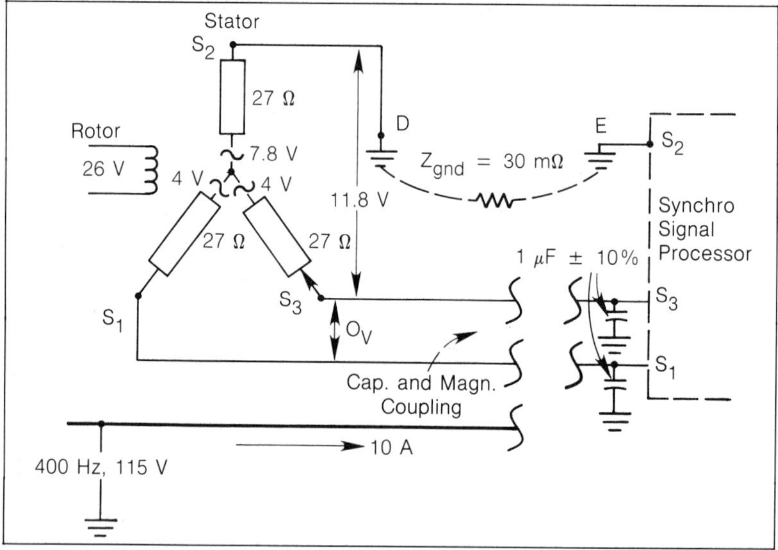

Figure 6.23—Example of a Synchro Installation Showing Three Potential EMI Problems

(a) Crosstalk

In a typical harness, power wires of few millimeters in diameter can be running at few centimeters from the synchro wires. This corresponds to a coupling capacitance of about 10 pF/m and a mutual inductance of 0.3 μH/m. Assuming the victim and culprit wires are running parallel over 3 m, both coupled noises can be calculated:

Capacitive Crosstalk:

$$V \text{ coupled} = 115 \text{ V} \times RC\omega$$
$$= 115 \text{ V} \times 27 \text{ } \Omega \times (3 \times 10^{-11} \text{ F}) \times 2\pi \times 400$$
$$= 0.23 \text{ mV (negligible risk)}$$

Inductive Crosstalk:

$$V \text{ coupled} = M\omega I$$
$$= 10 \text{ A} \times (3 \times 0.3.10^{-6} \text{ H}) \times 2\pi \times 400$$
$$= 22.6 \text{ mV} = \text{EMI error}$$

(b) Common Z Coupling

The aluminum and the joint resistance represent about 30 mΩ, in which 10 A of 400 Hz power return are flowing. So,

$$V \text{ error } (S_2) = 30 \text{ m}\Omega \times 10 \text{ A} = 300 \text{ mV}$$

This corresponds to an angular error of

$$\frac{300 \text{ mV}}{210 \text{ mV}} = 1.4°$$

(c) Unbalance Created by Filtering Capacitors

Assuming each capacitor is at its worst tolerance and that the stator arms impedance represents about 27 Ω at 400 Hz, the voltage to ground of points S_1 and S_3 is given by:

$$V_{S1} = 4 \text{ V} \times \frac{X(1 \text{ } \mu\text{F} + 20\%]}{jX(1 \text{ } \mu\text{F} + 20\%) + 27 \text{ } \Omega} \cong 3.99 \text{ V}$$

$$V_{S3} = 4 \text{ V} \times \frac{X(1 \text{ } \mu\text{F} - 20\%]}{jX(1 \text{ } \mu\text{F} - 20\%) + 27 \text{ } \Omega} \cong 3.97 \text{ V}$$

So, the differential error voltage at null position between S_1 and S_3 is:

$$3.99 - 3.97 = 20 \text{ mV} = \text{EMI error}$$

The solutions are:
1. Stop using chassis return for S_3 and use a dedicated wire.
2. Twist the synchro wires together to decrease magnetic coupling, and separate them farther from the power wiring.
3. Use better tolerance for decoupling capacitors, or use a lower value, like 100 nF if application permits, or check that the out-of-phase nature of the capacitive 400 Hz leakage can be rejected by the signal detector.

6.4 Electro-Explosive Devices

An electro-explosive device (EED), also called a pyrotechnic device, uses an electrical initiator to ignite a small load of explosive (see Fig. 6.24). This process is commonly used in aircraft and space vehicles to drop auxiliary tanks, canopies, used thrusters and payloads. EEDs are also used in satellites to activate certain sequences of events like deployment of booms, panels, etc. They can be controlled by the pilot, or as in satellites, remotely controlled from the ground or even fired by an inboard sequencer. Some EED

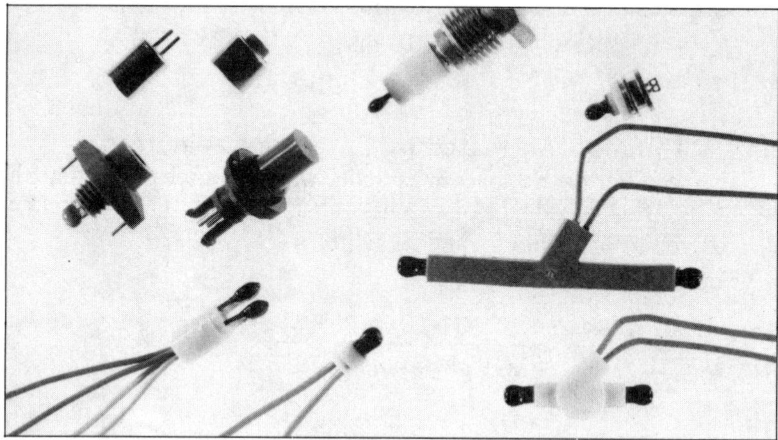

Figure 6.24—Various Types of Electro-Explosive Initiators, Used to Trigger an Explosive Charge for Civilian and Military Applications (courtesy of Davey-Bickford)

firing circuits can be quite elaborate, providing an electronic "safety catch," telemetry monitoring and the capability of interrogating the status of the EED.

EEDs require much the same EMI precautions as do low-level sensors, but the consequence of malfunction can be dramatic; if the EED input receives enough EMI current, it can unexpectedly fire and abort an entire mission or endanger human lives.

The U.S. Department of Defense has been actively engaged in determining the extent of radiation hazards and devising methods for controlling them. The problem has come to be known as radiation hazards (RADHAZ). The Navy has been involved in the RADHAZ problem in connection with ordnance programs known as Hazards of Electromagnetic Radiation to Ordnance (HERO). Initiated in 1958, HERO covers research and development with special attention to EEDs.

An EED can unexpectedly fire when unwanted energy couples to its input leads by radiation (see example in Fig. 6.25), conducted transients, capacitive or inductive crosstalk or static discharge. The two parameters are: the maximum **no-fire** current which is the guaranteed dc (or rms value for RF) current at which the EED will neither fire nor degrade with a reliability of 0.995 for a confidence level of 95 percent, and the minimum **all-fire** current which is the least dc (or rms) current which will guarantee firing with a reliability of 0.995 for a 95 percent confidence level.

All EED circuits should be designed with a safety margin. This margin is the difference in dB between the maximum expected ambient field and the field value which would induce the maximum

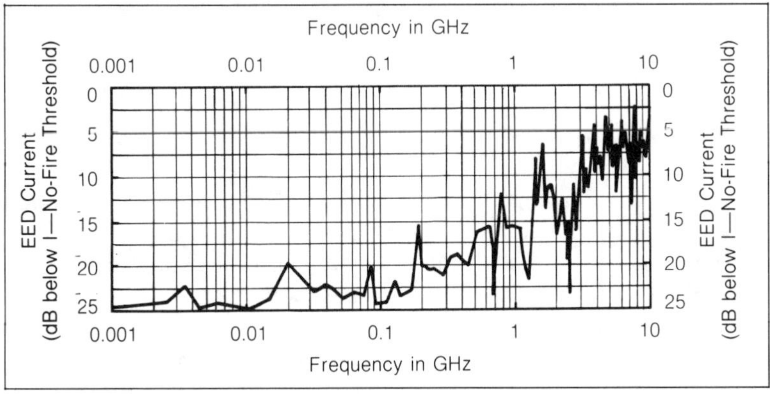

Figure 6.25—Sensitivity of a Bridgewire EED in a Weapon System to Radio Frequency CW Power (1,000 W/m2)[15]

no-fire current. This safety margin can be verified by bench tests, bearing in mind that to verify a 0.995 reliability with a 95 percent confidence a sample size of 650 devices is necessary, with all no-fires at the given level.

For EEDs exposed to nearby ship or airborne transmitters, a common value is that when the firing circuit is exposed to power densities of 100 W/m^2 (VHF-UHF) up to 1,000 W/m^2 (microwaves) there must be a 10 to 20 dB reserve below the no-fire threshold (NTF).

Space technology needs have led to more studies of EED responses to EMI. For instance, using infrared sensors integrated in the EED, an accurate monitoring of the bridge wire temperature under actual RF illumination has been conducted for both ground and in-flight conditions.[9]

RF power dissipated in an EED depends on the characteristic of a signal appearing across its leads, the impedance characteristics of both the EED and the leads and other factors. Electric initiators are classified under seven types: high-resistance wire, low-resistance wire, carbon (graphite) bridge, conductive film, semiconductor and spark gap.

The resistance wire and the carbon bridge are the most commonly used EEDs. Combinations of more than one of these types are used to obtain special characteristics for a particular application.

The wire-bridge detonator uses a fine wire of tungsten or other noble metal between two electrodes to form the bridge. Electricity flowing in the bridge heats the wire, igniting a spot charge which in turn sets off the detonator base charge. A typical low-energy bridge detonator is shown in Fig. 6.26a. The bridge resistance ranges from 2 to 5 Ω and is made of tungsten wire. The detonator is rated at a nominal 5,000 erg at less than 10 μs, with a capacitor charged to 50 V (one erg equals 10^{-7} J). The bridge wire is coated with an explosive **spot charge**. A typical firing energy is in the order of the millijoule.

The carbon-bridge initiator shown in Fig. 6.26b uses a colloidal graphite charge to form the electrical detonation circuit path. It functions like the wire bridge but is more sensitive. The lead wires inside the plug are coated with insulating varnish and twisted to attain the small separation at the face of the plug needed for graphite-bridge detonators. This detonator will function in 10 μs, with 300 V applied from a 2,200 pF capacitor (corresponding to 0.1 mJ).

The explosive mixture contains conductive material which forms the electric circuit. The heating effect of the current in this mixture causes detonation. The conductive material is a mixture of metals and graphite and is mixed with explosives. Conductive mixtures vary in resistance from approximately 1 to 800 Ω. Because of manufacturing difficulties with the wire-type bridge and its relatively low sensitivity, a deposited metal-film initiator is sometimes used. Titanium smears on a glass base have detonated the explosive with an average energy of 50 erg at 30 to 40 Ω resistance.

Arc, heat and shock wave are the three main techniques of electrical initiation of EEDs. Arcing exists at levels greater than 25 V and usually occurs in carbon-bridge and thin-film initiators. Hot-wire initiators as well as conductive-film and conductive-mix initiators operate by heating. The third type of initiation occurs in the exploding-bridge wire type of initiator. The shock-wave mode requires greater than 300 V to cause the formation of an intense shock wave which initiates a secondary high-explosive directly. Most

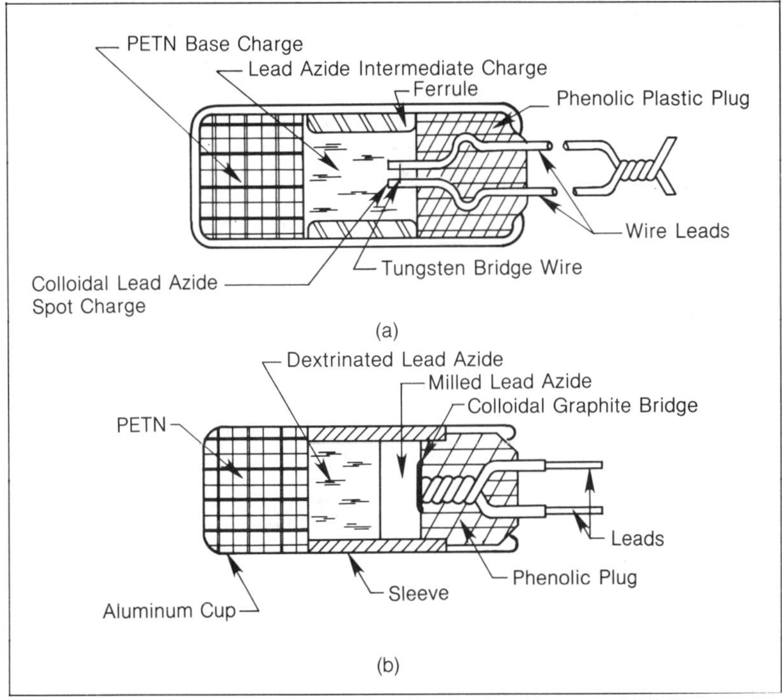

Figure 6.26—Electric Initiators: (a) Wire-Bridge Type and (b) Carbon-Bridge Type

initiators can operate under more than one mode, depending on the magnitude of the electrical stimulus.

In addition to the intentional sources, EEDs may also be activated by unintentional sources. The conventional sources are those designed into a firing circuit to supply a controlled amount of energy to initiate the EEDs. Some circuits use a capacitor bank which is charged slowly some time before firing time. Unintentional activating sources are those that couple sufficient EMI energy to result in inadvertent ignition. Both sources are listed in Table 6.6.

When RF-radiated energy is the cause of unintentional firing, the transmission path of a firing circuit becomes an important factor. Parameters which define the RF source and thereby influence the design of protective measures are field intensity or power density, type of modulation and duty cycle, frequency and polarization. Other important factors include electromagnetic coupling and thermal parameters. Here, RF waves propagating in free space become guided waves in a coupled firing circuit. The thermal parameter represents the mechanism by which RF initiation of EEDs take place by heating.

Mathematical models exist which have been used to predict the probability of firing an EED when the EMI sources are radars and nearby communications transmitters. Simplified models assume that the source transmitter antennas are boresighting the EED leads. Using prediction models similar to those discussed in Chapter 4, the effective area of the leads converts the arriving power densities to power available to ignite the EEDs. When superimposed on the known EED firing characteristics, the probability of initiation is predicted. A realistic model should address both pin-to-pin (differential) and pin-to-case (common-mode) types of EMI excitation.

The instantaneous power in an EED is:

$$P_{(t)} = V_{(t)}^2 / R_z$$

with,

$$R_z = \text{bridge wire resistance}$$

$$V_t = V_{peak} \text{ for quasirectangular firing pulse}$$

$$= V_c e^{-t/\tau} \text{ for capacitive discharge firing, where}$$

$$V_c = \text{initial capacitor voltage}$$

$$\tau = \text{firing circuit time}$$

Table 6.6—Intentional and Unintentional Sources of Energy which May Initiate EED Firing

Intentional EED Activating Sources	Unintentional EED EMI Activating Sources
Battery: Both wet and dry batteries are used. Extreme temperature variation, high accelerations and long storage time limit the use of wet cells. Capacity and weight limit the use of dry cells.	**Stray voltages:** Ground current loops, faulty connections, shorts and open circuits may produce voltages of sufficient magnitude in the firing circuit to cause ignition.
Thermal cell: Cells have been developed with solid electrolytes at room temperature. When required, a small thermite charge inside the cell is activated and the electrolyte melts, charging the cell.	**Static:** Charges built up by vibration, friction or the accumulation of a static charge on a person's body may actuate sensitive initiators
Generator: Mechanical-energy driven ac and dc generators are used. Size and weight limit their uses.	**Lightning:** An ungrounded weapon system may provide a path to ground through the EED or its firing circuit. Ionization voltages from lightning may induce currents great enough to fire an initiator.
Converter: Converted ac power by vacuum tubes or solid-state devices are used. The additional weight and space can often be better used either by using the available ac power directly or by using batteries.	**Transients:** Momentary surges of voltage or current in or near a firing circuit may induce currents or exceed the design limits of protective devices in firing circuits.
Electrostatic (Dust Generator): Whirling dust generators are used to develop potentials up to 5,000 V. Generation occurs only when the dust whirls with sufficient speed. High whirling speeds are obtained in most projectiles fired from a gun. Large amounts of energy may be produced with relatively small generators.	**Magnetic effects:** Magnetic fields of sufficient strength may induce a voltage in a wire.
	Thermoelectric effects: When two dissimilar metals at different temperatures are joined, a small thermoelectric voltage is generated, such as between a copper EED lead and an aluminum ground. Since such voltages are so small, ignition would be highly improbable.
	Test equipment: Some circuit test equipment used for checkout purposes inject voltages and currents into a weapons system that could possibly activate an initiator.
	Electrostatic: Whirling clouds of dust or steam may produce electrostatic charges to fire a sensitive detonator.

For this latter case, the constant firing energy can be approximated by:

$$E_{joules} \cong \frac{0.4 \; (V_c)^2}{R_z} \times \tau$$

Nearly all of the above math models yield pessimistic results for radar EMI sources and optimistic results for some communication transmitter sources. This occurs for radar models because the simultaneous superpositioning of a boresight condition, matching the polarization, ignoring the effect of EED lead twist or lossy material and underestimating the shielding effect of an intervening barrier (e.g., missile skin) result in a dangerously cumulative probability which, in effect, may exist arbitrarily close to 0 percent of the time. Thus, the recommended measures based on these models can be to shut down radars (to the chagrin of a ship's captain or a range safety officer). Some communication models, on the other hand, fail to recognize that an entire missile skin or aircraft can act as a pickup antenna and capacitively couple intercepted energy to the EED leads.

Among the additional parameters complementing the prediction process are personnel proximity or contact with the EED or its housing, structure of the EED container and ground plane, openings in the EED container, configuration of lead wires from control stations to EED or container and RF impedance of the firing circuit. As a result, some attempt has to be made to establish safe distances of separation between explosive and EMI emitter source. Table 6.7 illustrates one measure of RADHAZ protection.

Table 6.7—Explosive Safe Distances

Expolosive	Safe Distance in Feet	Source	Power*
Ammunition Blasting Caps	> 15 m (50')		3 W
	30 m (100')	Radar	3 W
	305 m (1,000')	Radar	2 kW
	1.6 km (1 mile)	Radar	100 kW
	30 m (100')	AM Radio	10 W
	305 m (1,000')	AM Radio	1 kW
	1.6 km (1 mile)	AM Radio	25 kW
	1.5 m (5')	FM Mobile	5 W
	3 m (10')	FM Mobile	25 W
	9 m (30')	FM Mobile	100 W

*For other power values, since power density varies to (1/distance)^2, the safe distance would vary to the square root of power.

The most dependable approach, so far, has been to measure by irradiation under moderate CW fields, the equivalent capture area of an EED and its firing circuitry once installed. This area Aeq. can be plotted for each frequency as:

$$\text{Aeq, m}^2 = \frac{P_{induced}, \text{ W}}{P_{incident,} \text{ W/m}^2} = \frac{P_{induced}, \text{ W}}{(E_{v/m})^2} \times 120\pi$$

Then knowing $P_{no\text{-}fire}$ and applying the safety factor, a safe value of $E_{v/m}$ or $P_{w/m}^2$ can be derived across the frequency.

6.4.1 EED Protection

It is a complicated task protecting EEDs from RF radiated energy while allowing the devices to be operable in the firing circuit. Basically, factors should be known concerning the RF generators, the transmission media, the coupling parameters to the EED and the EED firing characteristics as well as the type of susceptible material, its geometry and its location within a weapon system.

Ambient conditions, antenna configuration, polarization, type of modulation, frequency, power output and field intensity are pertinent factors in the problem of protecting a sensitive EED from unwanted ignition. Protective devices that have been developed for different EEDs include:

1. **Attenuators**—RF attenuating materials such as carbonyl iron powder, long-chain polymers, tantalum peroxide, magnesium dioxide and ferrite materials
2. **Shielding**—Scotch®-type metallic tape and twisted braid, complete metallic enclosures, etc.
3. **Thermoelectric Attenuator**—With a Peltier junction, dc will cause both Peltier and Joule heating, but ac will cause only Joule heating. The results are a 2:1 ratio of alternating current to direct current for the same heating effect. For the same current, dc has a heating ratio of 4:1.
4. **Transmission Lines**—lossy coaxial, parallel-wire, shielded parallel-wire and parallel strip lines
5. **Bypass Capacitors**—A certain amount of RF attenuation can be achieved by shunting the leads of the EED with a high dielectric constant bypass capacitor. The size is determined by the resistance of the SQUIB to be protected and the frequency to be bypassed.

6. **Relays**—Various methods have been devised to use a relay in the firing circuit in a way that requires a fairly heavy current for the relay to operate. Experimental work has been done on special coaxial relays that pass only dc and very low frequency ac.
7. **Filters**—RL and RLC dissipative filters have been used. The size, weight, mounting and effect on the firing time are limiting factors in the use of this type of protective device.
8. **Shorting Straps**—Usually a temporary "shorter" is applied during manufacturing to protect the initiator during loading, handling and shipping, and it remains in place until the initiator is armed. When the initiator is in the circuit, the temporary shorter is removed and a second one installed. The effectiveness of this device falls off at higher frequencies.
9. **Connectors**—a shielded cable connector to make ground contact before and break it after the two power contacts are made
10. **Fuses**—A fuse in parallel with the EED is limited by the one-time nature of its protection.
11. **Conductive (static dissipative) Plugs**—to avoid inadvertent firing by an electrostatic discharge

A measure of the effect of these various solutions can be summarized as: given the worst-case ambient in airborne equipment, the EED housing (filtering) decoupling must provide at least 70 dB of RF attenuation from 1 MHz to 12 GHz, decaying to 40 dB at 100 kHz. The immunity of the EED must be complemented by corresponding wiring and installation precautions such as:
1. the separation of firing circuits from other kinds of circuits. EED firing wires should not share shields, connectors, etc., with any other wiring
2. firing circuits which always should be balanced, twisted, shielded pairs and isolated from the EED can
3. the firing circuits should be isolated from each other

The case of accidental firing from electrostatic discharge (ESD) requires special attention. The discharge path can be differential (lead-to-lead) or common-mode (lead-to-case) through the explosive (Fig. 6.27). The breakdown voltage in the explosive being typically 0.8 to 1 kV/mm, this leads to practical arcing voltages in an EED of 1.5 to 5 kV.

Many solutions have been tried, such as building up the insulation of the can so that internal arcing is impossible. This is difficult to achieve practically and would put the protection under the threat of cracks, voids, etc., which can go undetected during manufacturing or operating.

Another attempted solution is the placement of shunting resistors between terminals and case. These resistors can work as either progressive "bleeders" or dissipators. ESD in airborne systems, however, can come from large stored charges with a source resistance as low as 1 kΩ. For instance, some standard ESD tests are made with a source resistance of 1,500 Ω. Therefore, the value of the shunting resistance has to be low enough to keep the lead-to-can voltage below the arcing voltage. This gives a first condition, i.e., a maximum value of the shunting resistance R_p, as shown in Fig. 6.28.

In addition, looking at all possible discharge scenarios, one case would be an ESD zap to one of the wires only. In this case a second condition has to be met, which is that voltage V_{AD} across the stressed resistance R_p does not allow for differentially firing the bridgewire BC (see Fig. 6.29). To achieve this, the divider network of $R_z/(R_p + R_z)$ must be of a sufficient ratio; that is, R_p must be larger than a minimum value.

The two conditions evolve in opposite directions. Therefore, there will be cases where a value of R_p can be found such as $R_{p\ min} < R_p < R_{p\ max}$. In other cases, the span will be too narrow to be safely met, and the two conditions even become contradictory if $R_{p\ min} > R_{p\ max}$.

For instance, assuming a bridgewire resistance of 5 Ω (the range

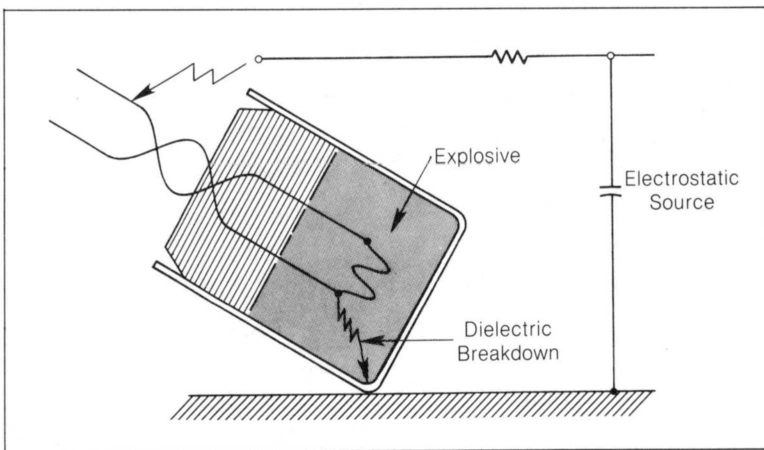

Figure 6.27—ESD Coupling to an EED

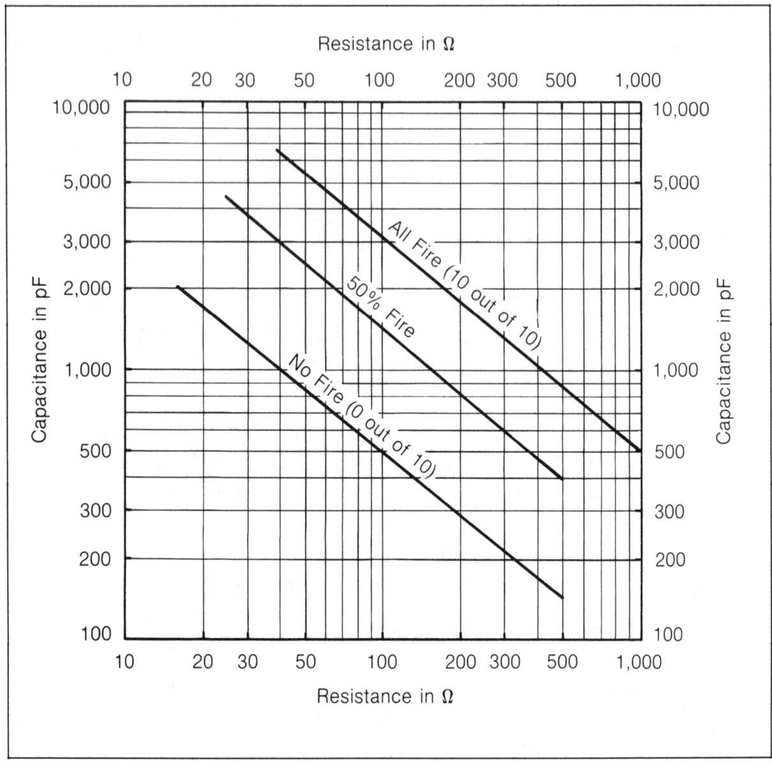

Figure 6.28—Results Obtained for Different Values of EED Lead-to-Ground Protection Resistances, vs. Storage Capacitor Values for a 5 kV ESD and a Bridge-to-Can Breakdown Voltage of 2 kV

is usually 1 to 10, with 3 to 5 being the most frequent) and a bridge-to-can arcing voltage of 2kV, the calculation of Fig. 6.29 shows that for a 15 kV ESD from a 1,500 Ω source resistance, the first condition (no arc) is met if:

$$R_p < 222 \ \Omega$$

On the other hand, the second condition dictates that, in case of a discharge to one lead only, the energy dissipated into R_z be less than the firing threshold.

If we take the EED with an "all-no-fire" threshold of 0.1 mJ and an ESD coming from a 100 pF/1,500 Ω source, the discharge time constant is 150 ns and the energy condition (based on the energy integral of an exponential decay through R_z) gives:

$$0.43 \, \frac{(V_{BCpk})^2}{R_z} \times \tau < 0.1 \times 10^{-3} \text{ J}$$

or,

$$V_{BCpk} < 88 \text{ V}$$

since, on Fig. 6.29:

$$V_{BC} \cong V_{AD} \, \frac{R_z}{R_p}$$

Solving for R_p shows that if $R_p < 222 \, \Omega$ per Condition 1, then Condition 2 is also satisfied.

However, if we now assume a larger electrostatic source such as

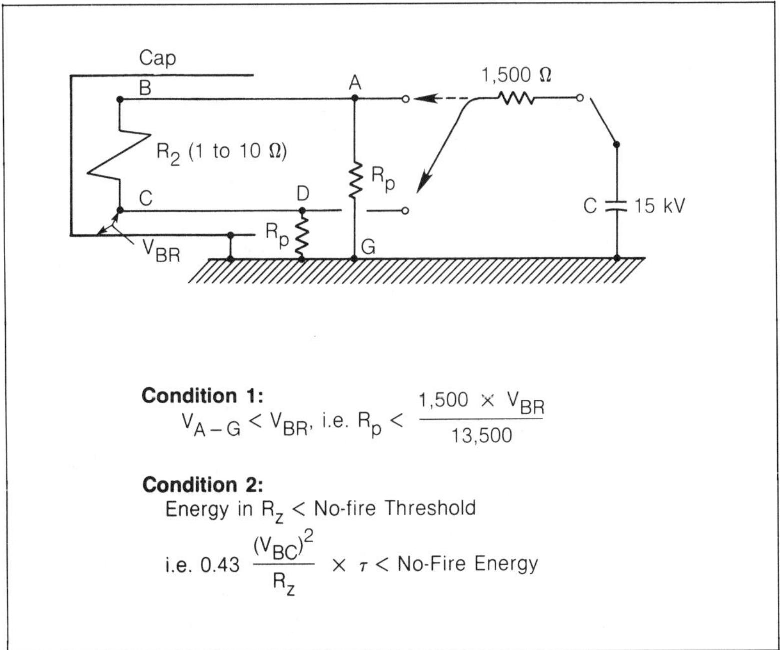

Figure 6.29—Equivalent Circuit to Calculate Adequate Protection Resistances R_p under Specific ESD Conditions

500 pF, giving a time constant (Through 1,500 Ω) of 750 ns, the condition for $V_{BC\ max}$ becomes:

$$V_{BC} < 39\ V$$

which corresponds to $R_p > 423\ \Omega$, a condition which conflicts with Condition 1 for no arcing. As a result, resistive shunts are more and more replaced by zener diodes, varistors or the like (see Chapter 7) which are integrated into the EED.

6.4.2 Thermal Parameters

An additional consideration to be given to EEd accidental firing is the thermal buildup under continuous EMI exposure. The bridgewire heating can be represented by:

$$\frac{d\theta}{d_t} + \frac{1}{\tau}\theta = P_{(t)}$$

where,

θ = temperature rise above ambient °C

t = time in seconds

τ = EED thermal time constant = R_{TH}/K_B in seconds

$P_{(t)}$ = time dependent power input in watts

K_B = heat coefficient of the bridgewire alone (recorded) in °C/W

R_{TH} = heat coefficient of the EED in °C/W, which includes lead wire and package heat losses

Usually K_B is rather large since it relates to the bridge's conversion efficiency of the applied pulse into heat. K_B is in the range of 1°C/W or more.

R_{TH} depends on the package size, and it dictates the temperature rise inside the EED. The smaller the R_{TH}, the lower the temperature rise for a given input. R_{TH} is usually several times smaller than K_B, leading to time constants in the subsecond range (typically 10-15 ms). Therefore, the EED cannot rapidly evacuate the bridgewire instantaneous thermal energy, and the device behaves as an integrator of energy, leading to thermal "stacking". As a result, when the EMI is a single event, or with a low repetition rate such as: pulse duration $< \tau$ and rep. period $> 10\tau$ the EED will cool completely between consecutive pulses, and the risk assessment should be based on the *energy* of a single pulse.

When the rep. period is $< \tau$, the *no-fire* prediction should consider the effect of induced RF currents as if they were dc currents having the same rms value.

6.5 Lamps

This section describes EMI causes and control in incandescent, fluorescent and gas lamps.

6.5.1 Incandescent Lamps

Once an incandescent lamp is energized, it is a fairly stable emitter of infrared and optical energy. Because of its relatively low temperature, relatively little RF energy is emitted. Consequently, incandescent lamps generally do not create EMI problems unless the supply line is noisy, in which case the filament will act as a miniature antenna even if the power line is otherwise shielded by a conduit.

For sensitive electronics or applications where extremely low radiated emission is required, specially shielded bulbs can be used (see Vol. 3 of this series on shielding).

The only serious problem with lamps is their large inrush current at turn-on, since the cold filament has a low resistance. Typically, a value of 10 times the normal current can be expected, but some bulbs take 20 times the nominal current when switched on. This creates more arcing and contact fatigue on the switch or relay controlling the lamp. On rare occasions, an incandescent lamp will develop a faulty filament which opens and closes, resulting in transient surges. The simplest EMI-control solution is to replace the lamp.

In the mid 1980s, RF-powered bulbs were introduced because the power efficiency and lifetime was better. In these lamps, a small oscillator is located in the socket to convert the ac mains into an

RF pulsed voltage. In the United States, the FCC has created regulations for the radio frequency emission of these devices. Their emissions are quite low above 1 MHz and, although limited history exists of their usage, they do not seem to have created interference problems. However, Fig. 6.30 shows that the first harmonic of their RF voltage produces a rather strong narrowband field in the low AM band.[10, 11]

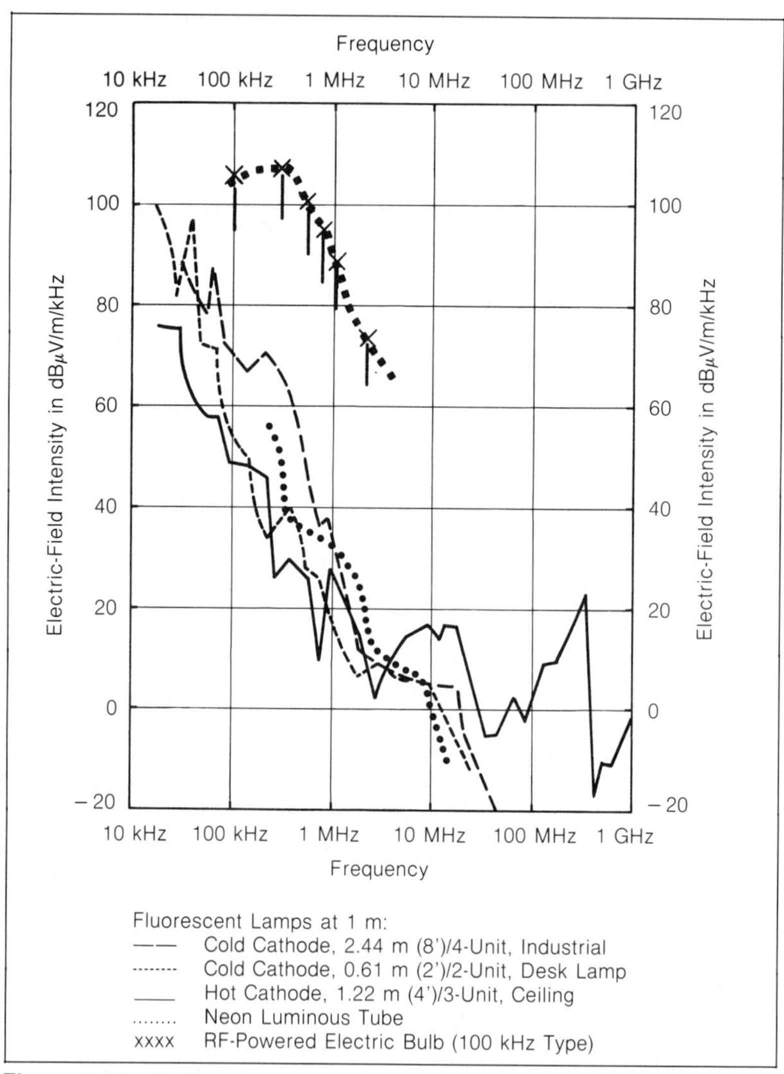

Figure 6.30—Radiated Lamp Noise (Fluorescent and RF Powered)

6.5.2 Fluorescent Lamps

Fluorescent lamps create both conducted and radiated RF noise. RF emission from fluorescent lamps is created by a column of gas being ionized and extinguished 120 times per second for a 60 Hz power mains supply. Thus, these transient surges result in broadband radiation from the bulb as well as conducted emissions back onto the power lines.

Measurements have shown that radiated emission levels are significant up to about 1 MHz, with some types exhibiting emissions up to 100 MHz or more. The fixture plays an important part in the radiation "efficiency." Clark has measured the radiated EMI from standard unmodified fluorescent units, plus some improved versions.[10] The cold-cathode units showed less emission above 1 MHz than the hot-cathode units. Measured values are reported in Fig. 6.30 where the legend indicates the length and number of tubes in each unit. The receiving antenna was located 1 m from the unit.

EMI is difficult to control economically in fluorescent lamps mounted in their fixtures. One technique is to shield the light-emitting area from the lamp-mount fixture with either conductive glass (this is expensive) or a wire screen. The ac lines feeding the fixture are also filtered to keep the conducted transients down to controllable levels. In Europe, most countries follow the CISPR* recommendations which stipulate maximum levels for fluorescent tubes (Fig. 6.31). Results of various tests also show that newer elec-

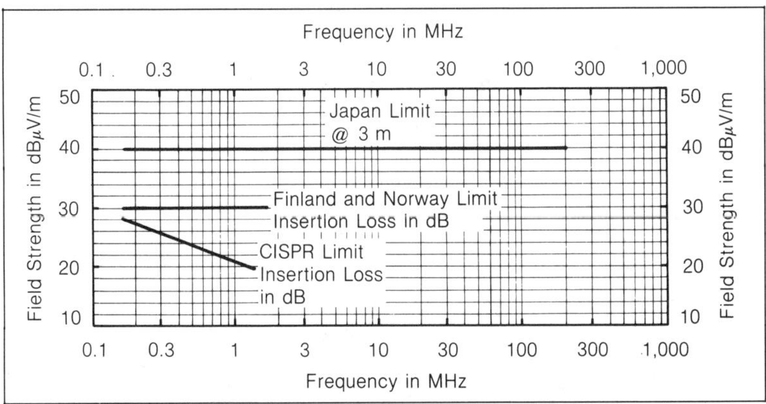

Figure 6.31—Interference Limits for Fluorescent Lamp Fixtures (Luminaires) (from CISPR Publication 15)

*Special Committee from the IEC on Radio Frequency Interference

tronic drives produce significantly more EMI than conventional magnetic ballasts (see Fig. 6.32).

Frequency	E-Field at 1 m dBμV/m in 10 kHz BW		Conducted EMI (10 kHz BW) dBμV, 50 Ω LISN/dBμA	
	Electronic Ballast	Magnetic Ballast	Electronic Ballast	Magnetic Ballast
30 kHz	100-140	60	80-125/70-115	70-75/35
100 kHz	80-110	40	80-100/60-105	60-65/22
300 kHz	70-90	40	60-85/45-80	55-60/20
1 MHz	60-80	30	55-75/30-55	40-55/18
3 MHz	40-70	30	45-65/25-30	40-58/20
10 MHz	< 40-55	< 20	40-50/<20	

Figure 6.32—Measured EMI from Eight Samples of Fluorescent Tubes, Four with Electronic Ballasts, and Four with Magnetic Ballasts. Conducted Levels Are Given Both in Voltages (with a 50 Ω/5 μH LISN) and Current (with MIL-STD 10 μF Bypass).

6.5.3 Gas Lamps

EMI emissions from gas lamps are somewhat similar to that of fluorescent lamps. The main difference is that gas lamps are energized from high-voltage transformers (typically furnishing 10 kV to the lamp) and the gas column remains ionized throughout the ac cycle. Thus a steady broadband, nontransient radiation takes place.

While EMI control of conducted and radiated emissions from gas lamps could employ the same techniques used for fluorescent lamps, this is rarely done. Most gas lamps such as neon signs are too big and cumbersome to shield. Filtering at the ac power mains may help mitigate some resulting EMI situations. Thus, little is done to contain EMI from gas lamps.

6.6 Motors and Generators

EMI aspects of electrical machinery are divided into four categories: interference reduction for large dc motors and generators, alternators and synchronous motors, fractional-horsepower machines and special-purpose rotating machines. Any rotating machine with sliding contacts should be regarded as a potential source of EMI because the switching and arcing processes of commutation cause rapid current and voltage changes that generate energy through a wide frequency range. Figure 6.33 compares the spectrum signatures of a small universal motor and a dc generator.[12] Figure 6.34 and 6.35 show measured EMI from larger motors.

6.6.1 Brushes

Brushes and brush leads are the most likely components from which EMI can be radiated or conducted. If a motor or generator is not adequately enclosed, the brushes and brush leads may require shielding. Provisions should be made in the original design of motors or generators for installation of capacitors at the brushes.

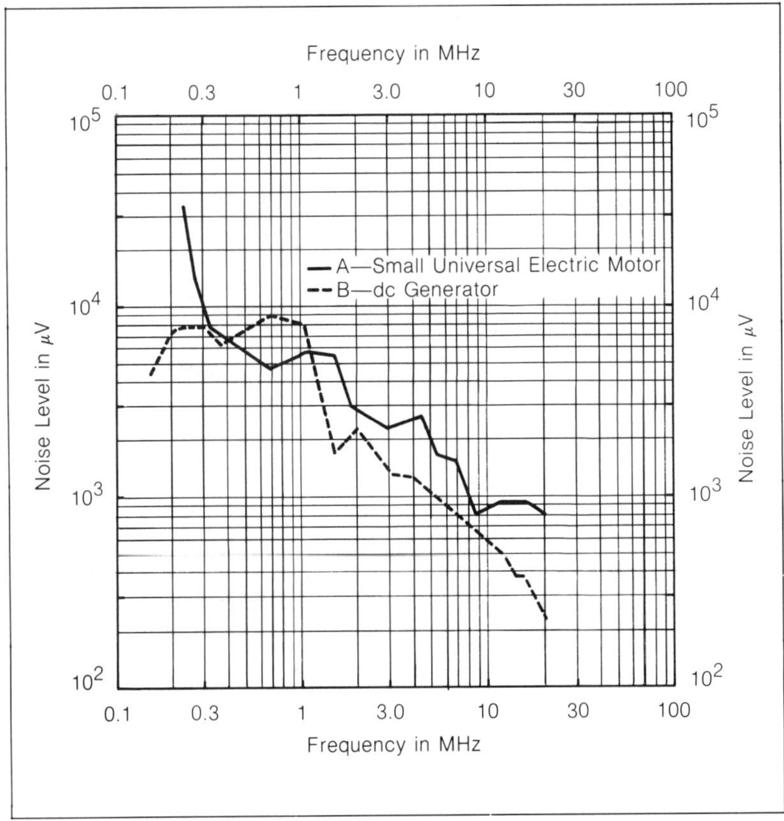

Figure 6.33—Spectrum Signatures of Radiated Noise from Small Universal Electric Motor and dc Generator, Measured in 1 kHz Bandwidth

Brush-generated interference may be reduced by incorporating the following in the design:

1. **Brush pressure**—EMI decreases at all frequencies with increasing brush pressure. Increased brush pressure, however, increases the rate of wear. The necessity for more frequent brush replacement is often a reasonable compromise for decreased interference.

2. **Current density**—EMI decreases with a decrease in current density. As the current density is increased, more heat is generated at the brush surface sliding on the commutator or slip ring. This heat causes the formation of a thick oxide film on the sliding metal surface. Rapid variations in the sliding contact resistance, resulting from irregularities in this oxide film, cause high-frequency transients that produce interference. To offset the heat increase, a somewhat larger brush surface area than necessary should be designed. Such

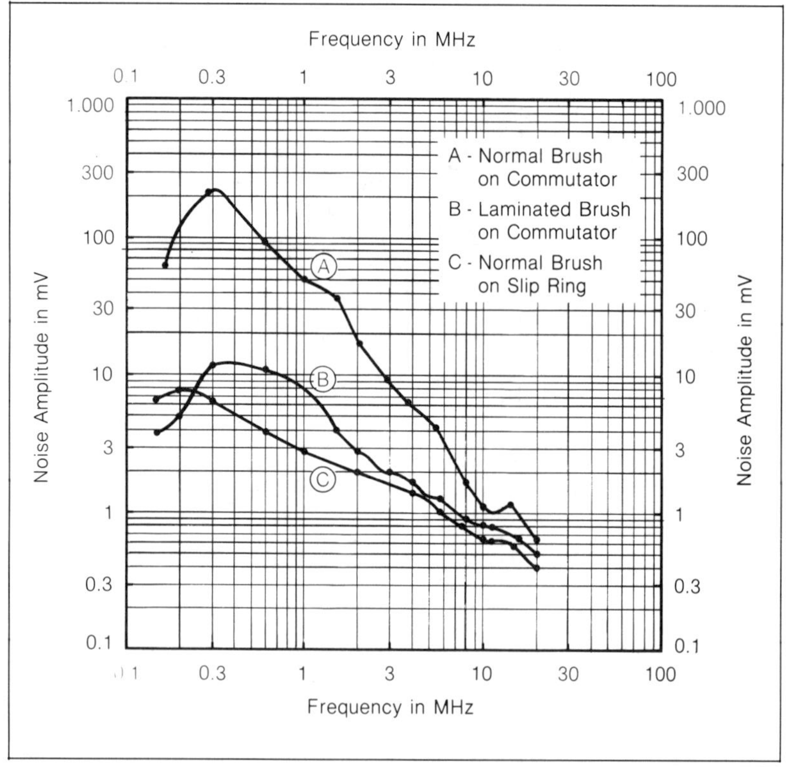

Figure 6.34—Spectrum Signatures of Different Types of Electric Motors

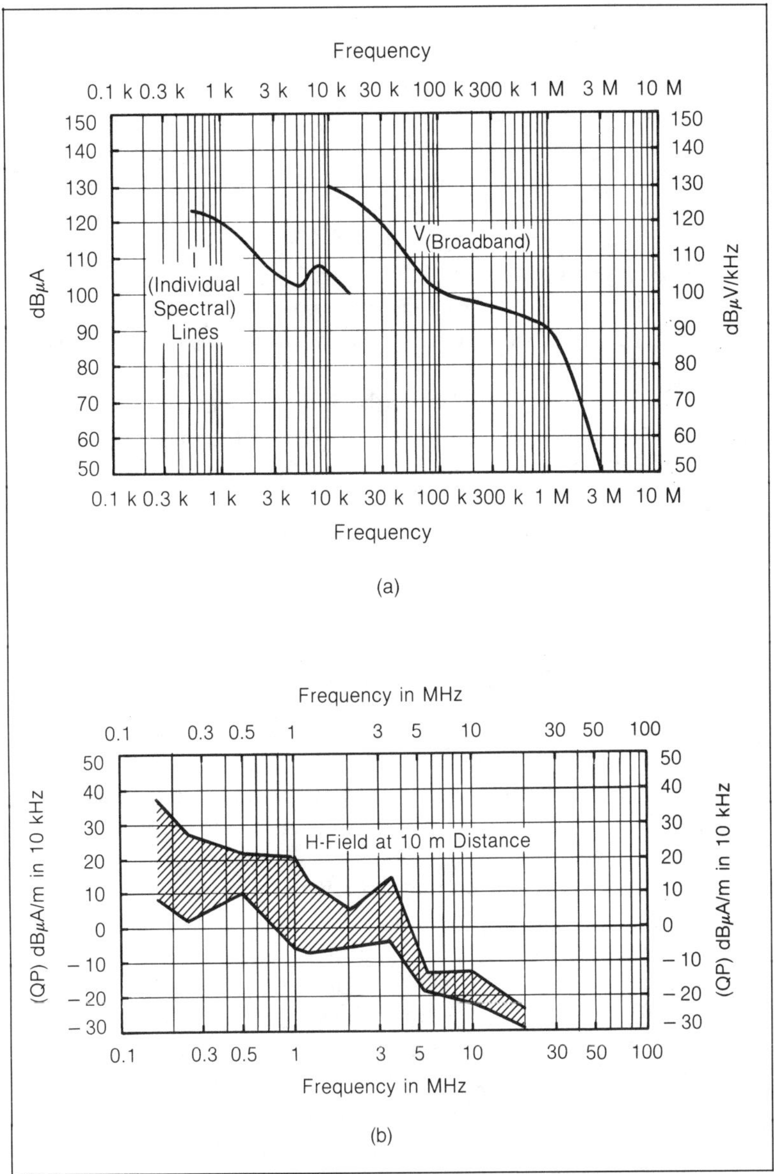

Figure 6.35—EMI from Large Traction Motors (a) Conducted EMI from 1,800 A Chopped Drive Car, with 500 Hz Chopping Frequency (Motor Unfiltered) (b) Magnetic Field Readings at 10 m from Cars. The envelope represents the extremes of various operating modes (accelerating, braking, coasting).

a design change will reduce heat and losses due to mechanical friction. On the other hand, if too low a current density is used, nonuniform grooves develop on the metal surface of the slip ring or commutator, and the increased friction frequently sets the brushes into a noisy chatter because of the wider brush surface area. The best tradeoff has been found with 8 to 9 A/cm^2 for carbon brushes and 11 to 14 A/cm^2 for metal graphite brushes.

3. **Brush resistivity**— Brush materials of low resistivity are low EMI generators. One example of such a brush is an electrographitic carbon brush with about 2 mΩ specific resistance in machines being used at less than 50 V. Low-resistance brushes are available with silver, copper or cadmium impregnated graphite. When used with a commutator, the resistance of the brush should match the requirements for good commutation. When used with slip rings, a wide choice of brush material is available because no switching action is involved.

6.6.2 DC Motors and Generators

Of all rotating machinery, dc motors and generators are the most serious offenders in generating EMI because they require commutators for their operation. Commutation is a switching action that is accompanied by interference-producing transients. When a switch is closed on a winding, the input impedance changes from almost infinity to zero. If the circuit contains inductance and capacitance, its voltages and currents cannot reach their normal values instantaneously because energy stored in the magnetic field of the inductance (or in the electric field of the capacitance) cannot dissipate instantaneously. Initially, the changing voltages and currents develop steep wavefronts which decay as a function of time. Since the bars of a commutator sliding rapidly past the contacting brushes produce a switching action, this causes extreme variations in impedance which, in turn, establish the series of voltage transients or pulses that cause EMI (See Section 6.1 on switches in general).

When designing a generator, measures can be taken to minimize the amount of EMI generated by commutator action. Reduction of commutation transients requires the use of design techniques to provide a smooth transition from one value of impedance to another within each armature coil. Interference produced as a result of commutation is reduced by six design techniques: interpoles, compensating windings, increased number of armature coils and commutator bars, laminated brushes, commutator plating and use of solid-state commutation.

A good way to improve commutation is by adding interpole windings. Interpoles counterbalance the self-induction of the armature coils during the commutation period and also reduce the induced voltage in the armature coils resulting from the coil, cutting fringing flux during the commutation period. The use of properly designed interpoles produces a rapid change in the armature-coil current at the beginning of the commutating period, reducing the steepness of the transient at the end of the commutating period.

Compensating windings produce the same effect as interpoles (but to a lesser extent) and help prevent field distortion. They also reduce cross-flux produced by armature coils. The use of interpoles and compensating windings lessens critical brush positioning requirements with respect to the commutator and provides EMF in the coils under commutation which oppose the EMF of self and mutual induction in these coils.

Increasing the number of coils on the armature (thereby increasing the number of commutator segments or bars) reduces interference by reducing the current broken per bar and the reactive voltage per coil. The largest number of armature slots in which the coils are uniformly distributed with respect to the commutator bars should be used, and the armature slots should be as shallow as possible. The use of short-pitch windings reduces interference by reducing the reactance voltage of each coil.

The break transients resulting from the switching action of the commutator can be smoothed out through the use of laminated brushes. These consist of brush materials of different resistivity, cemented together by nonconducting glue which provides insulation between adjacent brush segments. The ideal operation of laminated brushes is indicated on Fig. 6.36. Having the successive segments of the brush increase in resistance avoids the sharp current drop after the brush leaves the commutator segment. A more linear coil-current reversal results, thus reducing the break transients.

The segments of the laminated brush are insulated from one another by some suitable bonding material and electrically connected by the brush lead or brush spring. Circulating currents

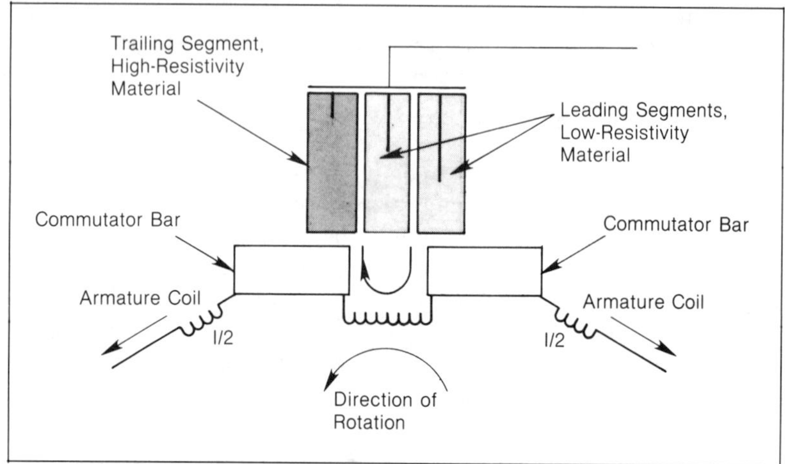

Figure 6.36—Commutation of an Armature Coil by Laminated Brushes

resulting from the self-inductance of the coil under commutation and from the coil-cutting fringe flux from the pole pieces must flow through the entire length of two brush laminations. The total resistance of this length is much greater than that presented by a direct path across the face of the brush (as would occur with a solid brush). Circulating currents are therefore reduced early in the commutation period, and desirable division of current through the two adjacent commutator bars is achieved. Good commutation can be achieved over a fairly wide range of brush positions, relative to the magnetic neutral, so that brush positioning becomes less critical and less dependent upon armature current.

The design of laminated brushes should include two or, at most, three laminations. The following criteria should be incorporated in the design:

1. The thickness of the leading-edge lamination of a two-lamination brush should be about 90 percent of the total thickness, and its resistivity should be as high as allowable for heat dissipation.

2. The resistivity of the trailing-edge lamination should be about 15 times that of the leading edge; this lamination should be thick enough to preclude mechanical weakness.

3. A thermosetting cement of 0.15 mm thickness should be sufficient to provide electrical insulation between the sections.

A cement that will preclude the formation of a smear of conducting particles from brush wear on the rubbing edge should be used; it should have a wear rate equal to that of the brush.

4. A brush with varying resistance characteristics from the leading edge to the trailing edge can be manufactured without the use of insulating separators and will act somewhat like a laminated brush.

After several hours in contact with a carbon or graphite brush, a copper commutator develops a layer of copper oxide mixed with carbon particles from brush wear. This copper-oxide film introduces unidirectional electrical properties (polarity effects) as in a copper-oxide rectifier. The oxide layer has a nonlinear resistance of higher value at the brush used as cathode than at the brush used as anode.

As a result, the cathode brush passes current in discontinuous high current density surges which cause EMI. Approximately 10 times as much interference may result from the cathode brush as from the anode brush. Plating the copper commutator with chromium to a thickness of about 0.025 mm will reduce the EMI level from a cathode brush to that of a relatively quiet anode. No adverse effects will result from the platings; the hard chromium surface prevents threading and grooving of the commutator. Wear rate and sliding friction of many brush materials on chromium are of the same order as those for copper.

Design features that improve commutation also reduce EMI. The most effective and economical technique is the installation of capacitors at the brushes. In generators, for example, installing capacitors (about 0.1 μF) at the brushes applies the remedy as close to the interference source as possible. The interference generated by the commutator and the brushes will be bypassed to the generator housing. The lead from the brush to the capacitor should be as short as possible, and the capacitor should be bonded to the generator housing to provide a low-impedance path to ground for the EMI currents.

Because of the combined interference-generating characteristics of the commutator and the brushes in a dc generator, an additional capacitor is installed at the output (armature) terminal. The preferred installation is a feed-through capacitor through the generator housing. The alternate installation is a 0.1 μF bypass capacitor mounted externally in order to maintain electrical contact with the generator housing and minimize the lead length between the terminal and the capacitor. Figure 6.37 illustrates the mounting of a bypass capacitor at the armature terminal. Since in vehicle/airborne applications the chassis is the power return, the shields as shown can only decouple to ground the electric field radiation from the hot wire. Magnetic-loop radiation can be reduced by running the hot wires as close as possible to the chassis.

Overall shielding is necessary to prevent the radiation of interference from within the generator. This shielding is afforded by the generator housing, which should be designed to provide maximum shielding effectiveness. Ventilation openings should be screened to prevent radiation of interference. No matter how perfectly a generator shield is designed, the shaft provides an exit path for interference because it must penetrate the shield. EMI should be bypassed directly to the generator housing by grounding the shaft through a brush riding on a special grounding slip ring (or riding directly on the shaft). This grounding will also eliminate bearing interference (bearing static or shaft current). Bearing interference

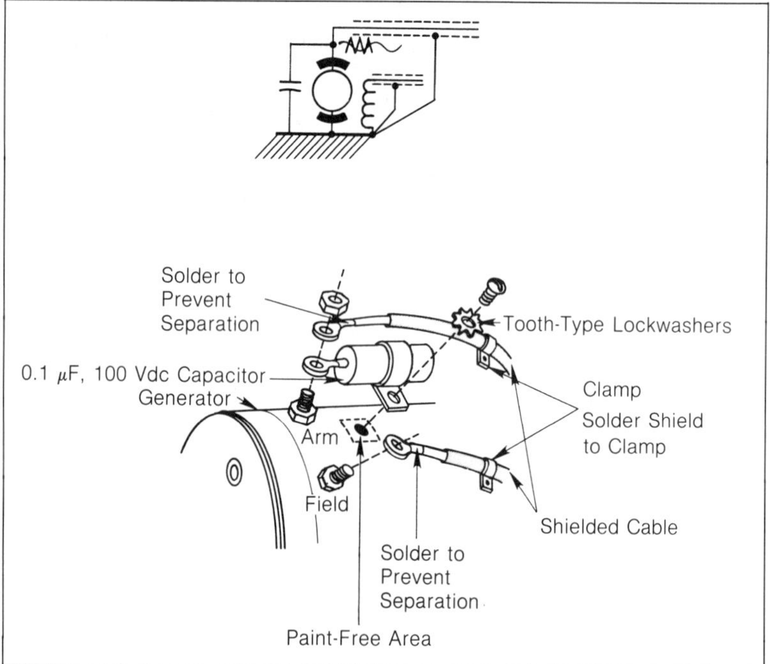

Figure 6.37—Installation of an EMI Bypass Capacitor on a Generator Armature Terminal

results from a periodic discharge of static electricity that takes place through the bearing and between the shaft and the housing. Eddy currents induced in the shaft and the housing by the flux lines in the motor can cause currents to flow through the bearing. These currents can also be caused by certain combinations of armature segments per pole, air gap and permeability inequalities, rotor eccentricities, insulation leakage or stray electric fields.

Another possible source of leakage from the generator shield is the inspection band. This band is in a detrimental place because of its closeness to the EMI-generating brushes and commutator; however, its function of permitting inspection of the brushes and commutator prevents its being moved to another location. To prevent leakage, the inspection band should be machined as closely as possible and should be wide enough to cover the inspection opening adequately with enough overlap to ensure good contact. The band should have machine screws spaced every 50 mm to permit secure tightening. Interference gasketing should be installed around the periphery of the opening. After removal of an inspection band, all contact surfaces on the band and the generator should be thoroughly cleaned before the band is put back into position.

The last shielding consideration for a generator housing is to ensure good contacts and low-impedance paths between the three sections of the generator: the two end plates and the main housing. This is accomplished by proper bonding and shielding practices (see Volumes 2 and 3 of this EMC handbook series). The design considerations for minimizing interference generated by brushes and commutation action in dc generators also apply to dc motors. Capacitors installed to the brushes bypass the generated interference to ground close to the source, providing an effective and economical means of suppression.

On some dc motors, an adjustable speed control is included in which the field leads are connected to an externally mounted rheostat. This arrangement necessitates breaking the shield continuity and therefore enabling interference generated inside the motor to be conducted out of the housing. Capacitors installed inside the motor housing connected to these leads, however, will bypass such interference to ground. Figure 6.38 shows a motor with four installed capacitors: one each for the two brushes, and one each for the two field leads. The feed-through capacitor should be mounted at the positive lead. A less acceptable interference-reduction technique for the same motor uses bypass capacitors at the brushes.

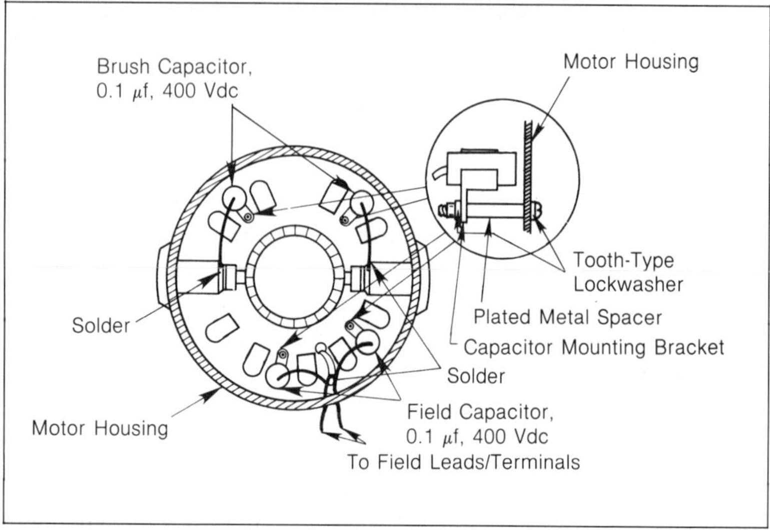

Figure 6.38—Capacitor Installation at Brushes and Field Leads in a dc Motor

6.6.3 Alternators and Synchronous Motors

Alternators and synchronous motors are very similar to dc generators and motors except that they supply or use ac and therefore have slip rings instead of commutators. Commutator interference is absent in these machines. There is, however, EMI from the brushes and from the generation of harmonics. Brush interference is lessened because most alternators and synchronous motors have the power windings in the stator and rotating excitation poles; heavy power currents need not be supplied to the rotor. Only the much smaller field currents have to be supplied through the brushes. Because commutation is not involved in the selection of brushes, a much wider choice in brush pressure, size and material is permitted.

In ac generators, the generation of harmonics and the resonant conditions that create interference can be minimized. Production of as pure a sine wave as possible (an important consideration in the design of alternators) is especially important when EMI reduction techniques are considered. A comparatively small harmonic content may be quite tolerable from all points of view except that of EMI.

In the reduction of harmonics, special attention must be given to:

1. **Flux distribution**. The most important factor determining the waveform of the generated voltage is the distribution of the magnetic flux around the periphery of the armature. Sinusoidal distribution, which produces the least amount of interference, may be achieved by chamfering the pole tips or skewing the pole faces.

2. **Symmetry.** In a perfectly symmetrical machine, all even harmonics disappear; therefore, special care must be exercised to construct identical pole pieces to make the yoke and armature perfectly symmetrical, to produce a uniform winding on the armature and to avoid all other irregularities.

3. **External connections.** In a three-phase alternator, the third harmonic and its multiples disappear at the terminals except when the machine is wye connected and has its neutral grounded. In this case, third harmonics are present in the voltage between any phase and neutral. This connection should be avoided or, if it is used, special attention should be given to the prevention of the third harmonic and its multiples.

4. **Distribution factor**. This should be chosen to eliminate the lowest harmonic not eliminated by any of the features mentioned in (2) or (3).

5. **Tooth ripples**. Their generation is greatly decreased by skewing through one slot pitch either the pole shoes or the armature slots. Tooth ripples may be eliminated altogether by making the number of armature slots per pole pair an odd number. The chord factors for the harmonics that are contained in the tooth ripples are then reduced to zero. Slip ring and brush materials should be such that interference is minimized. The design considerations applied to brushes and commutator surface materials in dc machines apply equally well here. The effects of brush bounce caused by vibration or irregularities of armature motion can be minimized by the use of two or more brushes per slip ring.

In addition to the interference generated as a result of the brush action on the slip rings and the harmonics present in the sine-wave output of an alternator, the exciter is a source of EMI. Because both the exciter (essentially a dc generator) and the ac generator are installed in a single housing, shielding considerations become a combination problem. Plating of the commutator, the use of proper brushes and brush pressure and the application of bypass capacitors are applicable to the exciter; the other design measures can be applied to the exciter as a separate unit. Although individually designed for interference reduction, the alternator and exciter each generate some interference. This residual interference is reduced by shielding and the use of bypass capacitors installed at terminal outlets.

Shielding of the alternator is incorporated in the design of its housing. Low-impedance paths between sections of the housing, provisions for bonding, and screening of all ventilating louvres must be carried out if the overall interference-reduction design is to be effective. As in dc generators, no matter how perfect the shield, a means of escape from the shield for the interference currents is provided by the alternator shaft which penetrates the shield. The same procedures for shielding dc generators therefore apply to alternators.

The alternator terminal outlets provide another means of leakage. They are prevented from radiating interference by the installation of capacitors. Bypass capacitors are installed inside the terminal strip and are connected to the terminal outlet just before the terminal breaks the shield. This arrangement removes EMI from the lead at the last possible point, preventing interference from coupling back into the lead and radiating from the terminals or from their connected wiring. Another type of installation is to mount feed-through capacitors through the terminal strip.

The problems of interference suppression for alternators also apply to synchronous motors since they have the same basic components as alternators. A synchronous motor will operate as an alternator and vice versa. An induction motor should be used instead of a synchronous motor whenever possible because of the lower EMI generated by induction motors.

The primary source of interference within a single-phase induction motor is the starting device. The starting winding is in series with a switch (or capacitor and switch) that is closed when the power is off. When the motor reaches approximately 80 percent of its rated speed, the switch is opened (either by centrifugal force or by a solenoid coil) and a single pulse of interference is generated. This switch should be placed in a shielded housing and the leads leaving the housing should be filtered.

6.6.4 Portable Fractional-Horsepower Machines

Portable fractional-horsepower machines include such equipment as portable electric drills and saws, certain household appliances, etc. Power is furnished by high-speed, lightweight, ac/dc or ac electric motors. Such equipment, using ac/dc motors (universal motors), is a major source of EMI because commutation is essential for its operation.

As in dc motors, an effective, economical method of designing for reduced commutator-brush interference is by installing capacitors at the brushes. In some portable ac/dc machines, restrictions of size and shape prevent the installation of capacitors at the brushes, and it is more feasible and economical to mount the capacitors in other parts of the equipment.

Installing the capacitors at the line side of the switch bypasses interference to the unit housing at the last point of exit to the power lines and prevents the interference from coupling back into an interference-free lead and being conducted by the power lines. If the filtered motor belongs to a 115 or 230 Vac equipment, then filtering capacitors must meet the safety aspects described in Section 2.2 of this book. If the mechanical design of the unit prevents the installation of capacitors on the line side of the switch, install them on the motor side. Shielding may be used to ensure that no interference couples back into the leads before they leave the unit.

6.6.5 Brushless Motors

In recent years a new concept in dc motors has been introduced, viz., brushless dc motors featuring solid-state commutation and the associated reduction of conducted and radiated EMI. The technique involves sensing the exact position of the rotor in relation to the stator by using two orthogonally located sensors, generally Hall-effect transducers.* If the rotor is constructed of a bipolar cylindrical magnet, the induced output of the sensors will be in phase quadrature as the rotor rotates.

---

*The Hall generator is a four-terminal device with dc input and an output proportional to the strength and direction of intercepted magnetic flux. The Hall generator output is also affected by the strength and direction of the current flow at its input.

Figure 6.39 shows a block diagram of the brushless dc motor. Figure 6.40 is a cut-away depiction. The dc source drives the Hall generators (HGs) through the feedback network and supplies power to the amplifiers. The sensed rotor position signals from the sin and cos HGs are each amplified by a push-pull amplifier creating four sine waves in quadrature which drive the coil windings of the stator. The amplifier outputs are rectified and applied as back electromotive forces to the feedback network which sets the value of the reference voltage applied to the HGs through another set

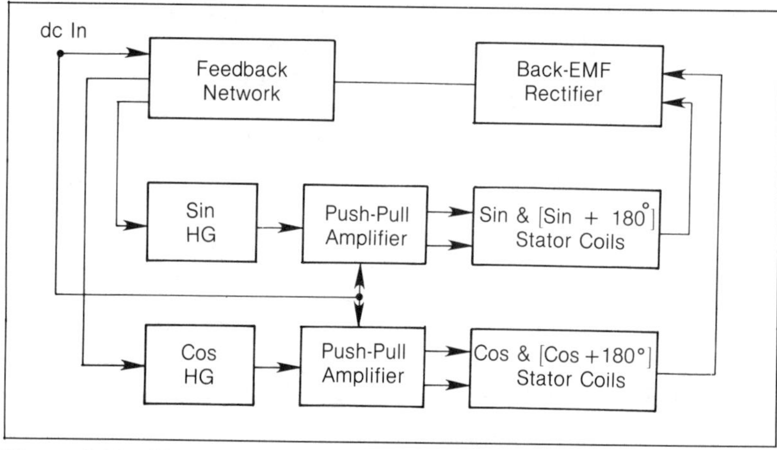

Figure 6.39—Block Diagram of the Brushless DC Motor

Figure 6.40—Small Brushless DC Motor with Integrated Electronic Drive (Courtesy of PAPST MOTOREN)

of amplifiers to close the loop. Load, temperature, speed and voltage compensation are automatically achieved in the process.

Presently limited in size, the brushless dc motor should prove to have a great future as a fractional dc motor where EMI control, long life (10,000 hours without maintenance, i.e., brush replacement) and operational parameter compensation are required.

6.6.6 Special-Purpose Machines

Special-purpose rotating machinery includes rotary inverters, dynamotors, motor generators and generators for electric arc-welding equipment. The function of conversion is common to most of this equipment: ac is converted to dc or higher frequency ac, or dc is converted to higher or lower voltage dc, or to ac.

A rotary inverter, which converts dc to ac, is basically a dc motor with added taps on the armature winding; slip rings are connected to these taps to provide the ac output. Interference is generated by both the ac and dc functions, commutator and brush action in the motor, and by brush action and harmonics in the alternator.

Figure 6.41 illustrates an interference reduction design technique for an inverter. The schematic diagram shows two feed-through capacitors bypassing EMI from the output leads of the alternator. The dc lead is shielded from the feed-through capacitor on the dc line. A capacitor shield is installed to prevent radiation from the terminal on the hot side of the capacitor. This shield also provides

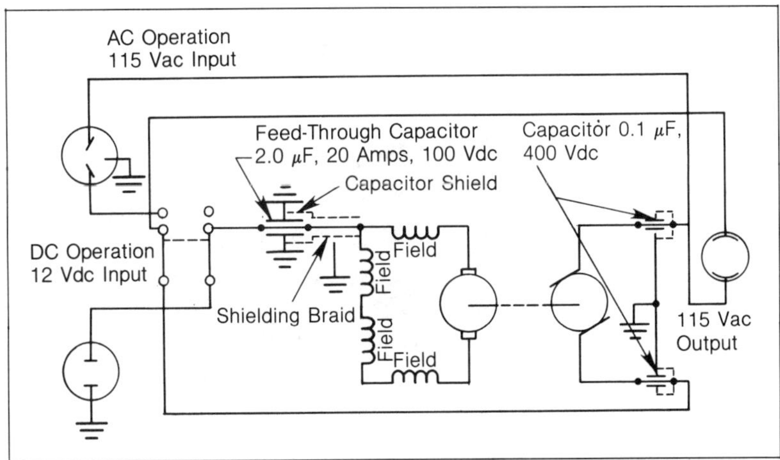

Figure 6.41—Rotary Inverter Showing Methods for EMI Suppression

a ground for the braid shielding. The ac output leads do not require shielding because the EMI generated by the alternator is much less severe than that generated by the dc motor. Bypass capacitors connected to the brushes in both the motor and the alternator should be included in the original design. The housing must adequately shield the unit with a feed-through capacitor mounted through the shield for connection to the dc input lead. The ac leads may not require suppression in addition to that provided by the capacitors at the brushes.

A dynamotor (a combination dc motor and generator with a single magnetic field) has an armature with two separate windings and two separate commutators, one at each end of the armature. It transforms low-voltage dc to high-voltage dc, or vice versa. The two commutators make this machine a major source of EMI. The suppression techniques for dc generators and motors apply to the dynamotor. Figure 6.42 illustrates a dynamotor with feed-through capacitors bypassing EMI to the housing on both the input and output leads. Complete shielding of the dynamotor prevents EMI from coupling through other paths.

The use of ac commutator motors should be avoided whenever possible. Universal motors come under this category, as well as repulsion motors and series ac motors. The performance advantage of these types is their high starting torque. Their EMI generation, however, is much more severe than that from other types of ac motors.

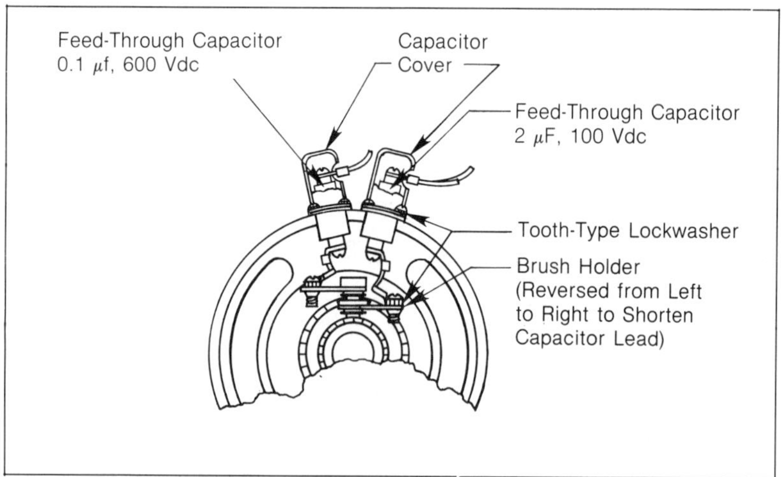

Figure 6.42—Dynamotor Showing Methods for EMI Suppression

High starting torque with ac motors can be obtained without increasing EMI by using capacitor-type starting, induction-run motors. These motors use a high capacitance for starting purposes only. Starting torques of 200 to 350 percent of full load torque are feasible with acceptable starting currents. Ratings from 1/8 to 10 hp are available.

Generators for electric-arc equipment require special attention only when connected to such a severe source of EMI as the electric arc. The generator can be driven either by an ac or dc motor to an engine. Little can be done to reduce EMI from the electric arc itself. The equipment should be located away from communication equipment, in buildings with good shielding characteristics. The leads, from the generator to the welding electrodes, can become very effective EMI radiators and should be adequately shielded.

6.6.7 Servo and Stepper Motors

Servo and stepper motors are used when a motion with precise speed, number of turns or shaft positioning is necessary, as with disk drives, robots, automatic machine tools (automated manufacturing systems), lifts and cranes, radar platform pointing and stabilization, artillery, etc. Their sizes range from miniature (a few watts) to motors in excess of a megawatt. More and more of these motors are driven by pulse mode, which allows for better efficiency since, for low duty cycles, a high torque can be generated with a large current even with a motor of modest continuous power rating. For instance, a motor which looks sized for 1 kW of continuous power can handle 300 V/30 A of pulsed power.

Pulse mode also allows for easy interface with digital command. The main EMI problem with servo drives and stepper motors is the H-field and power line transient created by the pulsed nature of their excitation (for large motors with wound rotor) or their powering (for small motors with permanent magnet rotor). The driving transistors usually have fast rise and fall times which can cause excessive power-line transients, even though the power wiring has been conservatively sized for the average current (but not for the actual dI/dt). The transistors also cause large magnetic fields near the motor and its supply wires.

This in turn creates EMI in nearby sensitive electronics, especially the tachometer or other sensor outputs which control the servo itself (see Section 6.3.4 position sensors). For instance, with the robot shown in Fig. 6.43, each one of the six motors needs an accurate, undisturbed position coding plus (generally) an acceleration/deceleration control to avoid position overshoot. Sometimes it

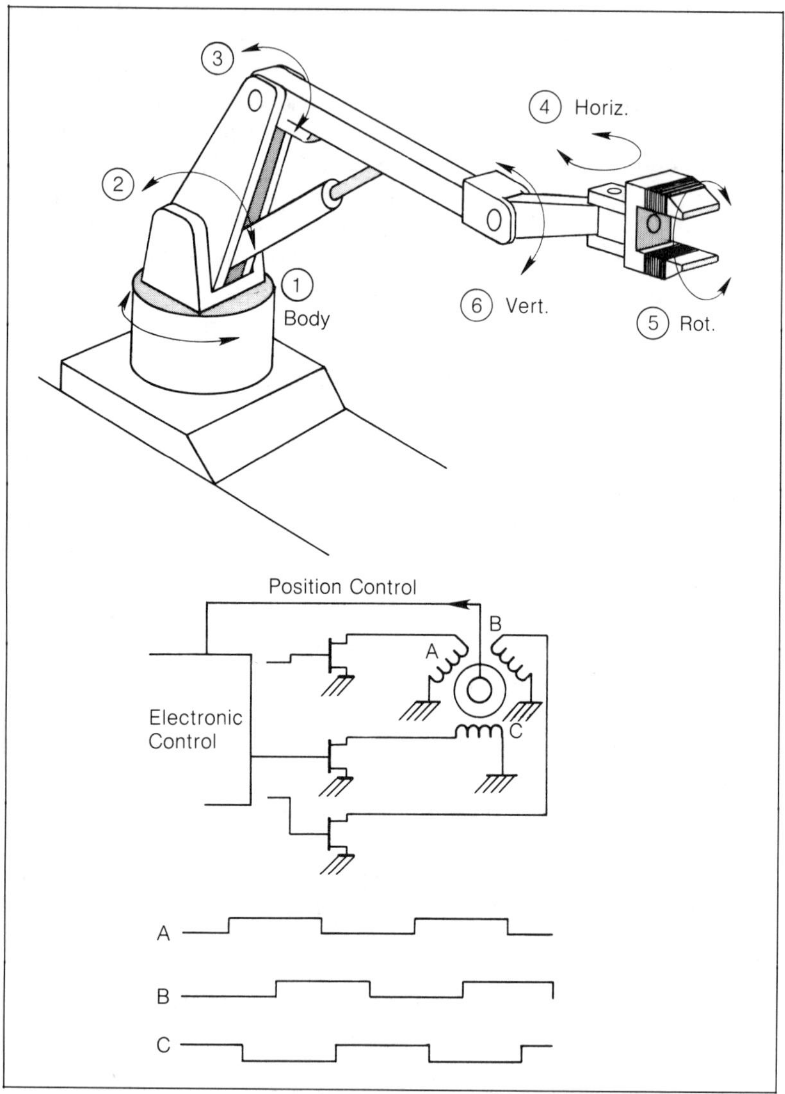

Figure 6.43—Typical Robot Equipped with Six Brushless Motors. Each motor needs a position feedback. This position feedback can be vulnerable to the large current pulses of phases A, B, C drives.

needs a "tactile" control to prevent the robot from damaging the object it is grasping.

The potential EMI problems in such motor applications are of two kinds: self-jamming and ambient noise. Self-jamming can occur because of the high current pulses (dI/dt can exceed 30 A/μs with fast power FET transistors) and high dV/dt which can both induce noise in the sensing circuits and feed-back cables. To make the servo motor immune to its own transients requires that the position monitoring be made unsusceptible to crosstalk, power-line transients and ground shift. Therefore, analog position sensors should be avoided and digital coding is preferred. For instance, an optical encoder with eight binary-coded sectors on a disk will give a resolution of 256 positions for one revolution of the shaft and benefit the better noise immunity of digital interface.

Immunity to ambient noise implies packaging, filtering and shielding precautions due to the harsh environment where robots are installed (welding, machine tools, steel mills, etc.) For instance, International Standard IEC-801 applicable to process control electronics, stipulates that such equipment should resist without malfunction the following ambients:
1. Radiated EMI = 10 V/m from 27 to 500 MHz
2. Power line induced transients = 2 kV/50 ns pulses and repetitive bursts
3. Signal line induced transients = 1 kV/50 ns pulses and repetitive bursts

These simulated ambients must not cause any alteration in the robot operation such as:
1. Single arm positioning error exceeding the normal tolerance
2. Cumulative arm positioning errors whose individual increments are tolerable but which add up after certain duration of the EMI transient injection
3. Sudden erring trips of the arm, which are potential threats to nearby workers

6.6.8 Motors and Generators Used in Vehicles

Motors and dc or ac alternators used in cars, trucks, light aircraft, recreational vehicles and boats pose specific EMI problems. For economic reasons they usually are not shielded, or at best they are poorly shielded and filtered. The regulator action is usually the "all-or-nothing" type. Transient disconnection of a battery terminal creates a "load-dump effect," and at engine turnoff the instant decay of the excitation field as well as instant stopping of the rotor (since the alternator is belt driven) creates a large dϕ/dt in the stator.

The onboard electronics have to function and survive with these disturbances on the battery bus. Figure 6.44 gives some examples

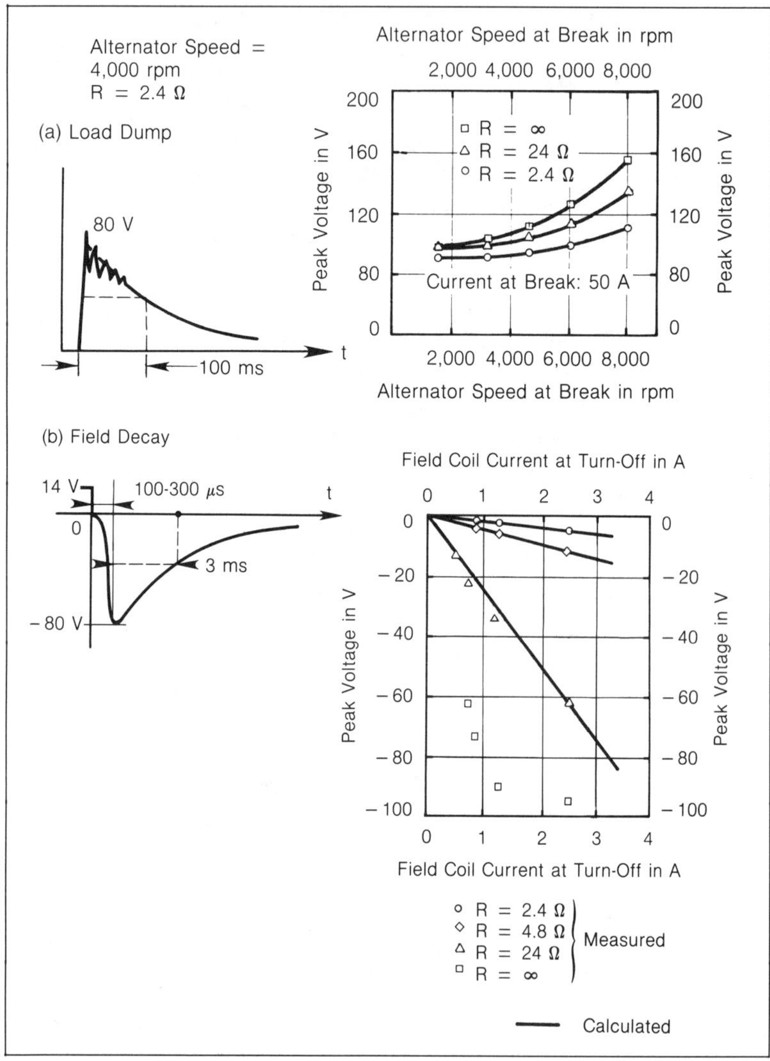

Figure 6.44—Values of Generator Transient on a 12 V Vehicle Battery Bus

of the "load-dump" overvoltage when a fully loaded alternator sees its load suddenly released (the resulting overvoltage is due, in part, to the reaction time of the regulator) and also the field decay transient on 12 V vehicle wiring.

The field decay overvoltage seen across a resistor R is equal to:

$$v_R = - \frac{V_B}{R_f} \times R \times e^{-\frac{(R_f + R)t}{L_f}} \qquad (6.19)$$

where,

$$V_B = \text{battery voltage,}$$
$$R_f = \text{resistance of field coil}$$
$$L_f = \text{inductance of field coil}$$

The peak value of v_R in the preceding equation is given at $t \cong 0$ by :

$$v_{R\ max} = - \frac{V_B}{R_f} \times R \qquad (6.20)$$

Figure 6.45 shows some examples of unsuppressed and suppressed motors and electromechanical sources in automobiles.

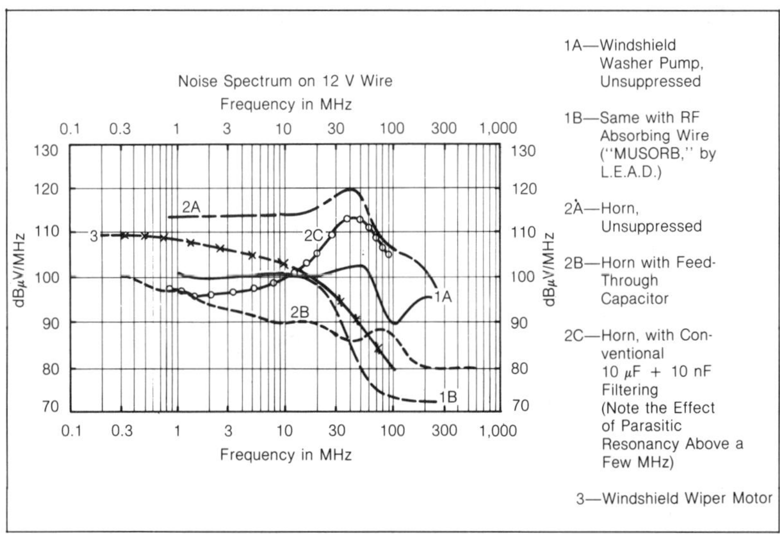

Figure 6.45—Examples of Noise Suppression in Automobiles: Noise Spectrum on 12 V Wire

6.7 References

1. Ott, H.W., *Noise Reduction Techniques,* (New York: John Wiley & Sons, 1976.)
2. Howell, K., "How Switches Produce Noise," (*IEEE Transactions on EMC,* August, 1979).
3. Mardiguian, M., *Electrostatic Discharge,* (Gainesville, Virginia: Interference Control Technologies, Inc., 1986).
4. Russell, B.D., "Switching Transients in Power Substations," (*EMC Technology,* July, 1983).
5. Notice No. K36, "VHF/UHF Relays," Omron Electronics, Inc.
6. Messer, L., "Retrofit of Military Aircrafts with New Electronics," *EMC Technology,* July, 1983.
7. Morrison, R., *Instrumentation Fundamentals and Applications,* (New York: John Wiley & Sons, 1984).
8. Morrison, R., *Grounding and Shielding Techniques in Instrumentation,* (New York: John Wiley & Sons, 1977).
9. Department of Defense, DoD Specification E-83578, "Explosive Ordnance for Space Vehicles," 1979.
10. Clark, "Interference Suppression of Fluorescent Lamps," U.S. Naval Civil Engineering Lab Test Report 166, October, 1961.
11. *Proceedings of the 1982 Lighting EMC Conference,* Berkeley, California.
12. Herman, J., *Electromagnetic Ambients and Man-Made Noise,* Chapter 3, (Gainesville, Virginia: Interference Control Technologies, Inc., 1979).

6.8 Bibliography

1. Freyman, D., (MBB), "Telemetry for EMC Tests of EEDs."
2. *Transient Voltage Suppression Manual,* (General Electric Company, fourth edition, 1983).
3. Melear, C., "Robots and Microprocessors Mix Well," (*Electronic Products,* July, 1985).

Chapter 7

Surge Suppressors

In terms of modern technology, there are three types of surge suppressors, or transient protection devices. They are the **semiconductor (variable resistor)**, the **gas discharge** and the **"crowbar."** All three types can be used against surges on power wiring, signal wiring, line-to-line or line-to-ground. However, their response time, energy handling capabilities and ON/OFF characteristics are different. They are all especially useful against high-energy transients, i.e., transients whose duration would prevent simple capacitive or inductive filtering. An EMI filter with 10 kHz cutoff frequency, for instance, is completely useless against a spike whose duration exceeds 30 μs.

Finally, there is the question of whether both types of devices should be mounted in DM or CM protection. A differential-mode (line-to-line) transient suppressor will offer no attenuation to common-mode transients, and vice versa. Total protection should include both DM and CM suppressors. Since they will be mounted on a high-energy input, they should be checked for safety compliance; however, a detailed discussion of that particular topic is beyond the scope of this handbook.

7.1 Avalanche and Varistor Devices

The first type of surge suppressors, the **semiconductor variable resistor**, behaves electrically like a zener diode, with an V-I curve like that in Fig. 7.1.

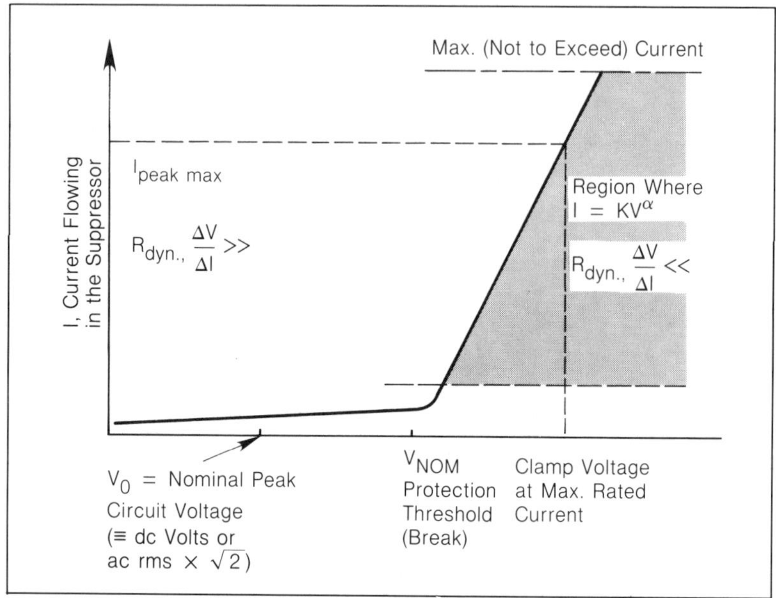

Figure 7.1—Basic Operation of Semiconductor/Variable Resistor Type of Suppressor

The α term on the curve corresponds to the rise of V-I characteristic beyond breakdown, i.e.:

$$I = K V^\alpha \text{ or } \alpha = \frac{\log I_2/I_1}{\log V_2/V_2} \tag{7.1}$$

The power-handling capability is limited by the Joule dissipation, i.e., $V_{clamp} \times I_{pulse} \times$ pulse duration because a clamp voltage is always present during the pulse current flow. The response time is usually very fast (< 100 ns). Therefore, the selection must be done according to the following steps:

1. Determine the maximum steady-state (high-line condition voltage of power mains in rms, peak ac or dc).
2. Determine the maximum transient voltage the victim circuit can withstand without damage.
3. Select the next higher rating (V_{nom}) of available surge suppressor.

4. Determine the maximum expected pulse current. Pulse current can be found in environmental statistics or from the surge withstanding specification. These tests are usually made with low-impedance simulators; if impedance is unknown, a default value of 50 Ω can be assumed.
5. Find the clamp voltage at maximum expected pulse current.
6. Check that (5) is less than (2).
7. Check that (4) does not exceed current-handling capabilities of the suppressor for the given pulse duration.
8. Compute the actual attenuation in dB if necessary.

Illustrative Example: An 8×20 μs surge is expected from lightning induction on a 120 V power line. The open circuit voltage is 2,000 V. **Mains and victim** impedance represent 20 Ω in this frequency domain. The victim input components (filter, power supply, etc.) cannot withstand more than 400 V without damage.

A metal-oxide suppressor is selected using nominal power line voltage + 10 percent for high line condition:

$$(120 \text{ V} + 10\%) \sqrt{2} = 185 \text{ V} \tag{7.2}$$

The next available varistor has a nominal V_{break} of 200 V. The selected model can handle 0.5 J, and the maximum non-repetitive peak current is 250 A.

Since the varistor will behave almost like a short above the break (very low dynamic resistance shown in Fig. 7.1), a coarse approximation of the surge current in the suppressor is:

$$I_{max} = \frac{2,000}{20 \text{ } \Omega} = 100 \text{ A} \tag{7.3}$$

The clamp voltage for 100 A per manufacturer's curve is 360 V, which is below our 400 V criteria.

A more accurate approach is to use a graphical method like Fig. 7.2. The load V-I curve is plotted against the suppressor V-I characteristics. The intersection determines the exact voltage and current across the device, i.e., 82 A in our example.

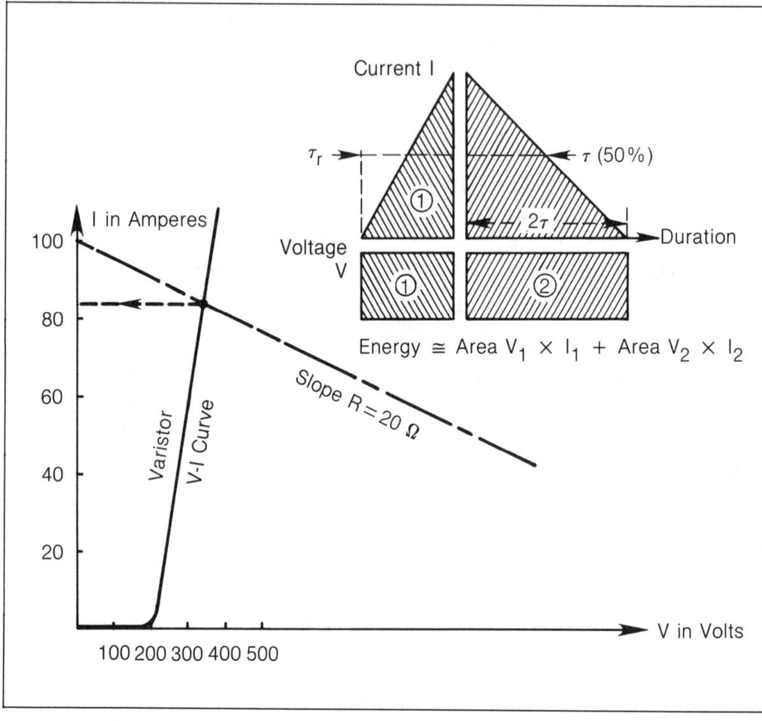

Figure 7.2—Graphical Method for Checking Varistor Operating Point and Piecewise Method to Calculate Varistor Dissipated Energy

The next check is to verify that I is below the I_{max} of the device. Since 100 A is below the 250 A maximum, this condition is met. Finally, the maximum energy has to be verified. An exact calculation would imply to solve for

$$E_{joules} = \int_0^T V_{(t)}I_{(t)}dt$$

with V and I mutually dependent. A coarse calculation is to assimilate the double exponential 8 μs/20 μs pulse to an addition of two triangles and to calculate the corresponding energy in the device:

$$E = 1/2 \times 360 \text{ V } (8.10^{-6} \text{ s} \times 82 \text{ A}) + 1/2 \times 360 \text{ V}$$
$$(2 \times 20.10^{-6} \text{ s} \times 82 \text{ A})$$
$$= 708 \text{ mJ} \tag{7.4}$$

This is slightly above the 0.5 J capability. So, to avoid the risk of stressing the device, a varistor with same characteristics but a higher energy capability should be selected. Figure 7.3 gives the shape of the pulse after suppressor insertion. The transient attenuation is then:

$$\text{V victim without suppressor} = 2{,}000\ \text{V} \times \frac{20\ \Omega}{20\ \Omega\ +\ 20\ \Omega} = 1{,}000\ \text{V}$$

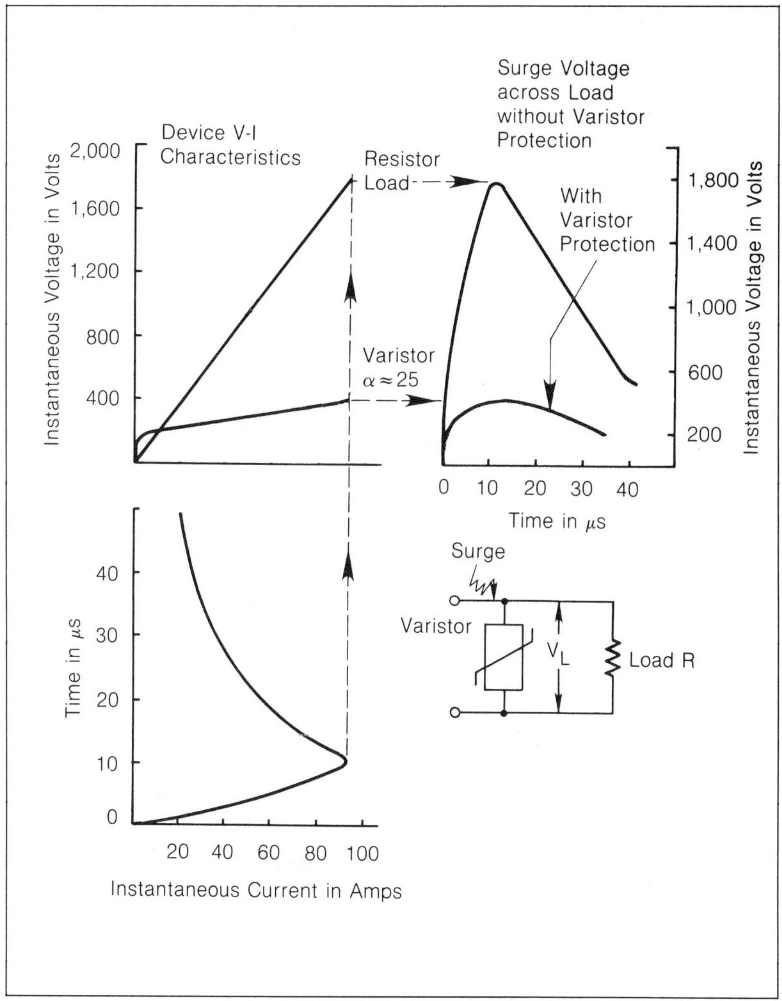

Figure 7.3—Comparison of Surge Voltage with and without Varistor

$$\text{V victim with suppressor} = 360 \text{ V}$$

$$\text{Attenuation} = 20 \log \frac{2{,}000}{360} = 15 \text{ dB}$$

Figure 7.4 shows another example of semiconductor transient attenuation for a sharp, short-duration transient and its effect in the frequency domain. Note that when a transient suppressor is mounted across an ac line, the actual threshold of clamping can vary with the phase position of the transient. For instance, with a 250 V varistor, the actual clamping voltage may vary by $\cong 100$ V depending on the pulse position (Fig. 7.5).

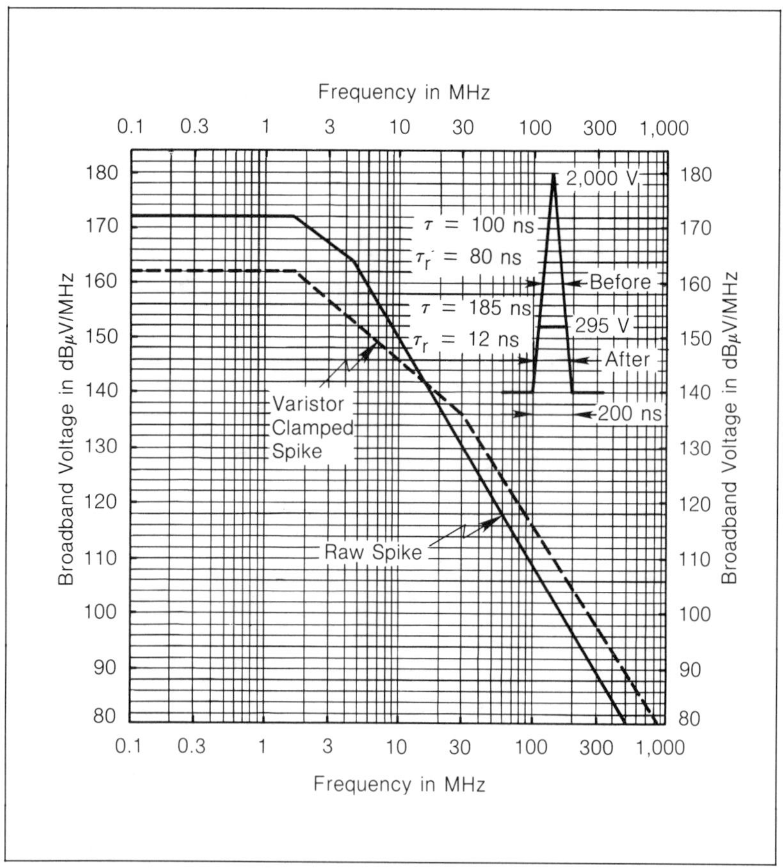

Figure 7.4—Spike Attenuation by Transient Protective Device

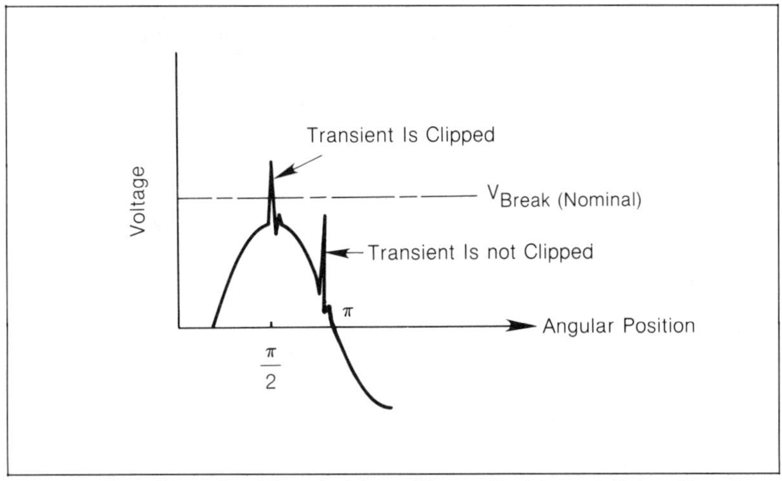

Figure 7.5—Clamping Threshold vs. Phase Position of Pulse Causing Threshold Inaccuracy

Depending on the application (power, signal, data processing, etc.) and the EMI threat (lightning, line transient, ESD, NEMP), different varieties of this first type of surge suppressor are used. Two examples of the first type of surge suppressors are metal-oxide varistors and special zener diodes.

The advantages and limitations of these suppressors are summarized in Table 7.1. Besides their limited power-handling capabilities, the other two limiting factors to watch when selecting varistors or zener arrestors are parasitic capacitance and lifetime.

Parasitic capacitance of such devices is usually large, in the 100 pF to 3,000 pF range. For power-line or telephone-line protection, this is generally not a problem. But a high-speed data link may not tolerate such capacitive loading.

If one considers a suppressor with 1,000 pF parasitic capacitance to be put across a line whose the two terminations in parallel represent 150 Ω, the bandwidth will be limited to about 1 MHz (3 dB bandwidth). This can be overcome to some extent by putting low-capacitance diodes (forward-biased for the expected pulse) in series with the suppressor. But this adds to the component count and the failure rate.

Table 7.1—Protective Device Comparison

Device	Carbon Block	Varistor	Zener (TransZorb)®	Clamping Diode	Gas Tubes
Range of Voltages	100-1,000	12-4700	3-200	5-400	90-1,000
Voltage Spec Increments	20%	10-25%	5-10%	10%	20%
Tolerance	20%	10%	5%	5%	15%
*Clamp Ratio @ 10A	Erratic	1.85 (13J)	1.65 No Spec	1.25 (15J)	2.55
Capacitance (200 V Device)	1 pF	100-300 pF	30 pF	30 pF	1.2 pF
Response Time	> 1 μs	10-1,000 ns depending on size	a few ns	< 100 ns	$\cong$ 1 μs
Price	Low	Medium to Low	Low	Medium	Medium
Steady State Power Dissipation Capability	High	Very low (1/2 W for 10 J Device)	Good to 50 W	Good to 10 W	Not Applicable
Failure Mode	Short or Gradual Shift	Gradual Shift in Breakdown	Open or Short Circuit	Short Circuit	Short
Other Properties and Notes	No Follow on Current Problem	Good for Mains Use		Ideal for SMPS Switch	After Ignition Enters Arc Mode with Low Voltage Drop- Requires Zero Current to Extinguish

*Defined as V_{clamp}/V_{break} measured for a particular unit.

Lifetime is specified by manufacturers as a function of the number of operations at a given current. For example, Fig. 7.6 shows a typical varistor pulse lifetime rating curve for several General Electric varistors. The end of lifetime is defined as a degradation failure which occurs when the device exhibits a shift in the varistor voltage (at 1 mA) in excess of ±10 percent of the initial value. This type of failure does not prevent the device from functioning, but the varistor will no longer meet the original specifications. For the type shown, the rated peak pulse capability (no-fail) is defined for a single 4,000 A, 8 μs/20 μs surge.

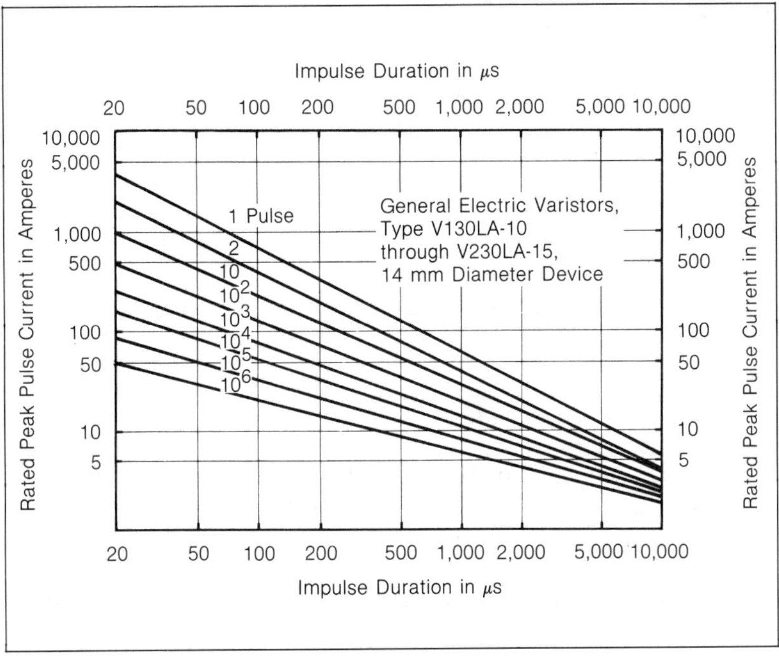

Figure 7.6—Typical Varistor Pulse Lifetime Rating Curve (Courtesy of GE)

Like any component, suppressors that frequently operate close to or above their maximum rating will fail more rapidly. The difficulty is that with an ordinary component, the current is generally a design parameter. With surges, the current obeys a probability distribution. Therefore, arrestors should always be conservatively rated. Depending on the application, arrestors can be specified to die in a short-circuit or in open-circuit mode when they fail. The first mode is generally preferred since the user is immediately warned that his protective device is not protective any more; in this case, the shorted line or a blown fuse will alert him. Fuses should never be mounted in series with the suppressor across the line (See Figs. 7.7a and 7.7b).

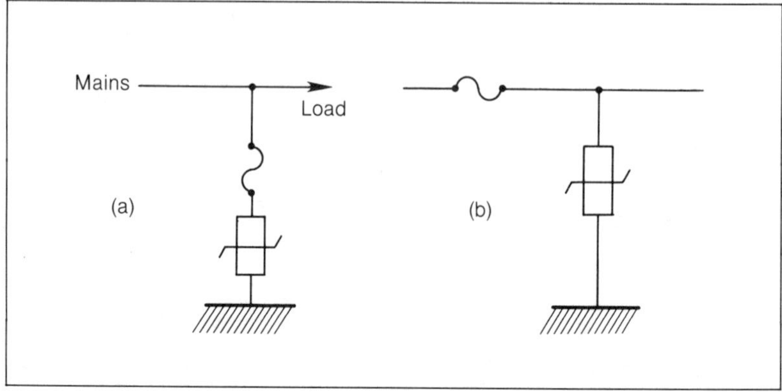

Figure 7.7—Addition of a Fuse to Protect the Surge Arrestor from Overstress. Example (a) is a wrong mounting since once the fuse is blown, the user is left unaware that he is no longer protected (unless an alarm circuit is provided with the fuse). Example (b) is preferable.

7.2 Gas-Discharge Devices

The **gas-discharge** surge suppressor behaves like neon tubes or thyristor devices in the breakdown region as shown in Fig. 7.8. After firing, the voltage drop of this type of suppressor is only a few volts and its power-handling capability is one order of magnitude greater than that for the semiconductor devices. Conversely, the gas-discharge response time is slower, and the first portion of the sharpest spikes can pass through the gas tube before it reacts. The selection steps are as follows:

1. Determine maximum steady voltage of the power mains.
2. Select the next highest rating (V nominal) of the available gas tube.
3. Determine the peak voltage and rise time of the transient (if unknown use 10 kV/μs.)
4. Find the threshold of the gas tube for the applied slope using the manufacturer's curve or default values shown in Fig. 7.9.
5. Be sure that the holdover current will not maintain gap arcing after the surge. If arcing continues, a power input fuse should also be selected to clear this follow-on current.

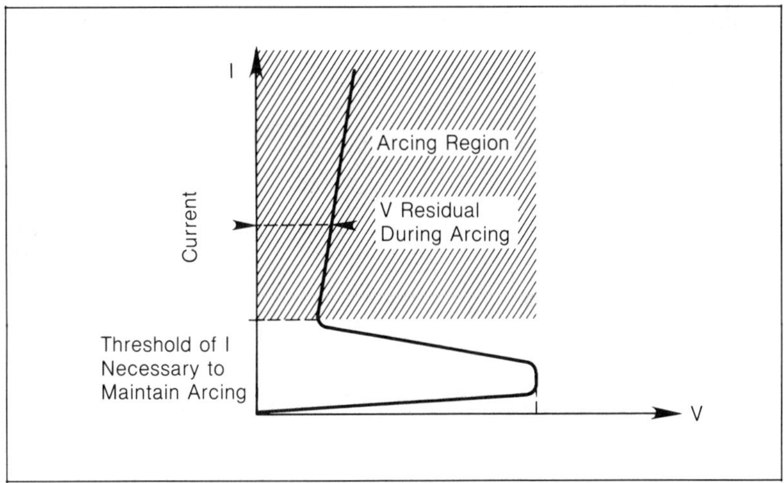

Figure 7.8—Illustration of Gas-Tube Transient Protector

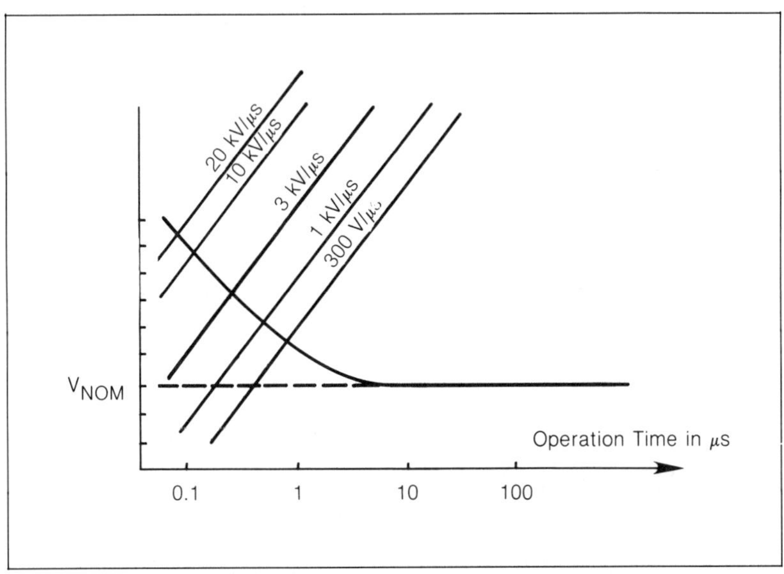

Figure 7.9—Default Values for Gas-Tube Threshold of Operation vs. Pulse Rise Speed (Gas Tubes between 100 V and 600 V dc Break-Over)

Illustrative Example: A 350 Vdc gas tube is mounted to protect a 220 Vac line. What protection is offered against a 3 kV/1 μs lightning induced surge? Since the slope is 3 kV/μs, the curve in Fig. 7.9 shows that the tube will react in $\cong 300$ ns, i.e., the first 1 kV front of the pulse will be allowed to pass. Then the tube will glow and the remaining pulse will be shorted. In this case, the maximum energy content associated with the pulse trailing edge will not enter into the victim (see Fig. 7.10), but it should be checked that the next component can resist to the 1 kV/300 ns peak that the tube has let pass.

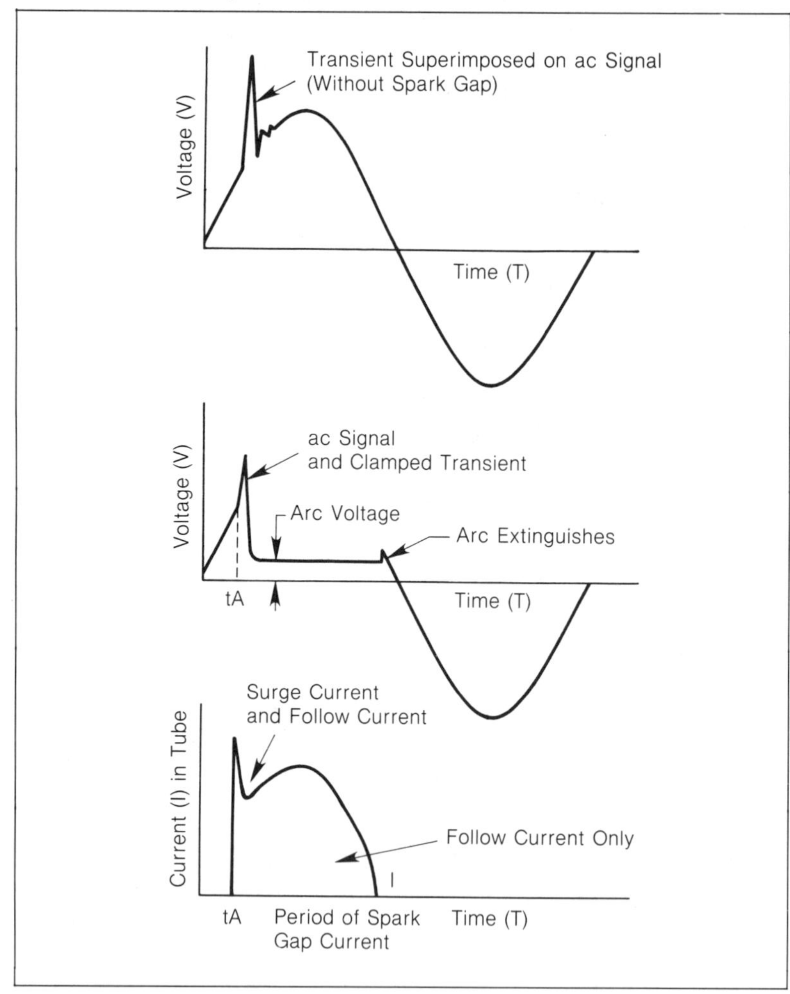

Figure 7.10—Illustrative Example of Gas Tube Attenuation

The gas tube has qualities and defects exactly reciprocal to those of avalanche-type suppressors:

1. It shows very small parasitic capacitance (1 to 3 pF).
2. There is practically no limitation in power handling, the only limit being heating of the leads and envelope.
3. It offers a slow reaction time.
4. A backward transient is generated when the tube arcs. The arcing is an abrupt dI/dT which can create problems in the very circuits the tube was intended to protect.
5. The line is shorted during tube arcing, therefore all intentional power or signal will disappear as well.

Sometimes, a hybrid assembly of gas tubes **avalanche suppressors** is used to combine their respective advantages (Fig. 7.11).

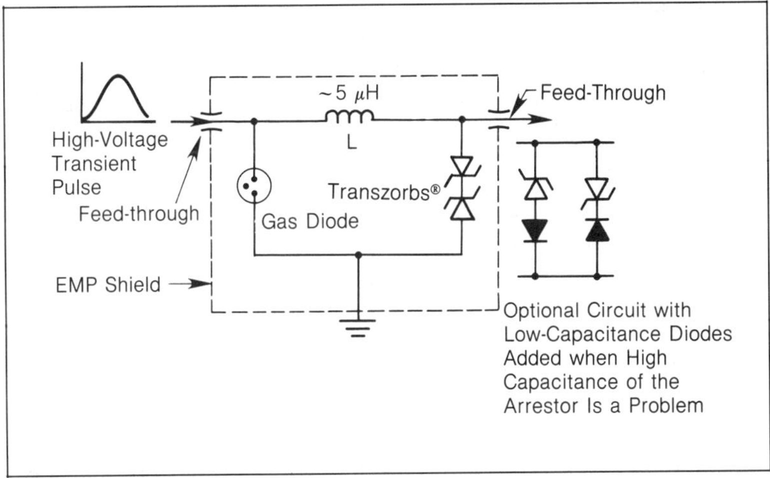

Figure 7.11—Hybrid Transient Protection Device (Typical)

7.3 Crowbar* Devices

A third variety of transient suppressor is called the "crowbar" type. Initially used as overcurrent protection for regulated power supplies, the concept has been extended to surge protectors. In these components (Fig. 7.12) a first detection is achieved by an avalanche device (zener or varistor) which triggers an SCR or any faster semiconductor switch. The response time of a crowbar device is usually very short since the zener can be integrated in the same chip as the SCR, whose characteristics can be optimized for high speed.

---

*The term "crowbar" is sometimes used also for gas tubes.

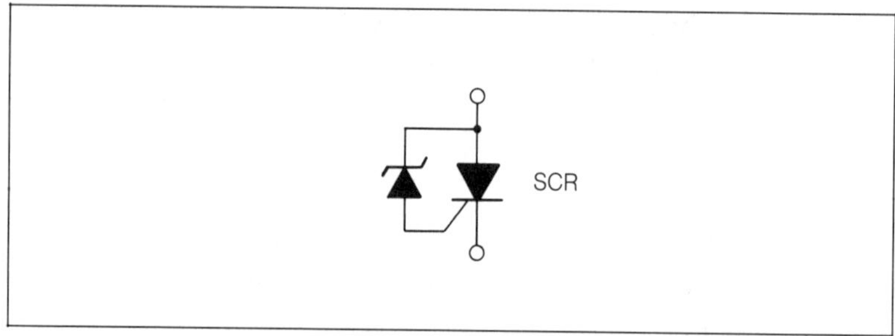

Figure 7.12—Typical Crowbar Device with a Combination of Fast Avalanche Clamp Plus an SCR

The main limitation of the crowbar is the follow-on current due to the SCR; that is, once the transient is over, the SCR might stay "on" because the normal line voltage sustains a current greater than the holding current. This can be overcome to some extent by carefully setting SCR characteristics. Another solution is a fuse, but this adds an additional component (something else which can go wrong).

7.4 Other Varieties and Applications of Surge Protectors

The principle of a device which has an infinite resistance at rest and becomes a short above a certain threshold is appealing for many applications. Besides power-line overvoltages and lightning-induced transients, surge protectors are used for NEMP hardening (see Fig. 7.13), ESD protection, etc.

Carbon block arrestors, with the variable resistor behavior of carbon, have been used for many years in the telephone industry. Although they are rugged and inexpensive, they tend to be abandoned because of the poor repeatability of their characteristics, especially after several operations.

Air gap arrestors (see Fig. 7.14) are used for high-voltage, overhead line protection, using the spark breakdown characteristic of two electrodes bent at a sharp angle. Although the setting of the breakdown voltage is not very accurate, it is good enough to provide coarse protection for lines carrying tens or hundreds of kilovolts.

Slow Rise Breakdown, Volts	Impulse Breakdown, Volts		Max. Impulse Discharge kA	Impulse Discharge Current, kA	Holdover Voltage	Arc Voltage	Insulation Resistance
	10 kV/µs	1 kV ns	8 × 20 µs	8 × 20 µs	Volts	Volts	MΩ
200 ± 50	< 300	< 1,000	10	5 (5 times each polarity)	140	25	> 10^4
300-500	< 550	< 1,000	10	5 (5 times each polarity)	150	25	> 10^4

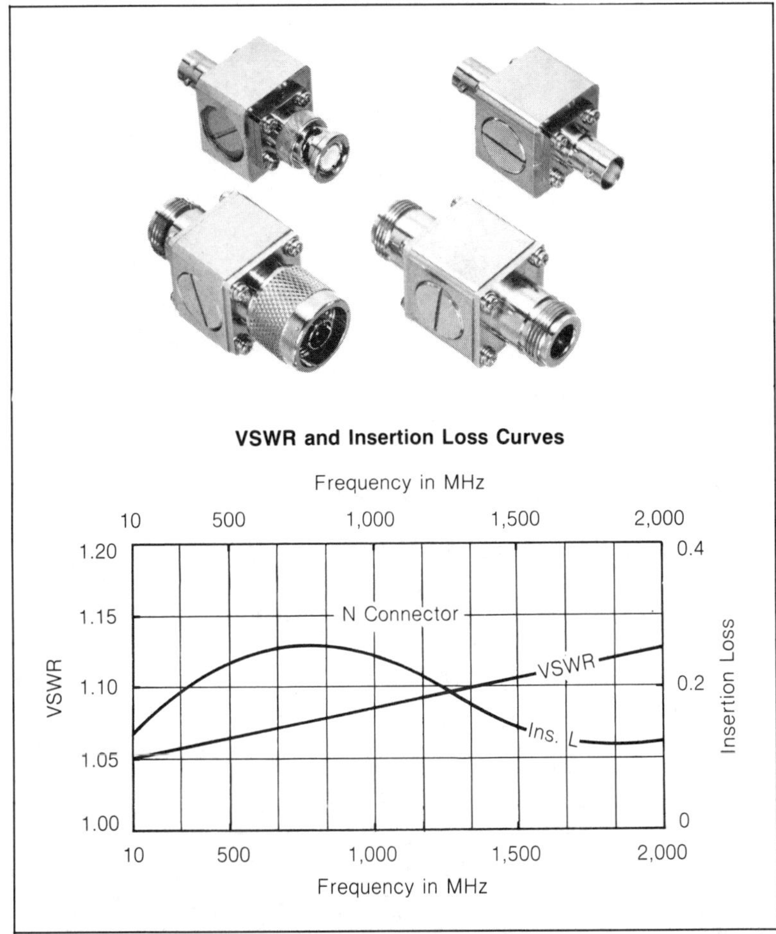

Figure 7.13—Example of Ultra-Fast EMP Protectors for Antenna and Video Cables, Made by Reliable Electric Co. The coaxial arrangement and special gas tube triggering allows for a response time in the tens of nanoseconds range.

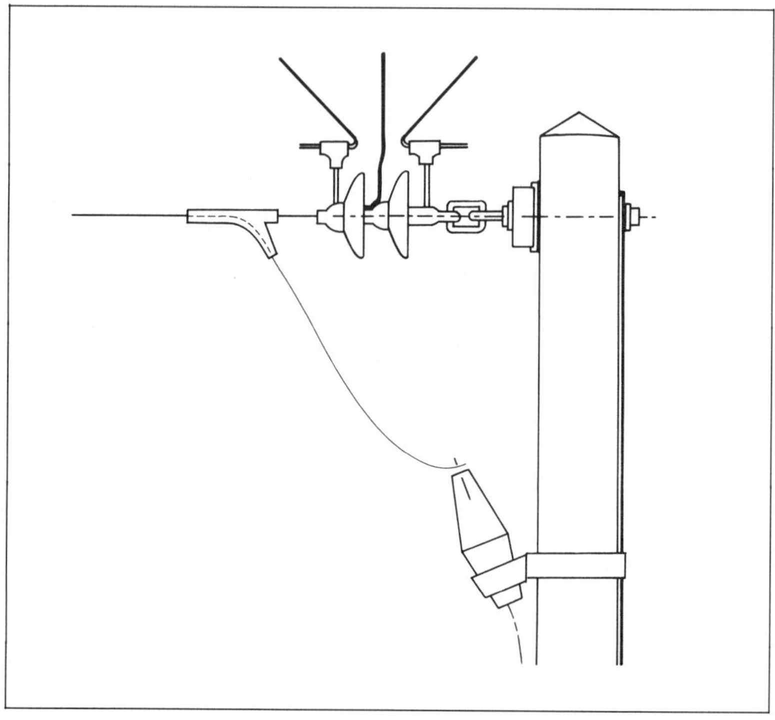

Figure 7.14—Air Gap Arrestor Used on Overhead Power Lines

A **high-voltage lightning** arrestor is an example of a solution to an EMI problem (lightning induction) which creates a new EMI problem: when arrestors operate, the sudden dI/dt and dV/dt arcing generates a line transient with much less energy but a faster rise-time than the lightning pulse itself. Also, damaged arrestors can create permanent EMI because of the corona effect (Figs. 7.15 and 7.16).

Varistor pins (see Fig. 7.17) are special mounts where the varistor has been fitted into a standard connector pin. Because of a high concentration of the ZnO powder, varistor pins with breakdown voltages in the few volts to 250 V range have been developed, with energy handling up to 3 J. Of course, the proper operation of these pins depends on a good connection of the varistor sleeve to a common ground area in the connector.

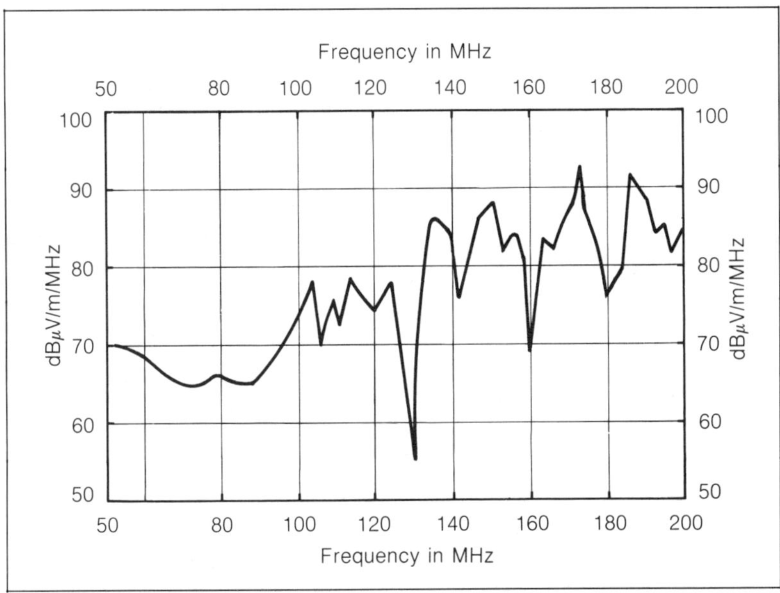

Figure 7.15—Partial Spectrum of Vertically-Polarized Peak Field from a Faulty Lightning Arrester on a 6.9 kV Test Line 4.57 m above Ground

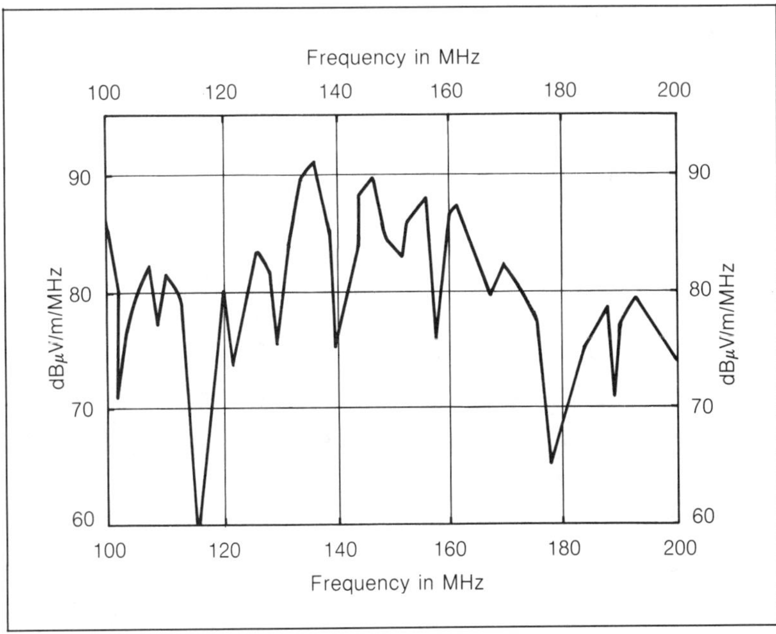

Figure 7.16—Partial Spectrum of the Horizontally-Polarized Component Measured under the Same Conditions as for Figure 7.15

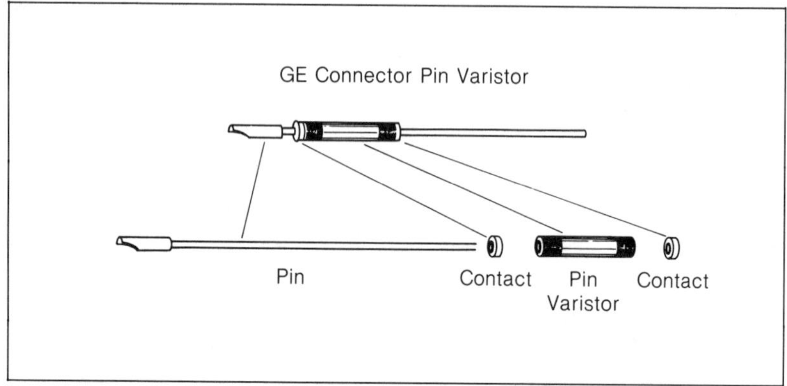

Figure 7.17—Detail of Varistor Pin for EMI Hardened Connector

Varistor Composite Material is a more recent concept.[1] Zener-like particles embedded with a certain dosage in a resin or other composite insulating material, give this material a voltage-dependent resistance. The material replaces the discrete element behavior of varistors by a distributed layer of "clamping material." The material absorbs fast pulses by capacitive effect, then dissipates the charge resistively. Compared to a standard varistor or zener, where capabilities are given in volts, amps, watts and joules, the characteristics of the composite are given in :

$$\text{\textbf{Breakdown Voltage}} = \text{V/mm (range} = 40 \text{ to } 4{,}000 \text{ V/mm)}$$

$$\text{\textbf{Maximum Current Density}} = \text{A/mm}^2 \text{ (range} = 10^{-3} \text{ A to a few A/mm}^2$$

$$\text{\textbf{Maximum Power Handling}} = \text{W/mm}^3 = \text{up to } 60 \text{ W/mm}^3 \text{ for a 30 ns pulse}$$

7.5 References

1. Malinaric, P.J., "Transient Suppressor Design with Varistor Composite Material," (*IEEE Transactions on EMC,* November, 1985).

7.6 Bibliography

1. EMP Protection Systems catalog, Belling Lee Intec, Ltd., Enfield, U.K.
2. Korn, S., *New Low Voltage GE-MOV,* Application Note 200.91, (General Electric Co., 1982).
3. Martzloff, F., Report 83CRD169, "Matching Surge Protective Devices to Their Environment," (General Electric Co., 1983).
4. *Transient Suppression Manual,* (General Electric Co., 1983).
5. "Transzorb Technical Notice," (General Semiconductor Industries, Inc.)

Chapter 8

Printed Circuit Boards and Layout

The printed circuit board (PCB) is the foundation for the entire electronic package. If the foundation is not right, the building will not stand. A multitude of EMC horror stories show that an in-depth look at PCB design would have saved thousands of dollars in testing, late fixes, retesting, etc., of a complete equipment. Therefore, look first at the PCB layout to accomplish the noise budget of Table 8.1.

Table 8.1—Example of Internal Noise Budget Allocation for Logic Circuits

Noise Source	Budget	TTL	ECL-10K
Power Supply	20%	80 mV	20 mV
Voltage Distribution IZ Drop	20%	80 mV	20 mV
Data Line Mismatch and Reflections:	20%	80 mV	20 mV
Crosstalk	20%	80 mV	20 mV
External Radiation Pickup	20%	80 mV	20 mV
Total Noise:	100%	400 mV	100 mV

8.1 Voltage Distribution and Zero-Volt Return

Power distribution is traditionally provided by supply and return traces. Their impedance (inductive reactance) is unimportant for slow-speed and low-power logic such as CMOS. Capacitor decoupling is not needed except at the connector input. As the logic speed increases, for TTL for example, considerably more care in layout is required because of increased common impedance coupling. High-frequency, ceramic-disc caps come to the rescue with one cap used typically to serve two DIP chips.

8.1.1 Decoupling

The so-called **decoupling** capacitor can be regarded as a reservoir which provides the inrush current that the logic device needs to switch in the specified time. The reason for this is that by no means can the long wiring from the power supply regulator to the chip provide the peak current without excessive voltage drop.

The value of the decoupling capacitor C close to the logic elements (chips) requiring the switching current I is:

$$C = \frac{I}{dV/dt} \tag{8.1}$$

where,

dV = voltage variation at capacitor output (supply rail sag) caused by the demand of a current I during the time interval, dt

dt = logic switching time

I = transient current demand of the logic family

Table 8.2 lists values of C for some popular logic families based on a maximum allowable V drop equal to 20 percent of the noise immunity level. Some guidelines should be followed to ensure that C works properly as a decoupling capacitor. Figure 8.1 shows a layout of power supply and return traces which are too far apart and therefore a poor design practice.

The problem, as shown in the equivalent circuit of Fig. 8.1, results in an inductance of about 5 nH for the cap leads (assuming they are cut very short) between the 5 V and 0 V traces and 5 nH for the chip and DIP pin leads. For a trace supply and return totalling 6 cm, the trace inductance is about 60 nH, so there is a total loop inductance of: 60 + 5 + 5 = 70 nH. Thus, the voltage drop from the capacitor to the IC resulting from the total loop inductance L is:

Table 8.2—Decoupling Capacitors for Some Popular Logic

Logic Family	Current Requirements		dV = 20% of NIL	dt = Rise Time	Decoupling C = $\Delta I/(dV/dt)$
	Gate Switch	Gate Drive			
CMOS	1 mA	1 mA	200 mV	50 ns	(500 pF)**
HCMOS***	10 mA	1 mA	200 mV	10 ns	750 pF
TTL	16 mA	8 mA	80 mV	10 ns	3,000 pF
STTL	30 mA	20 mA	60 mV	3 ns	2,500 pF
LSTTL	8 mA	11 mA	60 mV	8 ns	2,500 pF
ECL-10K	1 mA	6 mA	20 mV	2 ns	700 pF

*Calculation is based on only one DIP chip and one logic element driving a fanout of five gates. The noise immunity level is spread equally over five contributing noise sources: power supply, sag power distribution radiation pickup, common-impedance coupling, crosstalk and reflections due to impedance mismatch. (See Noise Budget of Table 8.1.)

**In the case of the standard (low-speed) CMOS, this value is conservative because the assumption that *the long wiring from the power supply cannot provide the peak current without excessive voltage drop* is not true. For rise times in the 50 to 100 ns range, even with 1 μH of supply leads inductance, the voltage drop would be acceptable. So unless protection against severe ambient EMI is required, this decoupling is not necessary.

***Although presumably "quieter" than TTL, high speed CMOS still exhibits a significant switching current due to its fast transition.

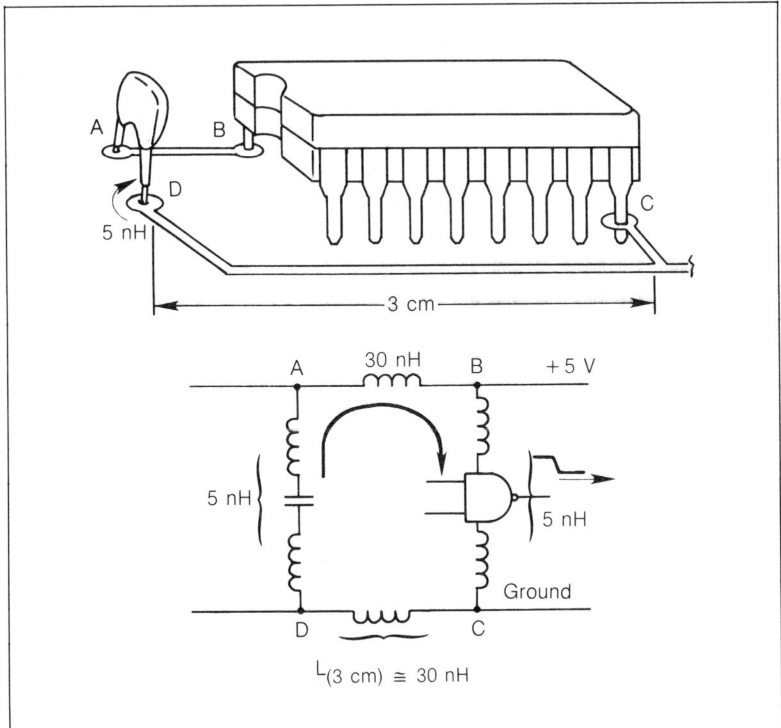

Figure 8.1—Effect of Inductance in Power Distribution

$$V = LdI/dt \qquad (8.2)$$

For Schottky TTL logic having a peak current demand of 30 mA/gate, the situation of Fig. 8.1 results in a voltage drop V:

$$V = 70 \times 10^{-9} \times (0.03/3 \times 10^{-9}) = 700 \text{ mV} \qquad (8.3)$$

This is far above the budget and even larger than the worst-case noise immunity of this logic.

To reduce this potentially dangerous situation, Fig. 8.2 illustrates how the power supply and return traces should be routed close together, thus reducing the series inductance by about 80 percent, (more or less). From Eq. 8.2, the voltage drop would become about 100 mV, or approximately 20 percent of the noise-immunity level, which is acceptable. Ultimately, the limitation of the cap high-frequency performance is predicted on the self-resonant frequency of the capacitance with its lead inductance. Figure 8.3 shows what the unavoidable limitations of discrete capacitors are. The critical frequency spectrum has been shown for three technologies: CMOS, standard TTL and Schottky.

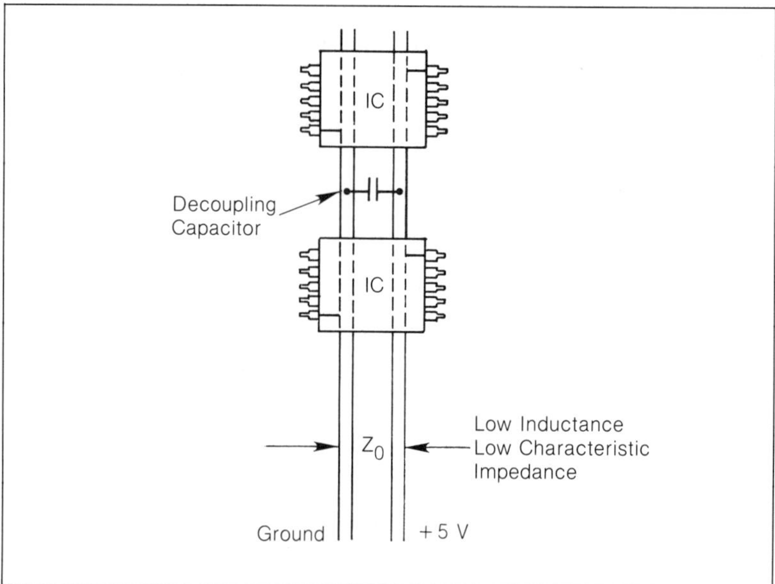

Figure 8.2—Recommended Power Rail Layout Having Low Series Inductance

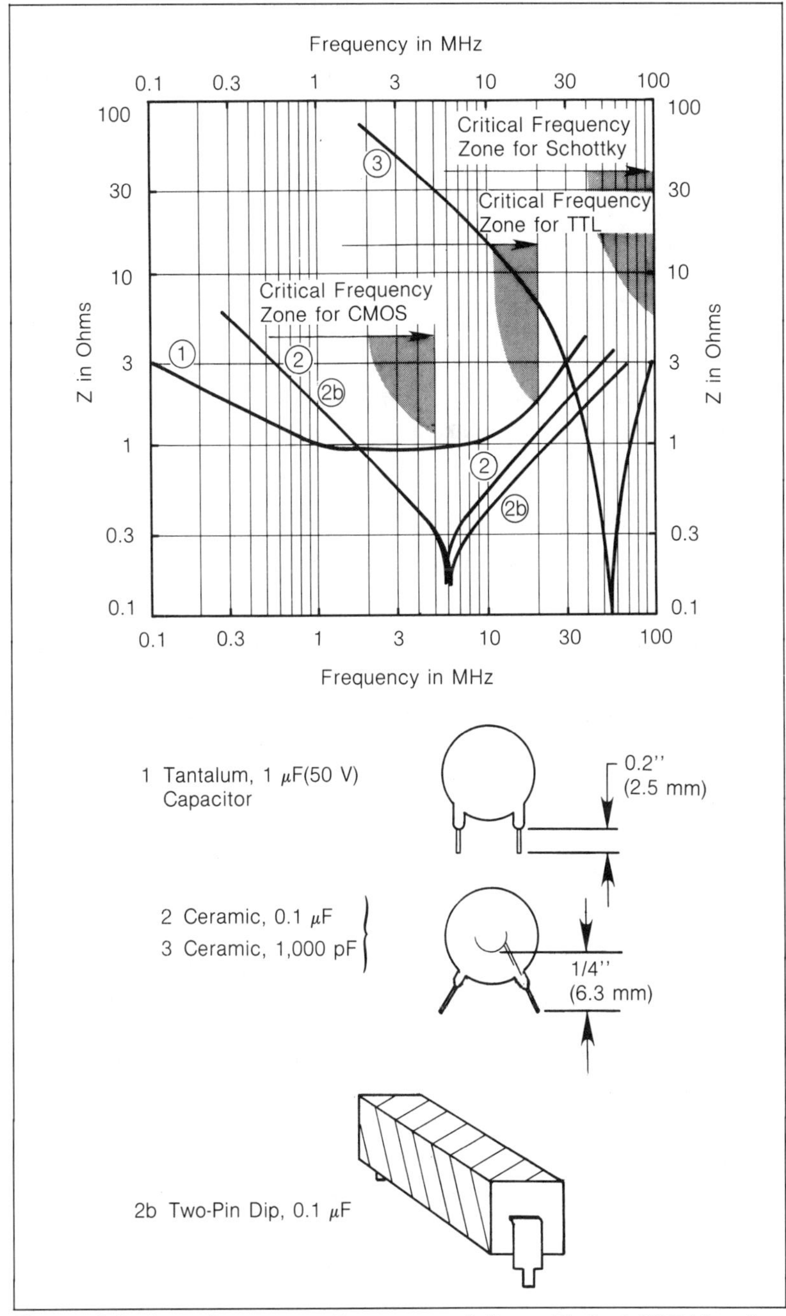

Figure 8.3—Impedances vs. Frequency of Some Typical PCB-Mount Capacitors

Notice that no conventional capacitor can work ideally for fast Schottky: they all behave like inductance and may even create **ringing** instead of decoupling. This can be reduced, however, by using either foil leads or, better yet, the new two-pin DIP package which is compatible with standard ICs for automated insertion. Such monolithic ceramic units are now available over a range from a few picofarads to about 10 μF. Thus, they are taking over some of the decoupling jobs formerly in the realm of tantalum units, which lose their capacitance at high frequency. Consequently, the needs for the 10 μF ceramic disc cap at the power supply PCB edge connector input will be eliminated by the use of a single monolithic ceramic two-pin DIP. Electronic Industries Association (EIA) (e.g., RS-198) and military specifications identify the basic classes of ceramic dielectrics and cap performance.

8.1.2 Other Methods of Voltage Distribution Cleanup

Excessive reduction of voltage sag on the power distribution traces by using decoupling caps is basically **cheating**. It could be stated that the distribution system was poorly designed in the beginning. So why compound the problem by using capacitors (band-aid engineering) which increase expense and reduce reliability? Other solutions to the problem which should be investigated are:

1. Use of a zero-volt plane with large $+V_{dc}$ traces
2. Raised power-bus distribution
3. Multilayer boards

Thin lines, far away from the return conductor, present a high inductance (typically 10 nH/cm). Table 8.3 shows the impedance versus frequency of typical PCB traces and planes.

If, for instance, two Schottky chips are **daisy-chained** on a length of 10 cm by a 1 mm supply trace, for a bandwidth of 100 MHz (which is the reciprocal $1/\pi\tau_r$ of the 3 ns transition for Schottky), the impedance is 72.5 Ω, corresponding to a voltage dip of:

$$30 \text{ mA} \times 72.5 = 2.1 \text{ V} \tag{8.4}$$

Table 8.3—Impedance of Printed Circuits

Impedance of Traces in mΩ or Ω

Impedance of Planes in Ω/sq

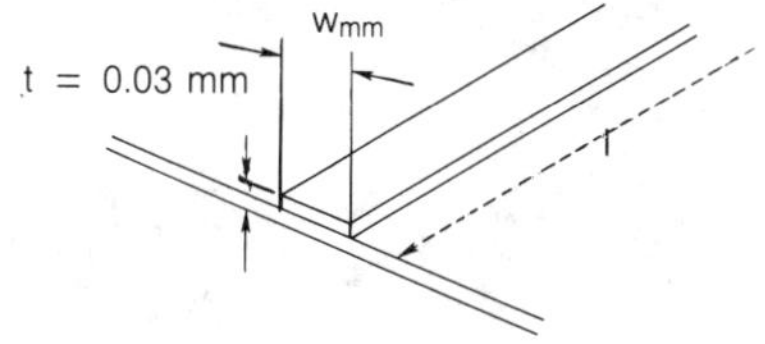

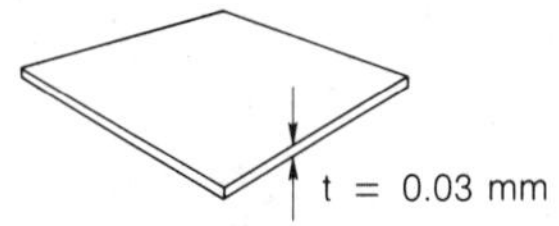

(Corresponds to 28.35 g
[1 oz] per sq. in
copper plane)

Freq.	w = 1 mm t = 0.03 mm				w = 3 mm t = 0.03 mm		
	l = 10 mm	l = 30 mm	l = 100 mm	l = 300 mm	= 30 mm	l = 100 mm	l = 300 mm
50 Hz	5.74 mΩ	17.2 mΩ	57.4 mΩ	172 mΩ	5.74 mΩ	19.1 mΩ	57.4 mΩ
100 Hz	5.74 mΩ	17.2 mΩ	57.4 mΩ	172 mΩ	5.74 mΩ	19.1 mΩ	57.4 mΩ
1 kHz	5.74 mΩ	17.2 mΩ	57.4 mΩ	172 mΩ	5.74 mΩ	19.1 mΩ	57.5 mΩ
10 kHz	5.76 mΩ	17.3 mΩ	57.9 mΩ	174 mΩ	5.89 mΩ	20.0 mΩ	61.4 mΩ
100 kHz	7.21 mΩ	24.3 mΩ	92.5 mΩ	311 mΩ	14.3 mΩ	62.0 mΩ	225 mΩ
300 kHz	14.3 mΩ	54.4 mΩ	224 mΩ	795 mΩ	39.9 mΩ	177 mΩ	657 mΩ
1 MHz	44.0 mΩ	173 mΩ	727 mΩ	2.59 Ω	131 mΩ	590 mΩ	2.18 Ω
3 MHz	131 mΩ	516 mΩ	2.17 Ω	7.76 Ω	395 mΩ	1.76 Ω	6.54 Ω
10 MHz	437 mΩ	1.72 Ω	7.25 Ω	25.8 Ω	1.31 Ω	5.89 Ω	21.8 Ω
30 MHz	1.31 Ω	5.16 Ω	21.7 Ω	77.6 Ω	3.95 Ω	17.6 Ω	65.4 Ω
100 MHz	4.37 Ω	17.2 Ω	72.5 Ω	258 Ω	13.1 Ω	58.9 Ω	218 Ω
300 MHz	13.1 Ω	51.6 Ω	217 Ω	39.5 Ω	176 Ω	6.39 m	300 MHz
1 GHz	43.7 Ω	172 Ω			131 Ω		

Ohms/Square	Freq.
813 μΩ	50 Hz
813 μΩ	100 Hz
817 μΩ	1 kHz
830 μΩ	1 kHz
871 μΩ	100 kHz
917 μΩ	300 kHz
1.01 mΩ	1 MHz
1.17 mΩ	3 MHz
1.53 mΩ	10 MHz
2.20 mΩ	30 MHz
3.72 mΩ	100 MHz
6.39 mΩ	300 MHz
11.6 mΩ	1 GHz

When changing from thin or very long traces to a copper plane as in Fig. 8.4, the immediate benefit is that a quasi-infinite plane has no external inductance; it has only resistance and internal inductance which increase as $\sqrt{F}$ above skin depth region instead of increasing with frequency as in the case of thin wires. For example, at the 100 MHz frequency of the previous example, a wide plane shows only 3.7 mΩ, i.e., the switching of 30 mA will cause only 100 μV drop in the common ground. Therefore, by **leaving as much copper as possible** rather than etching, a low-impedance is achieved for both the supply and return since they are close together. **This preferred practice can be expressed by this guideline: Make the PCB as optically opaque as possible by extending supply and ground return traces to large areas.** When copper planes have numerous holes in a grid pattern, impedance is increased by a factor shown in Fig. 8.5.

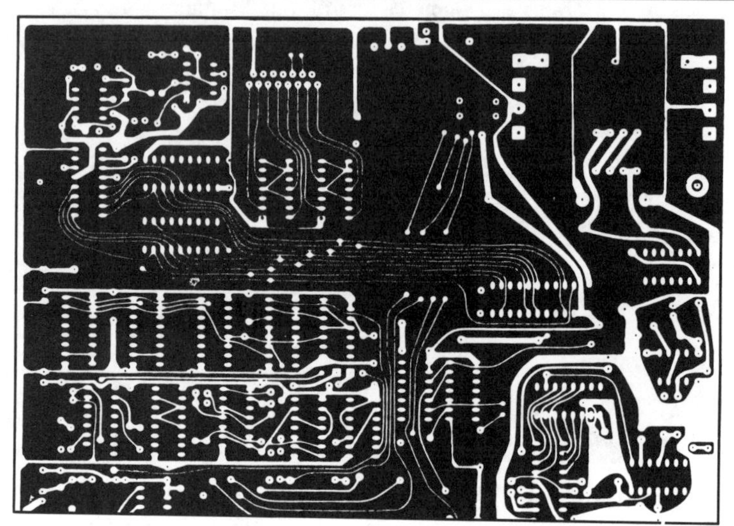

Too Thin 0 V and $+V_{dc}$ Traces: High impedances,
thin copper border around card will act as a pickup loop.

(a)

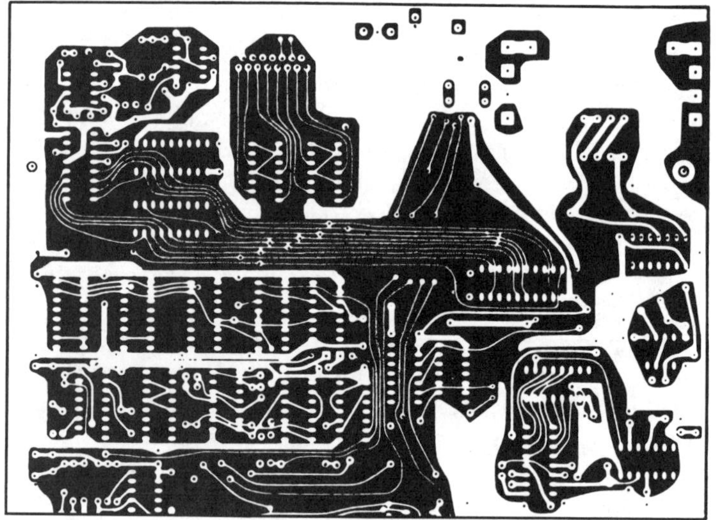

Ground & power traces have been enlarged: copper has
been left (not etched) wherever possible.

(b)

Figure 8.4—Enlarging Ground Areas in PCBs

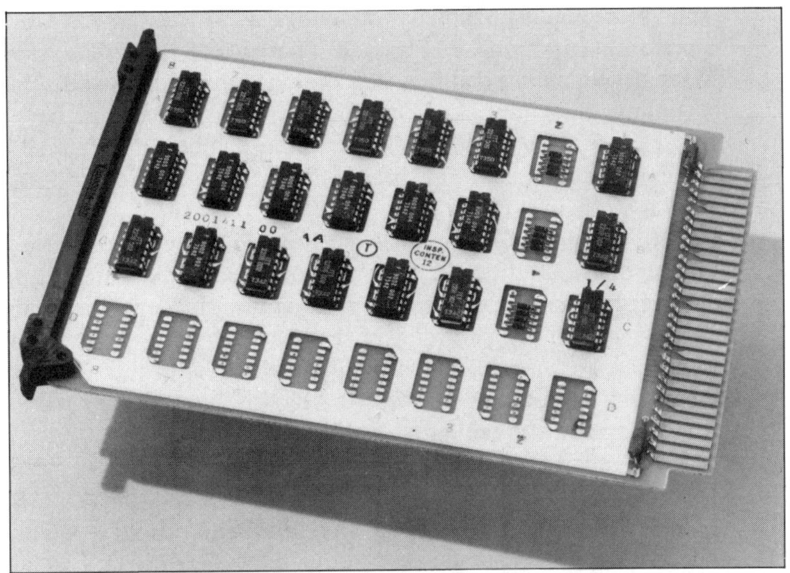

Figure 8.4A—PCB Ground Plane/Shield Retrofit • Insulated Copper Foil Connected to 0 V Traces as Much as Possible for Shielding and 0 V Return Improvement • Insulated Double Sided Sandwich Used for Power Busing and Shielding (One Sided + V_{CC} and 0 V Tabs at Each Dip Position

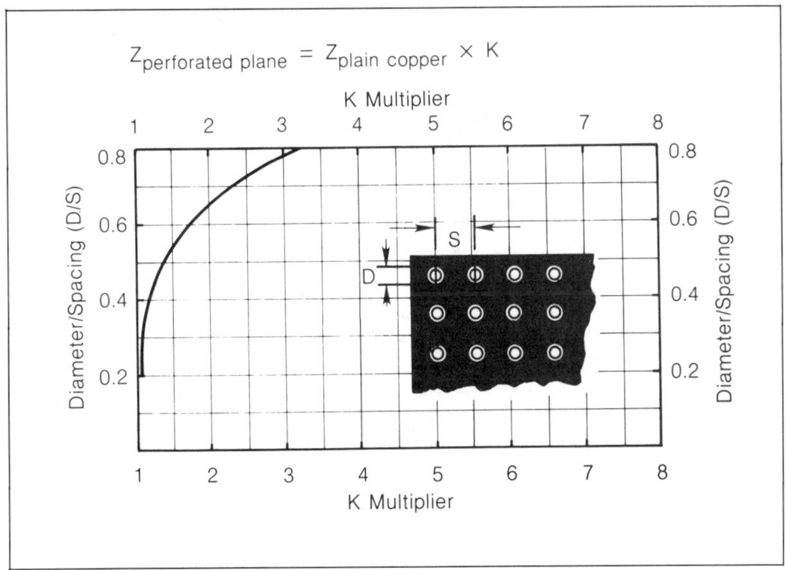

Figure 8.5—K Multiplier to Compute Impedance Increase of a Perforated Ground Plane vs. Plain Foil

When low-impedance power planes cannot be achieved, use flat buses for power distribution (see Fig. 8.6). These are better solutions for EMI than solid wires, and they also decrease the number of discrete decoupling capacitors. The flat buses provide (since they have a small L and large C) a lower impedance. Table 8.4 shows the characteristic impedance of three different transmission lines as a function of their geometry.

Thus, power distribution for serving logic will now be pictured as being distributed over a transmission line. Three different options are shown in Table 8.4. The third option corresponds to side-by-side traces on the same side of the board. If we set an objective of less than 10 Ω for Z_o, the third option would require D/W ratios of less than about 1.01 ($Z_o < 8$ Ω) which is impractical unless a discrete capacitor is added as in Fig. 8.2.

Because of the large area taken by a copper plane, the second option shown in Table 8.4 generally requires a multilayer board, which is discussed later. Thus, the first option in the table is sometimes used for single-layer, double-cladded PCBs. Physical realization of the first option is suggested in Fig. 8.6 for both

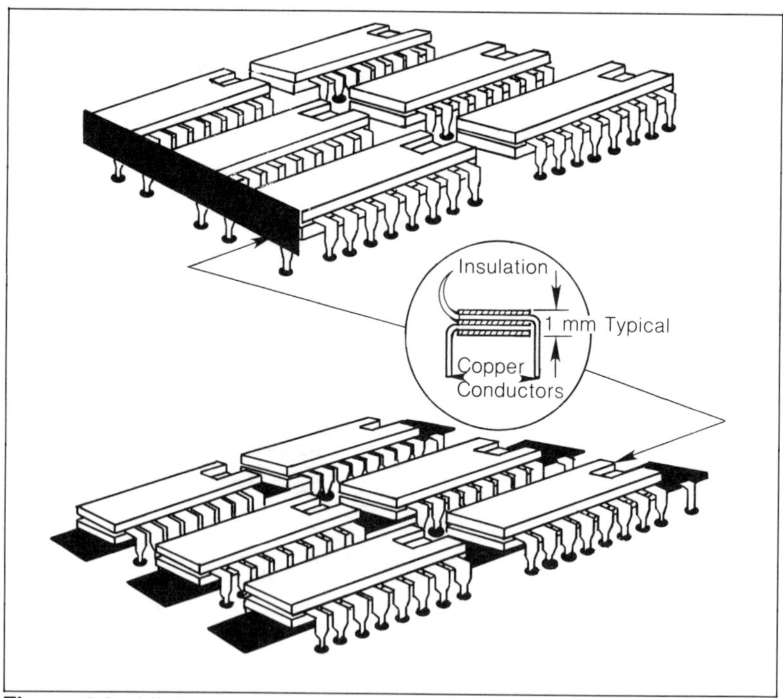

Figure 8.6—Minibus Power Distribution when a Ground Plane Is Not Achievable (Courtesy of Rogers/Mektron)

Table 8.4—Characteristic Impedance of Different Flat Conductor Arrangements

W/h or D/W	Z_{01} Parallel Strips (1)	Z_{02} Strip Over Gnd Plate (2)	Z_{03} Strips Side by Side (3)	Z_{04} Strip Line (2)
0.5	140	120	NA	60
0.6	124	108	NA	54
0.7	114	98	NA	49
0.8	110	95	NA	47
0.9	106	92	NA	46
1.0	104	90	0	45
1.1	98	84	25	42
1.2	92	80	34	40
1.5	86	70	53	35
1.7	80	66	62	33
2.0	72	60	73	30
2.5	60	56	87	27
3.0	54	48	98	24
3.5	48	43	107	21
4.0	42	40	114	20
5.0	34	34	127	17
6.0	28	28	137	14
7.0	24	24	146	12
8.0	21	21	153	11
9.0	19	19	160	9
10.0	17	17	166	8
12.0	14	14	176	
15.0	11.2	11.2	188	
20.0	8.4	8.4	204	
25.0	6.7	6.7	212	
30.0	5.6	5.6	227	
40.0	4.2	4.2	243	
50.0	3.4	3.4	255	
100.0	1.7	1.7	293	

Notes:
1. Mylar dielectric assumed, $\epsilon_r = 5$
2. Paper base phenolic or glass epoxy assumed, $\epsilon_r = 4.7$
3. Air assumed
4. For geometries corresponding to shaded areas, Z_0 shown can be higher than actual due to fringing capacitance.

$$Z_{01} = (377/\sqrt{\epsilon_r})\,(h/W), \text{ for } W > 3h \text{ and } h > 3t$$

$$Z_{02} = (377/\sqrt{\epsilon_r})\,(h/W), \text{ for } W > 3h$$

$$Z_{03} = (120/\sqrt{\epsilon_r})\,\ln(D/W + \sqrt{(D/W)^2 - 1}), \text{ for } W \gg t,$$

and $D \gg$ distance to nearby ground plane

$$Z_{04} = \frac{Z_{02}}{2}$$

horizontal and the more common vertical configuration.

For instance, the first option from Table 8.4 shows, for a bus width of 8 mm and a spacing of 0.2 mm, a 4 Ω characteristic impedance. The maximum voltage drop for one Schottky gate would then be:

$$30 \text{ mA} \times 4 \ \Omega = 120 \text{ mV}$$

which is tolerable. Finally, in extreme cases where a ground plane is needed but the wiring density does not permit it, **and** a multilayer is not feasible either, Figure 8.4A shows a **fix** option.

8.2 Analog/Digital Mix

When analog and digital devices are mounted on the same PCB, common-impedance coupling must be avoided because the 0 V return is polluted by fast current transitions. To reduce this problem, one can eliminate the common-impedance path(s) and extend the surface area of the ground conductor to lower its impedance. Each is illustrated in Figs. 8.7 and 8.8. Figure 8.7 shows an example of poor control of common-mode ground noise. The 0 V return is serving both an analog network (A/D converter) and the digital modules. This is not an acceptable EMI control practice.

In Fig. 8.8, the 0 V return is divided into separate analog and digital traces with increased width in both the supply and return paths to further reduce common-impedance coupling. Also, if possible, the analog amplifier is supplied from a distinct power supply output.

The 0 V return of the analog circuit no longer uses the noisy logic ground. **They are only commoned at the power supply 0 V terminal, or at the mother board ground plane**. Also, it can be observed that the input circuitry of the A/D converter is **no longer enclosed in the noisy loop** of the digital supply traces.

8.3 Physical Implementation
and Zoning

With the understanding of component behavior regarding noise and power distribution requirements, this section discusses general layout of wirewrapped, single-layer and multilayer boards.

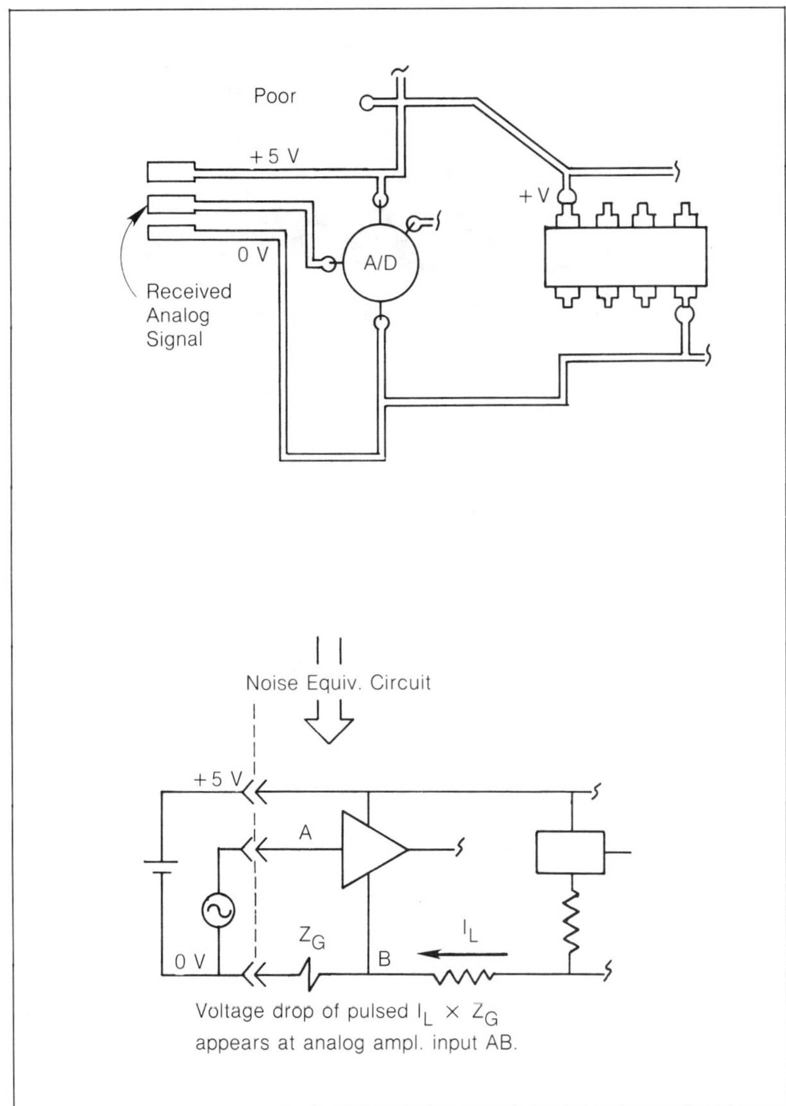

Figure 8.7—Poor Common-Mode Noise Control

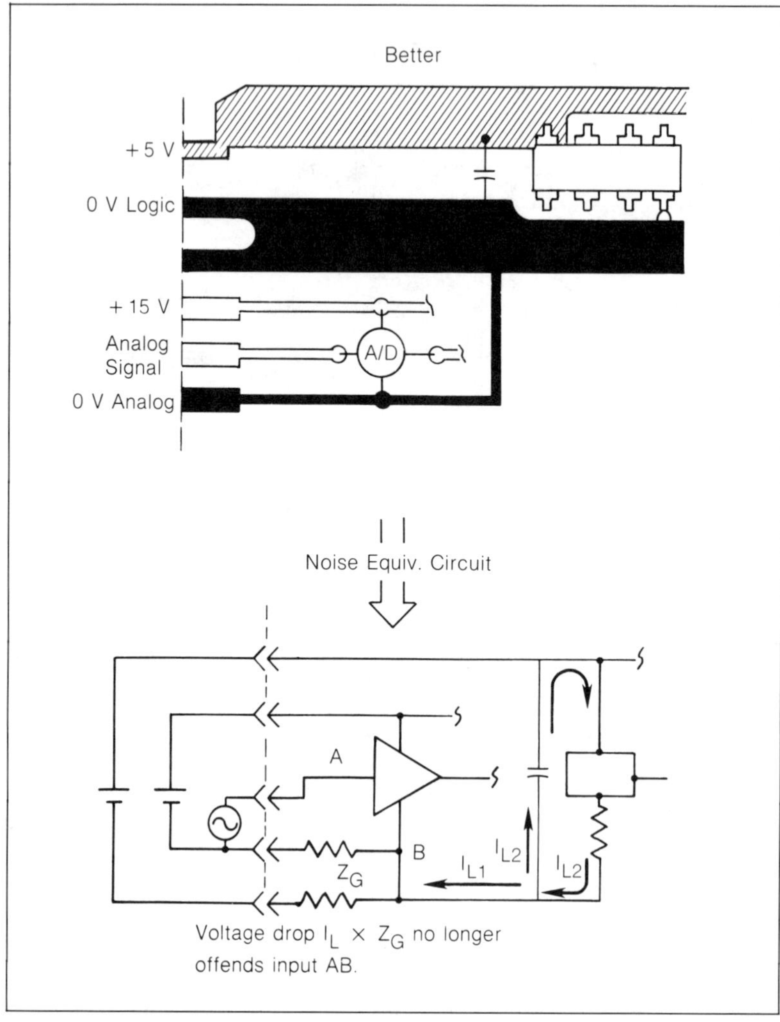

Figure 8.8—Improvement on the Common-Mode Noise Problem of Fig. 8.7

8.3.1 Wirewrap, Single-Layer Boards

For medium- (e.g., TTL) and low-speed (e.g., CMOS) logic, the wirewrap board is still popular for one-of-a-kind designs or the early stages of the logic **breadboard.** Its convenience (no artwork or tooling) and fast turnaround make the wirewrap board a popular choice. Figure 8.9 shows a typical wirewrap board with one side acting as a power-supply plane and the other side acting as the

8.14

power return and zero-signal reference return plane. Several pin holes are isolated and connected to either side so that pins can be mounted thereto to support discretes and interconnect wiring, i.e., the wirewrap.

In wirewrap boards, every fifth or tenth finger on the edge connector is typically earmarked for ground return connection. This provides a ground distribution system which limits the problem of common impedance coupling explained in Chapter 1 and Fig. 1.2a. Depending on the total load current and logic type used (see Section 4.4), a 1 to 10 μF tantalum capacitor is used at the connector input from the power-supply plane on one side to the ground distribution lines on the other side. This capacitor is bridged by a 0.01 to 0.1 μF ceramic-disc, high-frequency cap at the same connector finger location.

The discretes (resistors, diodes, caps, etc.) and ICs are mounted first to the wirewrap board pins. A ceramic-disc, high-frequency cap of 1,000 pF or other value (see Section 8.1.1) is traditionally used for every one or two DIP packages to provide decoupling. The Z1 wires are mounted next. In wirewrap parlance, Z1 wires are the longer runs and are defined as all wires greater in length than D/2, where D is the diagonal dimension of the wirewrap board. These wires are routed close to the ground plane. This helps control the characteristic impedance of interconnects by reducing the self-inductance of the wire. Since the longer runs, Z1, have the largest inductance, they are wired in first.

Figure 8.9—Typical Blank Wirewrap Board

Next, an option exists to reduce radiation, radiation pickup or crosstalk by laying down an X-Y grid over the Z1 wires as shown in Fig. 8.10. This grid matrix shield also helps control characteristic impedance of Z1 and return (the grid) wires, but could be excessive for EMI control for most applications.

Following either the Z1 interconnects or the above X-Y grid (if used), the shorter Z2 wires (Z2 < D/2) are laid out. Another option exists for an X-Y grid to cover all wires for the same reasons indicated above. The routing of Z1 and Z2 wires should be laid out for the shortest path and not in an X-Y directional manner which would otherwise increase crosstalk. Thus, random crossover of wirewrap wires is best for EMI control. A completed wirewrap board might then look like that shown in Fig. 8.11.

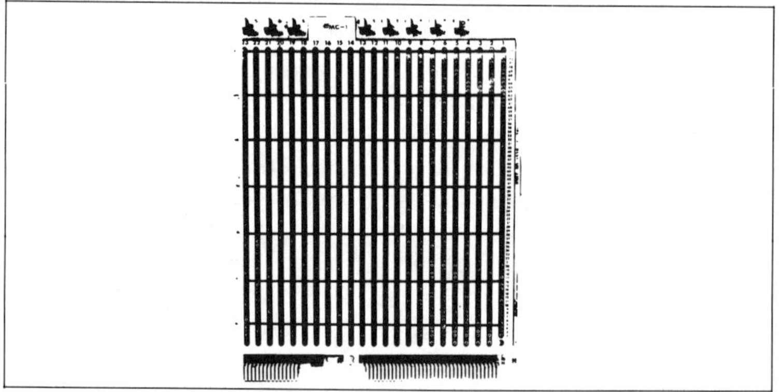

Figure 8.10—Controlling Radiation Pickup and Interconnect Impedance with a Topside X-Y Ground Grid Configuration

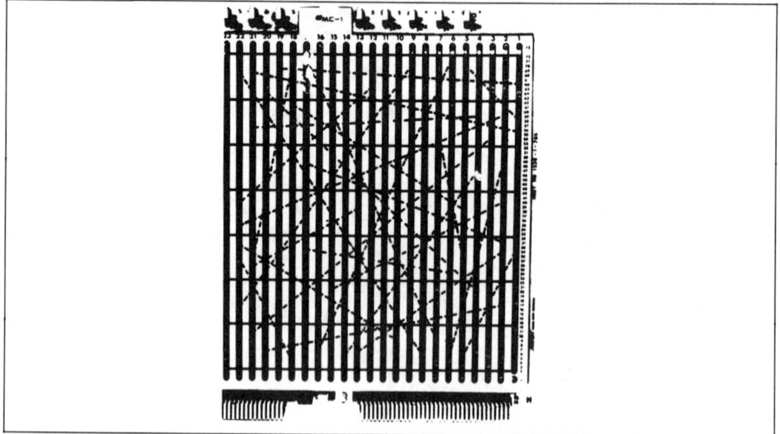

Figure 8.11—Completed Wirewrap Board

8.3.2 Single-Layer PC Boards

Printed circuit boards should be laid out such that the higher-speed devices (fast logic, clock oscillators, etc.) are located closer to the edge connector, and the lower-speed logic and memory, if applicable, are located farthest from the connector as shown in Fig. 8.12. This tends to compound common impedance coupling, radiation and crosstalk. Of course, the same could also apply to wirewrap boards as previously discussed.

If the logic speed is below TTL, then layout is relatively unimportant from an EMI point of view. The single exception is for **optical isolators, isolation tranformers or filters which should be located as close to the edge connector as possible.**

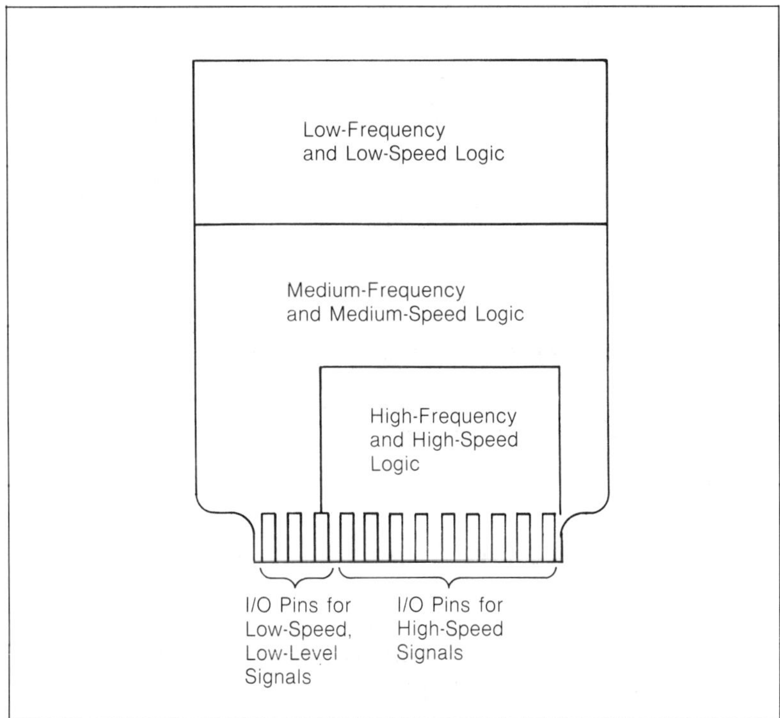

Figure 8.12—Functional Layout Guidelines

Regarding power distribution, the consequence of too much separation between power supply and return rails was discussed in Section 8.1. This is now illustrated in Fig. 8.13 in which a poor PCB layout scheme is used where the supply traces appears across the top and the return trace progresses across the bottom of the board. This results in undesired high self-inductance, greater circuit crosstalk and excessive traces radiation.

To prevent the problem depicted in Fig. 8.13 from occurring, use the PCB layout scheme shown in Fig. 8.14. Here the power supply and return traces are located close together on the PCB to form a transmission line which significantly reduces the distribution impedance. Vertical extension of the transmission line concept is possible via feeder traces on the top side of the board at the left and right, connected by plated through-holes. The layout in Fig. 8.14 also (1) helps in reducing circuit crosstalk, since fields are confined to a tighter configuration and (2) reduces overall radiation from the board since signal traces and their returns have lesser loop areas. Figure 8.15 is an extension of the EMC benefits over the layout in Fig. 8.14.

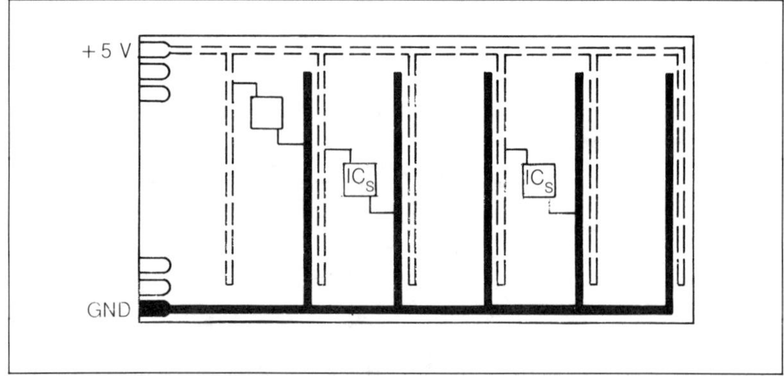

Figure 8.13—A Bad Layout Giving High Inductance to Power Distribution and Signal Return Paths

Similar to wirewrap boards, about every tenth pin on the edge finger connector of PCBs is typically planned for ground return as shown in Fig. 8.15. The regulated power bus at the edge connector is decoupled by a 1 to 10 μF tantalum capacitor, depending on the total load current. This capacitor is shunted by a 0.01 to 0.1 F high-frequency, ceramic-disc capacitor.

For the high-speed logic family, interconnect leads formed by microstrip traces above a ground plane constitute an impedance discontinuity at any abrupt change in direction, such as a 90° corner. To reduce the VSWR of this discontinuity, the right-angle corners are truncated by 45° as shown in the inset portion of Fig. 8.16. This practice is not necessary for TTL or lower-speed logic.

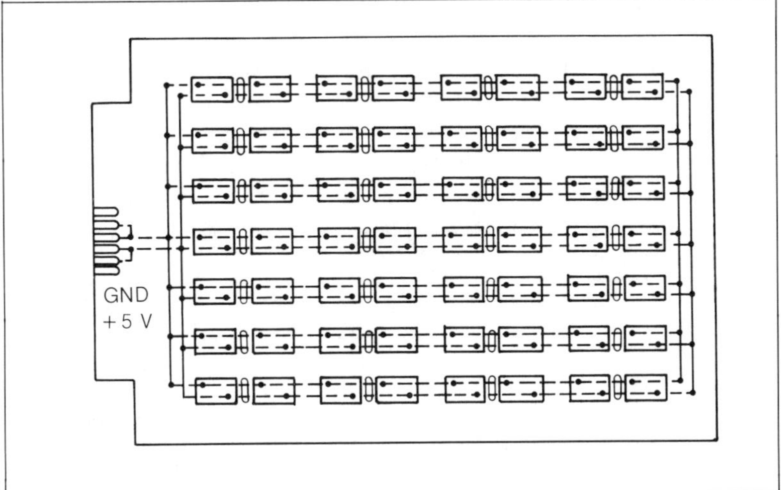

Figure 8.14—Better Layout than Fig. 8.13 to Reduce Power Distribution and Logic-Return Impedances, Trace Crosstalk and Board Radiation

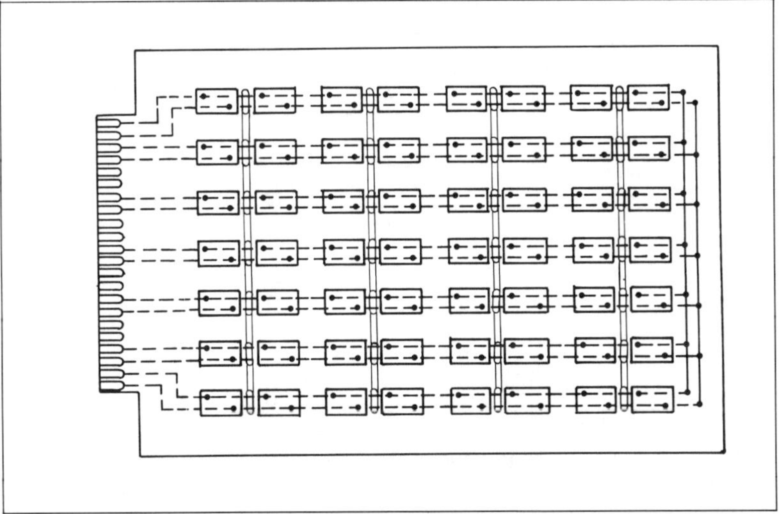

Figure 8.15—Better Layout than Fig. 8.14 for Further EMI Reductions

The problem with microstrip is that power traces take up both sides of a printed circuit board. Trace crossovers are virtually impossible, but can be easily eliminated if power is distributed via elevated-buses as described in Section 8.1.2. If elevated-bus power distribution is impossible or unwanted, then multilayer boards may be the only remaining option.

Connector areas must be treated as follows:

1. Power supply decoupling capacitors (serving the whole card) and I/O line filtering capacitors (for the lines which have not been cleaned up at the box interface) must be located very close to the edge fingers, using short and large traces.

2. Signal isolations transformers and optical isolators must be mounted close to their input lines, and avoiding parasitic coupling with their output lines.

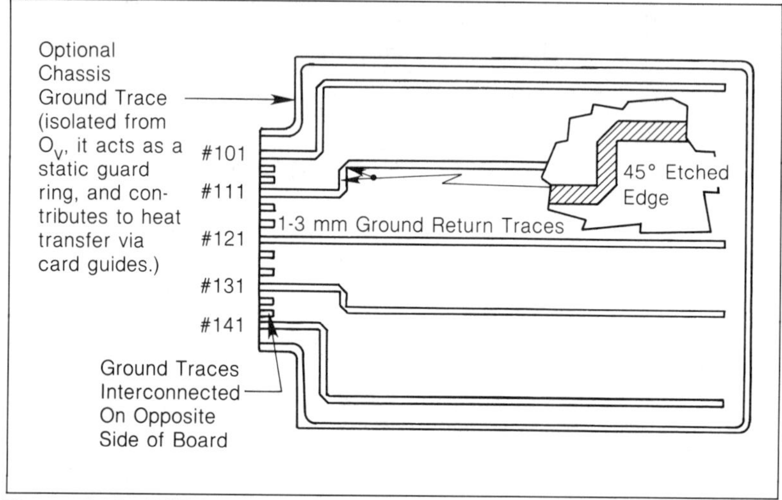

Figure 8.16—PCB Logic Ground Return Layout

8.3.3 The Ultimate Answer to PCB Noise Suppression: Multilayer Boards

It is generally difficult to successfully operate high-speed logic on a single-layer board because of common impedance coupling. While articles in trade journals describe such single-layer board achievements, they require considerably more attention to pertinent fabrication details. This often makes quality control and repeatability difficult during mass production. To avoid such problems, including excessive radiation and pickup, multilayer boards are recommended wherein the power supply and return and zero-signal reference are typically realized on separate one-ounce copper foil planes. A multilayer board is defined as two or more PCBs sandwiched together with the levels interconnected through plated through-holes.

There are usually n + 1 levels for an n-layer board as suggested in Figs. 8.17 and 8.18. Here a five-level, four-layer board configuration is illustrated. Level A is the component side and includes the interconnect traces. Together with the upper face of level B, they form a microstrip line. The upper and lower faces of ground return level B are electrically isolated by about 40 dB.* Thus, currents flow-

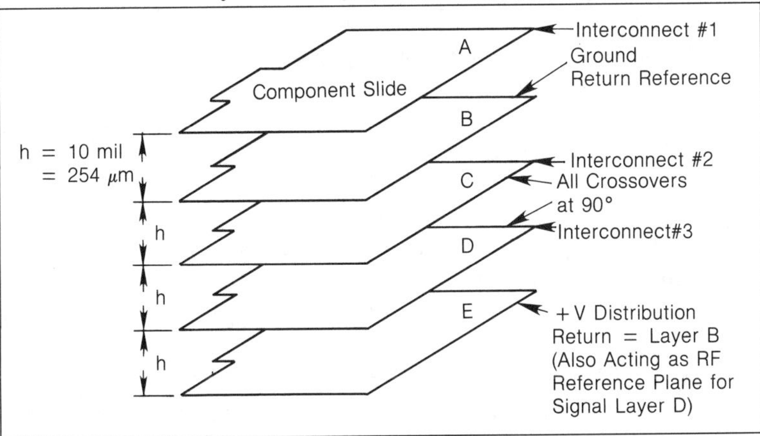

Figure 8.17—Multilayer Board for High-Speed Logic Impedance Control

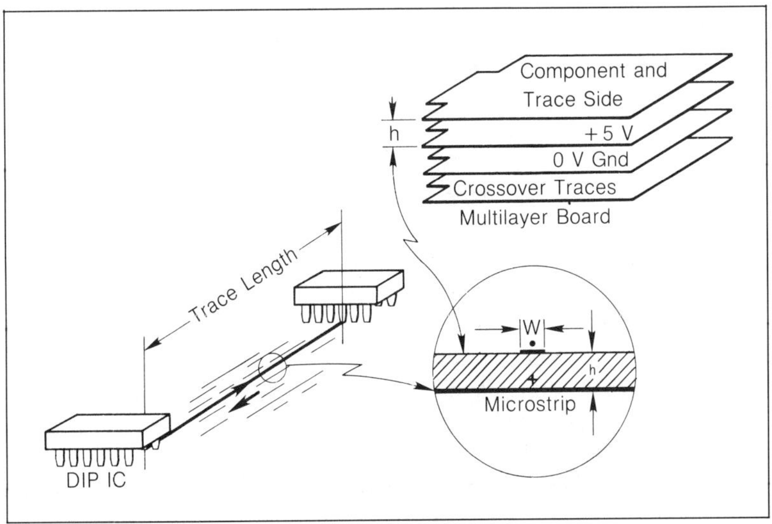

Figure 8.18—Radiation from Multilayer or Multiwire Circuit Boards

---

*The skin depth of copper at 100 MHz is about 0.2 mil ≈ 5 μm, and since one-ounce foil is about five skin depths at 100 MHz, the attenuation across the thickness is 8.7 dB × 5 = 43.5 dB.

ing on the lower face of level B are not seen on the upper face and vice versa. Level C contains more interconnect lines which are usually routed perpendicular to those of level A. This permits crossover on level A to be made on level C and vice versa. The lower face of planar level B forms microstrip lines with the interconnects of level C. Likewise, the interconnects of level D form microstrip lines with the planar level E. To avoid crosstalk, interconnect traces on levels C and D are routed perpendicularly.

Levels E and B form the power supply and return planes to facilitate low-impedance distribution. Depending on their spacing and dielectric constant, they provide a distributed board capacitance of the order of 10 to 1,000 pF/cm^2. Since the lower face of level E is also electrically isolated from its upper face, the plane of level E also serves as a shield to contain off-the-board radiation or pickup. Should remaining radiation or pickup from the traces of level A prove to be excessive, then level A can be topped with a plane similar to level E.

A variation of the multilayer board is Multi-Pac,* which is an assembly of stacked PCBs held together in a **sandwich** with press-fit contacts, as shown in Fig. 8.19. Some principal differences between multilayer and Multi-Pac features are:

1. Multi-Pac has via or pass-through holes from plane-to-plane, whereas with multilayer, holes must be drilled through all circuit layers, prohibiting circuitry in that area. Multi-Pac offers up to eight levels of circuitry, or solid copper sheets can be used in place of PC boards for high current capacity.
2. Controlled impedance with uniform board spacing is especially important for high-speed logic circuits.
3. Multi-Pac permits hybrid systems in any or all parts of the board. Unlike multilayer, Multi-Pac will allow additional circuit layers to be stacked on sections of the backpanel or daughter board where additional density is needed.
4. Because Multi-Pac is a stack of discrete PC boards, any board can be changed up to assembly time when the contacts are pressed in place.
5. Contacts can be removed with Multi-Pac, and circuit layers can actually be accessed for changes or repairs.
6. Multi-Pac meets MIL-STD-202, Method 103, Test B for humidity and MIL-STD-101 test for salt spray.

---

*Off-the-shelf, press-fit stacked boards are also available from:
HADCO, 12 B Manor Parkway, Salem, NH 03079 (USA)
ELFAB, 15 Sutton's Park, London Road, Earley, Reading, RG5 1A2 (UK)

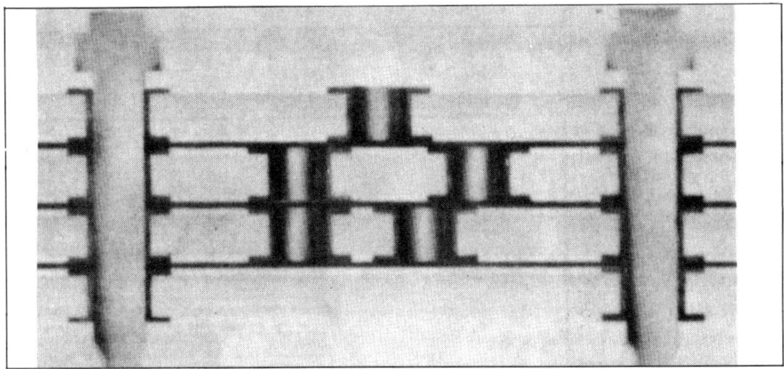

Figure 8.19—Edge Elevation View of Multi-Pac Board (Courtesy of Multi-Pac Corp.)

8.3.4 Multiwire® Boards and Other PCB-Derived Techniques

A Multiwire® circuit board, while resembling a PCB, is a customized pattern of insulated #34 AWG wires laid over an adhesive-coated substrate by a high-speed, numerically controlled machine as shown on Fig. 8.20. The polyamide insulation, rated at 2,000 V breakdown, makes it possible to cross the wires as close as 0.07 mm clearance without shorting out. Since crossover capacitance is less than 1 pF, crosstalk is negligible. Figure 8.21 illustrates the general fabrication approach which permits high density packaging applications. Thus, circuits which would require many levels of wire wrapping or several layers of PCB can be reduced to a single board of 1 mm thickness, more or less.

Uniform wire width and spacing reduces variations in applications requiring multiple supply voltages and ground planes, for controlled characteristic impedance (e.g., high-speed logic).

Since a Multiwire board has internal power and ground planes, the need for decoupling capacitors is in the same range as for multilayer. However, if in some specific areas a voltage is distributed by wire, its nominal 55 Ω characteristic impedance still requires the traditional use of high-frequency decoupling caps, mostly eliminated on multilayer PCBs.

From the transmission line point of view, such carefully-controlled 55 Ω impedances are ideal for high-speed logic interconnects to control backporching and other waveform problems. From the designer's and fabricator's points of view, the Multiwire process dif-

8.23

fers significantly from the traditional steps involved ranging from circuit concept to completed board.

Any fabrication process utilizing the Multiwire technique must be accomplished by or through the Multiwire Division* which owns the design and manufacturing rights to the product. A customer may be licensed to design and manufacture such products, but he must provide the following information:

1. Board overall configuration and fabrication data
2. Ground and voltage-distribution definition with edge connection finger locations and dimensions
3. Component locations and designations
4. A **from/to** net list

Other techniques have also been developed to match the flexibility of point-to-point wiring and the HF performances of PCB. For example, there are:

1. The SWITCHWIRE®, which uses an isolated nickel wire, point soldered to the circuit pads. The manufacturing speed is medium and can be done by an automatic machine. Typically suited for prototypes or small quantities, the wiring density can reach the equivalent of a eight-layer board.
2. The K6 and EMAFIL® use enameled copper wire, which is drawn continuously without interruption along all the points connected to the same potential, i.e., it tracks the schematic. At each pad, the enamel is vaporized and the copper is soldered on the side of the contact pad, therefore it is a versatile technique to switch later to a surface-mount process. The process allows for direct interconnects rather than a rigid X-Y pattern.
3. The SPEEDWIRE® from VeroBoard, where boards are equipped with pressure terminals which allow daisy-chaining of insulated wire and automatically strip out the insulation without cutting the conductor. SPEEDWIRE® is comparable to Multiwire.

---

*Multiwire Division, Kollmorgen Corporation, 31 Sea Cliff Av., Glen Cove, New York, 11542

Figure 8.20—Multiwire Machine Head Laying Down Insulated Wires in an Adhesive Substrate

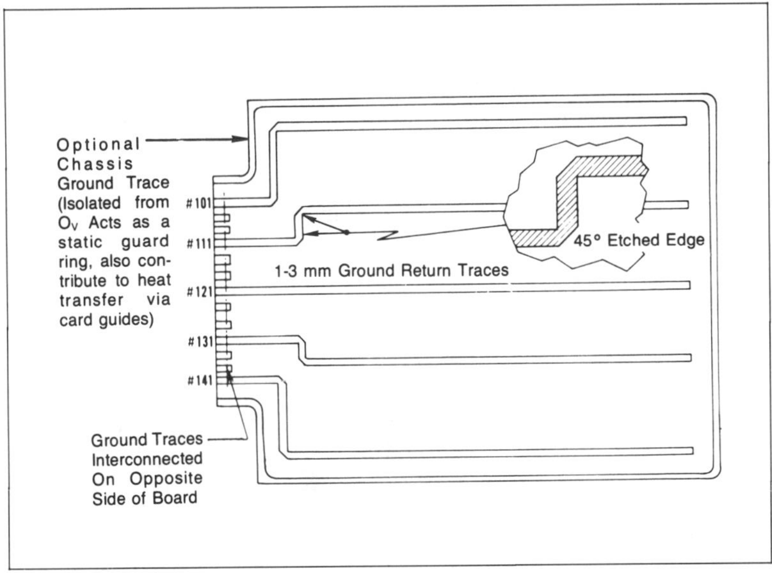

Figure 8.21—Expanded View of Typical Multiwire Board Realization

8.3.5 Reciprocity Between Susceptibility and Emission

What has been suggested up to now to make the PCB immune to self-jamming and ambient EMI is totally reciprocal with regard to radiation. A well-designed PCB, with minimum loop sizes, all signal traces running close to their returns or, better, above their return plane, will not radiate excessive levels and therefore will not cause the equipment to exceed FCC, VDE or MIL-461 radiated emission limits.

In addition to this, some precautions can help reduce emissions:

1. Select a slower (CMOS) or quieter (ECL) logic whenever possible.

2. Use low-profile chip sockets, or use direct chip soldering. Figure 8.22 shows that the radiating loop area for 1 chip can easily reach 0.4 cm^2, i.e., 25 chips represent a 10 cm^2 area! In the figure, the radiation is proportional to the current sunk from the capacitor, the loop area and the frequency F (or even to F^2 in the far-field).

3. Leadless chip carriers and surface-mount components yield less parasitic resistance, capacitances and inductances. This not only mitigates noise but allows for clock rates up to 4 GHz while the DIP package's limitation is 500 MHz (Fig. 8.23).

4. Watch for long signal traces which run all around the card. On Fig. 8.24, for instance, the run F-G-H-I-J-D-E is a potential radiating area.

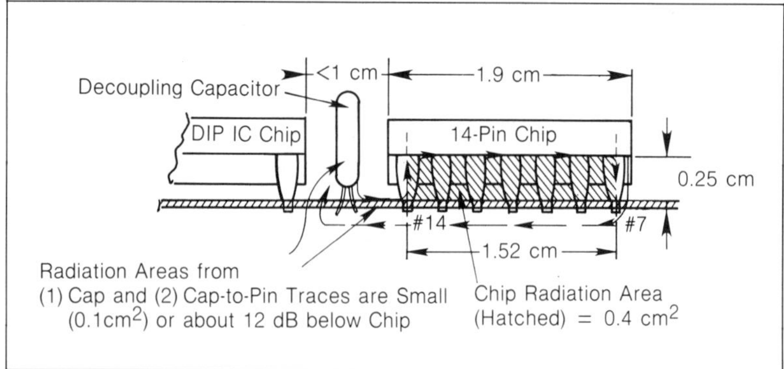

Figure 8.22—Radiation Area from 14-Pin Chip during Gate Switching Due to Instantaneous Supply Current Only (*For this aspect, flat packs and leadless packages exhibit less radiation.*)

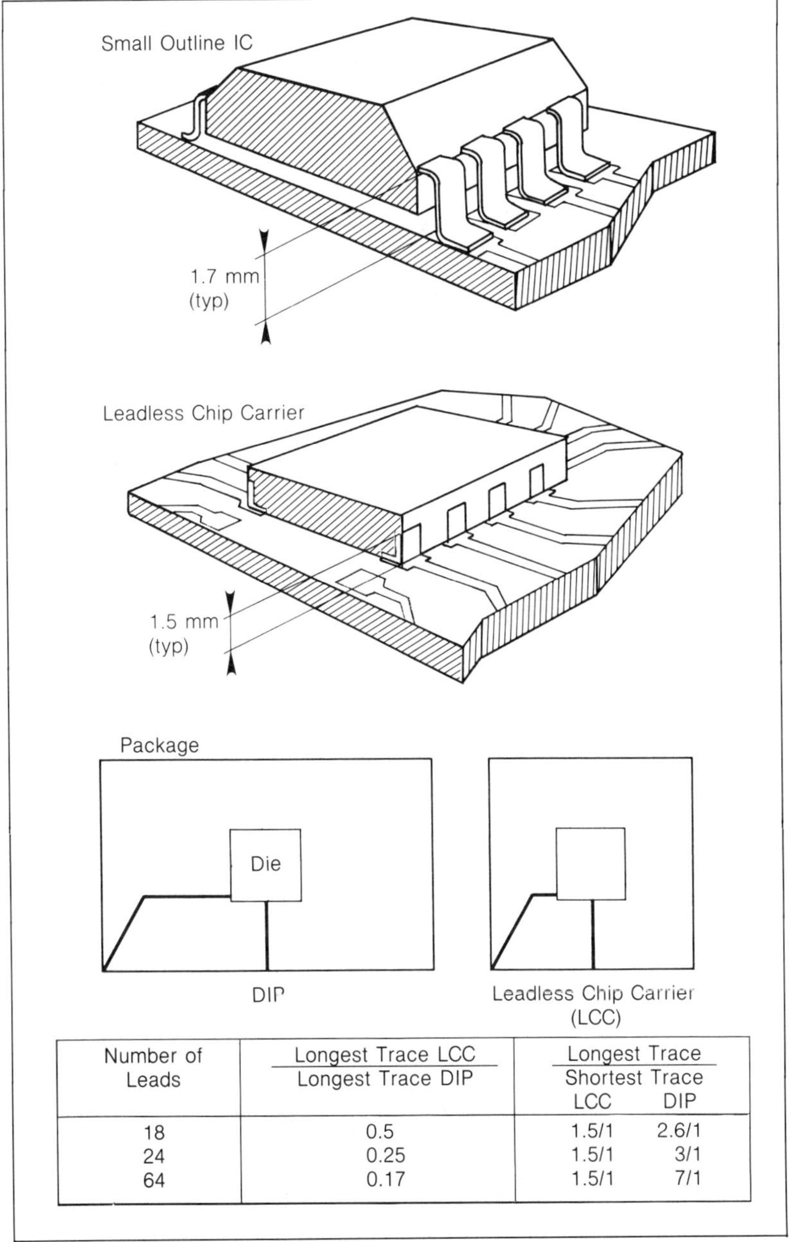

Number of Leads	$\dfrac{\text{Longest Trace LCC}}{\text{Longest Trace DIP}}$	$\dfrac{\text{Longest Trace}}{\text{Shortest Trace}}$	
		LCC	DIP
18	0.5	1.5/1	2.6/1
24	0.25	1.5/1	3/1
64	0.17	1.5/1	7/1

Figure 8.23—The use of surface-mount components (SMCs) provides a minimum of 2-to-1 reduction in the chip-to-traces area. This lessens both radiated emission and susceptibility problems and decreases parasitic inductance (*see Sec. 8.1*).

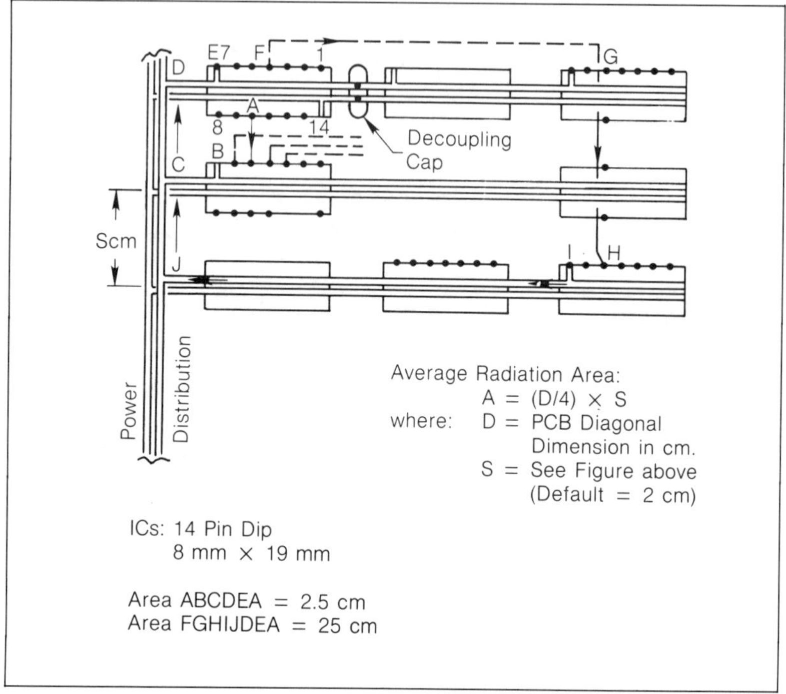

Figure 8.24—Radiation Loop Areas from Double-Sided, Single-Layer Board

8.3.6 Advantages of Surface-Mount Technology (SMT)

The benefit brought by SMT is shown in Fig. 8.25 where radiation levels of standard packaging are compared to those of surface-mount components (SMCs). Radiation from a PCB is made by two contributors: the traces and the chip/decoupling capacitor teams. Curves (1) and (2) reveal that with conventional package, traces radiation for the card shown are about 12 dB above chip radiation and are, therefore, the dominant mode.

Changing DIP packages to SMCs reduces chip radiation from curve (2) to curve (3). However, only a small overall benefit would be seen since traces, not chips, were the primary culprits. This small benefit is due to the 45 percent (or −5dB) reduction in card size brought by the lesser real estate of SMC.

The full benefit is seen when, in addition, the card becomes a multilayer (curve 4). In this case, the trace radiation area reduces to the trace length times the trace-to-ground plane height. This

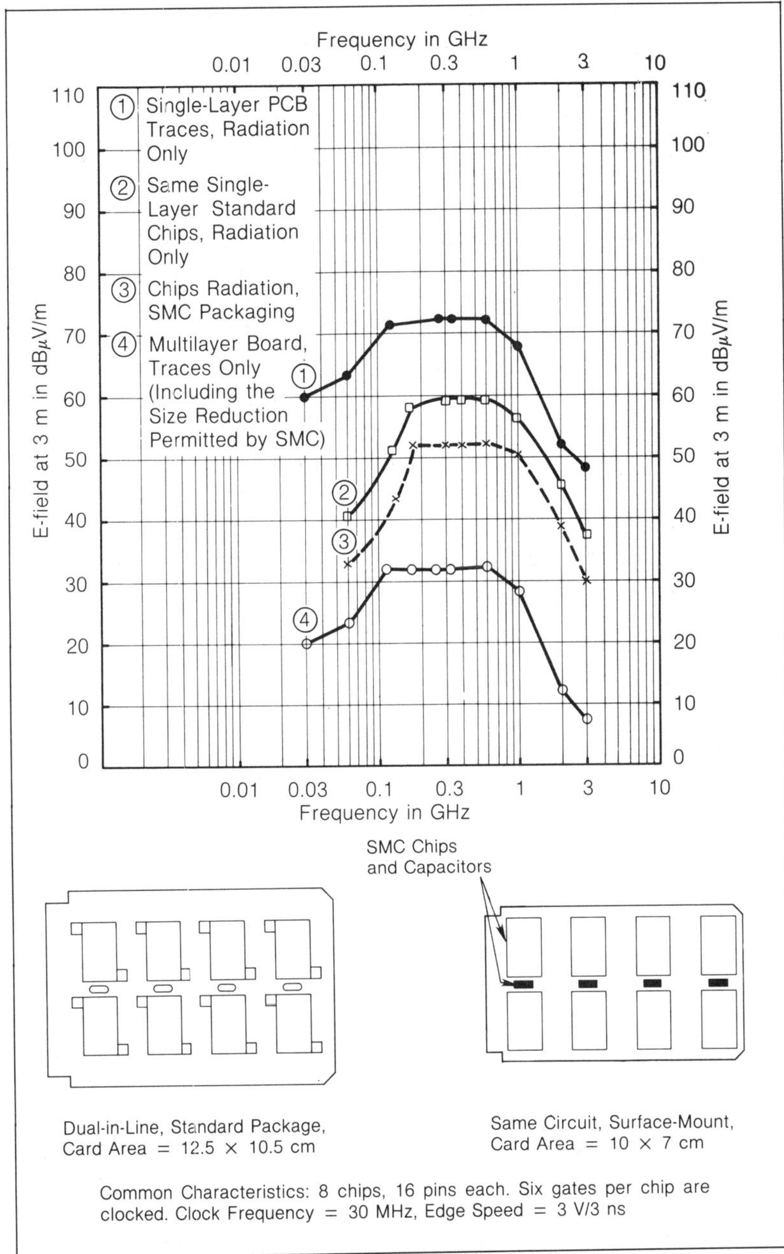

Figure 8.25—Radiation Comparison of a Single-Layer PCB (No Ground Plane) with Standard DIP versus the Same Card Changed to Multilayer (Less Trace Radiation) with SMC Packages (Less Chip Radiation, 45 Percent Decrease in Card Size)

reduction is so large that chip radiation now becomes the dominant mode, and this is why it has to be reduced by SMC packaging for both the chips and their decoupling capacitors. In other words, SMC and multilayer are complementary.

Even further reduction would be achieved if the packages used the center pin locations for $+V_{cc}$ and ground instead of opposite ends (see Section 4.4), and if the components were mounted on both sides of the board, as surface mount permits.

The SMC improvement can also be accounted in terms of the lesser die-to-leads trip length. The lead spacing which is typically 2.54 mm (0.1") with standard DIP can reduce to 1.27 mm or even 0.50 mm (0.02") with leadless chip carriers. Also, mounting "die-down" (or "cavity down") decreases the vertical run length. Because of the form factor of SMC being closer to unity than with DIPS, there are both an absolute decrease in lead length and a decrease in the ratio between the longest and the shortest leads. The former means a reduced lead impedance, allowing faster edge speed and clock rate for the same noise budget. The latter means a more even spread of delays between signals, allowing better timing. On the other hand, crosstalk is generally not reduced (it could even be aggravated) because the shrink in length is accompanied by a shrink in spacings.

So far, the only apparent limitations of the SMT approach are reliability and on-card testability (there is virtually no room nor feedthrough pins to give access to automatic test probes).

8.4 Summary

A successful EMI design starts at board level. For **Wirewrapped and single-layer PC board layout (one or two-sided):**

1. Allocate about 10 percent spaced connector pins to PCB ground lines.
2. Route dedicated ground lines and V_{dc} with traces > 1 mm wide.
3. Dedicate ground lines for high-gain analog circuits.
4. Landfill open areas with ground plane (or do not etch away; when viewed by transparency, the board should look as opaque as possible).
5. Check for crosstalk on long signal runs: if crosstalk budget (20 mV max. for ECL, 0.1 V for TTL and Schottky, 0.2 V CMOS) is exceeded, increase wire spacing or add a grounded guard trace between.

6. Implement necessary changes in computer aided design or design automation ground rules.
7. Decouple V_{dc} at connector with:
   a. 1 μF tantalum capacitor
   b. 0.01 μF HF ceramic disk or monolithic capacitor (0.001 μF   for high-speed logic).
8. Decouple V_{dc} for every 2 DIPs with 0.01 μF HF ceramic disk capacitor (0.001 μF for high-speed logic)
9. Consider raised power distribution for high-speed logic.

Multilayers, at extra cost, provide excellent noise reduction which might eliminate the need for extensive box shielding. The arrangement with voltage and ground planes **outside** and signal layers **inside**, is preferable. When signal traces are adjacent, check for crosstalk (as in a single-layer PCB).

8.5 Bibliography

1. White, Donald R.J. and Mardiguian, M., *EMI Control Methodology and Procedures,* (Gainesville, Virginia: Interference Control Technologies, Inc., 1985).
2. *MECL System Design Handbook,* (Motorola Technical Information Center).
3. Keenan, K., *Digital Design for Interference Specifications,* (Florida: TKC Corporation).
4. Minibus System Technical Catalog, (Rogers Corp., 1980).
5. Plonski, J.; Buck, T.; and Smookler, S., "Multiwire Boards Compete with Microstrip *and Stripline Packaging*" (Technical Report by Kollmorgen Corp.).

Chapter 9

Backplanes and Motherboards

After a correct PCB design has been established, the next step is to proceed with the backplane or motherboard design. The same techniques used in wirewrap and PCBs are used in backplanes. By definition:

1. A **backplane** is a printed circuit or wirewrapped board performing **interconnection** between the plugged-in PCBs.
2. A **motherboard** is a large printed circuit on which smaller PCBs (daughters) are plugged and which also has components (bulky discrete devices, power supplies, etc.).
3. A **planar board** is a large unique board which contains a composite of integrated circuits, discrete and electro-mechanical devices and the I/O connectors.

One problem with motherboards is that dimensions are larger and therefore all noise mechanisms are aggravated by one order of magnitude. For instance:

1. Since long parallel runs **(highways)** exist from one card location to another, they create **crosstalk**, which is a lesser problem with smaller PCBs.
2. Lines become **electrically long** and may require impedance matching, etc.
3. Because backplanes may be carrying hundreds of interconnect lines which are in a switching state during any given strobe gate, the propensity for **radiation** is severe.

One traditional way to suppress this by way of wirewrap backplanes is to lay down a grounded X-Y grid matrix on top of the wirewrap backplane, as shown in Fig. 9.1. A better method, which also improves quality control and lowers cost, is to replace the wirewrap backplane with a PCB backplane. This method usually leads to multilayer backplanes for the same reason that multilayer PCBs are used over the dual-sided, single-layer boards.

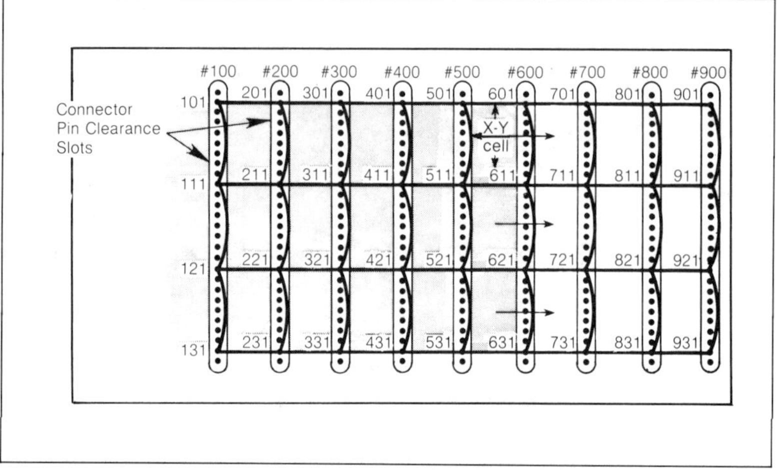

Figure 9.1—X-Y Ground-Grid Matrix in Backplane Wiring

A good way to start, **before all PCB layouts are frozen,** is to do the connector pin assignment and trace routes at the **motherboard**. By doing this, forbidden or dangerous vicinities can be avoided.

For instance, organize the runs by families so that:
1. No high-speed clock traces run close to sensitive traces (analog, alarm, reset) or to wires going to I/O connectors (interfacing outside world).
2. No high-speed clocks or data wires run without the protection of 0 V trace next to them.
3. About every 10th connector pin is a 0 V pin.
4. The $+V_{dc}$ distribution runs close to the 0 V traces or **plane.**

9.1 Crosstalk between Traces

Crosstalk occurs when a wire carrying fast signals is running close to another wire. The **culprit** wire induces, by mutual capacitance and inductance, a certain percentage of its voltage into the **victim** wire. Crosstalk increases when culprit and victim wires run close to each other, the frequency of the culprit signal increases (or the rise time is faster), the victim has a high impedance and the culprit and victim are far from the 0 V return. Crosstalk is expressed in dB by the formula:

$$\text{Xtalk}_{dB} = 20 \log(\text{V induced in victim/V culprit}) \qquad (9.1)$$

So -20 dB of crosstalk means that for 1 V of culprit voltage, 0.1 V peak will appear on the victim. More explanations and math models for crosstalk can be found in Ref. 1 and in Volume 8 of this EMC handbook series.

Table 9.1 gives average values of capacitive crosstalk (the one which predominates because of the high dielectric constant of epoxy) for 1 cm of trace length. For longer runs, the crosstalk increases accordingly. The procedure to apply in using the table is the following:

1. Select the geometry corresponding to the culprit-victim cross section.
2. Define culprit critical frequency or rise time.
3. Find the corresponding crosstalk per unit length (cm).
4. Apply length correction $= 20 \log(l_{cm})$ (do not use above length $l > 1/4$ culprit signal wavelength).
5. Apply impedance correction if $Z_{victim} \neq 100 \ \Omega$ by computing:

$$20 \log \frac{Z_{victim}}{100 \ \Omega} \qquad (9.2)$$

Table 9.1—Capacitive Crosstalk between PCB Traces for $Z_{victim} \cong 100\ \Omega$

W/S (Ccv,pF/cm) Freq.	W/h = 5 (Cvg = 2 pF/cm) Zo ≅ 34 Ω				W/h = 3 (Cvg = 1 pF/cm) Zo ≅				W/h = 1 (Cvg = 0.35 pF/cm) Zo ≅				W/h = 0.3 or no gnd return below traces (Cvg < 0.1 pF/cm) Zo ⩾ 120 Ω				Culprit Pulse Rise Time
	0.1 (0.02)	0.3 (0.04)	1 (0.08)	3 (0.16)	0.1 (0.03)	0.3 (0.06)	1 (0.1)	3 (0.2)	0.1 (0.05)	0.3 (0.10)	1 (0.20)	3 (0.28)	0.1 (0.08)	0.3 (0.12)	1 (0.30)	3 (0.40)	
1 kHz	−158	−152	−146	−140	−154	−148	−144	−138	−150	−144	−138	−136	−146	−142	−134	−132	Culprit
3 kHz	−148	−142	−136	−130	−144	−138	−134	−128	−140	−134	−128	−126	−136	−132	−126	−122	Pulse
10 kHz	−138	−132	−126	−120	−134	−128	−124	−118	−130	−124	−118	−116	−126	−122	−114	−112	Rise Time
30 kHz	−128	−122	−116	−110	−124	−118	−114	−108	−120	−114	−108	−106	−116	−112	−104	−102	10 μs
100 kHz	−118	−112	−106	−100	−114	−108	−106	−98	−110	−104	−98	−96	−106	−102	−94	−92	3 μs
300 kHz	−108	−102	−96	−90	−104	−98	−94	−88	−100	−94	−88	−86	−96	−92	−84	−82	1 μs
1 MHz	−98	−92	−86	−80	−94	−88	−84	−78	−90	−84	−78	−76	−86	−82	−74	−72	300 ns
3 MHz	−88	−82	−76	−70	−84	−78	−74	−68	−80	−74	−68	−66	−76	−72	−64	−62	100 ns
10 MHz	−78	−72	−66	−60	−74	−68	−64	−58	−70	−64	−58	−56	−66	−62	−54	−52	30 ns
30 MHz	−68	−62	−56	−50	−64	−58	−54	−48	−60	−54	−48	−46	−56	−52	−44	−42	10 ns
100 MHz	−58	−52	−46	−40	−54	−48	−44	−38	−50	−44	−38	−36	−46	−42	−34	−32	3 ns
300 MHz	−55	−46	−39	−33	−51	−43	−36	−30	−44	−36	−30	−28	−38	−32	−24	−22	1 ns
1 GHz	−53	−43	−33	−27	−50	−40	−30	−24	−42	−29	−22	−18	−32	−24	−16	−12	0.3 ns
3 GHz	−52	−42	−32	−26	−49	−39	−29	−23	−40	−27	−19	−15	−31	−21	−13	−9	0.1 ns
10 GHz	−51	−41	−31	−25	−48	−38	−28	−22	−38	−26	−18	−14	−30	−20	−12	−8	0.03 ns

NOTES:

- Epoxy glass is assumed ($\epsilon_r \cong 4$).

- Xtalk given as 20 log ($V_{victim}/V_{culprit}$ per cm of parallel run. For other lengths, add 20 log (lcm).

- For $Z_{victim} \neq 100\ \Omega$ (10 to 300 Ω), add 20 log ($Z_{victim}/100$).

- Clamp to 0 dB, Xtalk cannot be positive.

- Example of some typical values: W = 20 mils (0.5 mm) S = 30 mils (0.75 mm) for single layer, h = 30-40 mils (0.7 to 1mm). For multilayer h = 5 mils (0.12 mm).

- In regions above 1 GHz, where Xtalk approaches 0 dB, lesser actual voltages may appear due to strong "loading" of the culprit by the victim.

Table 9.1 —(continued)

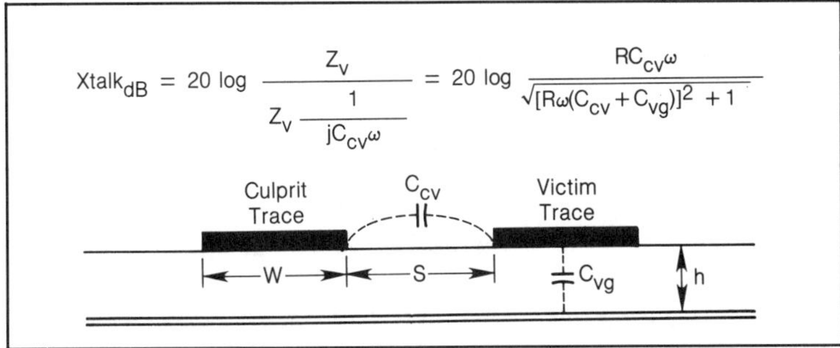

$$\text{Xtalk}_{dB} = 20 \log \frac{Z_v}{Z_v \dfrac{1}{jC_{cv}\omega}} = 20 \log \frac{RC_{cv}\omega}{\sqrt{[R\omega(C_{cv}+C_{vg})]^2 + 1}}$$

Illustrative Example: Two traces have a 10 cm parallel trip.
Trace width W = 20 mils (0.5 mm)
Trace separation S = 20 mils (0.5 mm)
No ground plane
Culprit = Schottky logic, 4 V swing, 3 ns rise time
Victim = Schottky logic, noise margin typical 1 V, worst case 0.4 V (input **low)**
Victim impedances = one gate **low** output impedance (about 50 Ω) on driving side, parallel with one gate **low** input impedance (about 1,000 Ω) on load side so total victim impedance = 46 Ω
Xtalk = −34 dB (table value) + 20 $\log(l_{cm})$ (length correction) + 20 log(46/100) (impedance correction)
Xtalk = −34 dB + 20 dB + (−7 dB) = −21 dB $\cong$ 9%

$$V_{victim} = 4 \text{ V} \times 0.09 = 360 \text{ mV} \tag{9.3}$$

This is slightly less than the **worst-case** signal-line noise margin, so normally no problem can be expected. However, if several adjacent culprit lines or fanout of several devices are running close to the victim trace, capacitive crosstalk may build up enough to upset the threshold of the receiving gate. We should also consider this value as too high compared to a noise budget like the one of Table 8.1.

9.2 Impedance Matching

With the longer dimensions of motherboards, termination of interconnect lines which are longer than twice the one-way delay is essential to avoid pulse ringing with high-speed logic. Correctly terminated lines will:

1. Reduce line noise by eliminating reflections
2. Reduce crosstalk
3. Avoid backporching when two-way delay $2\,T_d$ of line exceeds the signal rise time.

Given that the velocity of propagation is:

$$C = 3.10^8 \text{ m/s or 30 cm/ns in free space}$$
$$= 30/\sqrt{\epsilon_r} \text{ cm/ns in a dielectric with constant } \epsilon_r$$

The propagation delay T_d in a line of length l_{cm} with $\epsilon_r = 2$ to 2.5 is: is:

$$T_d = 1\,\frac{\sqrt{\epsilon_r}}{30} = 0.05\,l_{cm}, \text{ for Td in ns}$$

When T_d exceeds $T_r/2$, or when:

$$l_{cm} > 10\,T_r \text{ (for fanout} = 1\text{), and}$$

$$l_{cm} > 10\,T_r/\sqrt{N} \text{ (for fanout } N > 1) \tag{9.4}$$

it becomes imperative to terminate the line in a resistance equal to the line characteristic impedance. Figure 9.2 shows the problem of line matching. The decision of line matching versus rise time can be made by using the graph in Fig. 9.3 or Table 9.2, which has been computed for ECL. In a situation like Fig. 9.2, for instance, the line impedance Z_o is about 100 Ω while the input impedance Z_L of the driven gate was seen as about 400 Ω.

Coming from the source and arriving on this discontinuity, the 800 mV ECL pulse causes a reflection given by:

$$\rho_L = \frac{Z_L - Z_0}{Z_L + Z_0}$$
$$= \frac{400 - 100}{400 + 100} = +0.6 \tag{9.5}$$

So, a reflected pulse with an amplitude of:

$$0.6 \times 800 = 480 \text{ mV}$$

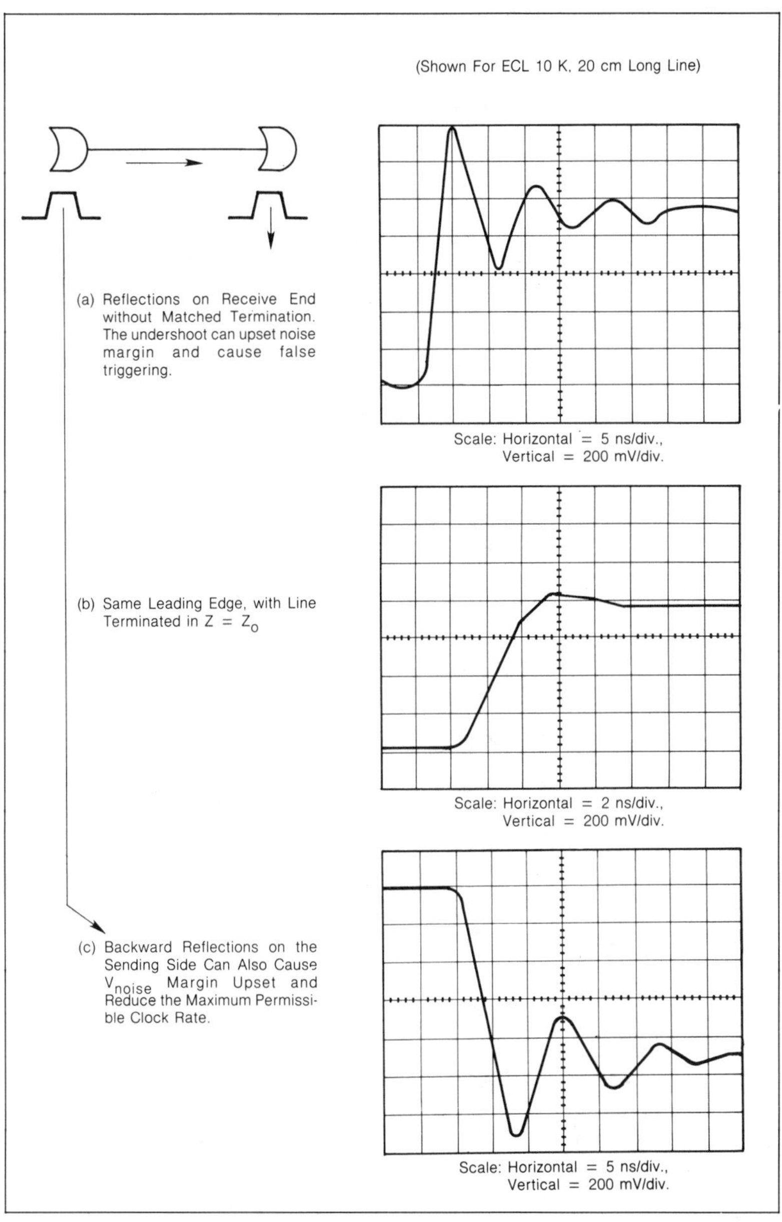

Figure 9.2—Examples of Termination Problems with Fast Logic

is going backward. The plus sign of the coefficient indicates a voltage addition, so the gate input "sees" actually a voltage equal to:

$$800 + 480 = 1,280 \text{ mV}$$

The 480 mV reflected wave travels back towards the source where, unless the source impedance matches the line impedance, it goes through another reflection with a coefficient:

$$\rho_G = \frac{Z_G - Z_0}{Z_G + Z_0} \qquad (9.6)$$

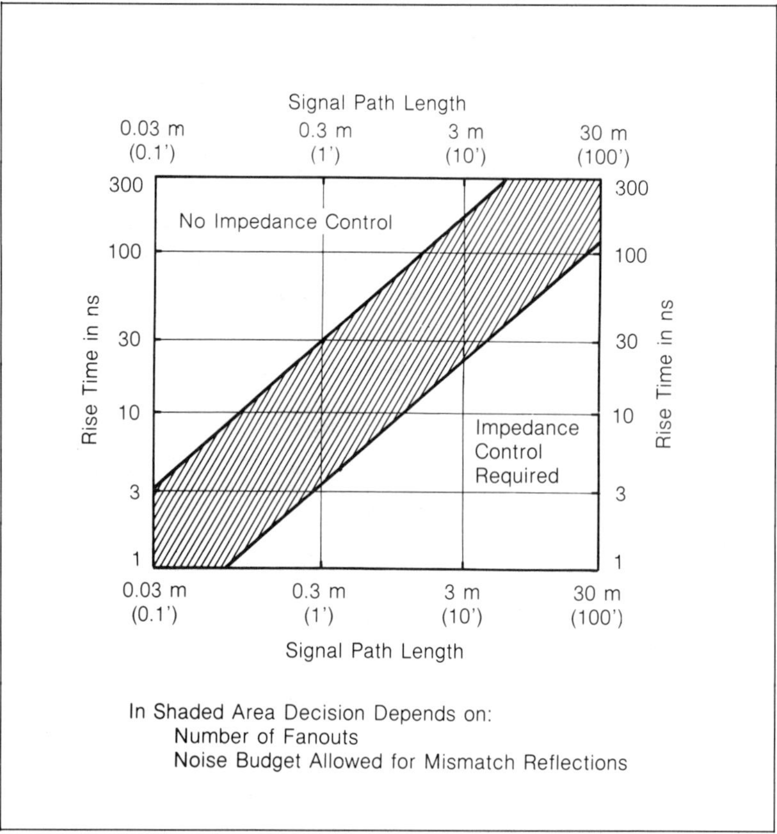

Figure 9.3—Rise Time/Signal Path Length Impedance Decision

Table 9.2—Maximum Unterminated Line Lengths versus Characteristic Impedance and Fanout (Computed for ECL-10K to Maintain Reflection Less than 35 Percent Overshoot, 12 Percent Undershoot)

	Characteristics Impedance Z_0 in Ohms	Maximum Line Length in Centimeters			
		Fanout = 1	Fanout = 2	Fanout = 4	Fanout = 8
MICROSTRIP: Propagation Delay 0.148 ns/in (58 ps/cm)	50	21.1	19.1	17.0	14.5
	68	17.8	15.7	12.7	10.2
	75	17.5	15.0	11.7	9.1
	90	16.5	13.7	9.9	7.6
	100	16.3	13.0	9.1	6.6
BACKPLANE: Propagation Delay 0.14 ns/in (55 ps/cm)	100	16.8	13.7	9.7	7.1
	140	15.0	10.9	7.1	4.8
	180	13.2	9.1	5.3	3.3

(From MOTOROLA ECL Application Note)

Since the generator impedance Z_G is generally lower than Z_0 of the signal line, this coefficient shows a negative sign which means that the new reflected wave is subtractive in terms of voltage.

As evidenced by Fig. 9.2, the forward and backward reflected pulses are "playing ping-pong" with a period equal to $2 \times T_d$. The corresponding oscillations at each end are decaying due to both the line losses and the reflection coefficients being smaller than one. But the first oscillations can upset the "0" or "1" margin of the driven gate. The sequence of events of a digital pulse mismatch is shown in Fig. 9.4. In time frame 9.4a, the load has not yet seen the coming pulse. In 9.4b, the reflected pulse travels backward toward the source. In 9.4c, the reflected wave has nearly reached the source-to-line interface. Finally, in 9.4d, the reflected wave sees a mismatch at the source, causing a new reflected pulse to travel toward the load. This pulse subtracts from the wave in Fig. 9.4c if $Z_g < Z_o$.

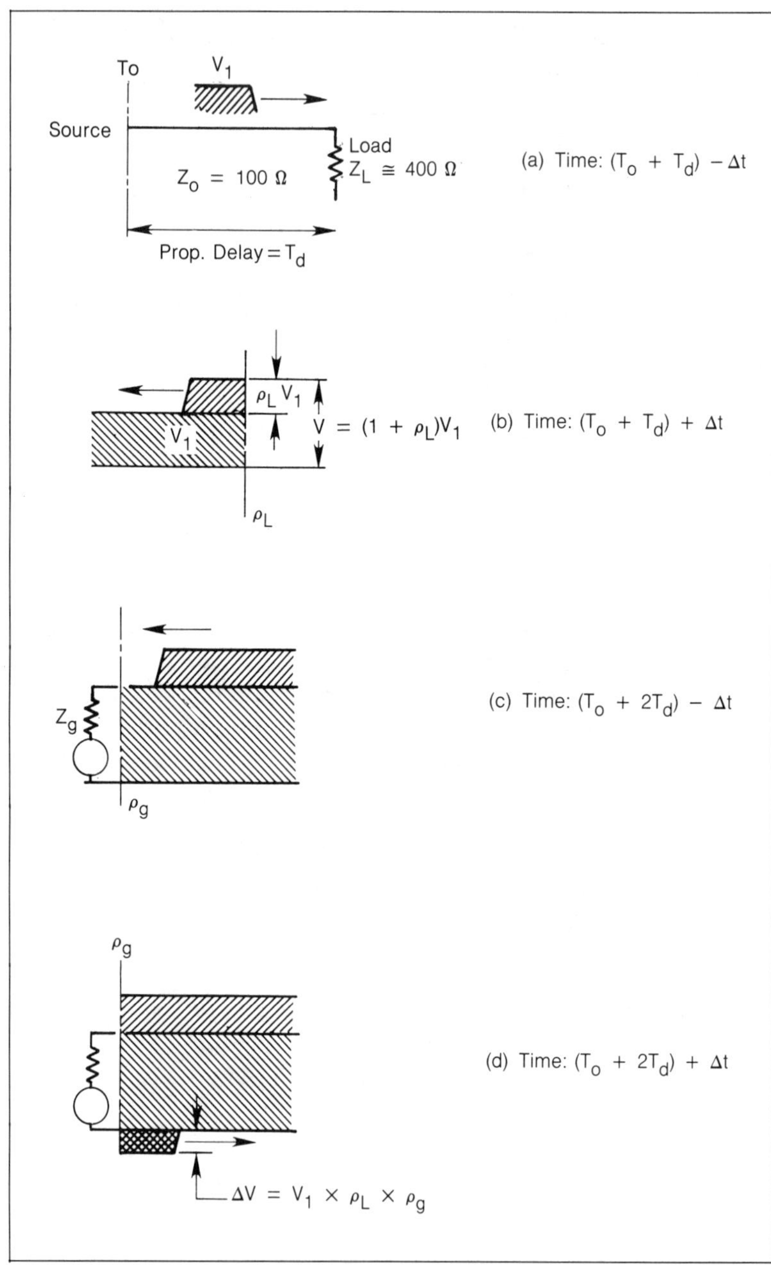

Figure 9.4—Sequency of Events with a Digital Pulse and Mismatched Terminations (Line Losses Not Shown)

9.3 Connector Areas at the Motherboard Interface

The I/O connectors area must continue the PCB-motherboard, noise-free concept. Preferably, if high speed rates are concerned, the connectors areas must respect the impedance matching like the PC traces and the cables, which means that alternate signal zero-volt pins may have to be provided at the connector to avoid discontinuities in characteristic impedance (see Fig. 9.5).

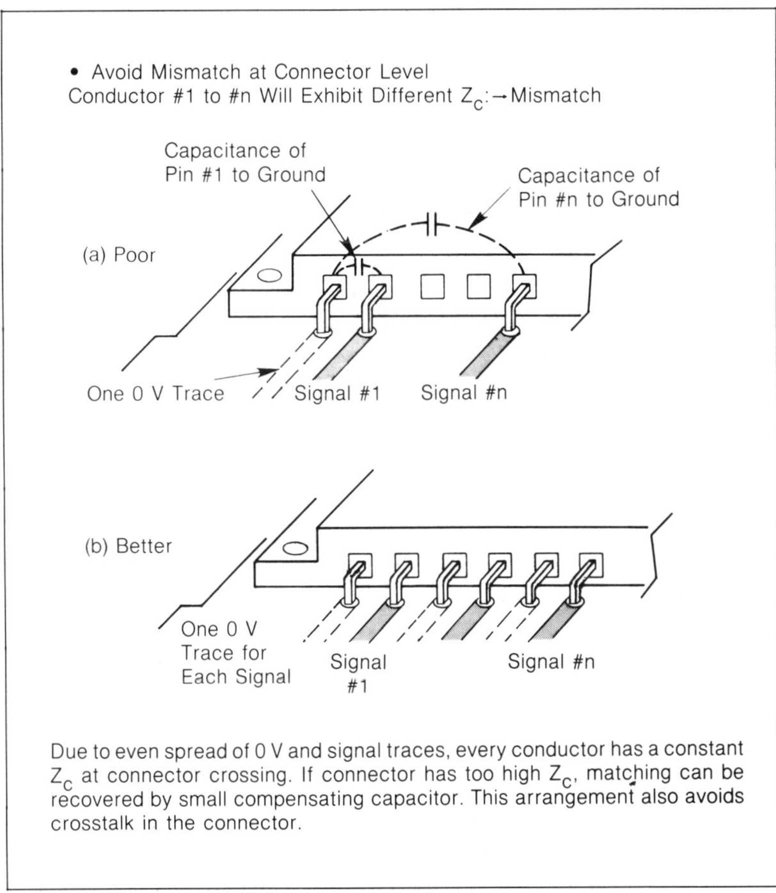

Figure 9.5—Interconnections of PC Board to Motherboard or I/O Cable for High-Speed Circuits

In any case, an extension of the board zero-volt traces or plane underneath the connector area is recommended (Fig. 9.6). It allows the most direct connection of all the I/O signal ground returns and makes it easier to achieve a direct coupling of noisy lines at connector level by discrete or **array** capacitors.

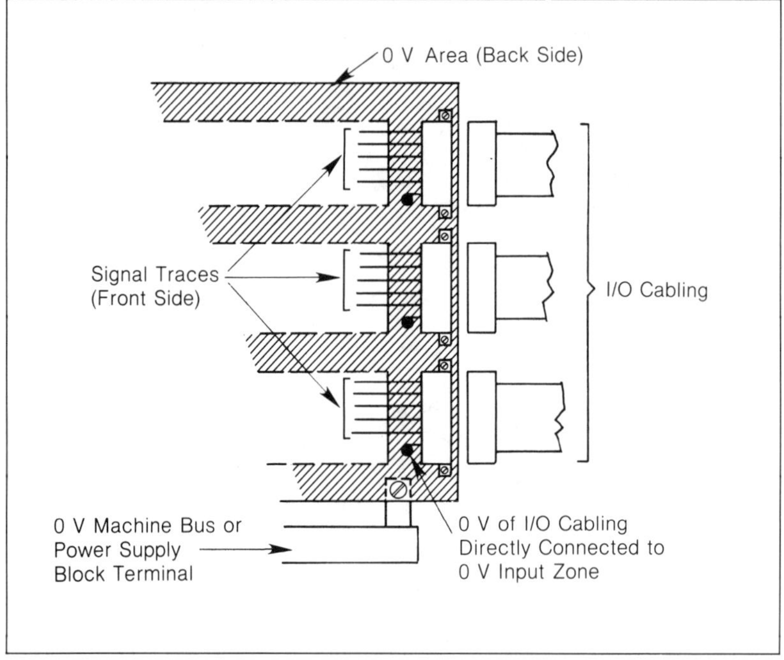

Figure 9.6—Treatment of Motherboards and Planar Boards at the I/O Interface

9.4 Summary

For successful EMI design of motherboards:
1. First do the pin assignment of the motherboard connectors.
2. Avoid close parallel runs of high-frequency clocks and signals with sensitive I/0 lines. A single clock wire can contaminate dozens of other lines by crosstalk.
3. Always **pair** a signal trace with its return trace, or better, with the ground plane.
4. When line becomes electrically long, implement impedance matching.

5. For noise elimination, the best techniques, in increasing order of efficiency:
    a. Wire-wrapped backplanes with a ground plane
    b. Double-sided PCBs with a ground plane on one side
    c. Multilayer boards

9.5 References

1. Metzger, G. and Vabre, J.P., *Electronique des Impulsions,* (Paris: Masson & C^i.e., second edition, 1975).

9.6 Bibliography

1. Motorola Application Note AN-556, "MECL Interconnection Techniques."
2. Southard, R.K., "High-Speed Signal Pathways," (*Electronic Engineering,* September, 1981).
3. Summers, J., "Interconnecting High-Speed Logic," (*Electronic Engineering,* March, 1985).
4. Wakeman, L., "Transmission-Line Effects Influence High-Speed CMOS," (*Electronic Design News,* June 14, 1984).

Chapter 10

Optoelectronics

Optical coupling, where light becomes the signal transmitting media, is the ultimate answer to many isolation and environmental immunity problems, EMI/RFI being just one of them. Optical transmission is used in optical readers and encoders, object counting and proximity sensing, optoisolators and fiber optic links. In terms of EMI, the benefit of optical coupling is the discontinuity created in electrical paths such as ground loops. Common impedance and earth potential differences become harmless. (See Volume 2, *Grounding,* of this handbook series).

An optical coupling system is based on:

1. A light source which converts an electrical signal into a corresponding light signal (except in object counting, rotational encoding, etc., where the light source does not need to be modulated by an electrical signal)
2. An optical transmission media: air, liquid or fiber
3. A detector which converts the received modulated light into a corresponding electrical signal

As far as EMI is concerned, the first and last items are the most relevant.

10.1 Light Sources

Light sources for optoelectronics can be incoherent (incandescent, gas discharge, electroluminescent, LED) or coherent (laser diode). Of the incoherent sources, the most used are incandescent (for optical interrupters and presence sensors) and LEDs. The latter are widely used in optical couplers and fiber optic systems. Not mentioning the wavelength and other optical features, the relevant electrical parameters for a light source are:

1. Its radiation efficiency which is:

$$\eta = \frac{\text{light emitted in photons}}{\text{input diode current in electrons}} \tag{10.1}$$

2. Its rise time

The first term will dictate the characteristics of the driving amplifier, the second term will relate to the available bandwidth. Table 10.1 summarizes some electrical features of optical coupler sources.

Table 10.1—Essential Characteristics of Light Sources

	LED (Broadband, Incoherent)	LASER (Narrowband, Coherent)
Typical Drive Current	10-50 mA	a few A to 15 A
Mode (Pulsed or Continuous)	P, CW	P
Max. Pulse Duration	Any	2 to 0.1 μs
Duty Factor	Any	< 1%
τ_r	300 to 5 ns	< 1 ns
Corresponding Bandwidth	1 to 60 MHz	> 300 MHz
Efficiency η	0.1 to 3%	5 to 40%

The LED incoherent sources are easier to drive, have better stability and longevity and are ideally suited for opto-isolator or low/medium-range fiber optic links.

Because of their narrow optical bandwidth, laser sources allow for less dispersion in the fiber and therefore are used for long-range and high-data-rate fiber optic links.

Two critical EMI and noise considerations with sources are:

1. Transforming a signal into light does not free the designer from taking EMI precautions on the side of the driver. The light source does not "filter" EMI from the wanted signal. If the

designer neglects EMC precautions at the driving amplifier, the signal will be corrupted and EMI will be carried in the form of light.

2. LED and laser sources are highly nonlinear devices. Therefore, while ideal for digital or two-state signals, they are more difficult to use with low-level analog signals when accuracy is important. In addition, their nonlinear nature gives way to possible audio rectification and RF demodulation problems in the presence of intense EMI ambients (see Chapter 4).

10.2 Light Detectors

Devices used to convert light into an electrical signal may operate in two modes:

1. Photovoltaic, where the device generates a voltage when exposed to light
2. Photoconductive, where the device behaves as a resistor whose value varies with incident light

In both modes, solid-state detectors are photocells, avalanche photo diodes, phototransistors and pin diodes.

Just like optical sources, detectors have an efficiency criterion which relates the output electron flow (current) to the input photon flow (light). Principal parameters for light detectors are shown in Table 10.2.[1]

When used in the photoconductive mode, the photodiodes need a reverse dc polarization to improve the sensitivity and make the diode work near the reverse breakdown knee. Just for active devices (Chapter 4), where sensitivity was defined as when $S = N$, the threshold of sensitivity for a photodetector can be defined by the "dark current," i.e., the current flowing through the detector circuit when there is no light.

Dark current increases with temperature, following the approximate rule of doubling the current for every 10°C increase. This dark current can be translated into noise equivalent power (NEP). The NEP in $W/\sqrt{Hz}$ is the optical power which would create an input current equal to the dark current if the detector were ideal. Multiplying NEP by the square root of the detector bandwidth gives the minimum discernable signal (MDS).

To ensure good signal processing with some margin, the receiver should be excited by an optical power higher than the MDS. This concept corresponds to the S/N ratio with conventional amplifiers. Since the current generated by the photodiode or phototransistor is small, it needs to be followed immediately by a preamplifier whose intrinsic noise adds up to the detector noise itself. Just as for light sources, the designer is not exempt from taking EMI precautions at the post-detector level, especially because of the high impedance and gain of the preamplifier.

Table 10.2—Common Detector Terms and Definitions

Term	Symbol	Definition
Sensitivity	$\rho(\lambda)$	The ratio of output current to the incident optical power measured in amperes/watt at a given wavelength
Noise	V_n	Extraneous voltage generated by a detector
Johnson or Thermal Noise	V_J	Noise due to random motion of electrons in a resistance element. The root-mean-square Johnson noise can be approximated as $V_J = \sqrt{4KTR\Delta f}$. Boltzman's constant, K, is 1.38×10^{-23} joules/kelvin, T is the absolute temperature in degrees kelvin, R is the resistance in ohms, and Δf is the electrical bandwidth of the circuit in Hz.
Noise Equivalent Power	NEP	The incident radiation in watts required to produce an output signal equal to the detector noise
Detectivity	D	The reciprocal of NEP
Specific Detectivity	D*	Detectivity normalized for a detector area of 1 cm^2 and a bandwidth of 1 Hz. It is equal to $D\sqrt{A\Delta f}$, where A is the detector active area in cm^2 and Δf is the bandwidth in hertz. When reporting D*, the wavelength or the black-body temperature and chopping frequency at which the data are taken must be specified.
Time Constant		The period required for a detector to reach 63.2 percent of its final output value following the application of steady-state incident radiation. The time constant equals $1/2\,\pi f$, where f is the chopping frequency in Hz at which the frequency response begins to roll off at 3 dB/octave.

10.3 Limitations in Linear Operation of Optical Links

An emitter/detector pair is a perfect device for switching and counting as well as for any two-state application. But being non-linear in nature, the current-to-light-to-current transfer function cannot be relied upon for accurate low-level analog applications. Performance can be improved to some extent by AGC circuits and matched photodiode feedback compensation. Temperature and aging also cause the transfer function to fluctuate.

10.4 Optical Isolators

Also called **optical barriers** or **optocouplers**, optical isolators play a large role as galvanic isolation devices in industrial controls, telephones, digital line receivers, isolation amplifiers (see Section 4.3.4), medical instrumentation, solid-state relays, etc., as well as any application where one or several of the following objectives are important:
1. Isolation of different voltages between circuits
2. Prevention of undesirable coupling from power circuits to control circuits
3. Protecting people from hazardous voltages
4. Elimination of dc or ac ground loops
5. Interface between two circuits having different ground references
6. Contactless/bounceless/sparkless switching, etc.

Figure 10.1 shows some basic types of optical isolators, and Fig. 10.2 shows three common applications. As illustrated in Fig. 10.2a, the triggering of triacs with optoisolators is accomplished relatively easily. The diode bridge converts the ac to the dc required by the SCR. Figure 10.2b depicts how excellent isolation between a patient and monitoring equipment can be achieved by optoisolators when used in medical electronics. Finally, Fig. 10.2c shows an IBM line receiver application.

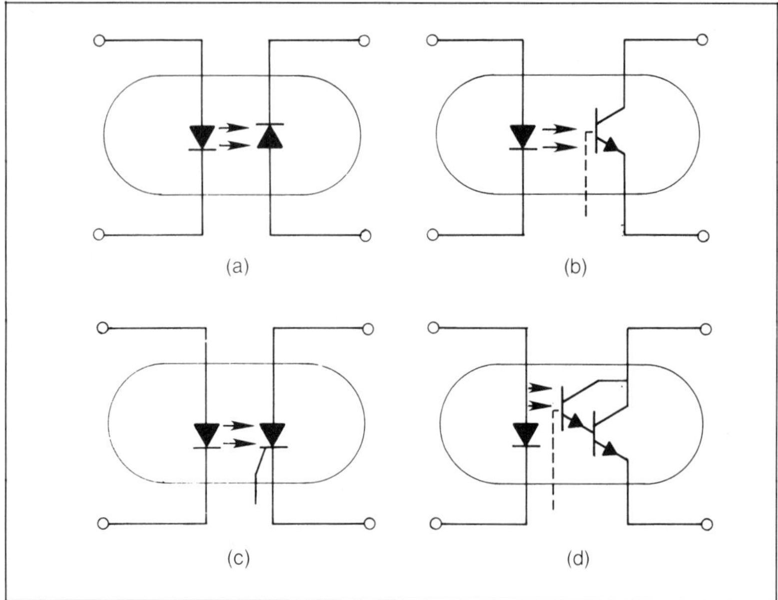

Figure 10.1—Basic Types of Optical Isolators: (a) LED Photodiode (b) Phototransistor with or without Base Terminal (c) LED Photo-SCR and (d) LED Photo-Darlington.

The dynamic and EMI characteristics of optical isolators follow:

Rise and fall time delimits the maximum useful bandwidth of the optical isolator. The simplest and cheapest optocouplers with only a diode/phototransistor pair typically have transition times in the 10 to 100 μs range for a digital-type signal. By adding Schmitt triggers and positive feedback amplifiers, the switching speed can be improved greatly. By 1986, modern optocouplers with a 30 ns transition time corresponding to a 10 MHz bandwidth were available.

Input/Output dc isolation is defined by at least two parameters: the isolation resistance R_{iso} and the maximum withstanding dc voltage V_{iso}. The former is sometimes defined by the maximum input-output leakage current under a given voltage like 1 or 3 kV. Typical values are:

$$R_{iso} = 10^9 \text{ to } 10^{11} \ \Omega$$

$$V_{iso} = 500 \text{ to } 5,000 \text{ V}$$

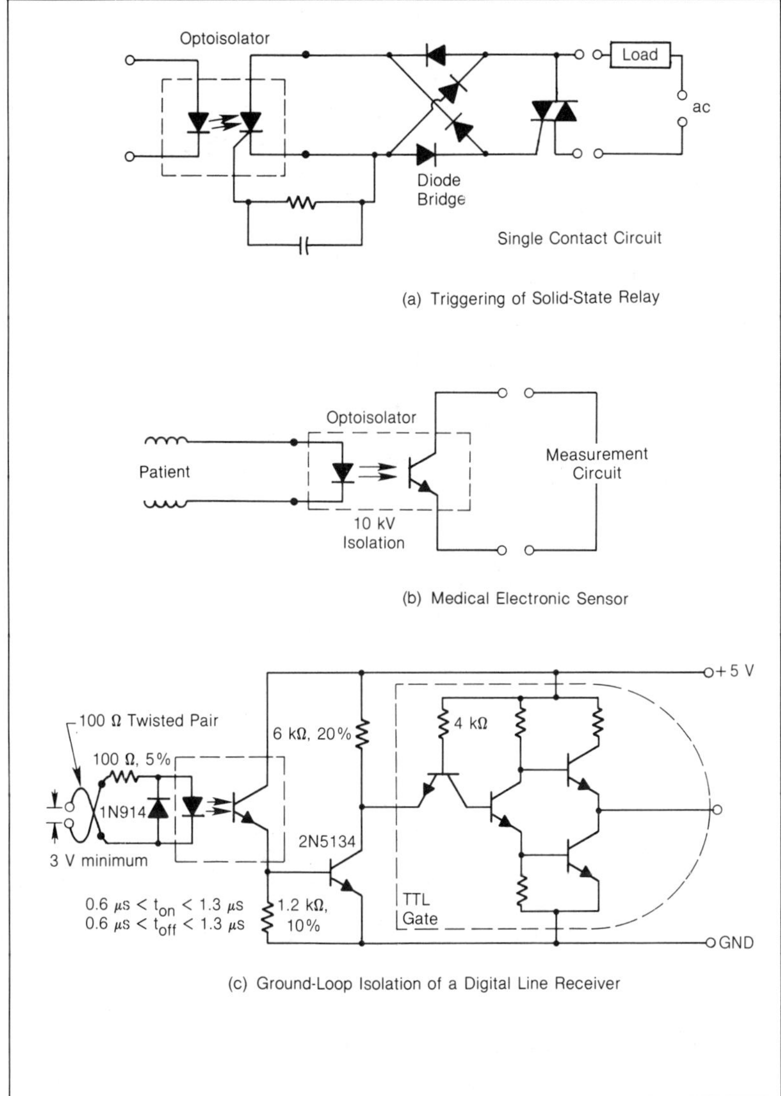

Figure 10.2—Examples of Optoisolator Applications

However, for extreme environments, optical isolators with up to 15 kV V_{iso} are obtainable. This voltage should normally be guaranteed between input and output pins or between any pin and the device's can, whichever is less.

Input/output capacitance (Fig. 10.3) consists of two capacitances: the internal LED-to-phototransistor (or other detector) capacitance which typically ranges from 0.1 to 2 pF, and the input-pin-to-output-pin capacitance which depends greatly on the package style and which ranges from 0.3 to 3 pF. The combination of these two capacitances dictates the ac isolation since they bypass the R_{iso} above a certain frequency. Each one participates in its own way to the common-mode rejection of the optical isolator, as explained in the next paragraph.

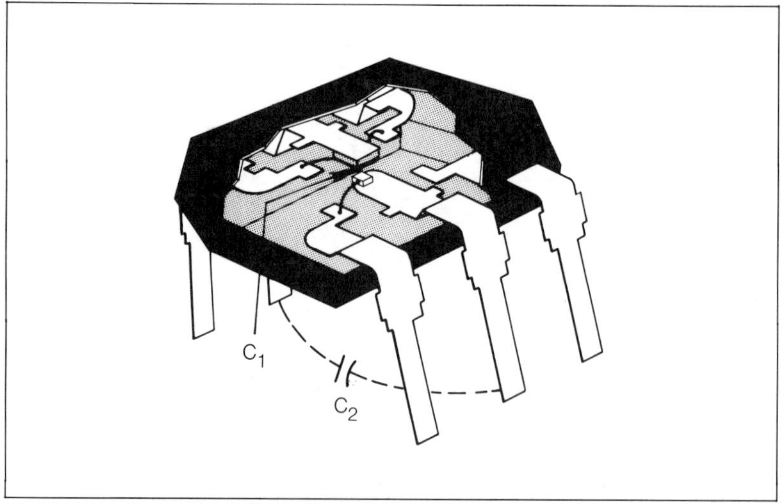

Figure 10.3—Parasitic Capacitances in Optoisolators. Input/Output parasitic capacitance consists of intrinsic LED-to-Photodetector capacitance C2 plus package & leads capacitance C1. It can be reduced by internal Faraday thin mesh. it can be aggravated by careless input/output wiring isolation.

Common-mode rejection (CMR) or transient immunity is important to describe the device's immunity to EMI, but it is the one criterion which is usually the most poorly documented (if documented at all) by manufacturers' technical data sheets. Some

of them label it as maximum input transient voltage without specifying the rise time. Some of them define a slew rate in V/μs. Some
of them do not specify it at all. In the few well-documented data
sheets available, the CMR is described as the maximum slew rate
in V/μs of CM voltage which can be sustained with the output
voltage staying in either a "high" (> 2 V) or "low" (< 0.8 V) status.

Figure 10.4 shows that the transient immunity is indeed a two-
mechanism phenomenon:

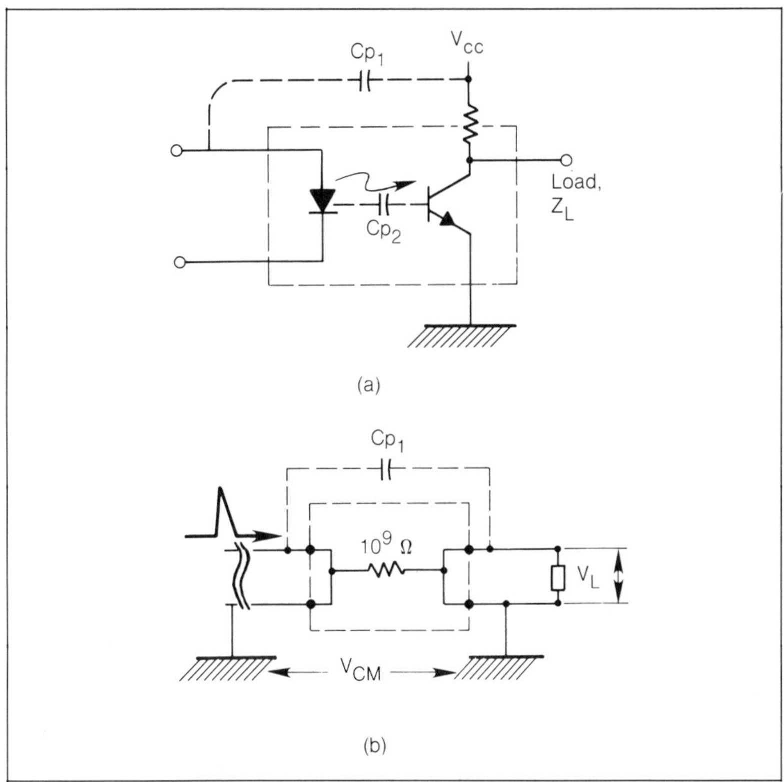

Figure 10.4—Equivalent Circuits of Optical Coupler CM Transient
Response (continued next page)

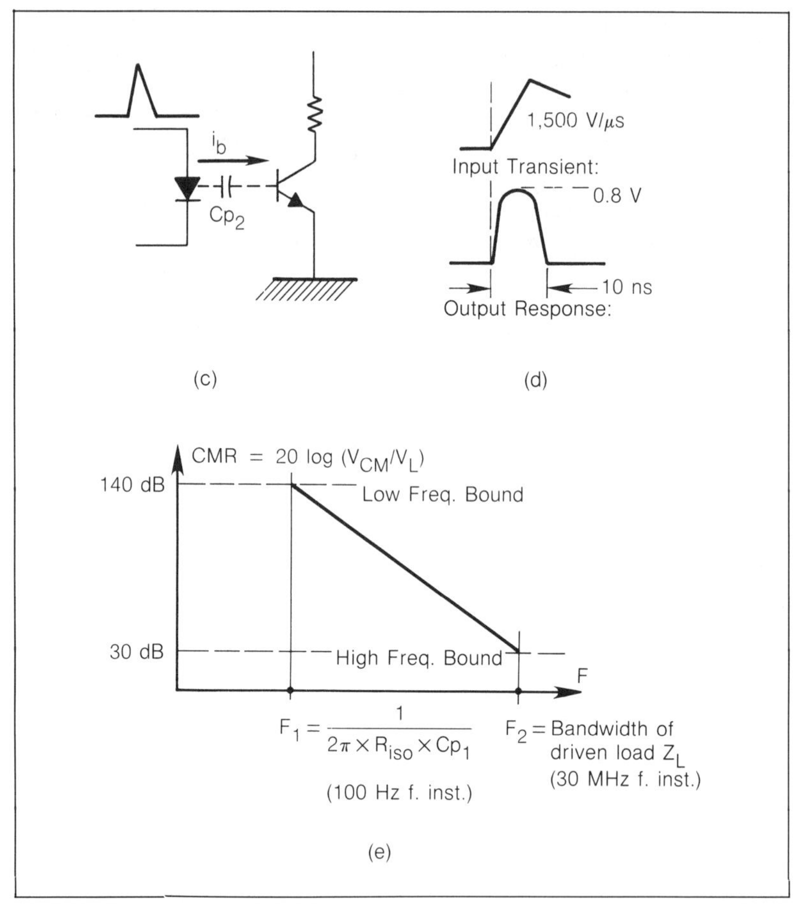

Figure 10.4—(continued)

1. **The input-lead-to-output-lead capacitance** (Cp_1) is a capacitance which bypasses the R_{iso} of the device, i.e., the device becomes a leaky barrier and some percentage of the EMI voltage appears across the load Z_L **without the optical isolator playing any active role in this transfer.** For instance, if we take Z_L as 1 kΩ (a typical dynamic resistance of a TTL gate input in the low-to-high transition region), this value is shunted by the optical isolator output resistance (typically 50 to 150 Ω), so the optical isolator output virtually looks like a 50 to 150 Ω load. The low-frequency bound of CM rejection is given by:

$$CMR_{max} = \frac{Z_L}{Z_L + R_{iso}} \qquad (10.2)$$

Assuming that: $R_{iso} = 10^9$ and $Z_L = 100$

this gives: $CMR_{max} = \dfrac{100}{100 + 10^9} = 10^{-7}$ or 140 dB

This CMR starts to deteriorate as soon as the reactance of Cp_1 by-passes the 10^9 Ω. Assuming $Cp_1 = 1.5$ pF (another typical value for optoisolators), this will occur for:

$$F_1 = \frac{1}{2\pi RC} \cong 100 \text{ Hz}$$

Beyond this frequency, the CMR will degrade at 20 dB/dec. Therefore, it would theoretically reach 0 dB for:

$$F_2 = 100 \text{ Hz} \times 10^{\frac{140}{20}} = 1 \text{ GHz}$$

However, when the frequency reaches the cutoff frequency of the load (typically the input of a digital gate, or a line receiver, comparator, etc.), the load itself, by its input capacitance, starts to have an ac noise rejection which improves at the same rate that the CMR degrades. For instance, with a TTL-type load beyond 30 MHz, the CMR will stay flat to a value computed by:

$$CMR_{(30 \text{ MHz})} = 140 \text{ dB} - 20 \log\frac{30.1^{-6} \text{ Hz}}{100 \text{ Hz}}$$

$$= 30 \text{ dB}$$

2. **The internal LED-to-detector capacitance** (Cp_2) exists because of the physical proximity of the LED and photodetector (on the order of one to few millimeters) and is aggravated by a resin lens used to channel the light and improve the overall efficiency. This resin has an $\epsilon_r > 1$ which aggravates the capacitance Cp_2.

   This capacitance can make the optoisolator **electrically triggered** if a high enough dV/dt exists across the optical barrier, like a transient between any or both input leads and the local ground. This noise current, $Ip = Cp_2\ dV/dt$, becomes a base current into the phototransistor.

Assuming that this one has a gain of 100 and a collector current of 1 mA when conducting, a base current of 0.01 mA will turn on the transistor.

From the previous equation, the voltage step which can cause this base current is:

$$dV/dt = 0.01\ \text{mA}/Cp_2 \tag{10.3}$$

If Cp_2 equals 1 pF,

$$dV/dt = 10^{-5}/10^{-12} = 10^7\ \text{V/s or 10 V}/\mu\text{s}$$

A simple electromechanical switch can cause spikes faster than this. Thyristors and other semiconductor switches cause transients in excess of 100 V/μs. Static discharges induce transients in the range of 100 V/ns; therefore, many real-life transients can upset an optoisolator even though its immunity based on dc data might seem impressive. Quality optoisolators are characterized against this parasitic turn-on, where the isolator is becoming an active device in the transmission of EMI. A good brand of modern isolator can resist up to 500 V/μs or even 3 kV/μs. However, these values are generally given for 25°C and degrade rapidly with an increase in temperature.

For instance, assume there is an optoisolator specified for 1,000 V/μs of CM transient immunity and a TTL type output. This means that when the output is at a **low** status, it takes at least a 1,000 V/μs spike to cause the output to exceed 0.8 V for more than 10 ns (the typical TTL minimum transition time). Therefore, the shortest pulses to cause an undesired response are a 10 V pulse with 10 ns transition time or a 100 V pulse with 100 ns transition time. For pulses having less than 10 ns transition (i.e., a bandwidth exceeding 30 MHz) the ac noise rejection of TTL as well as the time constant of the phototransistor will naturally improve the rejection by the same rate (20 dB/dec) as the capacitively injected base current increases, giving an overall flat CM rejection.

If we calculate the rejection of the above example for the worst-case pulse of 10 V/10 ns and consider that of the 0.8 V output, half (0.4 V) is due to the V_{CE} saturation of the output transistor, we come up with:

$$\text{rejection} = 20 \log \frac{10 \text{ V}}{0.8 - 0.4} = 28 \text{ dB (for TTL)} \qquad (10.4)$$

This is broadly in the same range as the CMR due to C_{p_1}, although one should compute each of them separately and retain the lower figure.

Transient immunity can be improved in several ways:

1. Decoupling of the phototransistor base. This is efficient but has a corresponding adverse effect on bandwidth.
2. Increasing the separation distance between the LED and the optical detector. This can go as far as installing a short piece of fiber optic to act as a light guide. This is an effective solution, but it increases the size of the device.
3. Inserting a thin metal mesh (optically transparent) between the LED and phototransistor. The principle is the same as Faraday-shielded transformers. The shield must be tied with a low-impedance conductor to the common ground on the detector side (see Fig. 10.5).

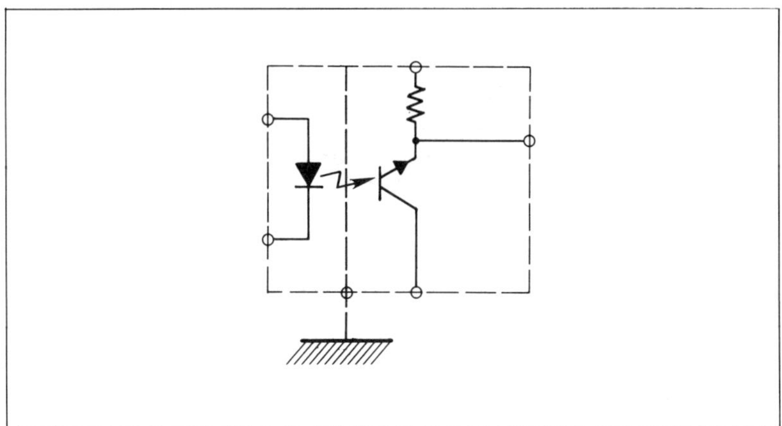

Figure 10.5—Optoisolator with Faraday Shield

Input ac and dc impedances should be considered because, being nonlinear, the LED input cannot be assimilated to a simple RC network. However, to avoid the LED overrun if the input signal exceeds the forward or reverse break voltage (Fig. 10.6), the input of an optoisolator is always driven in current mode, i.e., an input resistance is used. The useful range of I_F current is between 1 mA and 100 mA, corresponding to a V_F of 1.2 to 1.3 V. Thus, for a given range of input signal, the designer selects a series resistance R_S such as:

1. For the minimum input signal amplitude, V_{min}, the upper limit is:

$$R_{S\ max} = \frac{V_{min} - V_F}{I_{min}} \tag{10.5}$$

I_{min} being the minimum current to drive the optoisolator with the desired output response, i.e., amplitude and transition time (response time degrades when I_F decreases), considering also the worst-case temperature drift for V_F.

2. For the maximum input signal amplitude V_{max} the lower limit is:

$$R_{S\ min} = \frac{V_{max} - V_F}{I_{max}} \tag{10.6}$$

I_{max} is the maximum current compatible with the diode safe operating area (or maximum power dissipation $V_V \times I_F$) and the diode aging (for instance, 10 percent brightness degradation for $I_F = 60$ mA after 10,000 hours).

Figure 10.6 also shows the LED being shunted by its own junction capacitance, typically in the 30 to 100 pF range.

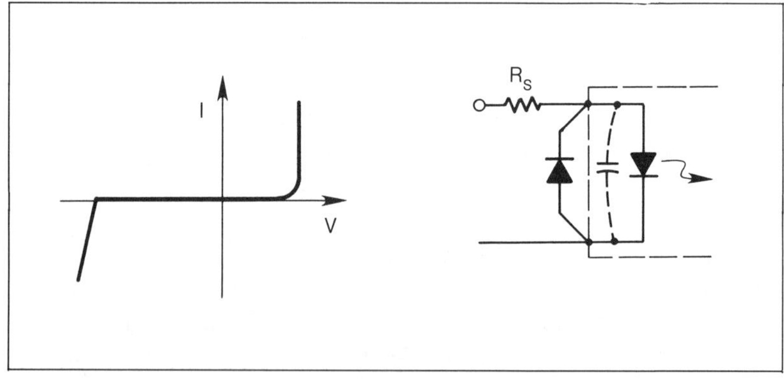

Figure 10.6—Characteristic of a Typical LED

If for instance, a 220 Ω series resistance is used to provide a 10 mA forward current for a 3.5 V input signal and a V_F of 1.3 V, a 50 pF parasitic capacitance will start to shunt the diode V_F at around 15 MHz, which gives the practical bandwidth of this LED input for a differential signal. As seen in Fig. 10.6, a reverse voltage protection can be provided by a silicon diode across the LED.

In some cases it is desirable to set a definite threshold for the LED voltage. This is done by shunting the LED by a resistor, the value of which is determined by the applied voltage, the series resistance and the desired V_F.

All these considerations are necessary to predict the behavior of the LED input in the presence of differential EMI. Although ground loop interference generally prevails in the EMI problems dealt with by optoisolators, EMI coupled differentially into the two wires of the cable pairs should not be overlooked. For instance, if computer cables are running in the same conduit with power cables over few meters, a 1 kV/μs transient on these cables can induce several volts per meter differentially in the signal pairs, which is enough to drive the optoisolator into an erroneous triggering. In this case, the ground-loop isolation provided by the optical barrier is without effect on the interference, and the solution resides in more conventional shielding and separation of the cables.

Finally, another EMC aspect of optical isolators is the way they are mounted. A few picofarads of input-output coupling are very easy to aggravate by careless wiring practices. Two signal wire pairs spaced by 3 mm in the same cable way already represent 5 pF/m of coupling capacitance. Input wires or traces must be kept away from their output counterparts. Figure 10.7 provides a comparison of pin-to-pin isolation of DIP optoisolators to that of a standard logic gate (from Ref. 2). In both cases, the devices were shut off, so the coupling is mainly due to the lead arrangement.

An optoisolator must be mounted as close as possible to the I/O connector. Return conductors (even if called "ground") for the input signal should be floated and distinct from the ground conductor of the detector side.

An optoisolator with an external base connection for the photo transistor (or SCR) should be treated carefully since this base lead can be very susceptible; it should be filtered and kept away from possible noise paths.

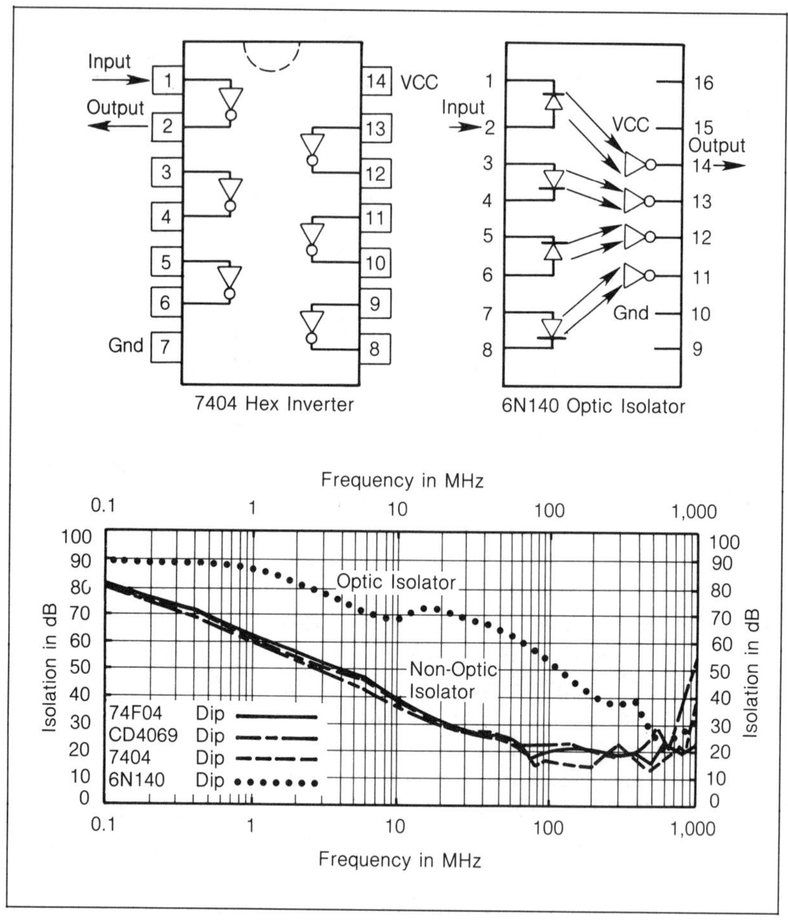

Figure 10.7—Input Pin to Output Pin Isolation of Optoisolators in DIP Package vs. That of a Standard Logic Gate. In the optoisolator package (measured die-removed), the improvement is due to the opposite, instead of side-by-side, pin location.

10.5 Fiber Optics

Fiber optics are the ultimate solution to electrical isolation and low path loss transmission where the transmitting media over the whole link, whether meters or hundreds of kilometers, is a glass or plastic fiber. Performance and construction details of fiber optic systems are covered in many books (like Ref. 1) and will not be addressed in this volume, the goal of which is essentially EMI.

Fiber optics are often promoted as the panacea for all sorts of interference problems and, indeed, they are a good solution. However, the emphatic advertisements about the virtually flawless fiber optic systems should not hide the sneaky ways EMI can defeat the best design by creeping in the back door. Some of the most relevant features of fiber optics with their possible limitations are enumerated in the following paragraphs.

1. **A fiber optic system is immune to ambient EMI fields.** A fiber optic link is obviously insensitive to electromagnetic ambient. Although theoretically possible, it would require megavolts per meter of field to have a discernable influence on the flow of photons.

   However, as already stated for optoisolators, this does not preclude the designer from taking EMC precautions at the transmitter and receiver. If improper shielding, insufficient CM rejection, crosstalk or other spurious effects are already resident in the transmitting (or receiving) equipment, they will either corrupt the electrical signal **before** it is converted into light, therefore carrying EMI in the form of light (unless optical heterodyning and mixer techniques are used) or they will corrupt the electrical signal after it has been recovered from the optical beam. This is especially true with the high-impedance, high-gain amplifier used to process the weak output of the phototransisitor.

2. **A fiber optic system does not radiate EMI**, therefore making the system "naturally" compliant with FCC, CISPR/VDE, military standards and TEMPEST specifications. As long as the cable was playing the major role in the emission process this is also true. However, all the electronics which drive the light sources (especially for the fast pulses used with laser transmitter diodes), their associated PCBs, etc., still radiate, and the usual shielding techniques are still required for the transmitting box.

   For instance, a digital current-to-light driver for high-speed logic can switch 100 mA/5 ns per channel, which requires the same precautions to avoid intra-box EMI (self-jamming) as in any electronic equipment. To permit a containment of EMI up to the point where electricity is converted into light, optical drivers and receivers are available with metallic housing and connectors, as in Fig. 10.8.

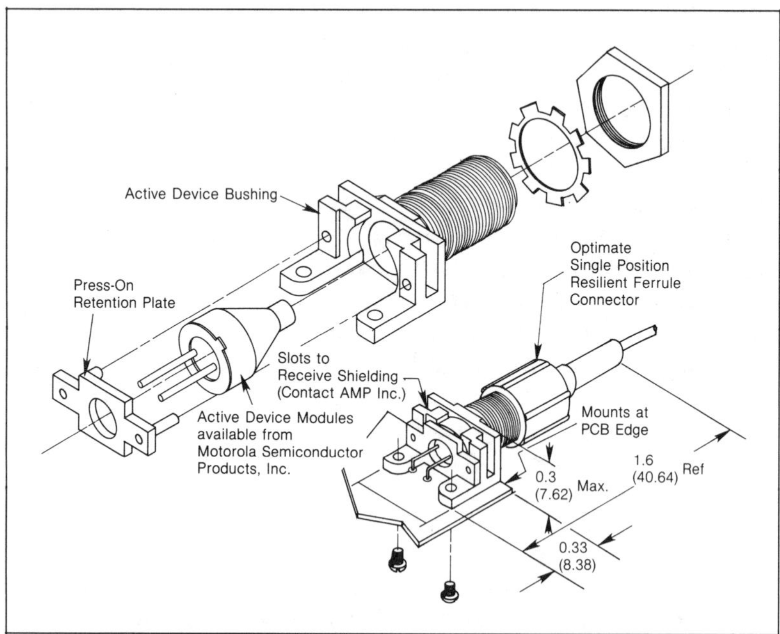

Figure 10.8—Shielded Fiber Optic Connector to Eliminate EMI Leakages up to the Fiber/Machine Interface (Doc. AMP, Kit Part No. 227240-1)

3. **A fiber optic system is immune to grounds or earth voltage differences.** Because of the perfect galvanic isolation, systems with high differences between their ground potentials can be interfaced without EMI problems.

4. **Such a system is also resistant to eavesdropping (TEMPEST).** Since no electromagnetic field is radiated, the capture of confidential information by electronic espionage is in theory impossible. However, what cannot be done electrically can be done optically if the spying party uses an optical tap (tee or optical coupler). Optical tapping can even be done without physically splicing the fiber, by using optical couplers based on the light leakage created when the fiber is sharply bent. Defense against this requires constant monitoring of the propagation characteristic using time-domain reflectometry (TDR).

5. **A fiber optic system is immune to lightning.** For the same reasons as all of the above, an optical link is naturally insensitive to the induced effects of lightning.

6. **It is immune to NEMP (nuclear electromagnetic pulse).** The intense electromagnetic field (for instance 100 kV/m with a 10 nsec/250 ns waveform) which accompanies a nuclear blast is a major threat to conventional electronics, while it has no effect on a fiber optic link. On the other hand, the ionizing part of the nuclear explosion can have a destructive effect on the fiber by permanently "obscuring" the fiber material (Fig. 10.9). A fiber link with an attenuation of 3 dB/km, for instance, could jump to 100 or 1,000 dB/km after irradiation. Intensive work has been done in the military area to develop fibers which are resistant to this phenomenon as well.

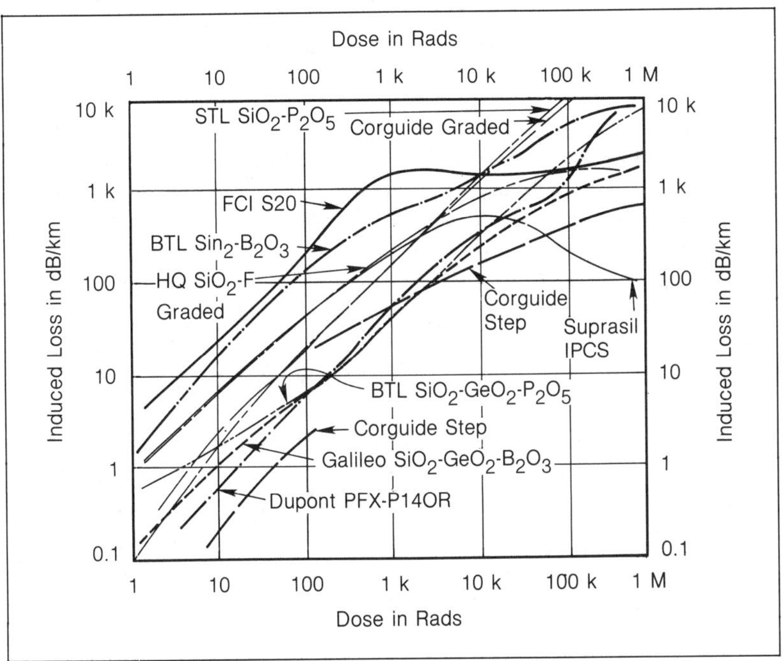

Figure 10.9—Optical Attenuation Induced by Ionizing Radiation

7. **A fiber optic system is insensitive to crosstalk.** Several fibers in a bundle cannot interfere with each other and they cannot pick up signals from nearby electrical cables, as would happen with conventional electronic wiring.

8. **Fiber optic links are lighter and thinner than hard-wired systems.** Due to the wide bandwidth of fibers,

multiplexing of many channels is possible. And since even with its mechanical protection the core size of an optic fiber is smaller than an equivalent coaxial or wire bundle, the overall weight and outside diameter of a fiber link is less than the cable which would carry the same data flow. Part of the overall savings comes from the fact that, because of its lesser attenuation, a long link needs fewer repeaters when it is a fiber optic type.

10.5.1 General EMI Precautions which Still Apply with Fiber Optics

Even though optic fibers provide unsurpassed immunity over the link itself, the digital modules at the driver and receiver side have not been designed specifically to provide high levels of EMI isolation, either for emission or susceptibility. Therefore, the whole transmitter, receiver and associated digital or analog processing electronics should be packaged with the same packaging precautions required for a conventional link. It could even be said that it takes more precautions if one wants to see the outstanding improvement brought up by the fibers. For instance, the fiber cable should pass through the shield via a metal tube which can eliminate extraneous interference.

Another problem may occur if the fiber cable needs special mechanical strength. Ordinary jackets are made of PVC. If more strength is needed, a Kevlar braid is used. And when extra strength is needed, some systems use a metallic braided armor or a metallic wire in the core. In this case, the galvanic isolation is compromised, and circulating currents between the two interconnected equipments can take place. Although they do not induce any noise by crosstalk, they can create the usual problems of CM current circulation at the transmitter or receiver electronics. The same is true for lightning or EMP induction which now can take place.

Finally, when repeaters are used over long-haul fiber links, the repeaters need to be powered. To this effect, copper wires are carried along with the optic fibers to supply the necessary voltages to the repeaters. The presence of these wires virtually defeats the electrical isolation which needs to be restored by shielded transformers and other common-mode rejection practices.

10.5.2 Component Configuration in a Digital Fiber Optic Link

The component configuration in a digital fiber optic system is shown in Fig. 10.10. The signals are first amplified and then transformed into optical signals in an input optocoupler. The input amplifier has the task of interfacing the electrical signals (current amplification) to the parameters of the optocoupler. The optical cable has connectors at both ends and can be any of the available types, depending on the attenuation which can be afforded.

At the output of the optical cable, the light is converted back into electrical signals (photocurrent) in an input optocoupler and the resulting photocurrent is finally amplified and processed in the output amplifier.

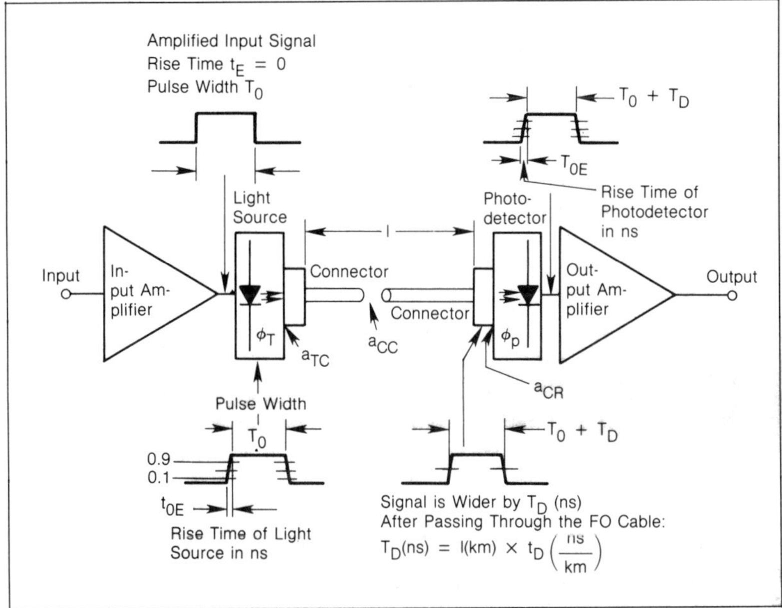

Figure 10.10—Digital Fiber Optic Transmission System

10.5.2.1 Form of Input and Output Signals

In the standard ITT range, the input to the transmitter is applied via a standard interface, normally TTL, so that the user may clearly understand what he needs to do in the external circuits to drive the transmitter. The convention used is that a TTL high turns the emitter **on**, whereas a TTL low turns it **off**. This format is suitable for nonreturn-to-zero (NRZ) codes provided that an LED or other continuous emitter is used in the output of the transmitter circuit and designs permit operation up to 20 Mb/s. Where long-haul applications are contemplated, it sometimes becomes necessary to resort to laser output transmitters.

The receiver consists of a photodetector upon which the incoming light signal impinges. The receiver is coupled to an amplifier, which is followed by a decision-making comparator, giving TTL high or low to the output depending on whether light is received by the photodetector or not. The photodetector can be either an avalanche photodiode (APD) or a pin-junction diode, and can be either dc coupled to the amplifier or ac coupled.

Transmitter and receiver units for optical cable links in information systems can be separately placed on PC cards for rack mounting or can be developed into integrated versions using existing ICs and placed on the same PC board.

10.5.2.2 Flux Budgeting

In order to establish the flux requirements for a fiber optic system, the characteristics of the receiver noise and bandwidth, coupling losses at connectors and transmission loss in the cable should be taken into consideration. In Fig. 10.10, the flux which the transmitter must produce is determined from the expression:

$$10 \log \frac{\Phi_T}{\Phi_R} = a_0 l + a_{TC} + a_{CR} + n a_{CC} + a_M$$

where,

Φ_T = flux (in W) available from the transmitter
Φ_R = flux (in W) required by the receiver
a_0 = fiber attenuation constant (dB/km)
a_{TC} = transmitter-to-fiber coupling loss (dB)
l = fiber length (km)
a_{CC} = fiber-to-fiber loss for in-line connectors (dB)
n = number of in-line connectors (excluding the end connectors)
a_{CR} = fiber-to-receiver coupling loss (dB)
a_M = margin (dB), chosen by the designer, by which the transmitter flux exceeds the system requirement.

For flux measurements, either a high-speed photodetector and oscilloscope could be used to measure the excursion flux, or an average-reading flux meter to measure $\Delta\Phi$.

10.5.2.3 Pulse Spreading and Rise Times Along the Optical Path

This description deals only with that part of the rise time introduced by the components of the optical fiber cable routes, excluding input and output amplifiers (Fig. 10.10). A rise time of $t_E = 0$ is assumed for the amplified input signal. The rise time is the time required for a pulse to rise from 10 percent to 90 percent of its peak value, and the pulse width is measured at half the peak amplitude. The total rise time t_r of the optical fiber route (excluding input and output) is:

$$t_r = \sqrt{t_{EO}^2 + t_{OE}^2}$$

where, t_{EO} = rise time of optical source
$\qquad t_{OE}$ = rise time of photodetector $\hfill$ (10.8)

On the other hand, the pulse width T at the output of the photodetector is given by:

$$T = T_O + T_D = T_O + 1 \times t_D \text{ [ns]} \qquad (10.9)$$

where t_D is the pulse spread (ns/km) on the optical fiber cable given in the data sheets for specified optical wavelength $\lambda \pm \Delta\lambda/2$ nm ($\Delta\lambda$ is the spectral bandwidth of a light source, electro-optical transducer). The value of t_D increases with spectral bandwidth $\Delta\lambda$.

Designing and budgeting of an optic link should consider the paramaters listed in Table 10.3. The calculation steps use the equations and definitions of Table 10.4.

Table 10.3—Some Equations for Practical FO Link Design

- Source Power

  - Power in dBm $= 10 \log (P_O\,\mu W/1{,}000)$

- Loss Equations

  - NA Loss: $dB = 20 \log \left(\dfrac{NA_r}{NA_t} \right)$, $NA_r < NA_t$

  - Diameter Loss: $dB = 20 \log \left(\dfrac{D_r}{D_t} \right)$, $D_r > D_t$
    (Area)

    Note: t = transmitter, r = receiver

  - System Loss: $dB = 10 \log (P_{in}/P_{out})$

- Detector Responsivity

  - $R = \dfrac{I_{out}}{P_{in}} \left(\text{in } \dfrac{\mu A}{\mu W} \right)$

- The Rise-Time/Bandwidth Equation

  - Rise Time $(t_r) = 0.35/BW$; $BW =$ Bandwidth

  - $t_r(\text{System}) = [t_2{}^2\,(\text{Transm}) + t_r{}^2\,(\text{Cable}) + t_r{}^2\,(\text{Receiv.})]^{1/2}$

Table 10.4—Fiber Optic Link Design Parameters

System-Defined Parameters	Standard Design Parameters
Link Length	Wavelength
Data Bit Rate (Bandwidth)	Emitter Output Power
Bit Error Rate	Coupled Power
Downtime Allowed	Intrinsic Loss in Fiber
Temperature Range	Fiber Dispersion
Radiation Environment	Detector Responsivity
	Receiver Sensitivity

Radiation-Dependent Parameters

Emitter Degradation
Induced Loss in Fiber
Fiber Recovery Rate
Induced Noise in Detector
Detector Degradation

Numerical Example:
Design of a Fiber Optic Link

In a fiber optic link system, the following elements are given:
Light source: LED
  Characteristics:
    1. Power output (at 100 mA): 125 μW
    2. Peak emission wavelength: 900 nm
    3. Output port diameter: 200 μm
    4. Numerical aperture: 0.7
    5. Rise time (t_r): 20 ns
Detector: Pin diode
  Characteristics:
    1. Minimum responsivity, ρ (at 900 nm), 0.4 μA/μW
    2. Input port diameter: 200 μm
    3. Numerical aperture: 0.7
    4. Min. input power for 10^{-9} Ber: 4 μW
    5. Rise time (t_r): 20 ns
Optical cable
  Characteristics:
    1. Length: 500 m
    2. Attenuation: 2.5 dB/km (at 900 nm)
    3. Numerical aperture: 0.5
    4. Core diameter: 200 μm
    5. Bandwidth-length product: 5 MHz $\times$ km

Losses due to splicing, misalignments, gaps, etc.: 5.67 dB
System loss margin: 3 dB

For Given Design Example Calculate:
Part 1
    1. Total losses in the system (including the loss margin)
    2. How much additional power is available at the input of the
       detector?
    3. The current at the output of the detector
Part 2
    1. The rise time of the fiber optic cable
    2. The link's bandwidth

Solution: Part 1

1. Total losses in the FO link   In dB
   a. Due to numerical aperture (NA)
      differences: 20 log (0.7/0.5)   2.92
   b. Losses of splices, misalignments, etc.   5.67
   c. Due to cable: 500 m (2.5 dB/km)   1.25
   d. System loss margin   3.00

                      Total losses   12.84 dB

2. $dB_{(System)} = 10 \log (P_{in}/P_{out})$
   $\phantom{dB_{(System)}} 12.84 = 10 \log (125/P_{out})$
   $\phantom{dB_{(System)}} P_{out} = 6.5 \ \mu W$

Therefore, the additional power available at the detector's input is
   $6.5 \ \mu W - 4 \ \mu W = 2.5 \ \mu W$

3. $I_{out} = P_{in} \times \rho = 6.5 \ \mu W \times 0.4 \ \mu A/\mu W = 2.6 \ \mu A$

Solution: Part 2

1. First, determine the cable bandwidth (BW):

$$BW_{(cable)} = \frac{5 \text{ MHz} \times \text{km}}{0.5 \text{ km}} = 10 \text{ MHz}$$

Then compute the rise time for the cable from the formula:

$$t_{r \ (cable)} = \frac{0.35}{BW} \text{ or } \frac{1}{\pi BW}$$

$$t_{r \ (cable)} = \frac{0.35}{10 \text{ MHz}} = \frac{0.35}{10 \times 10^6} = 35 \text{ ns}$$

2. $T_{r \ (System)} = [t_r^2 \ {}_{(Source)} + t_r^2 \ {}_{(Cable)} + t_r^2 \ {}_{(Detector)}]^{1/2}$ or:

$$T_{r \ (System)} = [20^2 + 35^2 + 20^2]^{1/2} = 45 \text{ ns}$$

Applying the formula again:

$$T_{r \ (System)} = \frac{0.35}{BW}$$

we obtain:

$$BW = \frac{0.35}{T_r} = \frac{0.35}{45 \text{ ns}} = 7.78 \text{ MHz}$$

10.5.3 Fiber Optic Adapters

Over the recent years, various manufacturers have introduced plug-to-plug compatible fiber links which can instantly replace a traditional wire cable (Fig. 10.11). The packages usually consist of a pair of electro-optical converters with a length of fiber from a few tens of meters up to a few kilometers. The connector block at each end contains an optical modem with simplex or full-duplex capabilities, corresponding to the most current standard interfaces like RS-232-C, ITT V-24 (up to 100 kbps), RS-422 (up to 10 Mbps), etc.

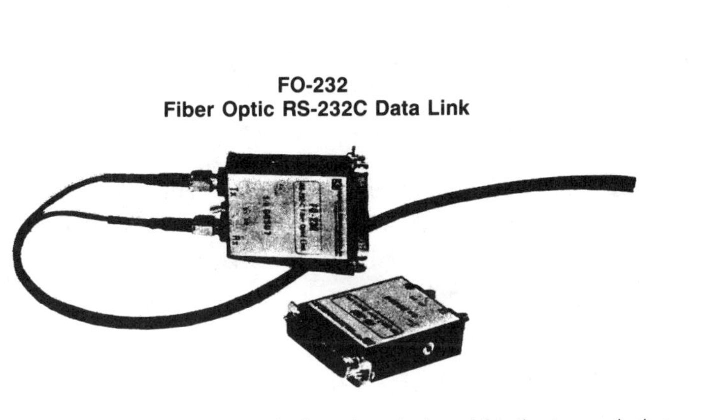

The **FO-232** is a fully-duplex optical modem designed for the transmission of asynchronous data via fiber optic cable. It replaces standard wire cable and eliminates EMI/RFI in compliance with the FCC regulations. The **FO-232** is plug-to-plug compatible with existing 25-pin connectors meeting EIA RS-232-C and CCITT V.24. The **FO-232** is capable of transmitting data up to 1.25 miles (2 km) at a full bit rate of 19.2 kbps or greater.

Figure 10.11—Example of Pluggable Replacement Fiber Optic Link to Substitute for a Conventional Cable (Doc: Light Wave Communications Co.)

The electrical connectors are pin-compatible with the corresponding standard pin assignment. All the user needs to do when he wants to immunize a data link is to disconnect the existing multipair cable and replace it with the fiber optic set. The only addition required is that each modem be powered by a 12 V supply which can be picked up from the host unit or made from a small pluggable power supply.

10.5.4 Radiation Hardening of Fiber Optic Links

The irradiation of an optical link during a nuclear explosion can cause severe transient or permanent effects. The most critical areas are the fiber itself and the detector, so the configuration of a hardened link is a trade-off between performance and immunity. For instance, avalanche photodiode detectors have better optical sensitivity than the PIN diode variety, but they cannot be used because of their vulnerability to radiation. As a result, hardened receivers generally have a lower sensitivity, which affects the system budget. All the same, radiation-resistant fibers generally have higher transmission losses under normal conditions than conventional fibers.

High doses of fiber irradiation can change the fiber attenuation from few dB/km to thousands of dB (Fig. 10.11). Research and testing have been conducted on phosphorous and other dopants which proved to decrease the damage by several orders of magnitude.

Finally, there are advantages in selecting modulation types which accommodate lower S/N ratios, such as frequency-shift keying (FSK) or phase-shift keying (PSK). This lower S/N compensates for some of the increased fiber losses.

10.6 References

1. Georgopoulos, C., *Fiber Optics and Optical Isolators,* (Gainesville, Virginia: Interference Control Technologies, Inc., 1983).
2. Belisle, K.M. and Jackson, C.K., "EMI Techniques for Decoupling and Isolation of Microcircuits," *Proceedings of the 1982 IEEE/EMC Symposium.*

10.7 Bibliography

1. "6N138 High Isolation Optoisolator Data Sheet," (Siemens Semiconductors).
3. Wall, J.A. and Posen, H., "Radiation Hardening of Optical Fibers Using Multidopants sb/p/Ce," *Proceedings from the 1980 Conference on the Physics of Fiber Optics.*

Chapter 11

Cathode Ray Tubes (CRTs) and Other Alphanumeric Displays

11.1 CRTs

EMI concerns with CRTs are mainly radio frequency emissions, transient disturbances created by flashovers, occupational hazards due to proximity to a high voltage circuit and susceptibility to ambient H-fields.

11.1.1 Radio Frequency Emissions

In terms of interference to radio and TV broadcasts, the most offending type of CRT application is the display of data in video display terminals (VDTs). Digital data sent to the tube to display alpha-numeric characters and pictures are high-speed switched square waves. Therefore, the CRT control circuits, their leads to the tube and the anode lead are efficient radiators in the VHF/UHF region.

The screen of a CRT is built from small dots or "pixels," arranged in horizontal lines, which are scanned by the electron beam. To display the desired figures, the electron beam is pulse modulated such that a digital "1" will create a white spot, and a digital "0" will leave a pixel dark. The refresh rate of the display is done at a power mains frequency of 50 or 60 Hz, while the horizontal sync is at 15 to 20 kHz.

In most VDTs, an elementary bit duration τ_b is less than 50 ns since the video bit rate is 20 MHz or more. Between two successive bits, the video signal generally returns to zero (RZ). This is obtained by combining the video signal (whose bit duration is τ_b) with a square wave of period T= τ_b, using a logic AND. This video-dot-clock is generally available in the VDT since it is used for the synchronization of the parallel to serial conversion.

Therefore, even though the initial digital message is random, the radiated spectrum contains discrete, narrowband components at odd multiples of the video-dot-clock, and these components are amplitude modulated by the random video message. This causes annoying interference, both radiated and power-mains conducted, to VHF/UHF radio and TV reception. But this also may lead to a possible reconstruction of private or confidential information at a rather great distance from the VDT. Experiments reported in Ref. 1 show that if a proper sync signal is fed into a modified TV receiver, a reproduction of the screen of a VDT from its radiated spectrum is possible as far as 50 m for unshielded VDTs. Beyond this distance, modulation of the narrowband components by the broadband signal becomes undetectable. It was also demonstrated by the same thorough study that the modulation index has to be brought down to less than 0.5 percent, i.e., 46 dB of shielding would be required, for the eavesdropping to become impractical.

Figures 11.1 and 11.2 show some conducted and radiated signatures of a VDT for both the "screen empty" and "screen full of text" conditions.[1] In Fig. 11.2, a difference of approximately 5 dB between narrowband emissions "full screen" versus "empty screen" is visible at some frequencies. A 10 to 15 dB difference between broadband profiles is also evident.

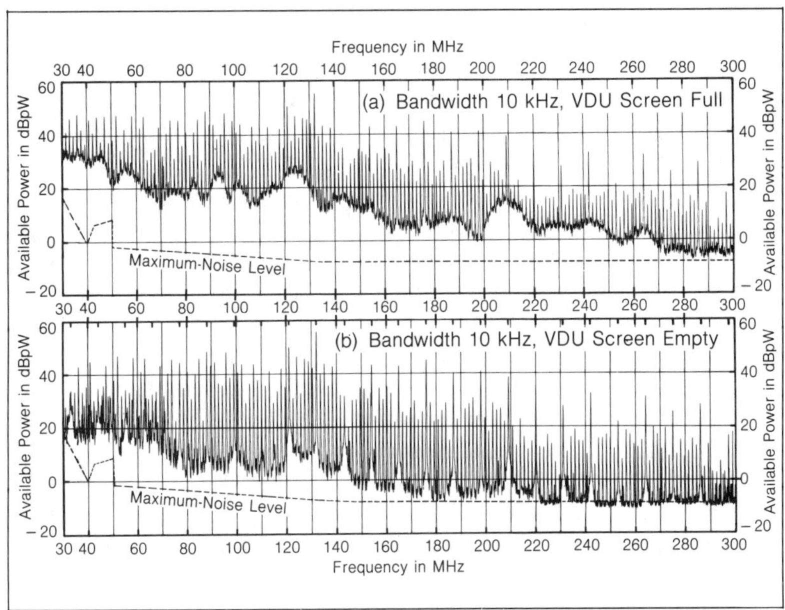

Figure 11.1—Maximum Interference Power Available on the Mains Power Cord, Measured by the CISPR Absorbing Clamp Method

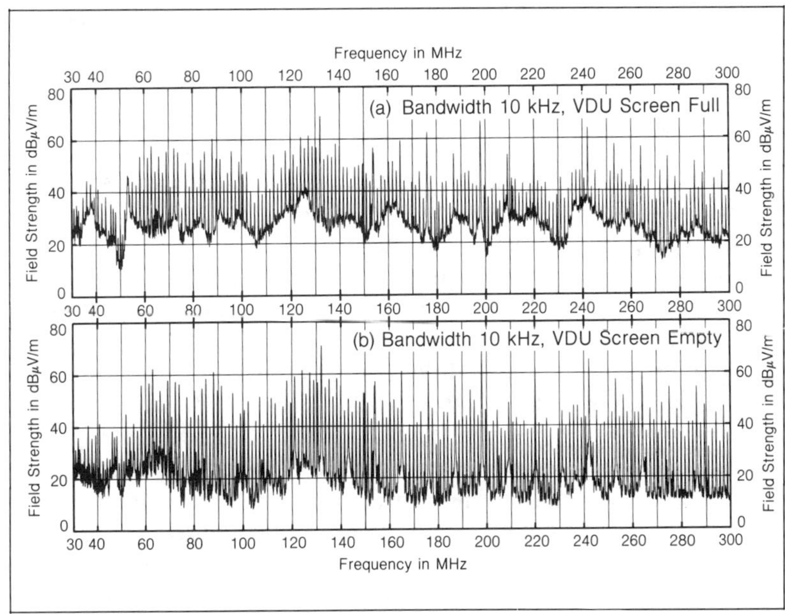

Figure 11.2—Field Strength in the Direction of Maximum Radiation at 1 m Distance (Horizontal Pol)

11.3

11.1.2 Flashover

Any CRT is prone to casual flashovers. Although not fully understood, these flashovers are believed to be due to secondary emission from the CRT gun starting an electron avalanche along the neckglass that causes a slight outgassing and finally ends up in a plasma discharge which restores normal condition.

Although quality control and cleanliness decrease this risk, flashover always has a finite probability of occurring at random intervals well separated in time. Taken from the study by Rhoades (Ref. 2), an equivalent circuit is shown in Fig. 11.3. The positive high voltage (HV) is connected by the anode button to the internal coating. The HV return is made through the external graphite film called "aquadag." The capacitance C_1 between the inner and outer coatings, with the glass as the dielectric, is about 1 to 2.5 nF. The distributed capacitance C_2 between the yoke and the internal coating is about 10 percent of C_1. The yoke return is common to the external aquadag. The high voltage (15 to 20 kV for black-and-white, 25 to 35 kV for color) creates a charge $Q=CV$ in the microcoulomb range and an energy ($1/2\ CV^2$) which can amount to one joule. Distributed resistance of the internal coating over the arc path is about 10 Ω. Some newer designs add a 400 Ω series resistance (discrete or distributed in the coating) to limit the amplitude and dI/dt of the arc current.

When an arc occurs, the current leaves the CRT pins, seeking a return path to the aquadag. The inductance of the path inside the CRT is about 0.4 μH, while a typical inductance for the external loop is about 0.3 μH. Thus the main arc path can be modeled as a series RLC circuit, driven by a spark gap. As the energy stored into $C_1 + C_2$ dissipates into the loop resistance, a 100 to 800 A ringing current will flow until the arc is extinguished.

Figure 11.3 shows a waveform of this current, with a 50 ns rise time. In a typical equipment, control electronics (sync, video signal, etc.), grounds and CRT drive ground are common. A suppressor is generally inserted to protect the drive circuits when the main arc current I_M returns to the aquadag connection (see Fig. 11.4). The transient voltage difference between point A and B can be in the 1,000 V range during the first rise of the arc current. This voltage causes some fraction of the I_M current to flow in all the parallel loops which are present across points A and B, like the controller to CRT cables, the dc and power mains filters and the stray capacitances. All these transient loops generate local electromagnetic field which can upset or even damage the control components. They also create power mains transients since one of the alternate return paths involves the equipment safety ground wire.

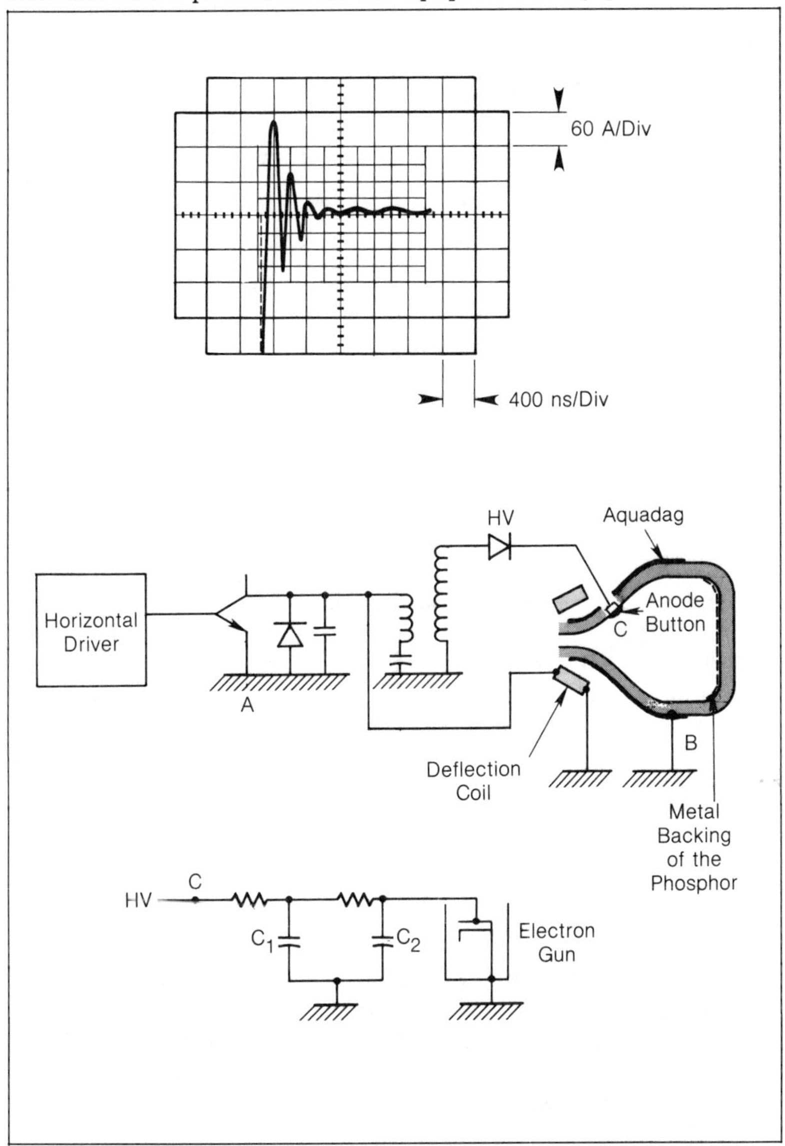

Figure 11.3—Simplified CRT Circuit and Equivalent Circuit of a Flashover. A typical flashover current with ≅ 50 ns rise time is shown on left.

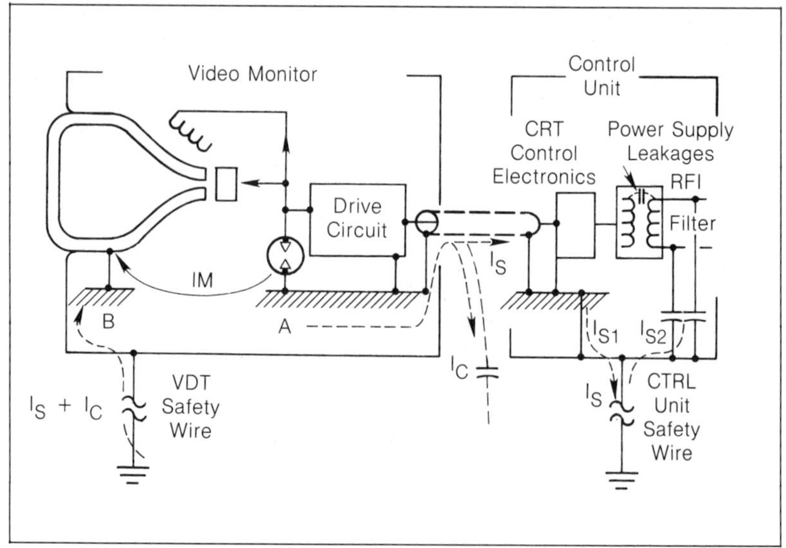

Figure 11.4—Main Return Path I_M and Alternate Escape Paths During a Flashover

The principal solution consists of confining the largest possible percentage of the external arc current return within the path AB, such that a minimal current flows into the "escape" paths. Making the connection AB as short and wide as possible, e.g., using a braid, will help reduce V_{AB}. Then the effort should concentrate on increasing the common-mode HF impedance of the escape paths. Noting that the impedance Z_{AB} is mainly inductive and represents, for the 50 ns rise time, a value of:

$$Z_{AB} = \frac{L}{dt} \cong \frac{0.3 \times 10^{-6}}{50 \times 10^{-9}} = 6 \, \Omega$$

We see that increasing the common-mode impedance of escape paths to several hundred ohms will reduce the escape current to a small percentage of the total arc current. Modern ferrites (see Sections 2.3.2 and 5.3 of this book) can achieve more than 100 Ω per turn above few MHz and provide the necessary RF damping on the leads between the CRT and the associated electronics.

11.1.3 CRT Radiation Hazards in the Proximal Region

Due to serious concerns expressed by labor organizations and national safety organizations (for instance, the U.S. Bureau of Radiological Health), several studies have been conducted concerning the potential hazard presented by long exposures (4 to 6 hours per day) to both the ionizing and nonionizing emanations from CRTs. This section will deal with the nonionizing radiations only.

The pulsating high voltages of the anode drive create a high impedance E-field. The pulsating currents in the deflection coils generate, in the proximal region, predominantly magnetic fields. In the early 1980s, the U.S. Bureau of Radiological Hazards (BRH) in Rockville, Maryland measured 34 VDTs for the near E-fields and H-fields. Both the crest value (in time domain) and the spectral distribution were measured for each field term.[3]

The results are summarized in Fig. 11.5 for the two most emitting samples of the survey. As expected from pulses having a repetition rate between 15 and 20 kHz, most of the spectral energy is concentrated within the fundamental and the next 3 or 5 harmonics.

The all-safe level for continuous exposure of a human being is generally considered to be < 1 mW/cm^2 or 10 W/m^2, corresponding to a plane wave E-field of 60 V/m (which becomes conservative in the near field where the wave impedance is greater than 377 Ω and the E and H terms are not in phase). From this we may derive that CRTs at and beyond 30 cm distance are safe by at least one order of magnitude. A potential hazard occurs if the operator's face (especially the eyes) comes within 5 cm of the CRT face or right above the flyback transformer for a long period of time.

11.1.4 Susceptibility of CRTs to EMI

The most annoying effect of ambient fields that plagues CRT displays is the distortion and "waving" of the picture in the presence of low-frequency H-fields. A typical VDT can start to show noticeable alteration in a magnetic flux density of 0.3 to 0.5 G at 50 to 60 Hz. Since the effect is inductive, it is aggravated by frequency increase, so the same tube can be susceptible to only 0.3 to 0.5 mG at 50 to 60 kHz.

11.7

(a) Peak Field Strength (Worst-Case Measurements from a Sampling of 34 Commercial CRTs)

Location	Distance in cm	E-Field V/m (Average of the 2 Highest Readings)	H-Field A/m
CRT Face (Center)	5	47	0.69
	30	1.35	0.04
Above Hv Transformer, Shield Removed	5	> 500	0.65
	30	9.5	.06
Shield in Place	5	0	0.06

(b) E- and H-Field Spectrums: Distance = 5 cm

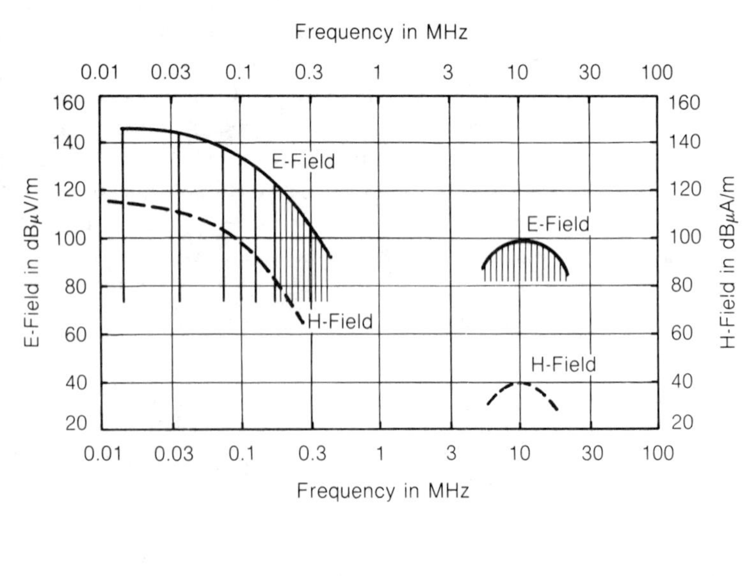

Figure 11.5—Electric and Magnetic Radiations from a CRT

11.8

Typical problems of this nature exist when VDTs are installed near cable runs with heavy currents, rotating machinery, ac-dc or dc-dc converters. CRTs with electrostatic deflection would be less prone to this effect and easy to shield, but most modern VDTs use magnetic deflection yokes because of the wide deflection angle needed by the shallow tubes currently in use. Low-frequency proximal H-fields are the most difficult to shield against. The solution is mainly a matter of cost.

For expensive applications where picture or text must be undisturbed in harsh, persistent H-fields, the tube is shielded by a high-permeability metal envelope as explained in the next section. For ordinary VDTs, the problem is generally circumvented by specifying a maximum H-field where the terminal is capable of working properly, thus exempting the manufacturer from the use of a costly shield. In case of a specific environment, however, the problem can be solved by a relocation or a reorientation of the VDT, since H-field sources are generally well localized and the field coupling is very directional. If this is impractical or insufficient, economics (and sometimes politics) will dictate if it is more advisable to shield the H-field source or to harden the VDT by a magnetic shielding "fix."

11.1.5 Shielding of CRTs

The previous sections have shown two reasons why CRTs need to be shielded: emissions and susceptibility. To shield against fields coming in or out, two approaches are used: shielding the components or shielding the whole VDT box. Figure 11.6 shows these different approaches.

Shielding the components requires that both the low-frequency magnetic sources (horizontal drive, flyback transformer) and the VHF/UHF sources (data pulses) be shielded. Shielding the whole box requires both a conductive cabinet and a full treatment of the CRT opening. The CRT opening represents one of the largest holes and is the most difficult to treat.

The disruption of the VDT shielding caused by the tube face is such that, if not considered in the initial design, it may become extremely difficult to achieve an acceptable product. The CRT opening can be treated by a screen mesh or a transparent conductive coating, making a good, continuous contact with the main housing shield where the tube is seated. These shields are fairly efficient against UHF/VHF fields, but insufficient against low-frequency H-fields. Against these latter, cone-shaped magnetic shields are available to fit all standard tube sizes (Figs. 11.6 through 11.8) if needed. They are fabricated from flat, unannealed Mumetal®, Co-Netic® or other highly permeable metal foil. After stamping, forming, welding, etc., the shield is annealed. From then on, the shield should not be given any bending or cold work of any kind.

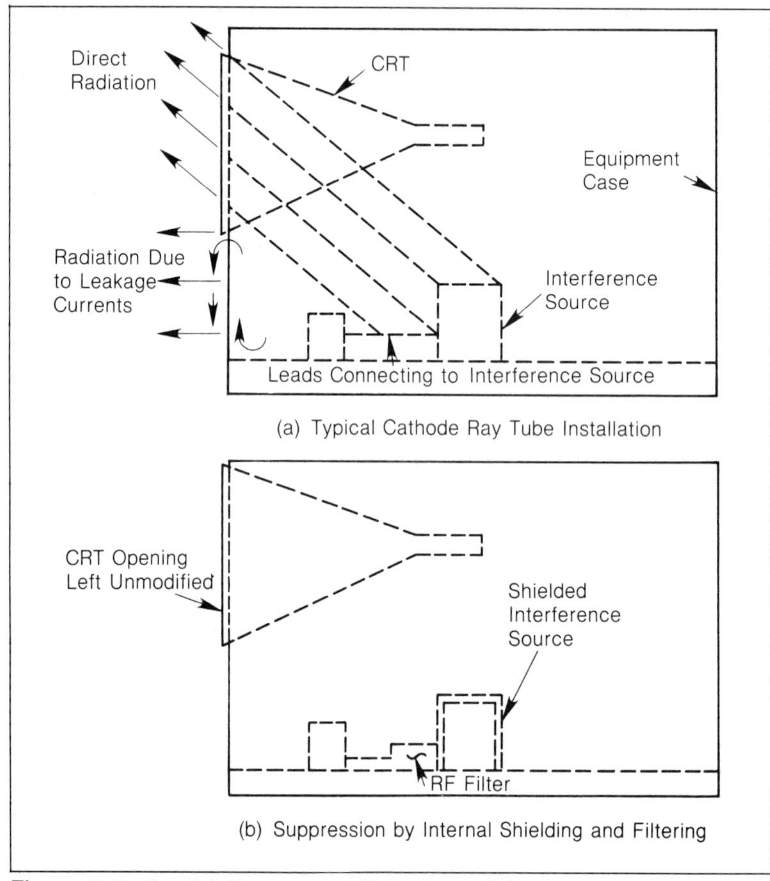

Figure 11.6—Possible Treatment of Cathode Ray Tube Openings (continued next page)

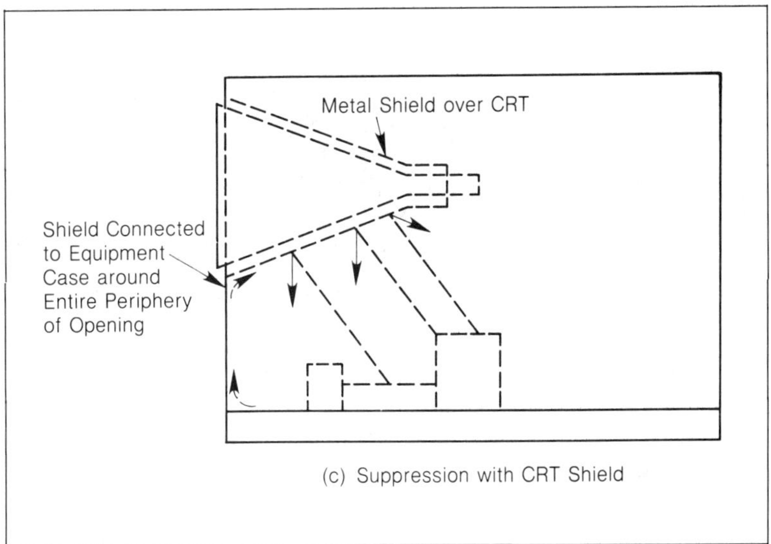

Figure 11.6—(continued)

Figure 11.7—Magnetic Shields for CRTs (Courtesy of MuShield Company)

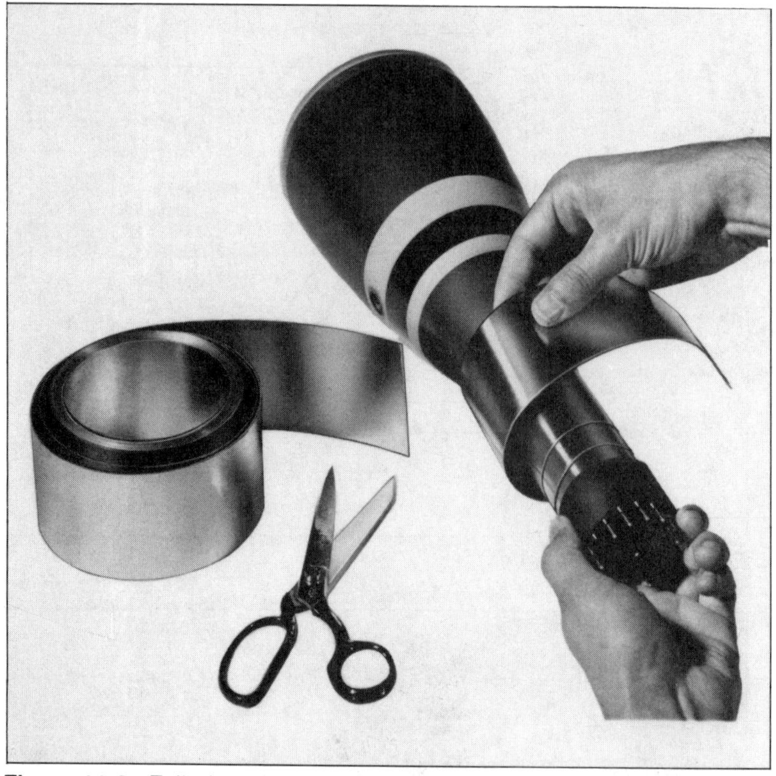

Figure 11.8—Tailoring of a Ductile Magnetic Shield to a CRT (Courtesy of Ad-Vance Magnetics Inc.)

The shielding needs for CRTs vary depending on the amplitude and frequency of the video and horizontal drive signals, but shielding requirements for a typical CRT are tabulated in Table 11.1, depending if the EMI specifications to meet are civilian or military.

Figure 11.9 illustrates the shielding effectiveness of thin coatings used on CRT faces, and Fig. 11.10 shows the relationship between light transmission and resistivity of gold coatings. Finally, Fig. 11.11 displays the shielding effectiveness of mesh-type CRT screens. More details on the design and performance of CRT shields are given in Volume 3, *Shielding*, of this EMC handbook series.

Table 11.1—CRT Characteristics

Video Pulses: 20 to 50 V, 10 to 30 MHz Yoke Drive: 3 A, 15 to 20 kHz		
dB Shielding Needed		
	10 to 1,000 kHz Range	**10 to 300 MHz Range**
	(Predominantly H-Field)	(E/M-Field)
Civilian Application (FCC and CISPR/VDE Class B)	NA	40 dB down to 10 dB
Military (MIL-STD-461)	50 dB down to 20 dB	70 dB down to 30 dB

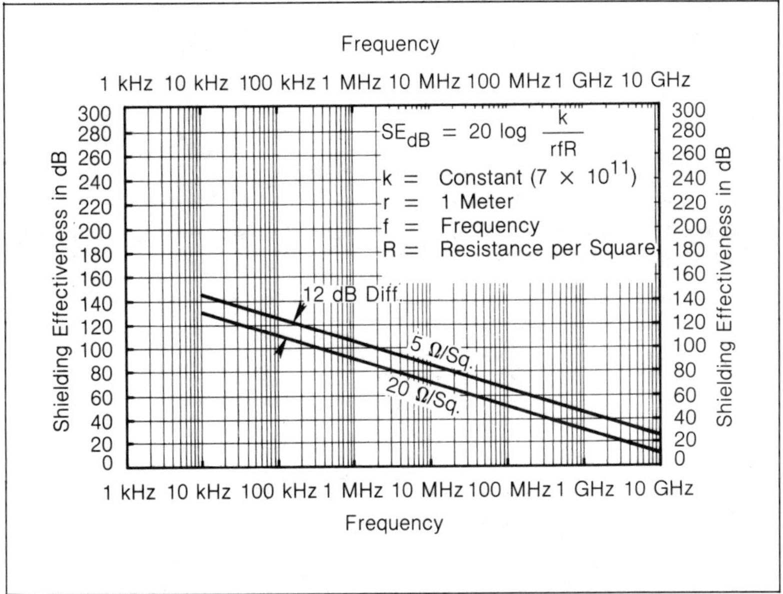

$$SE_{dB} = 20 \log \frac{k}{rfR}$$

k =   Constant (7×10^{11})
r =   1 Meter
f =   Frequency
R =   Resistance per Square

Figure 11.9—Shielding Effectiveness, Thin Coatings Used on CRT Faces (t << 2.5 Microns) (Courtesy of Tecknit)

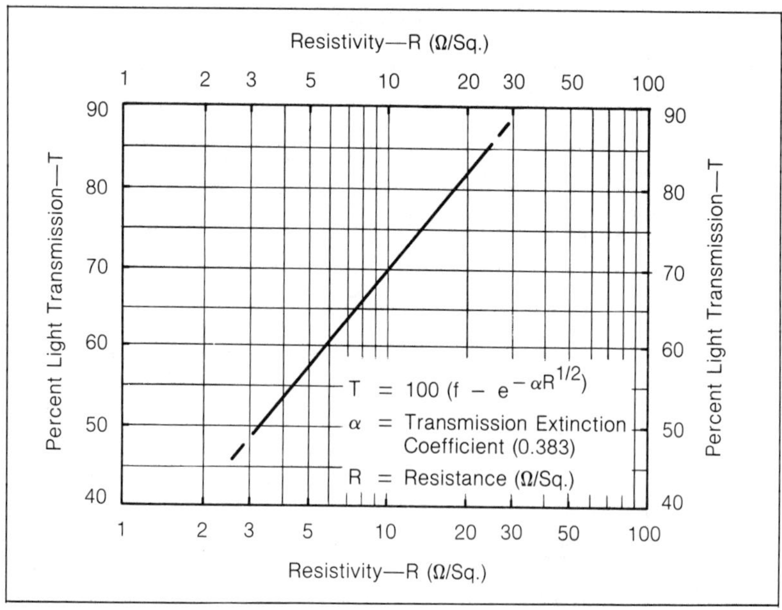

$$T = 100\,(f - e^{-\alpha R^{1/2}})$$
α = Transmission Extinction Coefficient (0.383)
R = Resistance (Ω/Sq.)

Figure 11.10—Light Transmission—Resistivity Relationship—Thin Gold Coatings (Courtesy of Tecknit)

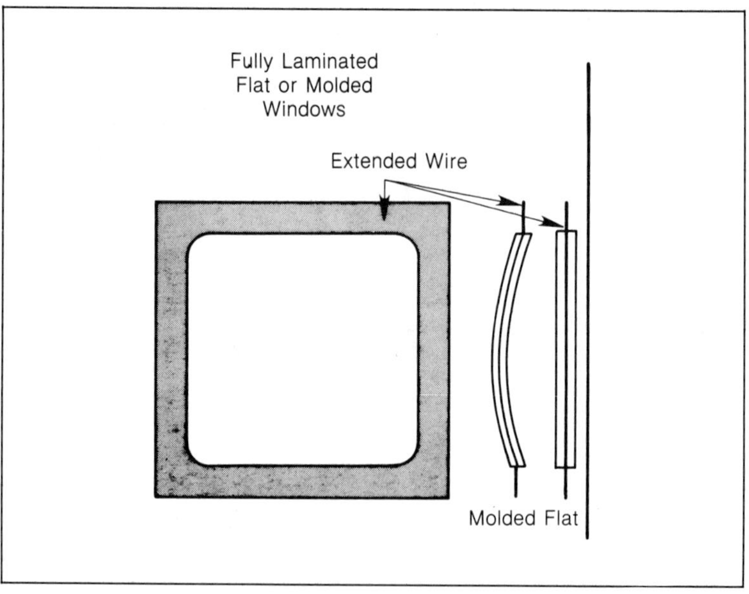

Figure 11.11—Example of a CRT Shield with 40 Wire/cm (100 Wires/inch) Mesh Embedded in the Transparent Acrylic (Courtesy of Dontech Inc.) (continued next page)

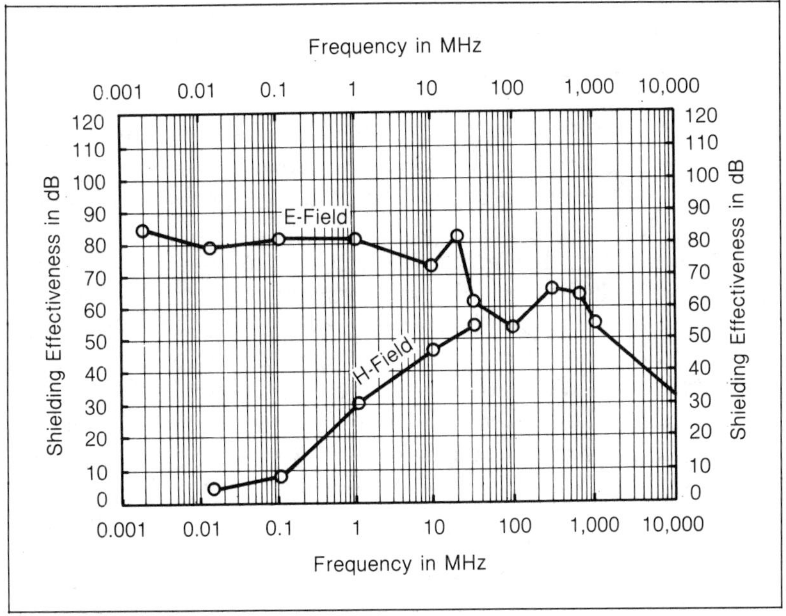

Figure 11.11—(continued)

11.2 Plasma, Liquid Crystal and Electroluminescent (EL) Displays

Plasma and liquid crystal displays are widely used alternatives to CRTs in many alphanumeric applications. Advantages include their flat shape, lower cost and absence of high voltage. These displays are rather unsusceptible to ambient EMI due to the relatively high signal amplitudes used to address both the plasma matrix (100 to 200 V) and the liquid crystal layers (10 to 40 V). Since the addressing speed is rather low, EMI emission is also rarely a problem. In a typical plasma display, the matrix elements are on for 150 μs and refreshed at a rate of 50 or 60 Hz.

LCDs and EL displays are controlled from low-power CMOS. In a typical EL panel, rows are driven by positive and negative 200 V, and columns are driven by positive and negative 60 or 90 V. The turn-on time per row is about 25 μs, with a refresh rate of 50 or 60 Hz as with CRTs. The rather low-frequency electric field is created by the 200 V pulses applied to the panel's thin layer with high dielectric constant, typically 3 pF per pixel.

The only case where such displays may need some shielding would be military or government applications where extremely low emission levels are required. It has been reported that even to meet the most stringent emission levels of MIL-STD-461, a shielding effectiveness of 0 to 20 dB in the 10 kHz to 30 MHz range is sufficient. In addition, the field emission is predominantly electrical, so even a moderately conductive screen produces a satisfactory attenuation.

A shielding problem may still exist with plasma or EL displays. Aside from the low shielding requirements of the display itself, the large opening it creates into its host equipment housing (except for displays on a separate panel, such as laptop computers) must be shielded to prevent EMI to and from the electronics. In a 512 × 256 matrix panel (25 cm × 15 cm), the fine resolution of 0.5 mm would degrade rapidly if a metal mesh was put on top. And a conductive transparent film would affect luminosity, which is already rather marginal with these low-power displays. Until conductive films with better light transmittance are found, the only viable solution is a fine-line mesh whose grid size is an exact submultiple of the column × row matrix. The wires can be located midway between pixels to reduce optical blockage.

11.3 References

1. VanEck, W.; Neessen, J.; and Rijsdijk, P., "Characteristics of EMI Field from Video Displays," *Proceedings of the 1985 Zurich EMC Symposium.*
2. Rhoades, W., "Characteristics of Unusual Power Mains Transients," *Proceedings of the 1985 Zurich EMC Symposium.*
3. "Study on Nonionizing Radiation from VDTs," (Bureau of Radiological Health, 1981).

Index

CMRR:
  parameters, 4.89-4.90
  see also common mode
coated conductors:
  EMI reduction, 3.24
coaxial cable:
  reducing EMI, 3.23
coaxial connectors:
  applications, 3.36-3.37
  selection, 3.37-38
collector-to-base:
  parasitic capacitance, 4.176
collector-to-sink:
  parasitic capacitance, 4.176-4.177
common impedance coupling:
  as coupling path, 1.4, 1.5
  contact impedance, 3.31
common Z coupling:
  position sensors, 6.39
common-mode:
  range, 4.65
  slew rate, 4.62
common-mode chokes:
  from inductors, 2.34
  principals, 2.34
common-mode induction or radiation:
  as coupling path, 1.4,1.5
common-mode rejection:
  analysis of, 4.65-4.70
  definition, 4.59
  of transformers, 5.5, 5.6
  ratio, 4.60-4.62
communications CE receivers:
  susceptibility, 4.8
communications-electronics receivers:
  susceptibility, 4.8
comparators:
  susceptibility, 4.85
  tests, 4.86
compensating windings:
  commutation, 6.61
components:
  complexity, 1.1
  cost, 1.1
  EMI problems of, 1.1, 1.3
  EMI properties, 4.38
  functional families, 1.3
  repairability, 1.1
  sensitivity to ESD, 4.151-4.154
  size, 1.1
composition carbon resistors:
  noise, 2.9
composition resistor:

equivalent circuit, 2.4
  resistance of, 2.2, 2.5
conducted interference:
  logic devices, 4.123
conductive (static dissipative) plugs:
  EEDs, 6.48
conductor:
  cause of EMI, 3.1
  definition, 3.8
  EMI reduction methods, 3.23-3.24
  inductance of, 3.8-3.15
  parallel lines, 3.18-3.19
  parameters, 3.8
  skin effect, 3.13, 3.14
  step front distortion, 3.15
  solution, 3.1
conductor straps:
  impedances, 3.19-3.20
connector contacts:
  faulty, 3.33
connector insertion loss:
  requirements, 3.33
connector shell:
  shielding effectiveness, 3.37-3.38
connectors:
  coaxial, 3.36
  crosstalk, 3.25-3.33
  definition, 3.24
  EEDs, 6.48
  noise, 3.25
  similarity to mechanical
    switches, 3.31
  types, 3.25
  use as connectors, 3.24
contact:
  bouncing, 6.24
  classification, 6.2
  in connectors, 3.31
  rating, 6.3
  types, 6.21
contact arcing:
  with load effect, 6.7
contact impedance:
  and insertion loss, 3.31
  causes, 3.32
contact plating:
  requirements, 3.32
  solving contact impedance, 3.32
contact protection:
  at switch, 6.9
contact resistance:
  solid-state relays, 6.32
continuous wave:

zero-crossover technique:
  application, 4.192
  limitations, 4.192-4.193
zero-crossover switching:
  EMI analysis, 4.185
zero-volt return:
  PCBs, 8.1
zoning:
  PCBs, 8.12-8.13

Other Books Published by ICT

1. Carstensen, Russell V., *EMI Control in Boats and Ships*, 1979.
2. Denny, Hugh W., *Grounding for Control of EMI*, 1983.
3. Duff, Dr. William G., *A Handbook on Mobile Communications*, 1980.
4. Duff, Dr. William G. and White, Donald R.J., Volume 5, *Electromagnetic Interference Prediction & Analysis Techniques*, 1972.
5. Feher, Dr. Kamilo, *Digital Modulation Techniques in an Interference Environment*, 1977.
6. Gabrielson, Bruce C., *The Aerospace Engineer's Handbook of Lightning Protection*, 1987.
7. Gard, Michael F., *Electromagnetic Interference Control in Medical Electronics*, 1979.
8. Georgopoulos, Dr. Chris J., *Fiber Optics and Optical Isolators*, 1982.
9. Georgopoulos, Dr. Chris J., *Interference Control in Cable and Device Interfaces*, 1987.
10. Ghose, Rabindra N., *EMP Environment and System Hardness Design*, 1983.
11. Hart, William C. and Malone, Edgar W., *Lightning and Lightning Protection*, 1979.
12. Herman, John R., *Electromagnetic Ambients and Man-Made Noise*, 1979.
13. Hill, James S. and White, Donald R.J., Volume 6, *Electromagnetic Interference Specifications, Standards & Regulations*, 1975.
14. Jansky, Donald M., *Spectrum Management Techniques*, 1977.
15. Mardiguian, Michel, *Interference Control in Computers and Microprocessor-Based Equipment*, 1984.
16. Mardiguian, Michel, *Electrostatic Discharge—Understand, Simulate and Fix ESD Problems*, 1985.
17. Mardiguian, Michel, *How to Control Electrical Noise*, 1983.
18. Smith, Albert A., *Coupling of External Electromagnetic Fields to Transmission Lines*, 1986.
19. White, Donald R.J., *A Handbook on Electromagnetic Shielding Materials and Performance*, 1980.
20. White, Donald R.J., *Electrical Filters—Synthesis, Design & Applications*, 1980.
21. White, Donald R.J., *EMI Control in the Design of Printed Circuit Boards and Backplanes*, 1982. (Also available in French.)
22. White, Donald R.J. and Mardiguian, Michel, *EMI Control Methodology & Procedures*, 1982.
23. White, Donald R.J., Volume 1, *Electrical Noise and EMI Specifications*, 1971.
24. White, Donald R.J., Volume 2, *Electromagnetic Interference Test Methods and Procedures*, 1980.
25. White, Donald, R.J., Volume 3, *Electromagnetic Interference Control Methods & Techniques*, 1973.
26. White, Donald R.J., Volume 4, *Electromagnetic Interference Test Instrumentation Systems*, 1980.
27. Duff, William G., and White, Donald R.J., Volume 5, *Prediction and Analysis Techniques*, 1970.
28. White, Donald R.J., Volume 6, *EMI Specifications, Standards and Regulations*, 1973.
29. White, Donald R.J., *Shielding Design Methodology and Procedures*, 1986.
30. *EMC Technology 1982 Anthology*
31. *EMC EXPO Records 1986, 1987, 1988*